ORIGINS of
INBRED MICE

Proceedings of a Workshop

Bethesda, Maryland
February 14–16, 1978

This medal was presented to the distinguished scientists
who have contributed so much to the research on inbred mice.
The sponsors are indebted to Van Cleef & Arpels
who designed and donated the medal.

ORIGINS of INBRED MICE

Edited by
HERBERT C. MORSE III

National Institute of Allergy and Infectious Diseases
National Institutes of Health
Bethesda, Maryland

Sponsored by

Cancer Research Institute, Inc.
and
National Institute of Allergy and Infectious Diseases
National Cancer Institute
Division of Research Services
National Institutes of Health

ACADEMIC PRESS   New York   San Francisco   London   1978
A Subsidiary of Harcourt Brace Jovanovich, Publishers

Academic Press Rapid Manuscript Reproduction

Copyright © 1978, by Academic Press, Inc.
ALL RIGHTS RESERVED.
NO PART OF THIS PUBLICATION MAY BE REPRODUCED OR
TRANSMITTED IN ANY FORM OR BY ANY MEANS, ELECTRONIC
OR MECHANICAL, INCLUDING PHOTOCOPY, RECORDING, OR ANY
INFORMATION STORAGE AND RETRIEVAL SYSTEM, WITHOUT
PERMISSION IN WRITING FROM THE PUBLISHER.

ACADEMIC PRESS, INC.
111 Fifth Avenue, New York, New York 10003

United Kingdom Edition published by
ACADEMIC PRESS, INC. (LONDON) LTD.
24/28 Oval Road, London NW1 7DX

Library of Congress Cataloging in Publication Data
Main entry under title:

Origins of inbred mice.

    Proceedings of a symposium held in 1978 in Bethesda, Md.
    Includes index.
    1. Mice—Genetics—Congresses. 2. Mice as laboratory animals—Congresses. 3. Inbreeding—Congresses. 4. Mammals—Genetics—Congresses. I. Morse, Herbert C.
QH470.M52074   599'.3233   78-10506
ISBN 0-12-507850-1

PRINTED IN THE UNITED STATES OF AMERICA

*This book is dedicated to John and Heather
for their daily reminders of the joys of life and learning,
and to Maya for her example and guidance*

Contents

Contributors   *xiii*

HISTORICAL PERSPECTIVE ON THE DEVELOPMENT OF INBRED MICE

Introduction   3
    Herbert C. Morse III

Introductory Remarks   23
    Helen C. Nauts

Introductory Remarks   25
    Richard M. Krause

Introductory Remarks   29
    Joe R. Held

Origins and History of Mouse Inbred Strains: Contributions of Clarence Cook Little   33
    Elizabeth S. Russell

Inbred Mice in Science   45
    Leonell C. Strong

The Creation of the AKR Strain, Whose DNA Contains the Genome of a Leukemia Virus   69
    Jacob Furth

Howard B. Andervont   99
    Margaret K. Deringer

Howard B. Andervont: An Appreciation   103
    Michael B. Shimkin

Biography as Related to Inbred Strains of Experimental Animals   109
    Walter E. Heston

Congenic Resistant Strains of Mice   119
    George D. Snell

Brief Autobiography Related to Work on Inbred Mice   157
    Margaret C. Green

A Brief Autobiography Relative to Work with Inbred
and Mutant Mice 167
   Earl L. Green

How It Began 179
   Clyde Keeler

INBRED STRAINS: MUTATIONS, BIOCHEMISTRY

Sources of Subline Divergence and Their Relative Importance for
Sublines of Six Major Inbred Strains of Mice 197
   Donald W. Bailey

Genetic Quality Control of the Laboratory Mouse (*Mus musculus*) 217
   Harold A. Hoffman

Biochemical Polymorphisms—Detection, Distribution,
Chromosomal Location, and Applications 235
   John J. Hutton

Genetic Control of Enzyme Activity 255
   Kenneth Paigen

INBRED STRAINS: VIRUSES

Genetic Regulation of Type C Viruses in Mouse Leukemia 281
   Frank Lilly

Chromosomal Location of C-Type Virus Genomes in the Mouse 289
   Wallace P. Rowe and Janet W. Hartley

Expression of XenCSA, a Cell Surface Antigen Related to
the Major Glycoprotein (gp70) of Xenotropic Murine Leukemia
Virus, by Lymphocytes of Inbred Mouse Strains 297
   Thomas M. Chused and Herbert C. Morse III

Biology of Mammary Tumor Viruses 321
   Robert D. Cardiff and Bruce W. Altrock

Diversity of Mouse Mammary Tumor Virus Genetic Information
and Gene Products in Rodents 343
   *J. Schlom, W. Drohan, Y. Teramoto, P. Hand, D. Colcher,
   R. Callahan, G. Todaro, D. Kufe, D. Howard, J. Gautsch,
   R. Lerner, and G. Schidlovsky*

INBRED STRAINS: CELL SURFACE ANTIGENS, MAPPING

Minor Histocompatibility Genes and Their Antigens 371
   Ralph J. Graff

Serological Definition of Cell Surface Antigens of Mouse Leukemia   391
  Lloyd J. Old and Elisabeth Stockert

Genes of the *Tla* Region: The New Qa System of Antigens   409
  Lorraine Flaherty

Recombinant Inbred Strains: Use in Gene Mapping   423
  Benjamin A. Taylor

INBRED STRAINS: SUBLINE DIFFERENCES

Differences among Sublines of Inbred Mouse Strains   441
  Herbert C. Morse III

Standardized Genetic Nomenclature for Mice: Past, Present, and Future   445
  Mary F. Lyon

The Biological Function of the LPS Gene   457
  David L. Rosenstreich

Effect of Macrophage Migration Inhibitory Factor on Peritoneal Exudate Cells of C3H/HeN and C3H/HeJ Mice   461
  Aldo Tagliabue, Luigi Ruco, James L. McCoy, Monte S. Meltzer, and Ronald B. Herberman

Different Hematopoietic Responses to Endotoxin in Different Sublines of C3H Mice   463
  Sallie S. Boggs, D.R. Boggs, and R.A. Joyce

Relationship of Endotoxin Toxicity and Response of Bone Marrow Derived Hematopoietic Cells   465
  Sallie S. Boggs, Robert W. Baker, Gretchen N. Schwartz, and Kenneth D. Patrene

Generation of an H-2 Restricted CML Response between AKR Substrains   467
  Marion M. Zatz

Derivation and Characteristics of the CBA/N Subline   471
  Donald E. Mosier

CBA/Ki Vs. CBA/St/Ki: Fifteen Years of Observations   475
  Annabel G. Liebelt

Differences between C57BL Strains at an Erythrocyte Antigen Locus   481
  Marianna Cherry, Donald W. Bailey, and George D. Snell

Subline Differences in Behavioral Responses to Pharmacological Agents   483
    Beatriz Moisset

Further Information on Subline Differences   485
    Thomas H. Roderick

WILD MICE: CLASSIFICATION, PROTEIN POLYMORPHISMS

Comments on the Relationship of Inbred Strains to the Genus *Mus*   497
    Michael Potter

Brief Review of European House Mice   511
    Joe T. Marshall

Genetic Heterogeneity of Spanish House Mice (*Mus musculus* Complex)   519
    Richard D. Sage

Biochemical Polymorphisms in Wild Mice   555
    Verne M. Chapman

The Extent of Allelic Diversity Underlying Electrophoretic Protein Variation in the House Mouse   569
    Francois Bonhomme and Robert K. Selander

Cytogenetics   591
    Orlando J. Miller and Dorothy A. Miller

WILD MICE: VIRUSES, T LOCUS, HISTOCOMPATIBILITY ANTIGENS

Population Genetics of T/t Complex Mutations   615
    Dorothea Bennett

Individual Mice of One Inbred Strain Produce Anti-H-2 Antibodies of Different Specificities   633
    Pavol Iványi and Paul de Greeve

Studies on Histocompatibility Antigens and Hybrid Sterility Gene in Wild Mice: A Short Survey   657
    Pavol Iványi

Characterization of H-2 Haplotypes in Wild Mice   667
    Jan Klein, William R. Duncan, Edward K. Wakeland, Zofia Zaleska-Rutczynska, Huei-Jen S. Huang, and Ellen Hsu

Four Major Endogenous Retrovirus Classes each Genetically
Transmitted in Various Species of *Mus* 689
    Robert Callahan and George J. Todaro

Index 715

Contributors

Numbers in parentheses indicate the pages on which the authors' contributions begin.

BRUCE W. ALTROCK (321), Department of Pathology, School of Medicine, University of California, Davis, California
DONALD W. BAILEY (197, 481), Jackson Laboratory, Bar Harbor, Maine
ROBERT W. BAKER (465), University of Pittsburgh, School of Medicine, Pittsburgh, Pennsylvania
DOROTHEA BENNETT (615), Laboratory of Developmental Genetics, Sloan-Kettering Institute for Cancer Research, New York, New York
D. R. BOGGS (463), University of Pittsburgh, School of Medicine, Pittsburgh, Pennsylvania
SALLIE S. BOGGS (463, 465), University of Pittsburgh, School of Medicine, Pittsburgh, Pennsylvania
FRANCOIS BONHOMME (569), Department of Biology, University of Rochester, Rochester, New York
ROBERT CALLAHAN (343, 689), Laboratory of Viral Carcinogenesis, National Cancer Institute, National Institutes of Health, Bethesda, Maryland
ROBERT D. CARDIFF (321), Department of Pathology, School of Medicine, University of California, Davis, California
VERNE M. CHAPMAN (555), Department of Molecular Biology, Roswell Park Memorial Institute, Buffalo, New York
MARIANNA CHERRY (481), The Jackson Laboratory, Bar Harbor, Maine
THOMAS M. CHUSED (297), Laboratory of Microbiology and Immunology, National Institute of Dental Research, National Institutes of Health, Bethesda, Maryland
D. COLCHER (343), Laboratory of Viral Carcinogenesis, National Cancer Institute, National Institutes of Health, Bethesda, Maryland
PAUL DE GREEVE (633), Central Laboratory of the Netherlands Red Cross, Blood Transfusion Service and Laboratory of Experimental and Clinical Immunology, University of Amsterdam, Amsterdam, The Netherlands
MARGARET K. DERINGER (99), Registry of Experimental Cancers, National Cancer Institute, National Institutes of Health, Bethesda, Maryland
W. DROHAN (343), Laboratory of Viral Carcinogenesis, National Cancer Institute, National Institutes of Health, Bethesda, Maryland

WILLIAM R. DUNCAN (667), Department of Microbiology, University of Texas Southwestern Medical School, Dallas, Texas

LORRAINE FLAHERTY (409), Division of Laboratories and Research, New York State Department of Health, Albany, New York

JACOB FURTH (69), Institute of Cancer Research and Department of Pathology, Columbia University, New York, New York

J. GAUTSCH (343), Scripps Clinic and Research Foundation, La Jolla, California

RALPH J. GRAFF (371), Waldheim Department of Surgery, Jewish Hospital of St. Louis, Washington University School of Medicine, St. Louis, Missouri

EARL L. GREEN (167), The Jackson Laboratory, Bar Harbor, Maine

MARGARET C. GREEN (157), The Jackson Laboratory, Bar Harbor, Maine

P. HAND (343), Laboratory of Viral Carcinogenesis, National Cancer Institute, National Institutes of Health, Bethesda, Maryland

JANET W. HARTLEY (289), Laboratory of Viral Diseases, National Institute of Allergy and Infectious Diseases, National Institutes of Health, Bethesda, Maryland

JOE R. HELD (29), Division of Research Services, National Institutes of Health, Bethesda, Maryland

RONALD B. HERBERMAN (461), National Cancer Institute, National Institutes of Health, Bethesda, Maryland

WALTER E. HESTON (109), 1380 Burgundy Drive, Fort Myers, Florida

HAROLD A. HOFFMAN (217), Comparative Pathology Section, Veterinary Resources Branch, Division of Research Services, National Institutes of Health, Bethesda, Maryland

D. HOWARD (343), Meloy Laboratories, Inc., Springfield, Virginia

ELLEN HSU (667), Department of Microbiology, University of Texas Southwestern Medical School, Dallas, Texas

HUEI-JEN S. HUANG (667), Department of Microbiology, University of Texas Southwestern Medical School, Dallas, Texas

JOHN J. HUTTON (235), Department of Medicine, VA Hospital and University of Kentucky, Lexington, Kentucky

PAVOL IVÁNYI (633, 657), Central Laboratory of The Netherlands Red Cross, Blood Transfusion Service and Laboratory of Experimental and Clinical Immunology, University of Amsterdam, Amsterdam, The Netherlands

R. A. JOYCE (463), University of Pittsburgh, School of Medicine, Pittsburgh, Pennsylvania

CLYDE KEELER (179), Central State Hospital, Milledgeville, Georgia

JAN KLEIN (667), Department of Microbiology, University of Texas Southwestern Medical School, Dallas, Texas

RICHARD M. KRAUSE (25), National Institute of Allergy and Infectious Disease, National Institutes of Health, Bethesda, Maryland

D. KUFE (343), Sidney Farber Cancer Institute, Boston, Massachusetts
R. LERNER (343), Scripps Clinic and Research Foundation, La Jolla, California
ANNABEL G. LIEBELT (475), Kirschbaum Memorial Mouse Colony, Northeastern Ohio Universities, College of Medicine, Rootstown, Ohio
FRANK LILLY (281), Department of Genetics, Albert Einstein School of Medicine, New York, New York
MARY F. LYON (445), MRC Radiobiology Unit, Harwell, Didcot, Oxon, United Kingdom
JAMES L. McCOY (461), National Cancer Institute, National Institutes of Health, Bethesda, Maryland
JOE T. MARSHALL (511), National Fish and Wildlife Laboratory, National Museum of Natural History, Washington, D.C.
MONTE S. MELTZER (461), National Cancer Institute, National Institutes of Health, Bethesda, Maryland
DOROTHY A. MILLER (591), Departments of Human Genetics and Development and Obstetrics and Gynecology, College of Physicians and Surgeons, Columbia University, New York, New York
ORLANDO J. MILLER (591), Departments of Human Genetics and Development and Obstetrics and Gynecology, College of Physicians and Surgeons, Columbia University, New York, New York
BEATRIZ MOISSET (483), Psychology Department, Temple University, Philadelphia, Pennsylvania
HERBERT C. MORSE III (3, 297, 441), National Institute of Allergy and Infectious Diseases, National Institutes of Health, Bethesda, Maryland
DONALD E. MOSIER (471), Laboratory of Immunology, National Institute of Allergy and Infectious Diseases, National Institutes of Health, Bethesda, Maryland
HELEN C. NAUTS (23), Cancer Research Institute, New York, New York
LLOYD J. OLD (391), Memorial Sloan-Kettering Cancer Institute, New York, New York
KENNETH PAIGEN (255), Molecular Biology Department, Roswell Park Memorial Institute, Buffalo, New York
KENNETH D. PATRENE (465), University of Pittsburgh, School of Medicine, Pittsburgh, Pennsylvania
MICHAEL POTTER (497), National Cancer Institute, National Institutes of Health, Bethesda, Maryland
THOMAS H. RODERICK (485), The Jackson Laboratory, Bar Harbor, Maine
DAVID L. ROSENSTREICH (457), Laboratory of Microbiology and Immunology, National Institute of Dental Health, National Institutes of Health, Bethesda, Maryland
WALLACE P. ROWE (289), Laboratory of Viral Diseases, National Institute of Allergy and Infectious Diseases, National Institutes of Health, Bethesda, Maryland

LUIGI RUCO (461), National Cancer Institute, National Institutes of Health, Bethesda, Maryland

ELIZABETH S. RUSSELL (33), The Jackson Laboratory, Bar Harbor, Maine

RICHARD D. SAGE (519), Museum of Vertebrate Zoology, University of California, Berkeley, California

ROBERT K. SELANDER (569), Department of Biology, University of Rochester, New York

G. SCHIDLOVSKY (343), Brookhaven National Laboratory, Upton, New York

J. SCHLOM (343), Laboratory of Viral Carcinogenesis, National Cancer Institute, National Institutes of Health, Bethesda, Maryland

GRETCHEN N. SCHWARTZ (465), University of Pittsburgh, School of Medicine, Pittsburgh, Pennsylvania

MICHAEL B. SHIMKIN (103), Temple University, Health Sciences Center, Philadelphia, Pennsylvania

GEORGE D. SNELL (119, 481), The Jackson Laboratory, Bar Harbor, Maine

ELISABETH STOCKERT (391), Memorial Sloan-Kettering Cancer Center, New York, New York

LEONELL C. STRONG (45), Leonell C. Strong Research Foundation, Inc., San Diego, California

ALDO TAGLIABUE (461), National Cancer Institute, National Institutes of Health, Bethesda, Maryland

BENJAMIN A. TAYLOR (423), The Jackson Laboratory, Bar Harbor, Maine

Y. TERAMOTO (343), Laboratory of Viral Carcinogenesis, National Cancer Institute, National Institutes of Health, Bethesda, Maryland

GEORGE J. TODARO (343, 689), Laboratory of Viral Carcinogenesis, National Cancer Institute, National Institutes of Health, Bethesda, Maryland

EDWARD K. WAKELAND (667), Department of Microbiology, University of Texas Southwestern Medical School, Dallas, Texas

ZOFIA ZALESKA-RUTCZYNSKA (667), Department of Microbiology, University of Texas Southwestern Medical School, Dallas, Texas

MARION M. ZATZ (467), National Cancer Institute, National Institutes of Health, Bethesda, Maryland

Historical Perspective on the Development of Inbred Mice

Herbert C. Morse III, Chairman
Richmond T. Prehn, Co-chairman

INTRODUCTION

Herbert C. Morse III

National Institute of Allergy and Infectious Diseases
National Institutes of Health
Bethesda, Maryland

"Happy are those who dream dreams and are ready
to pay the price to make them come true."
— L.J. Cardinal Suenens

"The introduction of inbred strains into biology is
probably comparable in importance with that of the
analytical balance in chemistry."
— Hans Gruneberg

The inbred laboratory mouse has become an indispensable tool in numerous areas of biological and medical research. Many fields of investigation either had their origin or altered their direction in response to the availability of this remarkable creature. Our current realization of the potential of the inbred mouse has developed over a period of more than 70 years and reflects the dedicated research of many skilled and imaginative scientists. Perhaps, because of the protracted period of time over which the utility of these mice became increasingly apparent and the attention given to striking developments arising from studies of recent investigators, the scientific community has never chosen to recognize the fundamental contributions made by the pioneering men and women responsible for the development of inbred strains, their worldwide distribution, and the early demonstrations of their potential for use in biomedical research.

On February 14, 1978, this oversight was partially corrected in the form of a historic awards ceremony paying tribute to those who made the most profound contributions to this field of science -- Drs. Clarence C. Little, Leonell C. Strong, Jacob Furth, Howard B. Andervont, Walter E. Heston, George D. Snell, Margaret C. Green, and Earl L. Green. The ceremony, attended by more than 200 students of various scientific disciplines, began with a series of talks, which comprise the first part of this book, by each of the honored scholars or their representatives. Dr. Elizabeth S. Russell spoke on behalf of Dr. Little who died in 1971. Dr. Margaret K. Deringer stood in for Dr. Andervont who was unable to attend for medical and personal reasons, and Ms. Jean Holstein spoke for Dr. Snell who contracted the "Russian flu" the weekend before the meeting. The ceremony closed with the presentation of medals (Fig. 1) and a monetary prize to each of the honored scientists. It was a grand occasion!

These collected talks, along with the presentation by another founding father of studies in inbred mice, Dr. Clyde Keeler, yielded new insights into the early history of mouse biology and should provide all students in this field with a perspective on this period of time which was not previously available. Additional references of major value in understanding the progress made in developing and using inbred strains are Dr. Keeler's 1931 monograph The Laboratory Mouse. Its Origin, Heredity and Culture (1), Biology of the Laboratory Mouse from the Jackson Laboratory (2), and articles by Staats (3), Heston (4), and Strong (5). Plans are also under way to publish a history of The Jackson Laboratory.

FIGURE 1. Recipients of awards for pioneering the development of inbred mouse strains. Back row: Margaret K. Deringer and John Andervont, recipients of the award for Howard B. Andervont; Robert A. Little, son of Clarence C. Little and recipient, along with Elizabeth S. Russell, of his award; Walter E. Heston; Earl L. Green; Ralph J. Graff and Marianna Cherry, recipients of the award for George D. Snell. Front row: Leonell C. Strong; Jacob Furth; Elizabeth S. Russell; and Margaret C. Green. (Photograph from the National Institutes of Health).

To those intimately familiar with the history of the laboratory mouse it will be clear that many gaps still remain in the literature on the people and events of early mousing. To name, at this time, others who made contributions of scarcely less importance than those of the award recipients would be invidious. I wish, however, to make two brief exceptions to this rule. These are to note the accomplishments of Dr. William E. Castle and Miss Abbie E.C. Lathrop, the former for his development of minds and the latter for her development of mice.

There are some remarkable similarities between the lives of these two people. They were born within one year of each other in small midwestern towns as children of school teachers. After obtaining their college degrees, both taught school for several years before moving on to other careers in Massachusetts. There, during the first decade of this century, they both became deeply involved in mammalian genetics and their pathways intersected. This fortuitous union resulted in the predominant use of the inbred mouse as a tool for studying "experimental evolution" for the next 70 years.

The story of William Ernest Castle's (Fig. 2) life is well known and his place in history secure. Castle was born in Alexandria, Ohio, in 1867. His interest in natural science, first expressed in his farm-boy childhood, was extended during his college years at Denison University under the tutelage of Clarence J. Herrick, professor of geology, zoology, and botany. Following graduation in 1889 he taught Latin for three years before moving to Harvard University to obtain a second A.B. degree and work under the direction of C.B. Davenport.

In 1895, Castle obtained his Ph.D. degree under E.L. Mark, having worked on the embryologic development of Ciona intestinalis. Two years later, he received an appointment to teach at Harvard where he was to remain until his retirement in 1936.

A listing of Castle's publications reveals an interesting gap between 1900 and 1903. It was during this time that papers by deVries, Correns, Tschermark, Davenport, and Bateson -- all related to the rediscovery of Mendel's work -- were published. These writings irreversibly altered his direction and from Castle's point of view, "Here was where genetics began" (6). Thereafter, the work of Castle and his students

FIGURE 2. Dr. William Ernest Castle, father of mammalian genetics, at age 93.

and the history of mammalian genetics in this country are inextricably linked.

The opportunity for Castle to extend his advocacy of Mendelian principles to advanced students of biology came about with the opening of the Bussey Institution for Applied Biology in 1908 (Fig. 3). From that time until its closing

FIGURE 3. The Bussey Institute of Harvard University. Here under the guidance of W.E. Castle and E.M. East, 40 students worked to complete their Doctor of Science degree. (Photograph from the Harvard University Archives).

(and his retirement) in 1936, Castle directed 16 students to completion of their Doctor of Science degree. Their names and the year in which they received their degree are shown as the "front" branches of the Castle family tree (Fig. 4). This construction originated from the contributions of Margaret M. Dickie on the occasion of Castle's 87th birthday and his investiture as an honorary trustee of The Jackson Laboratory (7). The "back" branches give a partial listing of the succession of visitors from the United States and other countries who benefited from Castle's formative influence in the development of mammalian genetics. These primary students have, in turn, been responsible for second and third

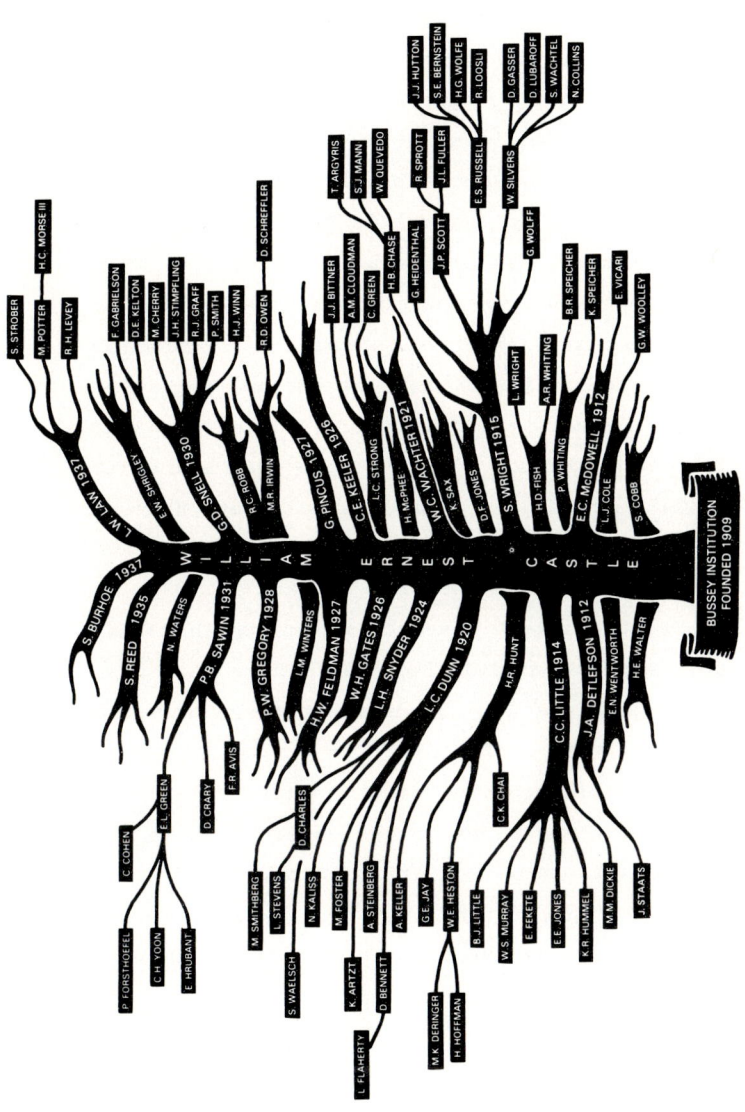

THE CASTLE "FAMILY TREE"

FIGURE 4. W.E. Castle's pervasive influence on mammalian genetics is illustrated by his intellectual family tree. Modified from (7).

generations of "Castlian" students, passing on the wisdom of the man and redeveloping the air of excitement which those at the Bussey experienced. This tree is far from being complete in the later generations, and I apologize to any who might feel slighted by not being included in its arborization.

Although only 13 of Castle's 246 publications were concerned primarily with mice, even a brief scanning of this family tree is sufficient to understand his pervasive influence on mouse genetics. Five of the award recipients from this meeting are found in branches of this tree -- Little, Strong, Snell, Heston and Earl Green -- as are a number of those who participated in the conference -- Keeler, Law, Russell, Bennett, Flaherty, Graff, Cherry, Potter, Staats, Hoffman, Hutton, and others. I am sure many of the readers of this book will be proud, as I was, to work out their place on this abundantly fruitful tree.

Following Castle's retirement from Harvard in 1936, he moved to Berkeley to become a research associate in mammalian genetics at the University of California. There he continued his lifelong interest in color inheritance of mammals in studies of horses with a special focus on the Palomino breed. After a brief illness, he died on June 3, 1962.

Miss Lathrop's contributions to the development of inbred mice are much less well known but are nonetheless of major historical import. I am not the first to draw attention to her work as she is mentioned briefly in Biology of the Laboratory Mouse (2), Biology of the Histocompatibility-2 Complex of the Mouse (8), in a paper by Potter and Lieberman (9), and she was the subject of an article by Shimkin in Cancer Research (10).

Miss Abbie E.C. Lathrop (Fig. 5) was born in Illinois in 1868 to parents, both school teachers, who had moved to Illinois from near Granby, Massachusetts. After receiving her degree from an unknown institution, she taught for several years before having to retire because of an illness which would later take her life, pernicious anemia. She moved to Granby around 1900 and, after a brief but futile fling at poultry farming, came upon the idea of raising small animals for sale as pets. "Her beginnings were small -- about as small as they could be, as she started with a single pair of waltzing mice which she got in this city [Granby, Mass.]. Her stock increased gradually and she advertised to buy mice. She received letters from people who thought she had them to sell, and she always made the effort to fill such stray orders. After she had sold 200 or 300 mice Miss Lathrop thought the resource of mouse farming as a business must be very nearly at an end, since the offspring from that number would be enough to supply pets for the entire younger generation, but the orders continued to come in" (11).

The source of the orders also changed drastically. Instead of receiving requests from mouse fanciers interested in obtaining creamy buffs, red creams, ruby-eyed yellows (very rare) or white English sables, she had orders for mice by the hundreds coming from research laboratories as far away as St. Louis and New York and as close as the Bussey Institute of Boston. It is more than likely that C.C. Little's classic studies on coat color genetics in mice, begun in 1907, were performed with mice obtained from Miss Lathrop's farm. Castle acknowledged, in a note sent to Dr. Michael Potter in 1958, that many of the mice used at the Bussey were purchased from Miss Lathrop (Fig. 6).

Abbie was initially confused by this increased interest in her mice, "but the truth came out that the mice are used in laboratories for scientific research as to the causes and treatment of various of the ills that human flesh is heir to. Cancer principally is greatly illuminated by the aid of these

FIGURE 5. Miss Abbie E.C. Lathrop at Granby. Redrawn from the Springfield Sunday Republican, October 5, 1913.

little creatures which are subject like people to the ravages of the fearful disease. In one of the cages of Miss Lathrop's mouse barn may be seen a lively little fellow with a lump upon his shoulder as big as a hickory nut. His days are numbered, for the cancerous tumor will strike a vital spot before very long and, with the delicacy characteristic of creatures low on the scale of life, he will promptly succumb" (11).

This article was written in 1913 and clearly reflects Miss Lathrop's, by then, substantial scientific experience in cancer research. At least five years earlier she had noted that her breeding program was being adversely affected by the development of "skin lesions," and mice were sent to

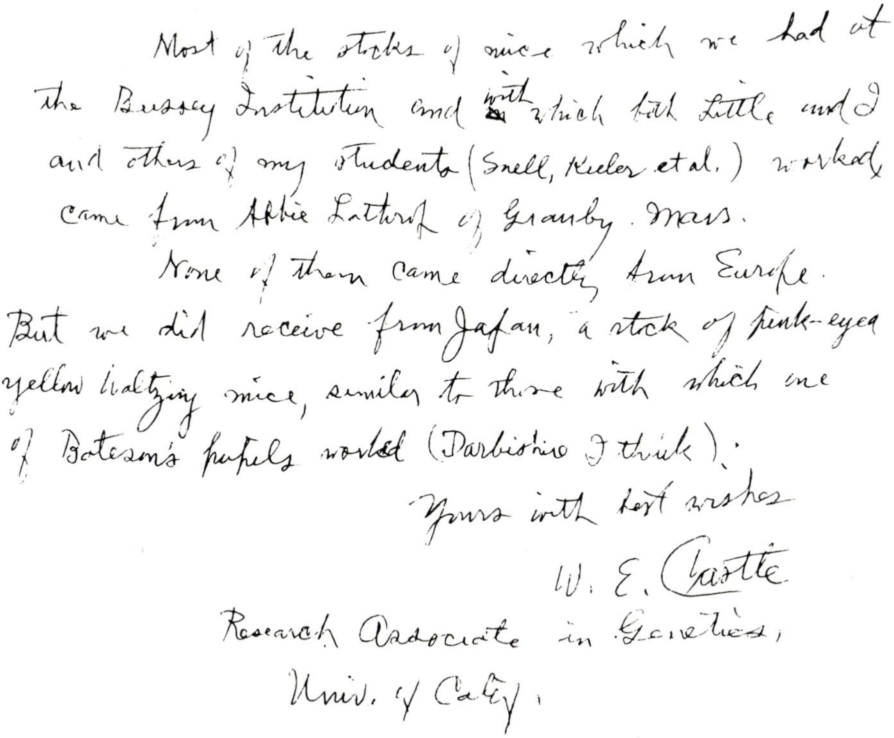

> Most of the stocks of mice which we had at the Bussey Institution and with which both Little and I and others of my students (Snell, Keeler et al.) worked, came from Abbie Lathrop of Granby, Mass.
>
> None of them came directly from Europe. But we did receive from Japan, a stock of pink-eyed yellow waltzing mice, similar to those with which one of Bateson's pupils worked (Darbishire I think).
>
> Yours with best wishes
>
> W. E. Castle
> Research Associate in Genetics,
> Univ. of Calif.

FIGURE 6. W.E. Castle's acknowledgment to Michael Potter that mice used at the Bussey were obtained from Abbie Lathrop. (Courtesy of Michael Potter).

a number of institutions purchasing mice to determine the cause of the malady. For her, the most significant recipient of these ailing mice was Leo Loeb, then at the University of Pennsylvania. His diagnosis was "cancer" and by negotiations, as yet unclarified, it was arranged that experiments designed by him would be carried out by Miss Lathrop in Granby (10). This union proved extremely fruitful and, between 1914 and 1919, resulted in the publication of ten manuscripts (listed in reference 10). Their studies demonstrated that a) the incidence of mammary tumors varied among different families of mice, b) the incidence of tumors in crosses between high and low tumor families was like that of the high strain, and

c) that pregnancy increased their frequency and that ovariectomy reduced it.

The careful records required to document these facts were kept in small hard-backed bound notebooks, now in the possession of Joan Staats at The Jackson Laboratory. A reproduction of one of the pages from these books is shown in Figure 7. These and other documents reveal that Abbie bred mice from the wild, foster-nursed mice, noted that the yellow coat color did not breed true, performed autopsies (Fig. 8, left), and had inbred mice to F_{12} by late 1915 (Fig. 8, right)! With approximately two and a half generations of breeding attained per year, this means she must have started these lines around 1910, only one year after C.C. Little initiated the inbreeding of his dilute brown non-agoutis. One wonders if this effort resulted from her contacts with Loeb or perhaps with Little.

It is quite remarkable to recognize that these scientific feats were accomplished during a period when the demands on her time to care for the business of selling mice and other pets were at their peak. From around 1910 until her death in 1918, Miss Lathrop's barn and sheds contained more than 11,000 mice, several hundreds of guinea pigs, rabbits and rats, and occasional ferrets and canaries. The mice were housed in light-tight wooden boxes, filled with straw, and were fed on a diet of crackers and oats. "Every 2 months 25 barrels of crackers must be laid in stock to supply their appetites, and a ton and a half of oats is their monthly portion" (11).

Children of the town were hired at seven cents an hour to periodically clean the cages and were allowed free access to the always fresh cracker barrels as a bonus. The cages were watered daily "in little jars which are first boiled as a protection against disease germs" (11). Of particular interest, in view of Dr. Earl Green's comments (this volume) on the primitive mouse watering provisions which he found in

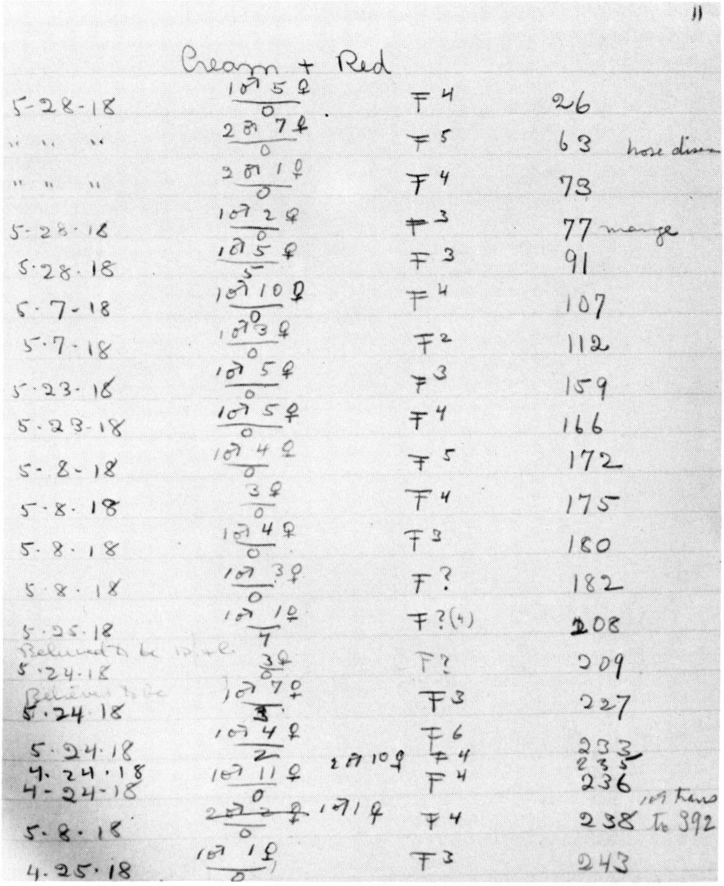
11

Cream + Red

5-28-18	1♂ 5♀ / 0	F4	26
" " "	2♂ 7♀ / 0	F5	63 nose dis[illegible]
" " "	2♂ 1♀ / 0	F4	73
5-28-18	1♂ 2♀ / 0	F3	77 mange
5·28·18	1♂ 5♀ / 5	F3	91
5·7·18	1♂ 10♀ / 0	F4	107
5·7·18	1♂ 3♀ / 0	F2	112
5·23·18	1♂ 5♀ / 0	F3	159
5·23·18	1♂ 5♀ / 0	F4	166
5·8·18	1♂ 4♀ / 0	F5	172
5·8·18	3♀ / 0	F4	175
5·8·18	1♂ 4♀ / 0	F3	180
5·8·18	1♂ 3♀ / 0	F?	182
5·25·18	1♂ 1♀ / 7	F?(4)	208
Believed to be [illegible] 5·24·18	3♀ / 0	F?	209
Believed to be 5·24·18	1♂ 7♀ / 3	F3	227
5·24·18	1♂ 4♀ / 2	F6	233
4·24·18	2♂ 10♀	F4	235
4-24-18	1♂ 11♀ / 0	F4	236
5·8·18	2♂ 2♀ 1♂ 1♀ / 0	F4	238 1♂ taken to 392
4·25·18	1♂ 1♀ / 0	F3	243

FIGURE 7. A page from Miss Lathrop's breeding records made in the final year of her life. (Courtesy of Joan Staats).

Ohio as a student, is the observation that "on some of the [mouse] cages, Miss Lathrop has rigged a device for giving them water. It is a bottle with a feed tube, from which a drop constantly hangs down into the cage, and a thirsty mouse has only to stand on his hind legs to quaff a cooling drink" (11).

Abbie's forerunner of The Jackson Laboratory production unit sold mice to laboratories across the country. The mar-

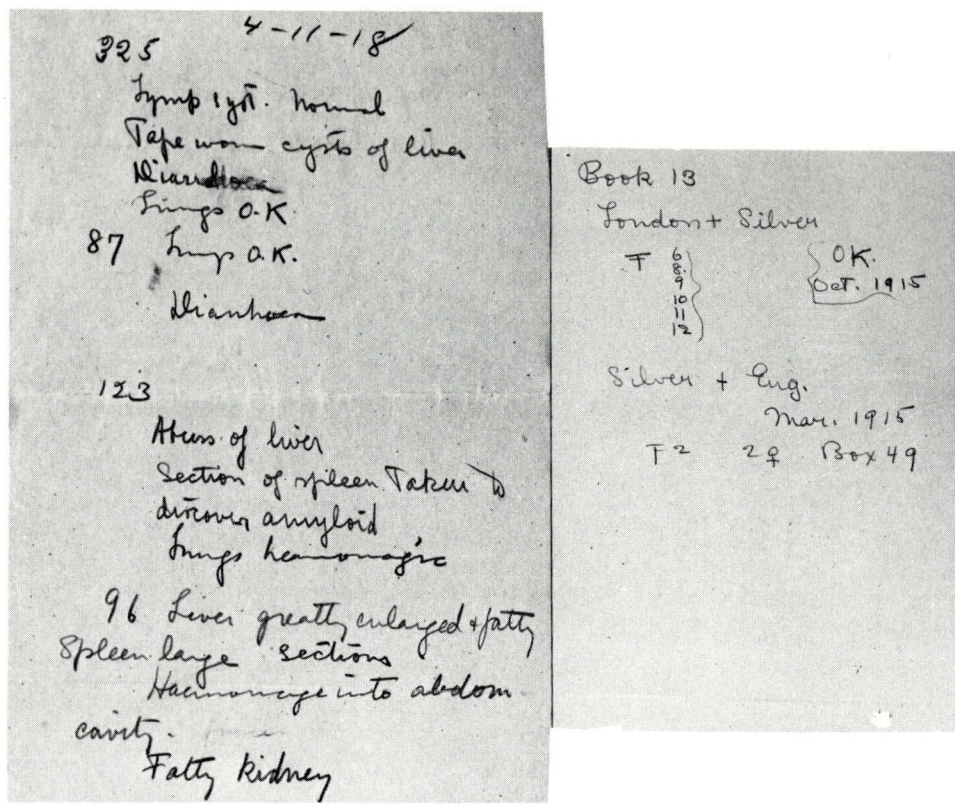

FIGURE 8. Autopsy reports prepared by Miss Lathrop (left side) and further documentation of her inbreeding through F_{12} by October 1915 (right side).

ket price varied according to the age of the mice, their color and other points, ranging from $10 to $20 per hundred. However, she continued to sell fancy mice as pets to the local children. In return for his 15 cents a child would receive a waltzing mouse or hooded rat and a brief lecture on pet care. "In schoolmarm fashion, she warned against neglecting the mice or handling them too much" (12).

Miss Lathrop actively pursued her business and scientific interests in her mice until her death at the age of 50 in 1918. Her grave in West Cemetery, Granby, is marked by a simple stone. The little white house by the side of the road near Five Corners in Granby is all that stands of a reminder of the Granby Mouse Farm (Fig. 9). From time to time, there have been suggestions that a marker be placed at the farm to honor Miss Lathrop's accomplishments, but this has never been done. Nonetheless, the memory of Miss Abbie Lathrop will not soon be forgotten for almost all of us involved in mouse biology have daily reminders of her contributions in the form of today's inbred mice. Their roots, in many cases, can be traced back to ancestors who spent their lives in the wooden boxes of Miss Lathrop's farm (Fig. 10).

These and other inbred strains shown in this figure were the focus of the rest of the meeting and this book.

The basic science areas considered in this publication touch upon a wide spectrum of disciplines including cancer research, immunology, biochemistry, hematology, pharmacology, virology, cytogenetics, theoretical and applied genetics, and zoology. It was clearly impossible to cover all fields of investigation in which the inbred mouse plays a major role, but this sampling of these fields provides a cross section of scientific endeavors which are rarely included in a single program or volume. This format provided some unique opportunities for cross-fertilization among scientists involved in highly divergent fields of research and will give the reader of this text a taste of some of the more recent and intriguing developments in murine biology.

Finally, I wish to express my gratitude to those who did so much in helping with the planning for this meeting and publication of this book. I am personally indebted to Drs. Michael Potter, Wallace P. Rowe, and Lloyd J. Old for the many hours they spent providing advice, support, and encouragement. I also wish to thank Dr. Richard Asofsky for his enduring understanding of and many contributions to my peri-

FIGURE 9. The Granby Mouse Farm as it appears today.

patetic wanderings in science and the development of this program and publication. This effort would have been impossible without generous financial support from the Cancer Research Institute through Mrs. Helen C. Nauts, the National Institute of Allergy and Infectious Diseases through Drs. John Seal and Richard Krause, from the National Cancer Institute through Dr. William Terry, and the Division of Research Services of the National Institutes of Health through Dr. Joe Held. Joan Staats of The Jackson Laboratory and Ellen DiCarlo of the Holyoke (Mass.) Transcript-Telegram provided invaluable assistance in piecing together Miss Lathrop's history. Dr. William B. Castle, W.E. Castle's son and my professor of hematology in medical school, was very helpful in directing me to sources of information on his father's life

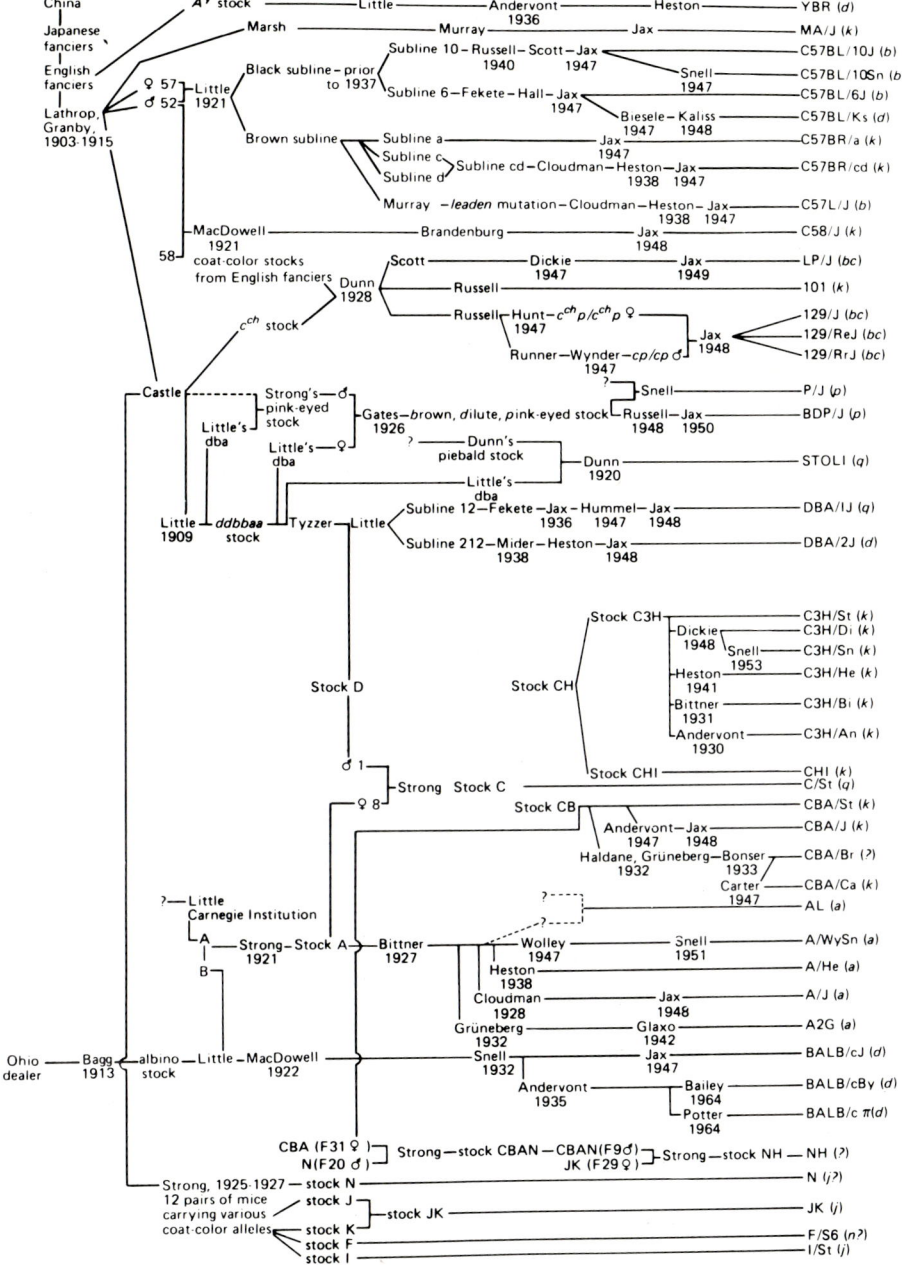

FIGURE 10. The genealogy of inbred mice.

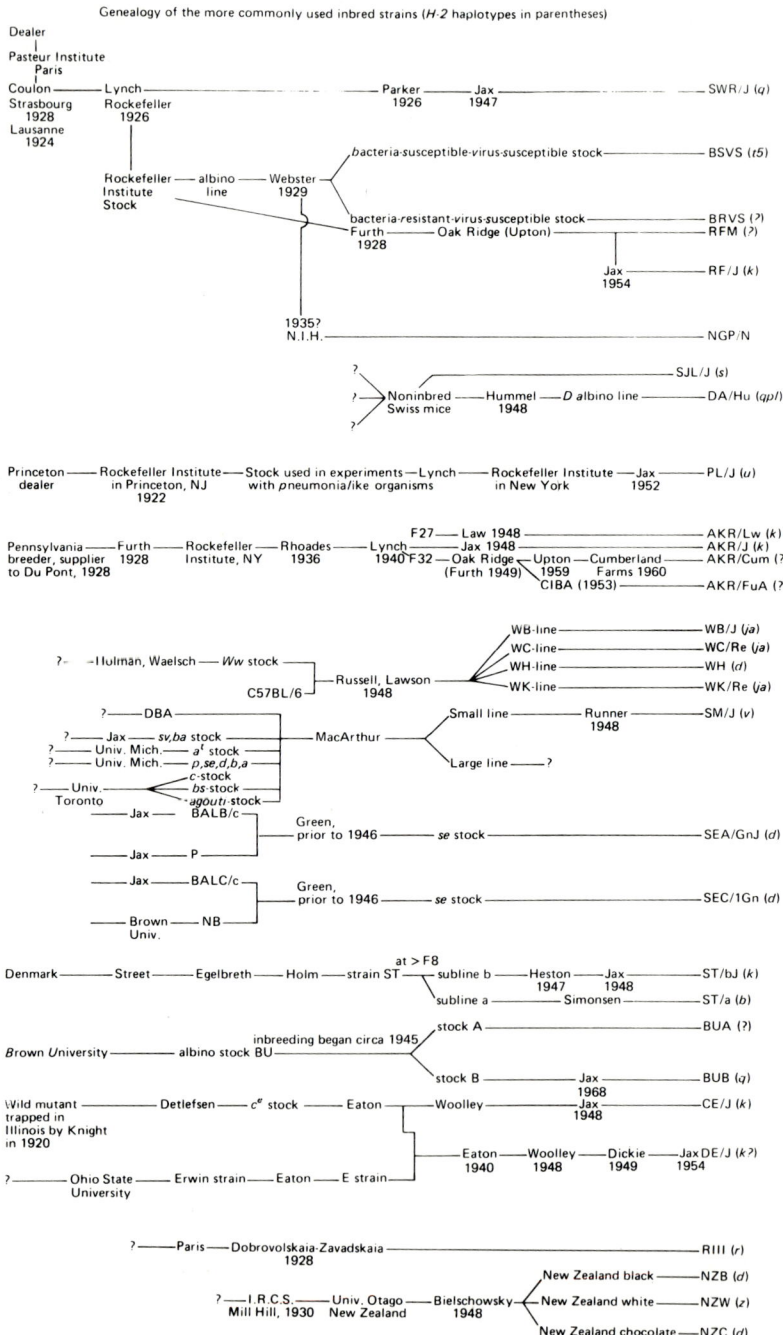

Genealogy of the more commonly used inbred strains (*H-2* haplotypes in parentheses)

*Based, in part, on Potter and Lieberman (1967), extended by Klein (1975); revised 1978.

and professional activities. Lastly, I wish to thank Academic Press for their courage in supporting this effort and to express my gratitude to Stoney Welsh for his multiple contributions to the development of the meeting and this book.

REFERENCES

1. Keeler, C.E. (1931). "The Laboratory Mouse. Its Origins Heredity, and Culture." Harvard University Press, Cambridge.
2. Staff of The Jackson Laboratory. (1966). "Biology of the Laboratory Mouse." Second Edition. McGraw-Hill Book Company, New York.
3. Staats, J. (1976). Cancer Res. 36:4333.
4. Heston, W.E. (1949). In "Lectures on Genetics, Cancer, Growth and Social Behavior," p. 9. Roscoe B. Jackson Memorial Laboratory, Bar Harbor, Me.
5. Strong, L.C. (1942). Cancer Res. 2:531.
6. Dunn, L.C. (1965). In "Biographical Memories." National Academy of Sciences 38:33.
7. Russell, E.S. (1954). J. Hered. 45:211.
8. Klein, J. (1975). In "Biology of the Mouse Histocompatibility-2 Complex," p. 30. Springer-Verlag, New York.
9. Potter, M., and Lieberman, R. (1967). Adv. Immunol. 7:91.
10. Shimkin, M.B. (1975). Cancer Res. 35:1597.
11. Springfield Sunday Republican, October 5, 1913, p. 12.
12. Holyoke (Mass.) Transcript-Telegram, July 12, 1968, p. 14.

INTRODUCTORY REMARKS

Helen C. Nauts

Cancer Research Institute
New York, New York

It is a privilege for the Cancer Research Institute to co-sponsor this workshop as a way to honor the scientists whose pioneering efforts gave us the inbred mouse. This experimental animal continues to be one of the most powerful resources available to science in the investigation of the diseases that plague mankind.

Not only have studies in inbred mice illuminated our understanding of a range of pathological processes, such as infectious diseases and metabolic and neurological disorders, they have given us penetrating insights into normal biological mechanisms. Nowhere is this more beautifully illustrated than in our growing comprehension of the intricacies of the immunological system.

But it is in the area of cancer that the inbred mouse has made its most enduring contribution to our growing understanding of this major disease of mankind. The role of viruses in the development of cancer, the influence of genetics on cancer susceptibility, insights into chemical carcinogenesis and the involvement of endocrine and nutritional factors are some of the landmarks of scientific inquiry into cancer made possible through the development of the inbred mouse.

The Cancer Research Institute, in the 25 years since it was founded in New York City, has helped to initiate, support, and coordinate research into immunological approaches to the control, and hopefully the eventual prevention of cancer. Much of the work that has been supported by the Institute, particularly in our International Fellowship Training Program, focuses on studies in inbred mice.

The Trustees and Scientific Advisory Council of our Institute join me in congratulating those who are honored today for their outstanding achievements. No award, however, can be sufficient to express the appreciation of the scientific community for the precious gifts each one of you has given us.

INTRODUCTORY REMARKS

Richard M. Krause

National Institute of Allergy
and Infectious Diseases
National Institutes of Health
Bethesda, Maryland

It is a special pleasure for me to participate in this Workshop on the Origins of Inbred Mice, and the National Institute of Allergy and Infectious Diseases is proud to take part in this program. I congratulate, also, the awardees who will be honored later this morning.

I welcome you all to the NIH. But apart from my official greeting, I welcome you from a very personal point of view. My research career began at the Rockefeller University, then the Rockefeller Institute for Medical Research. During the course of nearly a quarter century, I came to know Clara Lynch and to admire her contributions to the development and use of inbred mice in probing the genetic determinants of resistance and susceptibility to infections and tumors.

Clara Lynch joined Murphy's laboratory in 1920. She inbred her mice so that the strains became progressively more uniform. During the inbreeding, she noted that two strains of separate origin showed a significant difference in incidence of lung tumors. Subsequent breeding studies demonstrated that susceptibility to these tumors is genetically controlled.

But all was not tumor biology for Clara Lynch. Infection was then, and still is today, modulated by genetic forces. Lynch used her Swiss mice as one of the strains to show that susceptibility to yellow fever virus was influenced by hereditary factors. This work on yellow fever had practical consequences in the development of mouse susceptibility tests for use in epidemiological surveys. In later years, Lynch also examined the genetic susceptibility to tuberculosis.

Perhaps not all of you participating in this workshop know the origin of the Swiss mice. Lynch introduced them to this country from a laboratory in Lausanne. The original population was two males and seven females which Dr. Lynch brought into this country in a shoebox, kept in her stateroom on board ship. Science was lived at a more leisurely pace in those days. Incidentally, she arrived and proceeded through customs with no permit from the Department of Agriculture! Lynch, of course, exploited this strain and derivatives of these Swiss mice have been used throughout the world.

There was in the early years, I am sure, more than one shoebox in the origins of inbred mice -- a legacy of those honored and assembled here. And from those simple origins sprung an intricate genealogy. But how you do complicate matters with a whole new lexicon which rhymes in no one's ear except your own! You go on and on with BALB/c, MOPC, A/J, SWR -- sort of a New Deal bureaucracy for a latter day Orwellian animal farm.

I am most familiar with what is done in immunology with inbred mice. But the matter is not yet all of a piece. What exactly are the genetic determinants of immunity? There is a subtlety here that still escapes us. It is as if nature were deliberately evasive. But it will come. Spring is in the air -- even on a late winter day like this. "In March -- the

lawn is full of south," said Emily Dickinson, the Belle of
Amherst -- "the lawn is full of south."

My own work has taken me far afield and into immunogenetic studies with rabbits. Indeed, in May we will be sponsoring here at the NIH a conference on rabbit genetics. When I mentioned this to my good friend, Wally Rowe, he chortled with some amusement and said, "Just like rabbit geneticists, still running behind the mouse geneticists." But my students have moved with the times -- if that is what it is -- and they are mouse immunologists.

I have talked too long, but when asked to greet you at this Workshop on the Origins of Inbred Mice, it occurred to me that, in a manner of speaking, I was present at the beginning through my association with Clara Lynch.

INTRODUCTORY REMARKS

Joe R. Held

Division of Research Services
National Institutes of Health
Bethesda, Maryland

On behalf of the Division of Research Services, it is a pleasure to welcome you here this morning. We are pleased to have the opportunity to honor the men and women whose vision and foresight contributed so much to our present understanding of basic biology.

The initial reason for developing inbred strains was to study cancer. More than 100 years ago neoplasms occurring in animals were determined to be qualitatively similar to those found in humans. Shortly after the turn of the century, genetically defined animals became necessary for tumor transplant studies. It was observed that transplantation of spontaneously developing neoplasms failed except in closely related individuals. This led to the almost explosive development of inbred strains of animals in which the awardees played a critical role.

The Division of Research Services is committed to extending these pioneering efforts. An important mission of the Division's Veterinary Resources Branch (VRB) is to support NIH intramural research by providing healthy, genetically defined research animals. This was the primary reason for the Branch's organization in 1948 when it was known as the Laboratory Aids Branch.

The emphasis of VRB's programs has changed over the past quarter century. Initially, the emphasis was on large-scale animal production with minimal genetic definition. With the passage of time the demand by NIH investigators for more precisely defined animals has increased. Today, in terms of total numbers, the production of genetically defined inbred rodents is greater than that of outbred stocks.

One of VRB's roles in collaborating with the research institutes is particularly pertinent to the theme of this meeting, for it is within this organization that the various strains and stocks of research rodents are maintained and produced for the NIH. Included among these animals are practically all of those strains developed by the men and women to be honored here today.

In recognition of the usefulness of such animals to the biomedical research community at large, the NIH collection of rodents has been designated an international Genetic Resource by the World Health Organization. More recently, VRB has been designated by the International Committee of Laboratory Animals as a Resource Center for Nude Mice. The Division is committed to ensuring the genetic and microbiologic integrity of these animals. Later today you will hear about one of its quality assurance programs.

The change in the kind of research animal required is also reflected in the composition of the VRB-managed Genetic Resource. The initial group of 14 inbred mouse strains consisted of nearly all of the descendants of the original strains developed by these pioneers in the first decades of this century. Since then, the mouse resource has increased to nearly 100 inbred, congenic, and recombinant inbred strains.

The success of this resource could not have been achieved without the internal commitment by the staff to quality. Animal quality is dependent on health and genetic components. Both require appropriate management strategies.

The adaptation of gnotobiotic techniques to large-scale production has been one of the most important factors contributing to the development of the Genetic Resource. These techniques are effective in substantially reducing the risk of catastrophic loss of irreplaceable genetic material due to disease outbreaks. Maintaining and producing research animals under these conditions also results in a superior research tool since the experimental results are not compromised by the effects of subclinical or acute infections.

Gnotobiotic techniques have also had an enormous impact in extending the size of the resource. Introduction of new animals into an already established facility always involves a certain element of risk. By utilizing gnotobiotic technology, the degree of risk is substantially reduced. The Genetic Resource has been able to respond to the many requests by the research community for developing and maintaining specific animal models. Additionally, the use of gnotobiotic techniques has provided a means of protecting animal models with profound immunological defects. These newer models will further extend the value of the inbred strains developed by the awardees.

The Division of Research Services is proud of its role in continuing to build upon the foundations laid down by the farsighted individuals being honored here today. We are also pleased to be cosponsors of the workshop planned for these three days. It should afford a good opportunity for those scientists in attendance to exchange information related to the use of mouse models. We are also hopeful that communication will be continued in future meetings such as these.

ORIGINS AND HISTORY OF MOUSE INBRED STRAINS:
CONTRIBUTIONS OF CLARENCE COOK LITTLE

Elizabeth S. Russell

The Jackson Laboratory
Bar Harbor, Maine

It is a great privilege for me to be asked to describe the contributions of Clarence Cook Little to the origin, maintenance, and widespread use of mouse inbred strains. I wish he could be here to tell us in person about this fascinating business, for his recounting of the tales would add to the joy of all of us. "Prexy" -- I will have to call him by that name from here onward, since Clarence Cook Little is much too formal a title for such a friendly man -- was one of the world's best raconteurs, with a real relish for the interplay of human personalities.

I deem myself luckier than the other speakers this morning. They will inevitably feel constrained to be strictly scientific and objective and thereby will be overmodest in describing their personal contributions. Since I will be describing another person's role, it is appropriate for me to say that Prexy's character and personality, as well as his scientific acumen, contributed greatly to the origin and growing importance of inbred mice.

In preparing this account, I have asked many questions and received invaluable information from many people. I would like at this time to acknowledge special help from the direct memories of Elizabeth Fekete, Joe Murray, and George Snell; Joan Staats' invaluable archives and her Classified Bibliography of Inbred Strains of Mice; from the excellent biographical memoir on Little prepared by George Snell for the National Academy of Sciences; from the early scientific papers by the original Jackson Laboratory staff members; and from recent historical searches by Jean Holstein.

Three basic interests directed the scientific career of Clarence Cook Little: genetics, biological individuality, and cancer (1), but his influence went far beyond the scientific papers he contributed. Prexy was altruistic, charming, full of charisma, enthusiastic, idealistic, innovative, and in some ways unconventional and impractical but very persistent. He loved life. He had great powers of persuasion, but also appreciated and supported the efforts of his fellow man. Sometimes he may have placed too much trust in the altruism and idealism of others. He was a true "evangelist of science" and, to an extraordinary degree, his unique qualities fostered and shaped the history of inbred mice.

Clarence Cook Little was born in 1888 in Brookline, Massachusetts, where he grew up surrounded by a variety of animals, including dogs, cats, pigeons, and mice. His father introduced the long-haired dachshund to the United States, and both he and Prexy were much in demand as judges at dog shows. It is easy to see why he was attracted to biology when he was an undergraduate at Harvard College from 1906 to 1910. This was a heady period in the development of the new science of genetics. Following the rediscovery of Mendel's papers in 1900, biologists were eager to identify segregating factors (no one yet called them genes) and to check out the working of his laws on segregation and independent assortment in many different species. My uncle, George Harrison Shull, wrote a theoretical paper, "On the Composition of a Field of Maize" (2), about gene distribution in a cross-breeding species in 1908 when Prexy was a junior. Johannsen, in Denmark, had developed pure strains of beans by inbreeding through repeated generations of self-fertilization and was studying the interactions of genes and environment. The Bussey Institute of Harvard University, established in 1909 with William E. Castle as director, became a center for both plant and animal genetics, and at the Institute there was much discussion of genetic fixation by inbreeding. Sewall Wright, whom I regard as one of the deepest thinkers about principles of inbreeding and the types of characters fixed (3), was one of Prexy's fellow students and friends, as was E.C. MacDowell with whom he later worked at Cold Spring Harbor. Castle sponsored students working on genetics even earlier than 1909, and Prexy began working with him in 1907 on the inheritance of mouse coat color. Castle and Little published two papers from his undergraduate work (4, 5), the first establishing the pink-eye ($\underline{p}$) locus as distinct from dilute ($\underline{d}$) and albinism ($\underline{c}$), and the second deducing that dominant yellow ($\underline{A^y}$) is a lethal homozygote, not an easy conclusion in those days. That second paper, in particular, makes very exciting reading. No wonder the Jackson Laboratory has always welcomed student research trainees! Prexy also started in 1909 to mate brother-sister pairs to maintain his newly established homozygous dilute brown non-agouti stock. It would be very interesting to know when and how Prexy's major interest in his dilute brown non-agouti mice shifted from their coat color to their cancer incidence.

After graduation, he remained at the Bussey Institute to obtain his D.Sc. degree in 1914, and worked with E.E. Tyzzer at Harvard from 1914 to 1916. By 1915, he was arguing with Maude Slye in Science about the inheritance of cancer (6, 7). She followed incidence of cancer in long pedigrees of mice, but did not, of course, inbreed them, and never attached a

Mendelian interpretation to her results. From 1915 on, Little and Slye continued to argue about heredity and cancer, and I enjoyed sitting in on such a confrontation in 1936. Prexy worked on genetic control of acceptance of tumor transplants with Dr. Tyzzer, and their work resulted in two papers based largely on crosses between Japanese waltzing mice and mice from the "most homogeneous stock of common (house) mice that has been bred at the laboratory of the Bussey Institution. All the present animals are direct descendants of a single pair of closely related dilute brown (silver fawn) mice obtained in the spring of 1909. From the start the stock has been kept free from any outcross and has therefore an unbroken stretch of more than twenty generations of inbreeding" (8). Also in 1916, Little, discussing the relation of heredity to cancer in man and animals, philosophized that animal studies "have proved beyond question that hereditary factors play an extremely important part in determining the incidence of cancer in mice," but "none of the material thus far employed in investigations with spontaneous tumors meets the requirement of 'homogeneity of genetic constitution' needed to determine the 'method of inheritance' of susceptibility to cancer" (9). By 1916, he must already have been dreaming of using his DBA mice for studies of mammary tumor susceptibility.

Prexy was in the U.S. Army during World War I, and following his discharge late in 1918, he decided to accept a position at the Carnegie Institute of Washington at Cold Spring Harbor, rather than to return to Harvard to work with Tyzzer. He wanted to devote full time to research on genetics and cancer in mice, and felt this would be more possible in a laboratory devoted completely to research than at a medical school. His correspondence at this time reveals that Tyzzer had maintained the DBA inbred stock through the war years, for which we must all be grateful (Holstein, personal communication). Prexy built up his DBA colony at Cold Spring Harbor and also started the development of a new "family" of inbred strains, descended from mice received from Miss Abbie Lathrop, a mouse supplier from Granby, Massachusetts. She seems, by the way, from records now in the possession of Joan Staats at the Jackson Laboratory, to have been a very responsible person who knew a great deal about the characteristics of her animals. In his line C, descended from Lathrop's stock, Prexy mated female 57 and female 58 to male 52. The descendants of female 57 became the C57BL inbred line, the descendants of female 58, the C58 inbred line. Prexy himself inbred the C57 stock, and his colleague, E.C. MacDowell (a former Castle student), developed the C58 line. (Later on, J.M. Murray was responsible for inbreeding the portion of the pedigree that eventually became C57BR and C57L.)

During the decade of the 1920s, Little became involved in educational administration and innovation, but insisted at the same time on continuing and expanding his research on genetics and cancer. At this period he seems to have been a central figure in an informal group called "Mouse Men of America," whose members exchanged information and mouse stocks and got together at scientific meetings. The membership, which grew to approximately 100, included Prexy and his growing research group, Halsey Bagg, L.C. Dunn, John Gowen, and E.C. MacDowell (J.M. Murray, personal communications). Some of the present-day subline differences may possibly be attributed to that extensive interchange of stocks not yet genetically fixed.

Little left Cold Spring Harbor in 1922 to become president of the University of Maine. His laboratory was set up at the Agricultural Experiment Station, and during the three years at Orono Joseph M. Murray, William S. Murray, and Arthur M. Cloudman joined the group as graduate students working with particular mouse stocks. Bill Murray's responsibility was the DBA stock, Joe's the C57BR stock. In 1925, Prexy became president of the University of Michigan, but one of his stipulations for accepting the position was that his research laboratory should be continued and enlarged. Graduate students and mouse stocks moved to Ann Arbor, where they were soon joined by Leonell Strong and his mouse stocks.

Here I must insert an aside, to say that I first met Prexy in 1925, because his daughter Louise was a University High classmate and close friend. I remember attending a Christmas party at the President's House -- in 1927 I think -- and admiring from afar Prexy's son Bob, who is here today.

At both Orono and Ann Arbor Prexy introduced fine new programs but, particularly at Michigan, he seems to have attempted too much too soon, and he met with considerable opposition. Although he had administrative problems, research blossomed in his laboratory, his graduate students were trained in the medical school and the zoology department, and new members were added to the group: John J. Bittner, Elizabeth Fekete, and Charles V. Green.

Prexy's move from Orono to Ann Arbor had been promoted by a group of Detroit industrialists who were also Mount Desert Island summer residents. When administrative storm clouds gathered at the University of Michigan, these friends asked Prexy what he wanted to do next. He wanted to build a laboratory devoted entirely to research on mammalian genetics and cancer, far from the confines of academia. They promised to support this endeavor, and Prexy resigned from the University of Michigan in the spring of 1929. I'm pleased to say, however, that he presented a very inspirational graduation address to the University High School class of 1929.

With Prexy's guidance, the original building of the Jackson Laboratory was constructed in the summer and fall of 1929. Gradually the mice moved in, and all the researchers (Bittner, Cloudman, Fekete, C.V. Green, J.M. Murray, W.S. Murray, and L.C. Strong, and Prexy himself) were there by the spring of 1930. The research productivity of the first years at the Jackson Laboratory was very high -- it all must have been a good idea! For the purposes of this conference, it may be helpful to know how many and which inbred strains were then available at the Jackson Laboratory. I have used Joan Staats' Classified Bibliography to ascertain that the approximately 50 papers published between 1929 and 1936 reported work involving the following inbred strains: A, BALB/c, CBA, C57BL, C57BR, C57L, C58, DBA, N, and I; crosses between these strains, and interspecies crosses to Mus bactrianus and Mus wagneri; plus several different mutant genes. Two sublines were recognized in DBA; I know Elizabeth Fekete worked with DBA/1 and in 1937 George Woolley worked with DBA/212. C57BR/a and C57BR/cd were also distinguished. Little was directly responsible for the inbreeding of only the DBA and C57BL strains, but through the establishment of the Jackson Laboratory, he contributed greatly to the continuation and characterization of the whole array.

It was amazing that so much could be accomplished in the early years of the Jackson Laboratory, particularly as they were working under extreme financial difficulties. The year 1929 was not an ideal time for establishing what was intended to be a research institution supported entirely by private funds. Remember, there was no NIH, and no extramural granting capability way back then. I think we can attribute much of the success to Prexy's special qualities of persistence, charisma, enthusiasm, idealism, and interest in people. A particularly touching example is the publication in 1933 of the important paper on "The Existence of Non-chromosomal Influence on the Incidence of Mammary Tumors in Mice" by the staff of the Jackson Laboratory (10). That's team work! Incidentally, that publication was all based on DBA, C57BL (Prexy's own strains), and reciprocal crosses between them. When the supply of money dried up, as it did in 1932 and 1933, the staff voted to cut salaries back below "bare bones" budgets, lived as much as possible off the land, combined households, but continued to raise and study mice! That period of financial stringency had another effect on mammalian genetic research, which I feel has been very beneficial. The system I sometimes refer to as "Operation Bootstrap" was initiated then, I think following a suggestion by Bittner. Genetically controlled mice were supplied to outside investigators to support Jackson Laboratory research. That source of income pulled the Laboratory through a very tight time, and its continuation, gradually developing into more sophisticated arrangements, has become an important feature in the organization of biomedical research in the United States.

From all accounts I have heard, the Jackson Lab was a happy place in those early years despite the financial worries. Prexy's charisma held them together!

My assignment is to tell you about Prexy's contributions to inbred strains of mice, not to give you the entire history of the Jackson Laboratory, though the two are definitely related. By 1937, when I arrived at the Lab as an independent investigator, C.V. Green had died and Joe and Bill Murray and Leonell Strong had left for other positions (still mousy, I hasten to add). With the addition of new staff members George Snell, George Woolley, and Bill Russell, the research program became more diversified. After all, the genetic fixation that makes inbred strains useful for cancer research also fixes other characteristics that may well be studied to man's benefit. Prexy was interested in all the new undertakings, and very supportive. The Lab began to grow, with Prexy always responsible for obtaining funds. More formal student programs appeared, as well as visiting investigators, more mutant genes, and more inbred strains. We had become more systematic. One such effort you may want to hear about was carried out in 1937. From the beginning of the Jackson Laboratory, each staff member had maintained his own mouse colonies, though he would willingly supply a "breeding start" to others in or outside of the Laboratory, as geneticists have always done. Almost everyone had a colony of C57BL, and the incidence of minor anomalies such as small right eyes, malocclusion, and irregularities in position of the sacrum seemed to differ between colonies. Bill Russell had a huge pedigree chart tying together all of the C57BL colonies presently or recently at the Laboratory, and all of the stock which had been sent to outside investigators. There were 10 distinct sublines, many of them separated more than 20 generations previously. The big colonies in 1937 were C57BL/6, which Bill adopted, and C57BL/10, which George Snell maintained. The numbers were arbitrarily assigned of course. Within the sublines incidences of anomalies were similar, between sublines there were greater differences. Between 1937 and 1947, the Laboratory grew in physical size, number of staff members, and number of mice; it was a productive period, despite World War II, involvement in which led to the provision of large numbers of mice for study of tropical disease organisms.

Then came the drastic fire of October 1947 which resulted in destruction of most of the main laboratory and in loss of almost every mouse. That's when we came to appreciate the benefits of having provided inbred mice to other investigators.

Letter after letter came in, offering stocks to help us get
started again. Once more, Prexy's charisma kept the Lab
together and, unlike the crisis of 1932-1933, this time the
problem came at a particularly favorable time, if such can be
said of a catastrophe. This was the beginning of growth of
"big science," of research grants, of much more need for
inbred mice. The destruction of the Jackson Laboratory mouse
colonies, just when demand was increasing, led to much deeper
appreciation by others of the former availability of these
animal resources. In what I hope was a sensible plan, we
accepted the offers of mice most closely related to our former
stock, added some new special types that were generously
offered, built up a common colony from which both in-house
research colonies and animal resource colonies were derived,
and arranged to carry out recharacterization of the life
histories of mice from inbred strains in the common foundation
colony. Prexy once more was the enthusiastic leader of a
crusade with many devoted followers. All experiments in progress were lost, and some staff members and assistants had to
move to other laboratories for a short period, while those
who remained at Bar Harbor first piled on top of each other
at the Hamilton Station (site of behavior genetic studies of
dogs and of rabbit genetics), then sat at desks on the open
floor of a half-built animal house which was under construction at the time of the fire, and nursed the miniscule colonies
of mice which began to trickle in. It was a wild time, but
everyone seems to have loved it. And soon we had a new big,
beautiful (for that time) mouse wing, and then new office and
lab space and an auditorium. Phoenix rose from the ashes!
The times were favorable, yes, but Prexy, with his energy,
enthusiasm, power of persuasion, and vision, was the essential
driving force.

Of course I speak from the biased viewpoint of a long-time staff member of the Jackson Laboratory. But I genuinely feel I am being objective in my conclusion that without Prexy Little the development of inbred strains of mice would have occurred much more slowly, and their use would have been much restricted. We all owe him a tremendous debt of gratitude.

REFERENCES

1. Little, C.C. (1954). Genetics, Biological Individuality and Cancer. Stanford University Press, 115 pp.
2. Shull, G.H. (1908). American Breeders Association IV:296.
3. Wright, S. (1922). U.S.D.A. Bull. 1090, 63 p.
4. Castle, W.E., and Little, C.C. (1909). Science 30:313.
5. Castle, W.E., and Little, C.C. (1910). Science 32:868.
6. Little, C.C. (1915). Science 42:218.
7. Little, C.C. (1915). Science 42:494.
8. Little, C.C., and Tyzzer, E.E. (1916). J. Med. Res. 33:393.
9. Little, C.C. (1916). Sci. Mon. 3:196.
10. Staff of the Jackson Memorial Laboratory (1933). Science 78:465.

Clarence C. Little

INBRED MICE IN SCIENCE

Leonell C. Strong

Leonell C. Strong Research Foundation, Inc.
San Diego, California

Those of us who have been in the field so long must keep reminding ourselves that genetics is still a young science, but in terms of its power to illuminate biology, it is a young giant precociously strong and rich in promise. Some of us have known this for a very long time; for many, the importance of genetics to all biological research is a brand new revelation. The distance this science has traveled from its infancy might well be measured by Francis Bacon's analogy of knowledge as a river, narrow at its source and easy of survey, but broadening eventually into a mighty body scarcely to be encompassed. Those of us who began our labors at the source, where the stream was narrow, can readily testify that genetics is today, by comparison, a river in flood.

But I would remind you that the broader the science-river grows, the more pressing is the need to recollect and review the principles and practices lying closest to its source, for though simple and few in number, these constitute the laws, the foundation stones of a science that a tyro cannot afford to lose sight of. This is particularly true of mammalian genetics, which has forged tools now shared by nearly all disciplines in biology. Perhaps our collections and recollections in this symposium will serve a useful purpose in restating the origins, the nature, and characteristics of these living tools, the inbred mice.

The potential importance of the mouse as a model system for experimental research had been recognized as early as 1889 when it was found that malignant tumors could be successfully transplanted in the species. A period of highly naive optimism followed. It was supposed by some that, with such a splendid tool available, all the ails of mankind might be expected to yield swiftly to science. Particularly bright was the hope that the great riddle cancer might thus soon be resolved.

The "mousers," led principally by the students of E.E. Tyzzer and W.E. Castle at Harvard, and a few others, were tinkering with various aspects of mouse genetics. Halsey Bagg at Memorial Hospital in New York had a interesting albino strain under study, and C.C. Little carried the mouse model idea to Cold Spring Harbor, where he continued work on a partially inbred strain of dilute brown mice but lost his colony in a paratyphoid epidemic. Maude Slye at the University of Chicago completed a brilliant study on the dominant-recessive question of cancer in mice, almost proving the genetic link, but then drew an upside-down conclusion by misreading her data.

Optimism gradually faded as it was found that results from experiments with mice varied so greatly that an investigator often could not even verify his own observations, much less expect a second researcher in another laboratory to do so. Cancer research again fell into the doldrums.

I need not tell a gathering of geneticists the trouble with mice was their individuality. Each mouse had a unique genetic constitution that produced equally unique experimen-

tal results, and there did not seem to be much hope for
remedying these shortcomings. While the science of genetics
furnished the theoretical means of standardizing animals
through intensive inbreeding, nature's taboos against meddling thus with the order of things were held by such high
authorities as W.E. Castle and his school to be invincible.

It is certainly true that the natural scheme for preservation of the species discourages incest in an extremely
effective manner. In the grip of the debilitating ailments
that surface through the pairing of recessive genes, an
inbreeding species grows delicate and sickly, easy prey to
disease. For stragglers able to survive these hazards, a
barrier of sterility looms as the tenth generation of inbreeding is approached. Few scientists seriously believed that
a viable strain of inbreds could be continued, even if briefly
achieved.

In 1919, headed for a career in cancer research, I was a
graduate student at Columbia University under the tutelage
of the celebrated Nobel geneticist Thomas Hunt Morgan.
Although Morgan was skeptical about the possible genetic link
in the origin of cancer, I was impressed by the prevalence
of the disease in my own family and by other scattered evidence and was determined to investigate that link.

In looking for a place to begin my research, I was impressed with Professor Morgan's achievements in the application of statistical methods to biology through the use of
fruit flies. With them, Morgan et al. had, only a few years
earlier, discovered the sex-linked chromosome. I became
obsessed with applying these same techniques of quantitation
to cancer research through the science of genetics. But the
fruit fly would not serve my purpose; I needed a small
mammal in which cancer naturally occurs, one both plentiful
and cheap to maintain.

Except for its variability, the mouse was the best available candidate, among mammals second only to man in frequency and variety of spontaneous cancer. Regrettably, the frequency of occurrence was still all too rare; a single mouse with a spontaneous tumor was selling for $300 in laboratories on the eastern seaboard. The use of mice in the number for quantitative research necessitated a ready supply at minimum cost.

From the outset, I was leery of the transplantation technique. It seemed to me that the host factor must be of prime importance in any attempt to understand the nature of cancer, and with transplants the host was ignored or somewhat neglected. Nonetheless, lacking alternatives, I launched an experiment to measure the effect of castration on the growth of transplanted tumors.

Actual engagement in the work led me to question anew the pitfalls of the method. How, I wondered, can you distinguish between the mechanisms of cancer and the mechanisms of rejection? Also a matter for concern was the unpredictable course of transplants; sometimes they grew and sometimes they did not. Even with the most meticulous work, the success rate varied between 10 and 40%. If the tumor did grow progressively, the growth rates of the same tumor in a series would vary. Could the reason be genetic?

These first gropings spurred a change of direction; work on the effects of castration was set aside in favor of a plan to attempt genetic analysis of factors underlying susceptibility and resistance to transplanted tumors. It had been my intent to use some of C.C. Little's experimental dilute browns for my study. Accordingly, I found an opportunity to spend the summer of 1919 with him at the Carnegie Institute of Washington at Cold Spring Harbor on Long Island.

After a brief pause during which I married Katherine Bittner from my native Pennsylvania, a honeymoon residence was set up in a tent on the Cold Spring Harbor grounds where married students were being housed for the summer. The paratyphoid epidemic destroyed Little's mouse colony just as we were settling in. This forced a drastic change in plans. I was obliged, instead of setting to work with the dilute browns, to capture wild mice and start sorting out their hereditary traits through the tedious processes of mate, wait, select, and mate again. (As a note of historical interest, I recall that the best lair of wild mice was under Oscar Riddle's pigeon coop.)

Out of fear that contamination would occur in the blitzed mouse laboratory at Cold Spring Harbor, we kept the wild mice under the bed in the honeymoon tent. Since cages were hard to come by, any wooden box, fitted with a screen-wire cover, was pressed into use. (As I recall, wooden cheese boxes were considered prize finds.) The nutritional program for the mice consisted of bread scraps begged from the mess hall, combined with wild seeds gathered in the open fields and canned milk purchased out of a very slender budget.

Later that summer, Little learned that a pair of old dilute browns survived in Tyzzer's laboratory where the strain had earlier been started. Tyzzer obligingly shipped them to Cold Spring Harbor, but Little, seeing their advanced age, despaired of their breeding and gave them to me. I assigned them a box and put them into the tent where the wild mice were now multiplying at a rate gratifying to me and alarming to Katherine. By a visitation of pure luck, one last creative spark ignited in the old pair, but since the resulting litter contained only one female, this point of revival for the dilute browns was tenuous.

Meanwhile, my attentions had become engaged with the fascinating genetic events taking place among the evolving few scattered serving mice. Even before the analysis had been completed, it was obvious that susceptibility and resistance to transplanted tumors were indeed genetically controlled. The unfolding results cemented my distrust of transplantation as a means of tackling the cancer problem. Unknown variables in the host and its tumor were being added to unknown variables in another host and its reaction to the tumor, and these were being hopelessly compounded by adding the unknown variables of rejection mechanisms to the equation. Obviously, reliable data could not be obtained or even hoped for with the mice available.

These experimental results and deliberations were potent forces bending my mind to the task of remodeling Mus musculus. I could not see how any real progress could be made until such a tool was fashioned. The arguments against such a project had been running through my mind for some time. Maybe the much-feared sterility barrier was not as absolute as was assumed. If sufficient numbers of mice were used, it might be possible to squeak through the critical generations. As for the predicted fatal delicacy (to which the experts attributed the fate of Little's colony), it seemed to me that once all of the debilitating recessive traits had been bred out, there would be room to suppose that an inbred strain could be as hardy as needed, provided adequate standards of laboratory hygiene were observed.

Just how I would finance the project, which would assuredly take years, I had no idea. At that time there were no multi-million dollar grants or leviathan foundations to support research. Private means sustained some who wanted to do research, but we were as poor as the church mouse of proverb. Even the educational program was being completed on borrowed money. But obsession pays small heed to arguments of reason. Somehow, a way would be found to do something that needed doing so much.

Foundations for the development of a better mouse were laid in the summer of 1920. One of Halsey Bagg's albinos, which he claimed were inbred but which proved to be nothing of the sort, was chosen as "great white mother." Another albino, borrowed from C.C. Little and originating from a commercial colony at Storrs, Connecticut, was sire. The progeny were labeled simply, alphabetically, the A strain. Homozygosity was the chief aim. To her lasting credit, Katherine encouraged the enterprise and never once flagged in her encouragement and active help throughout what was to be an extraordinary ordeal in our lives.

In July 1921, doctoral work completed, we took a proliferating mouse colony to Annandale-on-Hudson in upper New York State, where I began my first teaching appointment in the biology department of St. Stephens (now Bard) College. It was an Episcopal institution and since the current rector lived on campus our entourage, now numbering three humans and about 400 mice, was installed in the vacant Episcopal manse at Barrytown-Four-Corners. Because no laboratory was available, the mice took up residence in the upstairs back bedroom of the manse. The intense musty-mousy odor that pervaded the house was no more of an inconvenience than many others endured during that period. The arrangement was at any rate convenient since it allowed Katherine, now expecting a second child, to care for Leonell, Jr., while continuing her services as chief caretaker and statistician for the mouse colony.

This serendipity was shattered abruptly just as winter set in. Word had reached local parishioners that the manse was being profaned by unspeakable creatures. An eviction notice for the mice was not long in coming, and they were transferred to the only alternative quarters, an abandoned chicken coop behind the biology building. By nailing several layers of tarpaper on the structure to keep out the cold and installing a galvanized metal floor to keep wild rats out, we transformed the chicken coop into the first Strong laboratory for mammalian genetics. Its only refinements were a single electric bulb powered by a wire strung from the biology building and a borrowed potbelly stove, which later proved to be a villainous piece.

A strict brother-to-sister mating system was adopted for development of the inbreds, with the exception of a few mother-to-son matings. The hardiest pair in each descent was chosen to continue the line. Gradually the numbers of non-shared genes were reduced by this means. As the inbreeding animals lost hardiness and vitality, I expanded the numbers of breeders to increase the margin of safety. Efforts to guard the mice against disease required stringent sanitation standards, and the work of caring for the colony became an increasing burden. Recording of pedigrees was another ever-expanding chore.

But the mice progressed. Soon they were eating us out of house and home quite literally. Often it was only the college garden that ensured food for our table, and Katherine had already been scolded by Bernard Iddings Bell, St. Stephen's president, for taking too freely of the vegetables. Fortunately recognition of my work had begun to spread. Through the interest of James Murphy's group I was invited in the winter of 1922 to lecture at Rockefeller Institute. As a result, Simon Flexner, institute director, made available a grant of $2,500 to support the project.

The project, however, kept expanding. Shortly after initiation of the A strain, an outcross had been made between one of these albinos and a survivor from Little's dilute browns. Cancer, although infrequent, was known to occur in both ancestral stocks. This hybrid cross was designed to test the idea that an increase in variability ought to increase the incidence of spontaneous tumors, the rationale being that cancer is just one more variable. Continued hybridizing in this stock proved the prediction true. Subsequent mating of a cancer-bearing mouse with a normal one produced the disease in the telltale 3:1 Mendelian ratio in the F_2 generation, a result that became the first laboratory proof that cancer is inherited -- and as a dominant trait contrary to Maude Sly's conclusion that the trait was recessive. This historic mating was the start of the so well known C3H high-tumor subline. At the same time, selection toward resistance to cancer, set up with this stock, produced the C12I, the CHI, and the CBA. The latter, selected for longevity, will still outlive any mouse in the laboratory.

It was impossible to pass up any interesting trait that appeared in the evolving mice, and thus a great many collateral lines were set up from mice showing characteristics that were thought to be of importance for further study. This haphazard expansion constantly strained the budget. By the third year of the work, application for funds was made to Columbia. Francis Carter Wood agreed to advance $200 in exchange for 800 mice of the evolving A strain, to be delivered over a year's time. To accept meant virtually giving away all of the animals needed for my own research program, but the situation was, as usual, desperate, so the deal was accepted.

As had been expected, a kaleidoscope of congenital defects began to appear in the mice in advanced generations of inbreeding. Cleft palate, cranial and skeletal malformations, blindness, and such lethal defects as spina bifida began to decimate the ranks of the mice as pairing of recessive genes opened the Pandora's box of hereditary disease and disability in the evolving strain.

By the winter of 1924, a critical stage of the inbreeding experiment had been reached. Numbers dwindled as the mice moved into the advanced generations and sterility became widespread. Even so, a handful of mice of the 7th generation were successfully mated and their littering was awaited with high expectation.

It was at this point that the coal-burning potbelly stove betrayed us. A student helper in the lab, charged with stoking the fire and banking it for the night, was in a hurry to get away for some social event on a night when I was lecturing. He banked the coal too soon, and poisonous gases escaped into the poorly ventilated shack, wiping out 80% of the mouse colony. Assessment of the damage the next morning showed that a lone pregnant female of the A strain had survived. By so slender a thread hung the "better mouse" ancestor of countless derived sublines used ever since in medical research throughout the world. It may be guessed that the days until that mouse littered were tense, anxious ones. Luckily, none of the sublines was entirely erased in the disaster, and in due time their numbers were multiplied to a safer level.

In the spring of 1925, I was offered an appointment at the University of Pittsburgh through George Gey's interest in my work. Higher pay and research facilities made it a tempting offer, but there were serious drawbacks to moving the mice, now in the most fragile stage of evolution. I decided to discuss the offer with Dr. Bell, who assured me that my best interests would be served by staying at St. Stephens. Thus I turned the offer down. It was therefore no small surprise to discover by way of a note tacked to the bulletin board at the end of the term that my appointment at St. Stephens was being terminated.

No satisfactory explanation for this unhappy turn of events was ever made. My elective classes had more than doubled in my second year of teaching, and my work had been pronounced satisfactory in every way. It was a dismaying situation, too late in the year to hope for an appointment elsewhere. Letters to nearly every university in the country elicited no offer. Without savings, survival for our family was problematical; for the mice it was impossible. Contemplating the awful prospect of killing off the inbreds and abandoning the work, I considered abandoning science too; perhaps I should become a missionary.

Help came from an unexpected quarter. Professor Castle at Harvard, chief scoffer at the folly of trying to establish a strain of inbreds, sent word that he would take the mice into his laboratory at Bussey Institute. Sadly, he could offer me no position. Dean Edsel of the Medical School was abroad, so no appeal for emergency funds could be made. A request to the Rockefeller Institute for a grant to support the work at Bussey was deferred because Simon Flexner was also abroad.

The Strong family stored its furniture and once more moved into a tent, this one on the grounds of Bussey Institute. With two small children sharing it, it was no honeymoon tent this time. When the winter cold drove us out of it, we took to sleeping on the benches in the institute auditorium. Our endurance was pushed to the limit when Katherine required surgery for a spinal ailment.

Mercifully, Simon Flexner returned from his travels and quickly approved funds for a one-year fellowship at Bussey Institute. During the stay at Bussey, the A strain, that ubiquitous pink-eyed mouse of laboratory fame was fully inbred. Individual mice were more alike biologically than identical twins.

For the first time, experiments with the mouse model system returned uniform, reliable, repeatable results with mathematical precision. As had been hoped, this blue-blooded mouse race proved the doom criers wrong by developing a new vigor once the threatening recessive genes had been bred out. They were more lustrous of coat, brighter eyed, and livelier than their wild ancestors.

Even more exciting for cancer research was the fully inbred C3H mouse. Not only had variability been eliminated, but in this subline, every female developed cancer of the mammary gland at approximately six months of age. By every known test of malignancy, the C3H is the most cancerous mouse in existence. Also in the complement of new mouse tools for quantitative research were the C12I and CHI with intermediate tumor appearance, and the long-lived CBA with low cancer incidence. Production of this cancer-resistant mouse had been designed as a means to apply comparative analysis between the cancer and non-cancer states.

Another crop of inbred sublines was begun in 1926 during the stay at Bussey. These included the F/St whiteface, valuable for a high incidence of leukemia in the older animals; I/St recessive; L/St, low incidence of mammary gland tumors, with lymphoblastoma, retinal opacity; and N/St with low tumor incidence and resistance to chemically induced tumors. Completing the lineup were Little's lost dilute browns, reconstituted from the ancient remnants of his colony and now fully inbred and pedigreed. It was my pleasure to present Little with a gift of breeding stock for his old line. I told him that I should not have saved them because it cinched his place in history as developer of the first successful inbred strain of mice. In turn, Little shared with me his newly established black C57 line, descendants of which still exist in my laboratory today.

Little had just become president of the University of Michigan and I accepted his invitation to help build a department of cancer research. The mouse colony, now commanding considerable attention in the scientific world, was moved to Ann Arbor in late June 1927.

Work toward stabilizing the inbred strains, and a half dozen experiments involving them, continued amidst a rising clamor from other investigators who wanted the mice for their work. When possible, a breeding pair was sent to anyone requesting them. Memorably, one of the first such pairs was a gift to Marie Curie. Keeping up with the demand, however, was far beyond the capacities of my small laboratory. When it was impossible to fill requests for the mice, there were grumblings that Strong was uncooperative. A few even complained that I was trying to restrict scientific material for my own selfish use. These charges were never justified. Few people realized that the inbreds had been created in the first place as a means of opening my own scientific line of inquiry into the cancer problem. I was glad to share the mice, but I had no intention of abandoning my career in cancer research to become a supplier of laboratory animals for others.

In 1930, along with others from the Michigan faculty, I joined C.C. Little in founding the Roscoe B. Jackson Memorial Laboratory at Bar Harbor, Maine. When the expected private support for the laboratory evaporated in the great depression, we turned to the sale of the inbred mice to other researchers as a means of supporting Jackson Laboratory. The numerous Strong inbreds plus Little's strains and a few others gathered from various sources comprised the inventory.

Intent upon building the institution and its trade, Little lectured widely on the virtues of the inbreds for research; his name became so closely associated with inbred mice that many assumed that he was their sole originator. The impression was further spread when Professor J.B. Haldane of London University paid a visit to the Jackson Laboratory to hear about our genetic wonders. I presented Haldane with a gift of the best breeding stock in my laboratory to take back to England. Apparently out of deference to Little as director of the institution that housed them, he introduced the mice into England as the "Little Inbred Mouse Strains." This error was not corrected until nearly a decade later when H.B. Andervont of the National Cancer Institute attempted to set the record straight at a Leeds genetics meeting.

The early confusion still continues in some quarters. Some think the credit belongs to Maude Slye; others believe the inbreds to be the work of John Bittner, a graduate student of mine who took some of them with him to the University of Minnesota; and still others continue to suppose their author was C.C. Little alone.

When I left Jackson Laboratory in 1933 to continue my cancer research at Yale, I left full stocks of breeders for all of my inbreds there, and these furnished a goodly part of the Jackson stock in trade until the great fire destroyed them. When the laboratories there were rebuilt and restocked, replacements for the Strong inbreds were obtained from various sources.

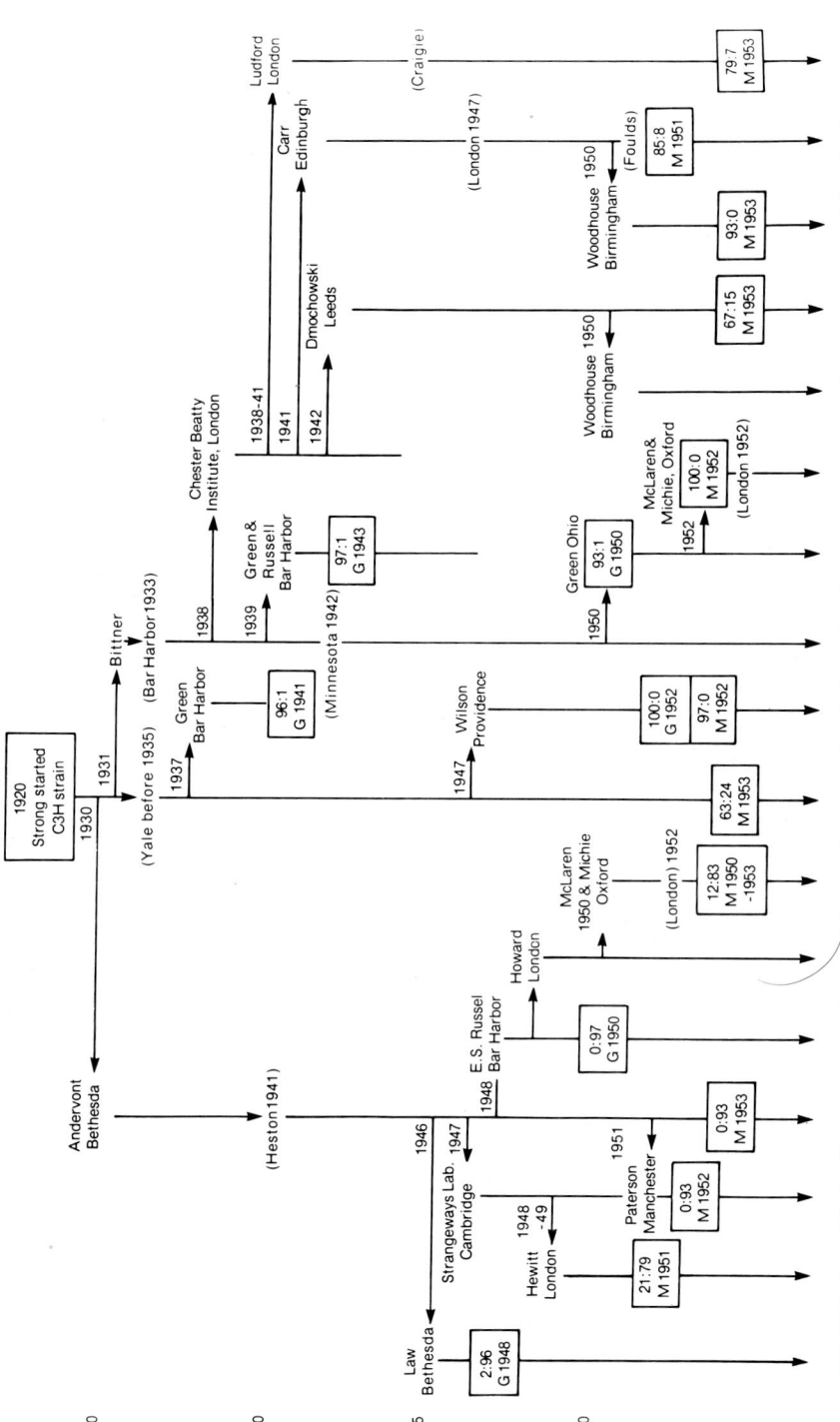

FIGURE 1. Distribution of Strong inbred mice to laboratories where colonies were established.

Figure 1 shows the distribution of Strong inbreds to other laboratories where mouse colonies were established for their propagation.

Many other sublines had their origin from a monumental cancer induction experiment carried out over more than a decade at Yale, in which injections of the carcinogen 20 methylcholanthrene into mice at an early age were continued in a line of mice for many generations. A cumulative effect of the carcinogen manifested in a greatly increased rate of mutation at many loci. As these mutants appeared, they were established as sublines. In all, 41 separate inbred strains have been in existence at one time or another. Most were studied and allowed to die out since it was impossible to support such sheer numbers. Others, thought to be of great potential value as research tools, were retained. Of these methylcholanthrene descents, only two remain in existence. These are the BRS subline, in which gastric lesions appear, and the polydactyly, a strain biologically unstable, bearing the pleomorphic gene LST. There is no demand for these strains, but I have thought them worth preserving for their potential value.

It has now been a very long time since the inbred mice made their debut as model systems for research. They were originally developed for the needs of genetics in relation to various aspects of cancer research, specifically to permit the application of quantitative analytical techniques. Their use in research has expanded immensely over the years.

For most, the procedures for their use have been sound, but in some instances it is obvious that errors in conclusions reported in the literature are traceable to insufficient training in the genetics of inbred mice on the part of researchers using them in other disciplines. There is, perhaps, time here to indicate caution. Only one illustration will be discussed.

In 1969, the pesticide DDT was listed as a carcinogen. The evidence for this conclusion appears to have been observation of an enhanced percentage of hepatomas obtained in the experimental mice. The observation led to the following conclusion: "The evidence for the carcinogenicity of DDT in experimental animals is impressive" (Report of the Secretary's Commission of Pesticides, 1969, p. 471).

Being a member of the commission, it was natural to inquire what strain of mouse had been used. It was replied that crosses had been made between females of C57BL/6 and males of either C3H or AKR, thus producing two hybrid stocks. This prompted a second question: Have all of the variables been considered in the final conclusion drawn?

The variables properly to be considered were as follows:

1. It is well known that spontaneous hepatomas are of frequent occurrence in C3H and CBA inbreds. These strains probably received this trait from a common ancestor, the C/St.
2. Hepatomas are more frequent in males than in females. Here, a C3H male had been used for the production of the hybrid.
3. Hepatoma tumors are inherited as a dominant trait and hence could be expected to occur spontaneously in the F_1.
4. The C3H mouse has an extremely high hereditary tendency to respond to chemical tumor induction with tumors of many types.
5. The heterosis obtained by hybridization, as in the case of the test animal used here, increases susceptibility to tumors in mice. Heterosis is a genetic phenomenon and its effect must be considered on the result obtained in any test involving hybridization, yet there is no mention of this variable in the case discussed.

In fact, there was no evidence of familiarity with any of these problems.

A more suitable mouse to test out the possibility that a given material is carcinogenic would be one with the lowest incidence of tumors, both spontaneous and chemically induced. These strains do exist and hybridization could also be used for any merits that process may be assumed to add to an experimental mouse.

Another approach that may be an even better test of carcinogenicity would be the use of two groups of test animals, one with high susceptibility and one with low. Then an average could be drawn. Perhaps "in mediocria firma" is a safer way.

When I resigned from the Pesticide Commission, the excuse of "insufficient time available" was true. But I think I might otherwise have resigned anyway, out of despair.

Had I remained purely a mammalian geneticist following the development of the inbred mouse strains, I believe I would have done several things to enhance their usefulness in all types of medical research. I certainly would have encouraged in every possible way the concept of using a minimum of two strains in the same experiment as a means to apply comparative analysis. I might wish that someone with the time to do so had followed up on the leads that were turned up in the application of this principle to cancer research in my own work and that of other investigators.

If the various inbred strains are arranged on a chart in order of their relative degrees of susceptibility and resistance to cancer, it is readily seen that they reflect a gradient from highly resistant through intermediate to high susceptibility (Fig. 2). This gradient appears to correlate with various levels of metabolic activity. High cancer susceptibility is accompanied by high levels of leukocytes and lymphocytes while the cancer-resistant state is associated with low levels of these two. A similar correlation is found in the degree of fluorescent pigments in the harderian glands. High fluorescence is present in high tumor susceptibility, whereas very weak fluorescence is found in the harderian glands of tumor-resistant strains. Perhaps more comprehensive analysis of base line metabolic levels in various strains of inbreds compared with the gradient of susceptibility and resistance would yield a valuable profile. Certainly such a composite could amplify our understanding of the peculiarities of individual strains.

A matter of considerable concern to me is the amount of variation in characteristics now to be found in the same inbred mouse strain obtained from different sources. In recent years, an investigator interested in verifying my cancer control studies attempted to obtain several hundred C3H mice from a large commercial source; he reported to me that he was informed they could no longer guarantee the incidence of spontaneous tumors in their C3H mice. This and similar reports of great variation among inbreds of the same strain designation prompted me to suggest that the original Strong colony, where characteristics have not changed more than a tenth of a percentage point in half a century, might serve as the norm against which variation in derived colonies might be measured. Those receiving the proposal failed to see the need for such a yardstick. Perhaps this is because we have lacked a definitive statement of the variation and divergence problem. It is to be hoped that a study of the type presently being conducted by Dr. Hilgers of the Loewenhoek Institute in Amsterdam may emphasize the need for some standard. Otherwise I do not see how we are to maintain quantitative results from inbred mice.

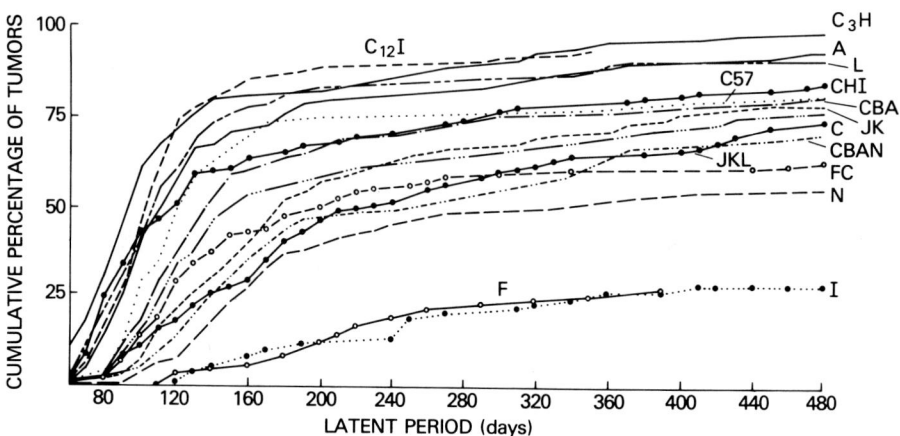

FIGURE 2. Incidence of methylcholanthrene-induced fibrosarcomas in 15 strains of inbred mice.

Another startling and threatening practice involving the integrity of standard designations for inbreds arises from the production of germ-free mice. It is a well known fact that foster-nursed C3H mice, lacking the milk tumor virus, will have low tumor incidence. The inclusion of the letter f in the symbol was long ago designated to distinguish this important fact. Yet I have recently learned that a major supplier of laboratory mice is marketing germ-free (that is, foster-nursed) C3H mice without the f symbol to distinguish them from standard C3H. At first it was assumed that this supplier might be reintroducing the MTV into the otherwise germ-free stock, but this proved not to be the case. In short, this supplier has for several years been in effect selling C3Hf mice as C3H -- by the thousands. Considering the naivete of many researchers without sufficient genetic training concerning the technicalities of the mice they use, we may legitimately feel some concern over the validity of experimental results being published where germ-free C3H mice were used with the mistaken belief that they were standard C3H high tumor mice. In defense of the practice, the supplier in question stated, "No one has ever complained."

If the inbred mouse is to retain its integrity and reliability as a tool for quantitative analysis in biological research, it is evident that the syndrome of characteristics for a given strain, identified by a given symbol, must adhere to a standard as unchanging as the international measuring devices for distance and time, so far as it is possible with living measuring devices, which the inbred mice assuredly are.

Secondly, broader dissemination of information concerning the genetics of inbred mice among researchers of other disciplines is essential if chaos is to be avoided. Ideally, a geneticist should be consulted prior to the commencement of any experiment where research animals are to be used, to make certain that the proper animal for the test is selected. Alternatively, an easy to use handbook as guide to the selection of the proper test animal for a given experiment needs to be compiled and made available to every researcher in the field of biology. Such a handbook should draw upon the expertise of geneticists deeply experienced with research animal models. Eventually, of course, we may hope for the addition of a more comprehensive course in genetics to the curricula of all schools where biology is taught, but you know and I know that such an enlightened move may be long in coming.

ACKNOWLEDGMENTS

In writing my part of the history of the inbred mouse and taking a place in a symposium on the subject, I desire, at this time, to record and to give my sincerest thanks to the several assistants and associates with whom it has been my pleasure and opportunity of profitably working. Without their labor, loyalty to me, and dedication to the problem at hand, beyond the dictates of duty, far less would have been accomplished.

Among these co-workers, four have shown that extra loyalty and dedication to the origin, continuation, and distribution of an adequate experimental animal worthy of being used in quantitative biological research in cancer and innumerable other scientific problems.

These four were: Harold Woodworth (35 years); Harold Spencer, Jr. (13 years); Fred Johnson (7 years); and Henry Matsunaga (18 years) -- total, 73 years.

For these services to science while working with me, I shall be forever with appreciative memories.

Leonell C. Strong

THE CREATION OF THE AKR STRAIN, WHOSE DNA CONTAINS THE GENOME OF A LEUKEMIA VIRUS[1]

Jacob Furth

Institute of Cancer Research
and Department of Pathology
Columbia University
New York, New York

This unique symposium is dedicated to the uses of inbred strains of mice. Among the developers of such strains, I am the only one who is not a mammalian geneticist. For me, the creation of the AKR strain was merely a means to get leukemic mice, and since their creation they have been widely used for understanding the development and control of the leukemias.[2]

In 1927, an anonymous donor made a large donation to the University of Pennsylvania for the study of this then mysterious disease, leukemia, with a free hand to spend it all. Dr. E.L. Opie surveyed the literature (1) and arrived at the conclusion that such research was long overdue. In chickens, Ellermann found a virus that caused all types of leukemias (2).

---

[1] The investigation was supported by Mr. Edward Mallinckrodt, Sr. (anonymously) and by numerous donors named in the publications cited.

[2] The medal given me carries the inscription "Venienti Occurite Morbo." (Treat the disease as you find it.)

The classical geneticists, Tyzzer (cf. 3) and his disciple, C.C. Little (cf. 4), found that mice were a storehouse of solid tumors. Two mouse geneticists had leukemia strains.

Dr. Opie then chose me, his disciple, to undertake research in this field, and at the start he was my invaluable consultant. My prior experience was limited to microbial genetics and immunology (under Otto Weil and Karl Landsteiner).

The beginning was tough. A letter to Ellermann requesting his virus was returned "Deceased; virus lost." In my laboratory, however, I was soon able to isolate several strains of avian "leukosis" viruses, some producing Ellermann's "erythromyelosis" (5), some lymphosarcoma (6), some the so-called "Marek's disease" [alias fowl paralysis, neurolymphomatosis (7, 8)].

Efforts to get a murine leukemia also met with unexpected difficulties. The two geneticists who happened to have leukemia strains failed to give us breeding pairs, while they benefited by reading the excellent survey of Dr. Opie (1) pointing to possible avenues of leukemia research. This created tough competition by possessors of leukemia strains.

We began inbreeding three stocks of mice; two obtained from a commercial breeder were named "A" and "S." "A" mice were purchased from a dealer in Pennsylvania who was the supplier of mice to the long extinct cancer research laboratory (supported by the DuPont family) where I learned that leukemia did occur in the "A" stock, albeit infrequently. Stock "S" was obtained from a dealer in New Jersey (with no knowledge of leukemia in the stock); "R" mice, a gift of the Rockefeller Institute, were known to be relatively husky, long-lived, and rarely developed neoplasia. I began inbreeding the first two stocks for leukemias; "A" for the lymphoid type of leukemia, "S" for the myeloid type, and "R" to be leukemia free. "A" ultimately yielded the AKR strain, the topic of this report. "S" yielded a good myeloid leukemia strain which was lost, and "R" yielded the strain known as Rf.

Leukemia-oriented Research During Genetic Inbreeding of Stock A Mice (1928-1937)

Mammalian genetics began in mice with inbreeding for such genetic traits as coat color and readily detectable abnormalities. The pioneering work of Tyzzer and his students, foremost among them C.C. Little (cf. 4), also included studies of spontaneous solid tumors, which happened to occur in their strains. Little was especially interested in them and, to enable expansion of work in mouse cancer genetics, he founded the Jackson Memorial Laboratory in 1929.

Nobody, to our knowledge, ever attempted to inbreed <u>de novo</u> a leukemia strain. The difficulties connected with the problem of inbreeding for leukemia were learned by us the hard way, after we began doing it in 1928, in spite of the advice of expert geneticists.

Leukemia is a diffuse neoplasm that originates in internal leukopoietic tissues and remains usually undetected until postmortem examination at about 8 to 16 months of age. At this age, living female siblings are usually infertile. Thus, to get a sufficient number of spontaneous leukemias, living male-female siblings have to be bred in numerous parallel lines. Further, a puzzling phenomenon was encountered in that numerous mice failed to express the leukemia genome, dying at 8 to 16 months of age, or even older, without any sign of leukemia.

Thus I would have been the victim of the dictum, "Publish or perish," were it not for the parallel approach to induce leukemia in mice with ionizing radiation, isolation of viruses producing avian leukemias, and analysis of the diverse types of leukemias and leukemoid reactions, as they were encountered.

We had no difficulties in isolating diverse types of avian leukemia. Much of the work on avian neoplastic viruses was novel and highly productive, but since our theme is the history of the AKR strain, they are not detailed here. They were helpful in introducing us to techniques of identifying the virus of the AKR strain.

The radiation research was initiated after reading a single report on the high frequency of fatal leukemias among roentgenologists as compared with that of physicians who were not roentgenologists (9). It was already known that most pioneer roentgenologists and some of their patients developed cancers of the exposed parts of the body (10), but a connection between radiation exposure and leukemia had not previously been made. Following this lead, we exposed large numbers of mice to various types of whole-body radiation by single or fractional doses, thus hoping to minimize the development of localized, solid, non-leukemic neoplasms. Leukemias appeared in a small to moderate number of mice after several months of radiation. Following the advice of Dr. Opie, we did not sacrifice the surviving animals after our objective (leukemogenesis) was attained but kept them until natural death. The numerous neoplasms that appeared afterward led us to digress later to radiation biology.

These non-genetic studies on leukemia were impressive enough in 1932 (four years after their initiation) to honor them with a gold medal for original research at the annual meeting of the American Medical Association. The studies thus far were published in 24 papers (18 on viral leukemia in chickens, five on murine leukemias, and one on a membrane filter device) with nine associates ranging from students to expert scientists.

I recall three amusing episodes in this relation. A current weekly, guided by appraisal of distinguished hematologists, expressed doubts about the conclusion of our prize-winning exhibit, stating, "What can he know about leukemia, he is an immunologist."

Shortly thereafter, after a key address at one of our annual conventions, the then leading oncologist, James Ewing (11), discussed my presentation. He accepted our work with mice, but he stated emphatically that from what he knew of human leukemia, it was not a neoplastic disease. Thereupon I wrote an article in the <u>JAMA</u> on the similarity of human and mouse leukemia (12). As I see now, this was good salesmanship, but not good statesmanship.

A similar episode occurred years later at an international meeting in St. Louis, after the explosion of the atomic bombs over Hiroshima and Nagasaki, but before the final "ballots" on their late effects were in. Here, too, leading clinical hematologists accepted our work on radiation leukemogenesis in mice but expressed strong doubts that ionizing radiations are leukemogenic in man. About two years later, biostatisticians of the ABCC (Atomic Bomb Casualty Commission) announced the development of leukemia in the exposed population in statistically significant numbers. Thus, I myself became a member of the ABCC and spent five years at Oak Ridge as a radiation pathobiologist. Among my young associates was Arthur Upton (13) who became the leading expert in radiation pathology. "Operation Greenhouse," a monumental study of an experimental atomic explosion, updates experimental work on the pathologic effects of ionizing radiation (14).

Spinoffs of these nine years of developing the AKR strain by inbreeding were the discovery of cell preservation in the living state by slow freezing (15), and proof that a single live lymphocyte but not crushed cells or fluid between them can transmit leukemia in very young adult mice (16).

Conventional Genetic Analysis of the AKR Strain After Homogeneity Was Attained (1937-1946)

The first up to date experimental studies on the genetics of spontaneous leukemia in AKR mice were made by Cole (17). In his studies the combined incidence of leukemia in Ak mice (including the f, g, h, and i sublines) was 71.2% in females and 67.0% in males. The difference in values in the four sublines of Ak did not appear to be significant (64.0 to 76.3% in females and 42.1 to 70.0% in males). In our low-leukemia strain Rf, the corresponding values were 1.9% in females and 1.2% in males.

During the war years, Dr. Clara J. Lynch, the geneticist of the Rockefeller Institute, became the custodian of the AKR sublines (18). She made the most thorough genetic study of several lines originating in my laboratories and continued inbreeding them between 1940 and 1946. She published her protocols in a comprehensive monograph in 1954 (18). In the United States, most AKR mice in current use are derived from her inbred lines.

In this monograph she adopted the appellation AKR, changing my system in which the original stock is designated by a capital and the sublines by small letters (e.g., Akr). In this review and in earlier ones she also used the appellation "R.I.L." Yielding to the now common usage of "AKR," I accepted this term. However, in citing earlier publications the appellations are those given therein.

The abbreviated tabulations of Lynch's data (Tables 1 and 2) give the best evidence indicating that her five lines were well inbred, and that the year to year fluctuation in leukemia incidence was due to some hidden non-genetic factor. Cole (17) carried his main five lines in five sublines which showed year to year fluctuations comparable to those of Lynch.

TABLE 1. Yearly Incidence of Leukemia During 1940-1946[a,b]

Year of birth	No. of lines pooled	% leukemia[a] in females	% leukemia[a] in males
1940-41	5	61.1	41.2
1942	5	62.8	47.6
1943	5	87.0	66.2
1944	5	94.3	71.4
1945	3	71.4	38.1
1946	3	56.3	26.3

[a] In mice 18 weeks or older.

[b] Abbreviated Table 1 of Lynch (18).

TABLE 2. Classification of Totals (Years 1940-1946) by Lines[a]

Line	% leukemia in females	% leukemia in males	P
A	74.6	53.7	.012
B	80.7	49.6	.001
D	70.5	45.7	.01
E	80.7	56.8	.001
F	75.8	53.7	.01

[a] Abbreviated Table 2 of Lynch (18).

"Homozygosity" was judged to be attained when the incidence of leukemia could no longer be increased and when transplantation in young animals with leukemic cell suspension usually yielded close to 100% takes in females and males. These are tenuous criteria, as we now know from the studies of Snell (19).[3] Further, neoplastic cells have an individuality of their own, different from that of the normal cell from which they arose.

The leukemias, induced or spontaneous, behave as neoplasms, lymphosarcoma and leukemia being extreme variants of the same disease. I suggested that localization of lymphosarcoma may be due to antigenicity of the lymphoma by lowering the host resistance. This reasoning followed the now forgotten pioneer work of James Murphy on the relation of normal lymphocytes in resistance to neoplasms (20). Murphy failed to bring convincing positive evidence to his thesis, namely, that induced lymphocytosis would confer resistance to tumorous growth. Later development of "homozygous" (isogenic) strains (19) placed the earlier work with transplanted tumors in proper perspective, and led to approaches on immunotherapy of spontaneous neoplasms, as will be discussed.

All Ak leukemias were lymphoid. They were found most frequently in animals dying between 34 and 45 weeks of age, the oldest when 84 weeks old.

Summarizing her findings between 1940 and 1949, Lynch concluded that the variability in leukemia incidence from year to year cannot be accounted for by selection among lines, by longevity, by the season of the year in which the mice were born, by the age of the mother at parturition, or by litter seriation. Impressed by our work on the role of the thymus in leukemogenesis (described below), she concluded that possible intercurrent disease affecting the thymus may influence the occurrence of leukemia.

Appraising work in her and in my laboratories in 1941-1942 on foster nursing (21), she concluded that the AKR strain was

---

[3] Mapping the cell surface antigens is the topic of Session IV (Amos, Graff, Old, Flaherty and Taylor).

susceptible to mammary tumor agents, but itself lacks such an agent or carries one of low infectivity.

Of the extensive early hybridization studies done in our laboratories (22), only two sets of abbreviated data are listed in Table 3. In the first by Schweitzer et al. (22), crossing mice of high mammary tumor-low leukemia strain C3H with Ak mice did not much inhibit leukemia development in the F_1 hybrids, while crosses between the low leukemia-MT-free Rf mice [Exp. of Cole (17)] caused marked inhibition especially when the Rf parent was female (Table 3).

Table 4 presents Lynch's (18) list of the principal sublines of AKR in the USA, derived from her AKR/M line. She discontinued breeding AKRs in 1946, leaving her AKR/M at the Memorial Hospital. Dr. Lloyd Old and other investigators at the Memorial Hospital, who will speak in coming sessions, will enlighten us about the fate of the AKR/M lines.

TABLE 3. Incidence of Leukemia in F_1 Hybrids, Crossing Ak Mice with Two Low Leukemia Strains

Strain	% Leukemia	
	female	male
Parental Ak	53	61
Parental C3H	0.3	0.3
F_1: Ak x C3H	51	48
C3H x Ak	28	39
Parental AK	67	71
Parental Rf	1	2
F_1: Ak x Rf	29	16
Rf x Ak	9	15

TABLE 4. Principal Sublines of AKR

Designation	Investigator	Generation	Location
AKR/M	Lynch	F_{18-19}	Memorial Hosp.
AKR/Lw*	Law	F_{27}	NCI
AKR/Fu	Furth	F_{32}	ORNL
AKR/Du	Dunning	F_{26}	Miami Univ.
AKR/Jax	Jax	F_{27}	Jackson Lab.

*Also named RIL/Lw.

The first major genetic contribution subsequent to the work of Lynch was made by Snell (19) who discovered that lines homologous by transplantation tests can differ genetically among themselves. The genetic studies on isogenicity nomenclature and other major developments in genetics will be topics in a session of this conference. The two editions of The Biology of the Laboratory Mouse, written by the staff of the Jackson Memorial Laboratories (3, 4), are informative textbooks for all students of the laboratory mouse.

Discovery of the Hormonal Role of Thymus in Leukemogenesis (1942-1946)

After inbreeding of the AKR strain was attained, the question was raised: Why the long latency of spontaneous leukemia (5 to 12 months) and when and where does the leukemic transformation of normal lymphocytes occur? The answer to this was first announced in 1942 at the conference on Gibson Island.[4]

---

[4]This was the original site of the now popular summer conferences named after Dr. Gordon who initiated them. They are now held in New England.

Until about five months of age, the lymphocytes and all hematopoietic organs of the AKR mice appeared normal, after which they underwent neoplastic transformation, first detected in the thymus by formation of isolated thymic lymphomas. From there they spread to many lymphoid organs. Sometimes only one, but usually both thymic lobes are involved. To probe this anatomic observation, thymectomies were performed (23, 24), the results of which are shown in Figure 1. This figure indicates that thymectomy prevented the development of leukemia, and thereby prolonged the life of AKR mice (24).

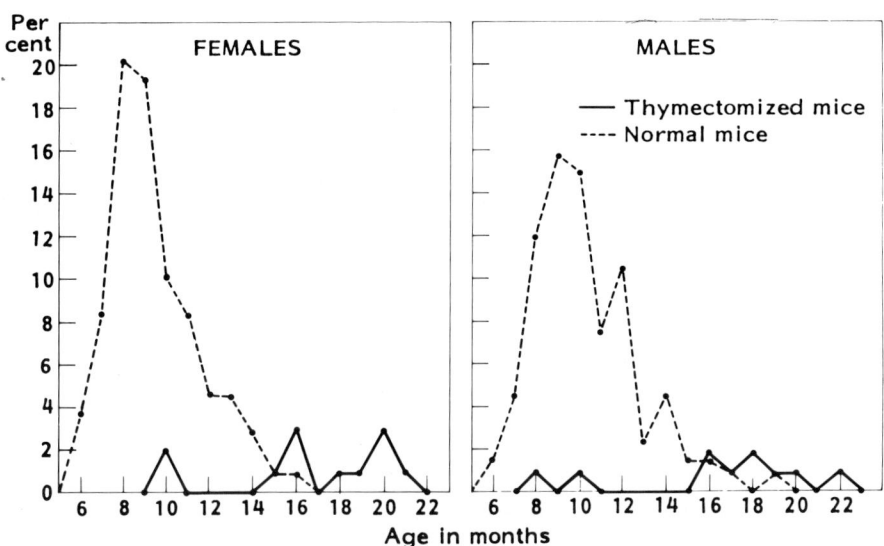

FIGURE 1. The effect of thymectomy on the incidence of leukemia and on longevity (24).

We noted frequently that when the incidence of pneumonia and other debilitating diseases was high in our colony, the leukemia incidence was low and the thymus was atrophic. The relation of "stress" and thymic atrophy was known to Selye (25). The relationship of adrenal glucocorticoid to atrophy of lymphoid organs was discovered by Dougherty (26), and this became related to stress. At this stage, two experiments were called for: the demonstration of thymic atrophy by some simple, noninfectious agent, and the exploration of whether thymic grafts will restore the leukemia incidence in thymectomized Ak mice.

As an uncomplicated stressful agent, underfeeding was chosen. This was motivated by well controlled systematic studies of A. Tannenbaum (27) who confirmed the statistical analysis of the Metropolitan Life Insurance Company indicating that undernourishment tends to prolong the life of people and reduces the likelihood of developing cancer. Probing this idea, Saxton duplicated the effect of thymectomy in careful, well controlled experiments by rigid, uninterrupted underfeeding (28). Underfed mice outlived even the matched controls which had been subjected to thymectomy and the leukemia incidence among them was very low (Fig. 2).

The findings on the relation of the thymus to leukemia (23, 24) were a veritable "breakthrough." (Until then the thymus was regarded as just another lymphoid organ.) They attracted numerous investigators, whose contributions soon outclassed those made in my lab. The state of knowledge in 1959 in leukemogenesis and neoplasia in general is well surveyed in The Physiopathology of Cancer (29). In Chapter 12 (30) on leukemia and thymus 581 articles are cited. H.S. Kaplan et al. (31, 32) reviewed the relation of ionizing radions to leukemia and the effects of thymectomy and thymus grafts. L.W. Law et al. (33, 34) reported on carcinogen-induced leukemia, and genetic and thymic factors, and D. Metcalf

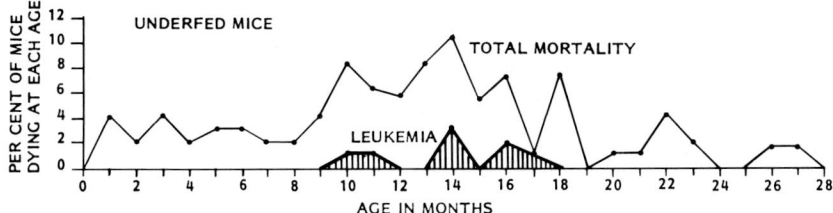
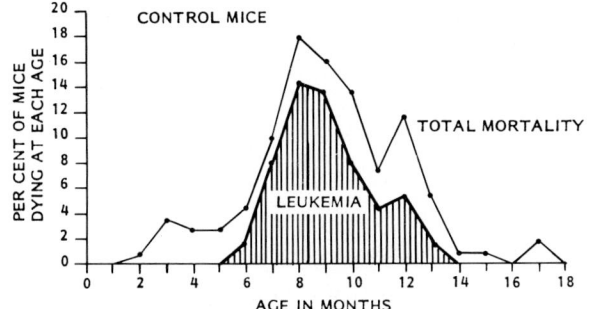

FIGURE 2. Inhibition of development of leukemia with increase of life span by underfeeding (28). Percentage of animals dying at each month of age. Lines enclosing shaded areas indicate percentage of animals dying with leukemia.

(35, 36) on the thymic origin of his LSF (lymphocytosis stimulating factor) and its relation to leukemia.

To put the discovery of the thymus in historical perspective, the following is worthy of note. In the third decade of this century, I witnessed a prestigious lecturer calling the lymphocyte a cell of mystery: most abundant in the body, seen everywhere, dies daily in large numbers in the intestinal epithelium, but it is unknown where it is born and what its function is. Lymphocytes are concentrated in certain organs, known as "lymphopoietic."

The recognition of the relation of the thymus in lymphomagenesis preceded that of its role in immunologic competence. Both are prevented by thymectomy and restored by thymic grafts.

We suggested earlier that the thymus is the common base for both. Other vital factors affect both, operating via the thymus. Foremost of these are "stress" (25) induced by adrenal glucocorticoids (26) which are related to ACTH (adrenocorticotropic hormone).

In 1964, a symposium was held on "The Thymus" (37), where it was recognized as a special functional organ, playing a significant role in immunologic responsiveness (37; page 99) in carcinogenesis as well as leukemogenesis (37; page 121). A landmark of knowledge on the structure, function, and role of the thymus was the subject of another symposium (38) held at about the same time with contributions of 72 authors.

Several Ciba Foundation Symposia were held on leukemia and allied disturbances as noted in AKR mice (39-42). They mirror the development of knowledge in areas of leukemia, thymus, leukemia virus, and immunology. The first, held in 1954, was titled "Leukemia Research" (39). The second, held in 1959, had the same title (40). The third in 1962 was titled "Tumor Viruses of Murine Origin" (41). The fourth, held in 1966, was titled "The Thymus: Experimental and Clinical Studies" (42).

In the first symposium Ludwik Gross reported on a filterable agent in Ak mice, allied to Bittner's mammary tumor virus. I attributed his success to the use of neonatal and C3H mice. These are leukemia-free but are sensitive to AKR leukemias as was reported earlier (22). His sustained work and that of others are well described in his book <u>Oncogenic Viruses</u> (43).

The fourth Ciba Symposium (41) on the thymus was masterfully organized by Sir Macfarlane Burnet at the Hall Institute. Members of the Hall Institute included Jacques Miller, Metcalf, Nossal, and other researchers on the thymus. Sir Macfarlane graciously invited me even though I was a renegade in thymus research (having been attracted in the early 1950s to derangements of homeostasis and neoplasia without carcinogens). The main topic of this conference was the structure of the thymus, its role in cellular and humoral immunity, the function of lymphocytes, and autoimmune disease. AKR mice were used or referred to in studies by several contributors. I presented here my last fragmentary work on the thymus. It was done with Ioachim and other associates in tissue cultures of the thymus. It aimed at detection of cells which secreted the leukemogenic factor. Cultured epithelial reticulum (Fig. 3) restored the lymphocyte levels and immunocompetence in thymectomized rats to which the AKR virus became adapted.

With great intensification of cell culture research resulting in the development of numerous isolated cell lines, it seems likely that the various cellular elements of the thymus will be grown in pure cultures. Their interrelationship, nutritional requirements, promoters, inhibitors, transforming principles, and hormonal secretions will be identified.

Inasmuch as in vitro studies will likely use inbred strains, attention is called here to the splendid volume Readings in Mammalian Cell Culture (44) and to the proceedings of conferences at the Alton Jones Cell Science Center of the Tissue Culture Association in Lake Placid, New York. Prominent in this area are Gordon Sato and his associates. One of their major discoveries is that replacement of serum by hormones permits the growth of cells in a medium defined by them (45). Most pure functional cell lines, now preserved and distributed in the frozen state, are derived from Sato's laboratory.

The first comprehensive book on thymic hormones, edited by Luckey and published in 1973 (46), describes the technique of thymectomy, disorders created by thymic deficiency in man, and isolation and characterization of several thymic extracts

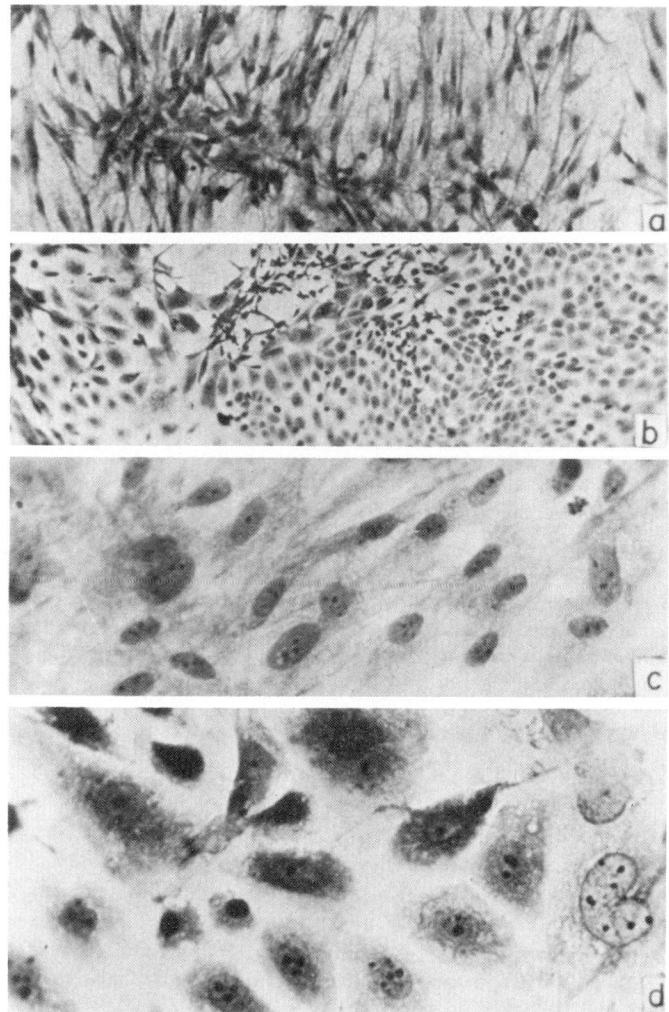

FIGURE 3. Appearance of epithelial-reticular cell cultures of the thymus at different magnifications: (a) x140, (b) x140, (c) x400, (d) x400.

with hormonal activity. Most known thymic hormones were isolated by A.L. Goldstein in collaboration with A. White (published in 14 articles in 1966-1972), while G. Goldstein et al., who called attention to an endocrine function of the thymus manifested in myasthenia gravis and lupus erythematosus, published in eight articles (1966-1970) as cited in this book.

A more updated book, The Biological Activity of Thymic Hormones, was edited by Van Bekkum in 1975 (47). In the introductory chapter Abraham White sums up the available experimental data, concluding that thymosin isolated by him and other hormones isolated by A.L. Goldstein fulfill five of six postulates essential for designating the thymus as an endocrine gland. The sixth essential postulate, the chemical synthesis of any of these hormones, has yet to be achieved. White postulated that the thymus produces a family of several hormones, each of which may alter the extent of expression of one or more of the immunological roles of the thymus in regulation of host competence. This book deals with the differentiation of thymic elements, their interrelationship, the diverse specific function of T lymphocytes, impairment of thymic function, administration of thymic extracts and thymosin to patients, mitogen, and mixed lymphocyte reaction for thymosin.

Inasmuch as my knowledge in the development of the diverse areas of research in which AKR mice are utilized is second-hand after 1966, I shall be brief and sketchy, citing original articles by major contributors as well as excellent semipopular surveys by them which began to appear in increasing numbers in the Scientific American articles written by them. A cursory review of a MEDLARS (Medical Literature Analysis and Retrieval System) search indicates that in a year covering 1974-1975 about 180 articles appeared on thymus hormones. In about the same year a literature search by the librarian of Jackson Laboratory (the main source of AKR mice in the USA) has shown that in about the same number of articles AKR mice were used. My attempt to break down these publications per area of research proved exceedingly time-consuming because the topics covered are overlapping and are limited to a segment of the area which the author investigated in depth.

Leukemia Virus Genetics: The Transcriptase of the RNA Virus of AKR (1970)[5]

The year 1970 was a landmark in molecular genetics and cancer research with the announcement of the discovery of the reverse transcriptase (RT) by Temin (48) and Baltimore (49). It triggered rapidly expanding research in molecular biochemistry and created a firm base for molecular genetics. It led to the discovery of many other enzymes linking the DNA-integrated silent viral segment of the AKR cell to its infectious RNA virus.

Temin (48, 50), studying a broad spectrum of tumor viruses, suggested that RNA tumor viruses are repeatedly evolving by emergence or escape from the genome of normal cells (protovirus hypothesis). The protovirus is sensitive to deoxyribonuclease and resistant to ribonuclease. (The infectious AKR virus is sensitive to ribonuclease.) By using labeled DNA copy in hybridization tests, he found that DNA contains some of the infectious virion sequences. On the basis of his phylogenetic and ontologic studies, he postulated that the primary event is the evolution of nucleotide sequences in the DNA. The emergence of infectious RNA virion is a secondary event.

---

[5]The genetic regulation of viruses is analyzed in Session III by Lilly, Rowe, Chused, Cardiff, Schlom, and Law.

In his Clowes lecture (48), Temin discussed how the genome of a normal cell changed into the genome of a cancer cell. The viruses that cause cancer have genes for cancer (now named T genes). Carcinogenesis by strongly transforming viruses involves the formation in infected cells of genes for cancer copied from a viral genome. Temin's protovirus hypothesis states "that the genes for neoplastic transformation arise in an organism as a result of misevolution of a normal system of DNA information-transfer."

In his 1975 Harvey lecture Baltimore sketched his strategy of RNA viruses (51). He contrasted three types of viruses: the polio virus, the virus of vesicular stomatitis, and RNA tumor viruses. They use different strategies for infecting cells. The strategy of the RNA tumor viruses is the use of reverse transcriptase in the virion to make a DNA copy which integrates into cellular DNA. This is illustrated schematically in Figure 8 of reference 51.

Once the proviral DNA is integrated into cellular DNA (49), it spreads the virus by inheritance (51). In the AKR mouse viral information has been shown to reside in two stable genetic loci and to segregate in genetic crosses as simple Mendelian traits (cf. Rowe et al., Science 178:860, 1972, and Chattopadhyay et al., Proc. Natl. Acad. Sci. 71: 167, 1974).

Leukemia Virus Immunology

Soon after the discovery of transplantability of leukemic cells, immunologic techniques were used to characterize both the leukemic cells and the agents producing them. In 1941 Elvin Kabat, who spent three highly productive years in my laboratories, disclosed that the plasma of leukemic chickens contained huge quantities of virus sedimentable at high speed (52). Avian leukosis, in contrast to murine and human leukemias, contains colossal quantities of infectious virus in the blood. Further, the serum of leukotic chickens contains virus neutralizing antibodies (53).

Leukemic cells, as neoplastic cells in general, differ from their ancestral normal cells by a genetic deviation, detectable by transplantation tests, one of the tools of geneticists. The most sensitive early immunologic techniques such as complement fixation and hemagglutination (if the virus adhered to blood cells) were not sensitive enough to detect small quantities of tumor viruses.

Medawar, in his Harvey lecture of 1957 (54), discussed known factors in the immunology of transplantation. As he predicted years earlier immunology offers negotiable pathways in the central regions of biology as genetics and embryology.

The discovery of the role of the thymus as an endocrine organ (23, 24) was soon followed by studies on the effect of thymic ablation at various periods of life. This led to research on immunology of the thymus (55, 56) summarized by J.F.A.P. Miller in 1966 (42, page 153). He has proven that the thymus directs the maturation of immunologic capabilities by means of a humoral mechanism.

Subsequently, numerous books and symposia on the role of the thymus in immunology have appeared. Some have already been cited; others are listed by Spiegelman et al. (57, 58). The great discovery by Karl Landsteiner and M.W. Chase (59, 60) on how to recognize small molecules (haptens), which themselves are not detectable by conventional immunologic techniques, tremendously advanced the usefulness of immunology in tumor-virus-related research, yet to be fully exploited.[6] (This

---

[6]The relationships between worldwide and domestic mice (genetic, virus content, immunogenicity) are topics of later sessions.

is reviewed in Scientific American in an informative, well illustrated article, "The Development of the Immune System," by M.D. Cooper and A.R. Lawton III in 1974.)

Thus the prophetic vision of Medawar materialized and became widely popular in the 1970s.

Retrospect: The New Strategy of Tumor-Virus Research

The history of the AKR strain is a good example of the old adage that problems never cease to exist. Solution of one usually brings to light many more new ones. Science is cascading, thereby leading to increasing specialization, often to such a degree that a team of different specialists is required to tackle each basic problem. Peyton Rous, when in retirement, listened to a lecture to which he was invited and whispered to me, "In what they say about Rous sarcoma I do not recognize the tumor I worked with." Presently I, too, find it difficult to comprehend many lectures and publications in which the AKR strain is a basic tool of research.

AKR is a uniquely stable strain, perhaps because of the nine years of successive brother and sister matings required to attain homozygosity. It took several years to identify its several cardinal features: onco-(leukemo)-genicity based on inheritance of leukemia genes, the expression of which (releasing an infectious RNA virus) is under control by some thymic hormone while other thymic hormones are related to immunogenicity.

The following are illustrative examples of the great diversification research has taken in the 1970s.

It has long been known that the genes are associated with a large number of proteins (basic histones and acidophilic proteins), but not their specific relation to the genes. The discovery of reverse transcriptase opened new vistas. Enzymes are proteins. Unlike nucleic acids, protein can be identified immunologically with ease and great precision. Many enzymes are needed for DNA-RNA function. Some are positive control elements wanted for activation or inactivation of the host genetic information. Others are negative control elements. There is also a cascade regulation, where several enzymes operate sequentially. The new strategy of the regulation of the viral genome calls for identification of these controlling enzymes. The background knowledge obtained from thorough studies of bacterial viruses to mammalian cells in general had to be adapted to cells with tumor viruses. For research on oncornaviruses, AKR is an excellent tool for formulation of new tumor and viral genome strategies. The discovery of reverse transcriptase broke the ground.

The structure and function of chromatin (i.e., nucleic acid associated proteins) was the topic of a Ciba Symposium in 1975 (61). By 1976 Maclean could speak in more precise terms on gene expression in eukaryocytes (62). The involvement of ribosomal (rs) RNA, transfer (t) RNA, and messenger (m) RNA in gene expression was recognized. His studies were done with pure cell lines grown in dissociated cell cultures. His Table 6 lists 14 hormones, whose actions are mediated by changes in levels of cyclic AMP active in multicellular organisms.

Analysis of virus-linked proteins (enzymes) led to the development of probes for identification of viruses in cells and information on gene function. First of these was linking leukemia viruses to a protein in the blood sedimentable at 60-70 g (58, page 11). He was one of the over 100 participants in an international symposium on the Molecular Approach to the Etiology of Human Cancer (58). In his article Spiegelman presented his molecular hybridization with radioactive probes, giving references to the historical research including the pioneer oncogenic theory of Huebner and Todaro (63, 64).

Research in immunology, both hormonal and cellular, vastly expanded. The book <u>Lymphocyte Differentiation, Recognition and Regulation</u> is a well organized, balanced review of some 2,317 references on this subject (65). "From investigations on the division of lymphocytes into the two great families, the regulatory role of T cells in antibody formation, the one cell one antibody rule, the genetics of immune responsiveness, the cellular mechanism whereby the immune system distinguishes "self" from "not self," the physiology of lymphocyte surface receptors for antigens, major discoveries have resulted" (citing the excellent review by G.J.V. Nossal, who is a scholar on this subject).

The <u>Thymus and Self</u> (66) is a monograph by Rygaard on the immunobiology of the mouse mutant <u>nude</u> discovered in 1968. It is born without a thymus and, therefore, forms a baseline for study and isolation of the normal thymus functions. This gave an opportunity to the author to review the state of knowledge about the immune system structure and function with emphasis on the apparent dichotomy of the immune system, supportive and inhibitory.

These are but morsels of the rapidly growing and diversifying AKR-related literature in the current decade.

ACKNOWLEDGMENT

I gratefully acknowledge the editorial assistance of Dr. Herbert C. Morse III and his assistants. Apologies are extended to many colleagues for omitting to cite their relevant work due to the limited scope of this article, and time and space limitations.

This sculpture of Ak mice crawling over bookends is a farewell gift from my colleagues at Oak Ridge National Laboratories. It was carved in cherry wood by a Cherokee whittler. It was presented to me with the statement (which touched me with surprise and jealousy) that the Ak mice would long survive me. How true!

REFERENCES

1. Opie, E.L. (1928). Medicine 7:31.
2. Ellermann, V. (1918). "Die übertragbare Hühnerleukose." Springer Verlag, Berlin.
3. Snell, G. (1941). "Biology of the Laboratory Mouse," 1st ed. The Blakiston Co., Philadelphia. (Gives reference to E.E. Tyzzer, C.C. Little and other pioneer workers on the genetics of tumor formation.)
4. Green, E.L. (1966). "Biology of the Laboratory Mouse," 2nd ed. McGraw-Hill Book Co., New York.
5. Ellermann, V. (1921). "The Leucosis of Fowls and Leucemia Problems." Gyldendal, London.
6. Furth, J. (1933). J. Exp. Med. 58:253.
7. Furth, J., and Stubbs, E.L. (1934). Proc. Soc. Exp. Biol. 32:381.
8. Furth, J. (1935). Arch. Path. 20:379.
9. Aubertin, C. (1931). Bull. Off. Soc. Franc. d'Electrol. Radiol. 40:218.
10. Furth, J., and Furth, O.B. (1936). Am. J. Cancer 28:54.
11. Ewing, J. (1940). "Neoplastic Diseases," 4th ed. W.B. Saunders, Philadelphia.
12. Furth, J., Ferris, H.W., and Reznikoff, P. (1935). J. Am. Med. Assoc. 105:1824.
13. Furth, J., Upton, A.C., and Kimball, A.W. (1959). Radiat. Res., Suppl. 1:243.
14. Upton, A.C., Kimball, A.W., Furth, J., Christenberry, K.W., and Benedict, W.H. (1960). Cancer Res. 20; No. 8, Part 2.
15. Breedis, C., Barnes, W.A., and Furth, J. (1937). Proc. Soc. Exp. Biol. Med. 36:220.
16. Furth, J., and Kahn, M.C. (1937). Am. J. Cancer 31:276.
17. Cole, R.K., and Furth, J. (1941). Cancer Res. 1:957.

18. Lynch, C.J. (1954). J. Natl. Cancer Inst. 15:161.
19. Snell, G. (1966). In "Biology of the Laboratory Mouse," 2nd ed. (E.L. Green, ed.), p. 457. McGraw-Hill Book Co., New York.
20. Murphy, J.B. (1944). Cancer Res. 4:622.
21. Barnes, W.A., and Cole, R.K. (1941). Cancer Res. 1:99.
22. Schweitzer, M.D., and Furth, J. (1939). Am. J. Cancer 37:224.
23. McEndy, D.P., Boon, M.C., and Furth, J. (1944). Cancer Res. 4:377.
24. Furth, J. (1946). J. Geront. 1:46.
25. Selye, H. (1950). "Stress." Acta, Inc., Montreal.
26. Dougherty, T.F. (1952). Physiol. Rev. 32:339.
27. Tannenbaum, A. (1959). In "The Physiopathology of Cancer," 2nd ed. (F. Homburger, ed.), p. 517. Hoeber-Harper, New York.
28. Saxton, J.A., Jr., Boon, M.C., and Furth, J. (1944). Cancer Res. 4:401.
29. Homburger, F. (ed). (1959). "The Physiopathology of Cancer," 2nd ed. Hoeber-Harper, New York.
30. Furth, J., and Baldini, M. (1959). In "The Physiopathology of Cancer," 2nd ed. (F. Homburger, ed.), p. 364. Hoeber-Harper, New York.
31. Kaplan, H.S., Marder, S.N., and Brown, M.B. (1951). Cancer Res. 11:629.
32. Kaplan, H.S., Brown, M.B., Hirsch, B.B., and Carnes, W.H. (1956). Cancer Res. 16:426.
33. Law, L.W. (1942). Cancer Res. 2:108.
34. Law, L.W., Dunn, T.B., and Boyle, P.J. (1955). J. Natl. Cancer Inst. 16:495.
35. Metcalf, D. (1956). Brit. J. Cancer 10:442.
36. Metcalf, D. (1956). Brit. J. Cancer 10:169.
37. Defendi, V., and Metcalf, D. (eds.). (1964). "The Thymus." Wistar Institute, Philadelphia.

38. Good, R.A., and Gabrielsen, A.E. (eds.). (1964). "The Thymus in Immunobiology." Harper & Row, New York.
39. Wolstenholme, G.E.W. (1954). "Ciba Foundation Symposium on Leukemia Research." J. & A. Churchill, Ltd., London.
40. Wolstenholme, G.E.W. (1959). "Ciba Foundation Symposium on Carcinogenesis. Mechanism of Action." J. & A. Churchill Ltd., London.
41. Wolstenholme, G.E.W. (1962). "Ciba Foundation Symposium on Tumor Viruses of Murine Origin." J. & A. Churchill Ltd., London.
42. Wolstenholme, G.E.W. (1966). "The Thymus: Experimental and Clinical Studies." J. & A. Churchill Ltd., London.
43. Gross, L. (1971). "Oncogenic Viruses," 2nd ed. Pergamon Press, Oxford.
44. Pollack, R. (ed.). (1973). "Readings in Mammalian Cell Culture." Cold Spring Harbor Laboratory, Cold Spring Harbor, N.Y.
45. Hayashi, I., and Sato, G. (1976). Nature 259:132.
46. Luckey, T.D. (ed.). (1973). "Thymic Hormones." University Park Press, Baltimore.
47. Van Bekkum, D.W. (1975). "The Biological Activity of Thymic Hormones." John Wiley & Sons, New York.
48. Temin, H.M. (1970). Perspect. Biol. Med. 14:11.
49. Baltimore, D. (1970). Nature (London) 226:1209.
50. Temin, H.M. (1974). Cancer Res. 34:2835.
51. Baltimore, D. (1976). "The Harvey Lectures." Academic Press, New York.
52. Furth, J., and Kabat, E.A. (1941). J. Exp. Med. 74:247.
53. Kabat, E.A., and Furth, J. (1941). J. Exp. Med. 74:257.
54. Medavar, P.B. (1957). "The Harvey Lectures." Academic Press, New York.
55. Miller, J.F.A.P. (1960). Nature 187:703.
56. Miller, J.F.A.P. (1961). Lancet 2:748.

57. Spiegelman, S., Burny, A., Das, M.R., Keydar, J., Schlom, J., Travnicek, M., and Watson, K. (1970). Nature New Biol. 227:563.
58. Spiegelman, S. (1976). Pure Appl. Chem. 47:11.
59. Landsteiner, K., and diSomma, A.A. (1940). J. Exp. Med. 72:361.
60. Chase, M.W., and Maguire, H.C., Jr. (1973). Int. Arch. Allergy 45:513.
61. Ciba Foundation Symposium 28. (1975). "The Structure and Function of Chromatin." Associated Scientific Publishers, Amsterdam.
62. Maclean, N. (1976). "Control of Gene Expression." Academic Press, London.
63. Huebner, R.J., and Todaro, G.J. (1969). Proc. Natl. Acad. Sci. USA 64:1087.
64. Todaro, G.J. (1973). Perspectives in Virology 8:81.
65.* Katz, D.H. (1977). "Lymphocyte Differentiation, Recognition and Regulation." Academic Press, New York.
66. Rygaard, J. (1973). Thymus and Self. John Wiley and Sons, London.

---

*Another comprehensive book, "The Lymphocyte," Vol. I, II, edited by J.J. Marchalonis, published by Marcel Dekker, New York, appeared in the same year (1977), which I had no opportunity to review.

Jacob Furth

HOWARD B. ANDERVONT

Margaret K. Deringer

Registry of Experimental Cancers
National Cancer Institute
National Institutes of Health
Bethesda, Maryland

It was with great pleasure that I accepted Dr. Morse's invitation to receive the award to be given to Dr. Andervont at this Workshop on Origins of Inbred Mice. I know of no one for whom I'd rather accept such an honor. It has been my privilege to know Dr. Andervont since August of 1942 when I joined the staff of the National Cancer Institute in Dr. Heston's laboratory. Ever since that time, Dr. Andervont has been my adviser, colleague, and friend.

Howard Bancroft Andervont, known affectionately as Andy by his family, colleagues, and friends, was born in Canton, Ohio, on March 8, 1898. He graduated in 1923 from Mt. Union College in Alliance, Ohio, with the degree of Bachelor of Science. His alma mater, Mt. Union, also conferred an honorary degree of Doctor of Science on him in 1942. Dr. Andervont did graduate work under the direction of Dr. Charles Simon at the School of Hygiene and Public Health of The Johns Hopkins University where he received the Doctor of Science degree in 1926. He was a fellow of the Carnegie Institution of Washington at Hopkins in 1926-1927 and following that was faculty instructor under Dr. Milton J. Rosenau in Preventive Medicine at Harvard University School of Medicine from 1927 to 1930. He became the first professional member of Dr. J.W. Schereschewsky's Office of Cancer Investigations of the U.S. Public Health Service at Harvard and moved with that group in 1939 to the National Cancer Institute in Bethesda. In 1947, he became Chief of the Laboratory of Biology, a position he held until 1961. At the time he relinquished that post he became Editor of the Journal of the National Cancer Institute. Dr. Andervont retired from the Institute in 1968.

Dr. Andervont's bibliography includes a total of 165 publications. This list attests to his many accomplishments. I shall mention only a few for I know that modest as he is, he would prefer it that way. His curriculum vitae states his "major contribution has been to determine the relative importance of genetic and environmental factors in the development of spontaneous tumors in experimental animals." His recognition of the importance of genetic factors led to the inbreeding and maintaining of large colonies of mice which he used in his research. Among these were strains C3H, BALB/c, I and RIII and, in addition, he was successful in producing a colony of wild mice.

Research on viruses was one of Dr. Andervont's earliest interests and he published extensively on the mammary tumor virus of mice. He was the first to demonstrate that this virus could be transmitted by the male parent. Many of his studies demonstrated the induction of cancer by chemical agents, in particular by polycyclic hydrocarbons.

I am grateful to Dr. Andervont for having supplied me with a nucleus of several of his inbred strains from which I established colonies. These strains were employed in various experiments, for example, in studies of the development of various types of tumors in substrain BALB/c and in studies of the development of mammary tumors in substrain RIIIeB, produced by the transfer of fertilized ova from strain RIII to strain C57BL. A number of other strains in my colony came originally from Dr. Andervont by way of Dr. Heston's laboratory.

Dr. Andervont's curriculum vitae also states that "exceptional contribution has been made in counseling investigators in cancer research." I am just one of the many individuals who has had the good fortune to be associated with Andy. I am sure that all who have received his wise counsel would share my pleasure in the award to him of this well deserved honor.

Howard B. Andervont

HOWARD B. ANDERVONT: AN APPRECIATION[1]

Michael B. Shimkin

Temple University
Health Sciences Center
Philadelphia, Pennsylvania

Howard Bancroft Andervont, Editor of the <u>Journal of the National Cancer Institute</u>, retired in March 1968. Son of Ernest Bancroft Andervont and Catherine Magdalena (Kuehn) Andervont, he was born in Canton, Ohio, on March 8, 1898. He had a fortunate, barefooted, stubtoed boyhood. He attended public schools and then Mt. Union College in Alliance, Ohio, where he helped support himself by working in steel mills. His biology instructor, Dr. Joseph Scott, awakened his interest in biological science. Following his graduation with a degree of Bachelor of Science in 1923, Scott recommended him to Dr. Charles Simon, of the School of Hygiene at Johns Hopkins University in Baltimore, Maryland. He was Simon's assistant in research and teaching on filterable viruses during his three years of graduate work.

In 1926 he received from The Johns Hopkins University the degree of Doctor of Science; his thesis was on the relationship between cowpox virus and vaccinia virus. He remained at Johns Hopkins for a year on a Carnegie Foundation Fellow-

---

[1] Reprinted from J. Natl. Cancer Inst. 40:XIII-XXV, 1968.

ship and then accepted an instructorship with Dr. Milton J. Rosenau, Professor of Preventive Medicine at Harvard University, Boston, Massachusetts. One of the members of Rosenau's department was Dr. J.W. Schereschewsky, assigned from the U.S. Public Health Service, who was instrumental in persuading the Surgeon General of the Public Health Service that cancer was a public health problem. Upon Rosenau's retirement in 1930, Andervont became the first professional staff member of Schereschewsky's Office of Cancer Investigations. The history of this group and an appreciation of Schereschewsky were published by Andervont in 1957.

Andervont remained with the Public Health Service for the rest of his professional career. The Office of Cancer Investigations at Harvard became part of the National Cancer Institute, created in 1937, and was moved to the National Institute of Health in Bethesda in 1939. Andervont led biological research at the National Cancer Institute until 1961, when he relinquished his post as Chief of the Laboratory of Biology and became Editor of the Journal of the National Cancer Institute.

Howard B. Andervont married Letha Marie Krabill on September 14, 1926. They have three children: Mrs. Barbara (Robert E.) Bowman, born in 1928; Mrs. Carolyn (James M.) Edie, born in 1930; and John D. Andervont, born in 1935.

Virus research was one of Andervont's earliest interests. He showed that herpesvirus of man could be transmitted to mice by intracerebral inoculation, and thereby provided the means of studying this human virus in an animal. It was but a short step to an interest in viruses as related to cancer and the development of a method of concentrating by adsorption on charcoal the Rous sarcoma virus that infects fowl. He made it possible in this way to obtain consistent production of viral tumors in fowl, when previously negative and poorly reproducible results were common. Fully a quarter of a century before the current strong interest in viruses as a cause of human cancer and at a time when most scientists regarded the existence of such a relationship with great skepticism, he developed his abiding belief in the importance of studying the viral etiology of cancer.

Andervont's early work on immunology of transplanted tumors led him to the realization of the importance of genetic factors, and the usefulness of genetically homozygous strains of mice in cancer research. Personally maintained and developed colonies of inbred strains became the material for all his subsequent research.

Following the discovery by Bittner of the milk-transmitted mammary tumor agent in mice, Andervont began a comprehensive, long-term study of this factor. He was first to convert a strain of mice which develop mammary tumors infrequently to a permanent strain with a high frequency of tumor development by foster-nursing the mice with milk containing the agent. He was first to demonstrate the transmission of the mammary tumor agent through the seminal fluid of male mice, to show that the agent was prevalent in low concentration in wild mice, and to demonstrate passive immunity to the agent in mice.

Another line of Andervont's investigations had its origin in the early studies by the Harvard group of the induction of cancer in animals by chemical agents, particularly polycyclic hydrocarbons. Andervont pioneered the development of biological methods of studying experimental induction of cancer with chemicals. He continued his interest in the mechanism of chemical carcinogenesis and was among the first to make a systematic investigation of tumors induced in different inbred mouse strains by chemical carcinogens, the inheritance of cancer susceptibility to various agents, and the possible correlation of the incidence of spontaneous and induced tumors.

Andervont's personal research is recorded in his 165 publications. They have been well recognized and have had deep influence on cancer research throughout the world. They demonstrate his lucid, thorough approach to a problem, clear presentation of the results, and conservative interpretations.

For 20 years Andervont was also administratively responsible for, and guided the research of, the largest group of investigators in the National Cancer Institute, whose work encompassed tissue culture, electron microscopy, genetics, radiation biology, cell physiology, cell biology, tumor virology, and etiology of spontaneous leukemia. Under his direction, research of extremely high caliber was accomplished, and discoveries of outstanding importance were made: discovery of the Stewart-Eddy polyoma virus which produces multiple tumors in a number of animal species, discovery of the Moloney mouse leukemia virus, development of quantitative techniques for the study of tumor viruses, the role of hormones in experimental induction of cancer, quantitative methods in the testing of cancer-producing chemicals, relationship of lung tumors of mice to factors associated with the chromatin of the sex cells, first cloning of a mammalian cell in tissue culture, development of more than a dozen tissue culture strains derived from single cells, development of a synthetic medium for tissue culture, and first conversion of normal to cancerous cells in the test tube.

Recognition of his commanding role in biological research on cancer during his long career has come from his colleagues and from the United States Government. He has been an active member of many scientific societies and of many committees of the American Cancer Society. He was president of the American Association for Cancer Research in 1955. For 15 years he has been a trustee of The Jackson Laboratory at Bar Harbor, Maine. In 1942, an honorary degree of Doctor of Science was conferred upon him by Mt. Union College, his alma mater. In 1961, he received the Distinguished Service Award of the Department of Health, Education, and Welfare, "for outstanding contributions to research, writing, and counseling in cancer biology and leadership in establishing and guiding the National Cancer Institute's Laboratory of Biology." In 1962, he was the recipient of the National Civil Service League Award.

Andy, as he is known to his many friends, is a modest man, and fulsome praise would embarrass him. He is an individualist, with his own convictions which he practices but does not impose on others. His belief in the freedom of science and in the paramount importance of personal ideas in the conduct of research has been the touchstone of his own endeavors and of the endeavors for which he has had administrative responsibility. All of his own reports are based on his own work; the animals he used were the animals he personally raised, examined, and evaluated. This personal involvement he never delegated to assistants, and the intimate knowledge of his media is evident in his publications. By the same token, the work of his associates was theirs alone. It would be unthinkable for him to add his name to publications reporting work in which he did not participate. He was free with his counsel when requested; he gave generously of his wisdom, when appropriate; he assumed responsibilities, when he considered them worthy.

Andy's relationships with his administrative heads reflected his ways in the laboratory. He was always dependable, straightforward, and, again, modest. His presidential address to the American Association for Cancer Research, published in 1956, demonstrates these qualities in his own words.

The home of Andy and Letha was always open to colleagues from all over the world, and to their many friends from many walks of life. The weekly poker parties became traditional from the Boston days, and the older group will always cherish the New Year's Eve gatherings.

Howard Bancroft Andervont is an epitome of a scientist. Andy, may your years be many, and may your tribe increase!

BIOGRAPHY AS RELATED TO INBRED STRAINS
OF EXPERIMENTAL ANIMALS

Walter E. Heston[1]

Fort Meyers, Florida

Soon after Congress passed the Bone Act establishing the National Cancer Institute in 1936, a Research Fellowship Program was set up to bring in a group of young investigators to complete the staff of this newly established Institute as soon as possible. These fellows were hired in 1938, and for two years, while the physical plant of the Institute was being built in Bethesda, they were placed for additional training in cancer research laboratories throughout the country.

As one of these research fellows, I was brought in as a mammalian geneticist and was fortunate to be placed for this two-year training period with Dr. C.C. Little at the Roscoe B. Jackson Memorial Laboratory at Bar Harbor, Maine. In no other two years have I learned more than in those two years at the Jackson Lab. There I received a good foundation in cancer research and, above all, I was well indoctrinated in the value of inbred strains of experimental animals, particularly mice, in medical research and especially in cancer research.

---

[1] Address: 1380 Burgundy Drive S.W., Fort Myers, FL 33907.

After choosing the genetics of lung tumors for my initial studies, I established a breeding colony of the high lung tumor strain A mice from animals received from Dr. J. J. Bittner who was on the staff of the Jackson Laboratory at that time. For a strain most resistant to lung tumors I chose strain C57L and established a breeding colony of this strain from animals received from Dr. Arthur C. Cloudman also on the staff there. Later I added a colony of strain C57BR from animals likewise received from Dr. Cloudman. In the spring of 1940, I was called in to the Institute in Bethesda and I shipped these colonies of individually identified mice along with the hybrids I had produced to Bethesda by Railway Express. Not one animal was lost in shipment. Subsequently, when I became involved in work on mammary tumors and hepatomas, I added strains C3H and C57BL to my breeding colonies from animals received from Dr. H.B. Andervont at the National Cancer Institute and still later other strains were added.

Since I was the one on the staff of the National Cancer Institute who specifically had received his degree in mammalian genetics, I assumed a special responsibility for maintaining my colonies of inbred mice and fostering the use of inbred strains in the research of the Institute. I had the full support of Dr. Andervont and my efforts were well received by my colleagues working with cancer. After all, one could not transplant experimental tumors without inbred animals. One day a few years later, however, in talking with Dr. R.R. Spencer, our director, it was decided that I might direct a comprehensive memorandum to the director of the National Institutes of Health, putting forth my ideas of the use of inbred strains in cancer research and also pointing out the value of the genetic approach and the advisability of using inbred strains in the research on other diseases carried out in the other Institutes of the National Institutes of Health. Accordingly, such a memorandum was formulated and sent. Within a few days I received from the director a half-page reply acknowledging me as a young investigator in cancer but implying that the other research of the Institutes was in good hands and was doing very well with the animals being used. This bothered Dr. Spencer more than it did me for he felt that he had been responsible for having put me into this somewhat embarrassing situation. Any misgivings I might have had about my brash actions were to be lifted a few years later when this director retired and a lecture series was set up in his honor. As the first speaker of the series a well known geneticist was selected.

One day in the early 1940s, Dr. Michael B. Shimkin who occupied an adjoining laboratory at the NCI received information that the inbred strains of guinea pigs at the USDA Beltsville Station were to be discontinued. These were of the original families with which Dr. Sewall Wright had done his monumental studies on the effects of inbreeding from which he derived the concept of the inheritance of threshold characters. These were the only inbred guinea pigs in existence and now, because of lack of space and financial support, they were about to be discarded. Dr. Shimkin immediately contacted Dr. Eaton who was in charge of the colonies, and Dr. Shimkin and I went to Beltsville and rescued the remaining breeding animals and brought them back to the National Cancer Institute where I set up breeding colonies. Aside from having a sense of their general potential value in medical research, we knew that they would be essential for any studies of transplanted tumors in this species and would be uniquely suitable for such studies as those later carried out with Robertson on the effect of ascorbic acid on the development of sarcomas. Unlike the mouse, the guinea pig can be made deficient in ascorbic acid. These families or strains included 2 and 13, and also family 35 that had a high incidence of polydactyly and family 32 that really should not have been called 32 for it had experienced some outcrossing. These latter two families were at such a low breeding level when we got them that it was impossible to get them going again. Years later, when he was in my laboratory, Dr. Sewall Wright poked his face in the cages and confirmed the identification of strains 2 and 13, not by their coat color, for both strains are tricolor, but by the shape of their noses. Strains 2 and 13 are still in existence and widely used.

In subsequent years inbred strains of rats were added to our breeding colonies from Dr. Dunning's colonies in Miami. These included strains A x C 9935, Marshall, and Fisher, all strains that Drs. Curtis and Dunning had been inbreeding for many years. In addition, we began inbreeding Osborne-Mendel rats and Buffalo rats that had been extensively used in medical research but up to that time had not been inbred. Years later, when a central animal breeding facility was set up at NIH with a geneticist in charge of the inbred colonies, all the guinea pigs and rats were transferred to this facility.

In the early 1940s, Dr. Margaret K. Deringer joined my laboratory, and about ten years later Mr. George Vlahakis joined us. To these two associates along with many other dedicated associates, technicians, and animal caretakers goes much of the credit for the work of this program. It was always the philosophy of the laboratory that probably the most important person on the team was the animal caretaker, and weaning and marking animals and making up new matings were considered the most important tasks of the laboratory and for this work I was always present until the day of my retirement.

Possibly of equal importance to our promoting the use of inbred strains of experimental animals was the part we played in the introduction of the use of the F_1 hybrid. During World War II, instead of being placed in the Armed Forces, I was instructed to remain at the NCI and to work along with Drs. Lorenz, Deringer, and Eschenbrenner on the Manhattan Project. Our principal work was concerned with the effects of long-term, continuous exposure to low dosage gamma irradiation. It was my suggestion that as test animals we use an F_1 hybrid because of its genetic uniformity, vigor, and long life. Since we already had a considerable amount of information on LAF_1 hybrids from our lung tumor genetic studies, it was decided to use this hybrid. The program was extensive, involving thousands of such hybrids, and some of it attracted considerable attention. At one time we sent a group of several hundred uniform and individually identified LAF_1's to the South Pacific for one of the tests of the atomic bomb. They made the round trip, arriving back in our laboratory for later observations of the effects of the irradiation without loss of an animal. The results of all of these studies demonstrated the value of the F_1 because the tumors that arose in these very old animals occurred at ages beyond which inbred mice would not have lived. It also set the stage for the use of this particular F_1 in subsequent radiation work. Unfortunately, this F_1 was a rather poor choice in that both parent strains are poor breeders. Nonetheless, the studies did much to open up the extensive use of the F_1 hybrid between two inbred strains as a genetically uniform, vigorous, long-lived experimental animal.

Work in our laboratory was also instrumental in introducing the use of the fostered strains. Bittner and others had shown that when newborn mice of high mammary tumor strains were foster-nursed by females of low tumor strains, few or none of them developed mammary tumors. For this reason, in the middle 1940s, we decided to start a colony of C3H mice foster-nursed by C57BL, the C3HfB, to be used in testing for the mammary tumor virus. We kept these animals in a healthy condition so that they lived to a very old age and, much to our surprise, the females were not free of mammary tumors as almost half of them developed tumors at an advanced age. While we initially thought that these tumors were due to the high genetic susceptibility of the females plus the hormonal influence from their having had many litters, the observation of particles in the tumors of this line by investigators in Dr. DeOme's laboratory at Berkeley showed that this strain had a line of mammary tumor virus that was somewhat different from the milk-transmitted line. These and other observations opened the way to subsequent extensive studies of many fostered lines from various combinations and the lines of virus they retained. The C3H with the A^{vy} gene, or C3H-A^{vy}, is the highest mammary tumor strain we have ever had and the fostered strain C3H-A^{vy}fB was particularly interesting in that it had an incidence of approximately 100% mammary tumors caused by a virus not transmitted through the milk but presumably genetically transmitted.

It was always the policy of our laboratory to send mice of any of our strains to any investigator who requested them. Since it was generally recognized that we had a rather clean colony, a reputation earned by our strict isolation of the colony, and possibly also since NCI always paid the shipping charges to insure prompt and safe delivery, we received many requests from researchers throughout the world. Consequently my wife was later to find out that my mice had become more widely known that I was. Much of the medical research of Japan today is done with strains of mice descended from my colony. This largely has been due to the efforts of Dr. Tatsuji Nomura in getting mice from us and setting up his own extensive production facility and even of his sister, Miss Michi Nomura, who first came to my laboratory to make arrangements for mice of our colony to be shipped to Japan. Through the years our mice were sent to establish breeding colonies in practically every country in the world and sent with almost no losses in shipment. In earlier days airline pilots took a special interest in our shipments of mice going to the far corners of the earth to be used in cancer research, often even taking them into the pilot's cabin to insure their safe passage.

Communication was always with the investigator who was to receive the mice to make sure that the breeding data did not get mixed up. An exception to this, however, was when Dr. Andervont and I sent C57BR and BALB/c mice to the U.S.S.R. These had to be transmitted through the State Department and, of course, they got mixed up, leading to a confusion of the Russian strains that did not get straightened out until about 15 years later when I visited the U.S.S.R. for the Cancer Congress in Moscow.

The day of retirement came for me on December 31, 1975, and a year later when all the research underway had been concluded, the Heston mice were no more. When word of my impending retirement got around, many came to me expressing concern about what was going to happen to my colonies of mice. It seems to me that it all worked out as it should have. I had always shared my mice with any who had asked for them so that any who were wanting to work with Heston mice had them. By this time even my associates had their own colonies. With my retirement there was clearly no longer a need for a colony of Heston mice. After all, of what value are a man's colonies of mice without the man.

Walter E. Heston

CONGENIC RESISTANT STRAINS OF MICE

George D. Snell

The Jackson Laboratory[1]

Bar Harbor, Maine

My subject today concerns the hunt for histocompatibility genes. This is a hunt that I was engaged in for nearly 30 years, from 1944 or 1945 to the end of 1973. Three kinds of traps are suitable for catching this particular game. These are the antibody trap, the linkage trap, and the congenic resistant strain trap. At one time or another I have used all three, but Dr. Morse tells me that my subject today should be primarily the congenic resistant strain trap. While engaged in this hunt I did not wear the typical outfit of the Maine hunter. The appropriate garb was not a bright red wool jacket but a laboratory gown, and the pervading odor not that of pine and balsam forests but of the mouse room. Yet I am sure that no Maine guide could have provided a more enthralling chase. Over the years, I was joined by many wonderful companions and friends. I cannot, in the short time I have, possibly do justice to their contributions to both the profit and the pleasure of the hunt.

---

[1]The Jackson Laboratory is fully accredited by the American Association for Accreditation of Laboratory Animal Care.

To explain how I became involved in the hunt for histocompatibility genes, I need to go back to my arrival at the Jackson Laboratory in 1935. At the time I joined the staff, virtually all the other investigators -- all six of them -- were engaged in studying the mammary tumor factor. Only Dr. Cloudman was actively engaged in work with tumor transplants. It thus happened that I was unaware of the important work on the genetics of transplantation which had been done by Dr. Little and various colleagues both before and after the founding of the Laboratory. My own interest at the time centered on the induction and analysis of translocations in the mouse, a project which I had started in 1931 at the University of Texas.

While this was an interesting subject, Dr. Little showed no enthusiasm for it, and I also felt it was reaching a point of diminishing returns. I therefore began devoting a great deal of thought and effort to finding some major undertaking that would warrant a long-term commitment. The project should be genetic, should use mice, and should offer the prospect of yielding some really clear-cut and basic information. The search involved both extensive reading and several years of trial and error in the laboratory. In the end, I was greatly influenced by three review articles. Two of them, by Woglom (45) and by Spencer (49), the latter at the time assistant chief of the National Cancer Institute, concerned immunity to transplantable tumors. These acquainted me with the thinking then current on the immunological aspects of transplantation. It was through these reviews that I became involved in the subject of immunological enhancement, a subject outside the scope of this paper, but one which occupied much of my effort until the genetic studies of transplantation became a full-time job. The inspiration for these studies came from Dr. Little's chapter on the genetics of tumor transplantation (48), a chapter in The

Biology of the Laboratory Mouse, which I edited as one of my jobs during this period. Through reading this chapter I became aware for the first time of the evidence that susceptibility and resistance to transplants are under genetic control, with a dozen or more loci each playing a part. Here, perhaps, was a clue to the sort of research project for which I had been looking.

The work of Little and his colleagues, while it showed that multiple loci are involved in transplant rejection, did not serve to identify individual loci. This was a major gap in the evidence. Gorer, in 1938, through the use of serological and transplantation techniques, had identified one locus concerned with the rejection of tumor transplants, but I did not become acquainted with this work until somewhat later. The problem was to find some way of separating out and studying individual loci. If this could be done, there would be a wealth of genetic material on which to work. Probably at this time I was not thinking about the function of the loci beyond their role in transplantation, but even with this limitation, the problem looked interesting.

What technique, or techniques, could be devised for rendering individual, transplant-influencing loci available for identification and study? Two methods suggested themselves. The first was the use of marker genes, a method of general applicability with which I was thoroughly familiar from prior Drosophila work and my own translocation studies with mice. The second was the production of congenic resistant, or CR, lines. Actually, there was a precedent for the production of congenic lines with a blood group difference in Irwin's studies of species differences in the Columbidae (46), but I was unfamiliar with this at the time and Irwin used the methodology in a context quite different from the one I was contemplating.

These two methods were outlined in a seminar given at the Jackson Laboratory, probably in 1944. About this time I devoted a good deal of effort to the mathematics of establishing congenic lines and to the tests which might be used to analyze them once they were produced. However, it was 1948 before the methods were published (50). J.B.S. Haldane added to this paper a much more elegant mathematical analysis of the development of coisogenicity than I had been able to produce. Figures 1 and 2 show the two mating systems, subsequently called the cross-intercross and cross-backcross-intercross systems (56), which were suggested as suitable for the production of CR lines. These figures, I think, are self-explanatory and require no comment.

Actually, in the 1948 paper, the expression isogenic resistant lines rather than coisogenic or congenic resistant lines was used. The term coisogenic was coined for comparable lines in Drosophila by Chovnick and Fox in 1953 (52). I adopted it in 1958 (10). Following a suggestion by Dr. Earl Green, the term congenic was listed as a possible alternative in 1961 (15). It is now in regular use. Perhaps more important with respect to terminology, the word histocompatibility was introduced for the first time in the 1948 paper.

Although the first publication concerning the hunt for histocompatibility genes did not appear until 1948, the hunt was actually started in 1945. My records show that the first backcross matings used to detect histocompatibility genes by the linkage method were made in that year, and the first transplantable tumor, kindly provided by Dr. Cloudman, transplanted to the backcross mice early in 1946. This work is outside the scope of this paper, but it developed into a sizable program, with seven linkage stocks carrying 18 marker genes being used. It led to a rediscovery of H-2, already discovered by Gorer (46), to a profitable joint study of this

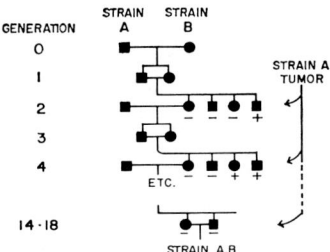

FIGURE 1. The cross-intercross or M system of producing congenic resistant strains. Reproduced from Snell and Bunker (20) with the permission of Transplantation.

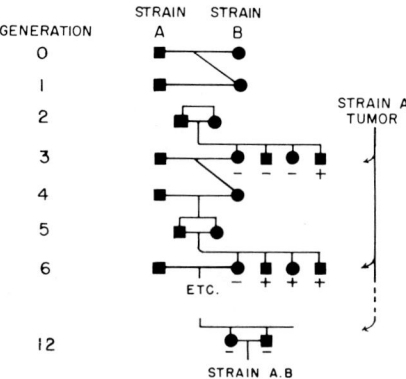

FIGURE 2. The cross-backcross-intercross or N system of producing congenic resistant strains. Reproduced from Snell and Bunker (20) with the permission of Transplantation.

locus with Gorer (50), and ultimately to the demonstration of eight H-2 alleles (53, 54). Allen (54) also used the linkage method to demonstrate one of the first recombinants within H-2.

In considering the production of CR lines, it is convenient to divide them into four groups based on variations in the method used. Although there was considerable overlap in the production and testing of these groups, they were started at different times, and I shall present them in chronological order. The CR lines in each group which reached the testing stage are listed in Tables 2 through 5. Table 1 lists the inbred strains which entered, either as first or second parent, into the production of the various CR lines. Table 6 gives, for each group, a summary of the number of lines started, tested, and still carried. The terminology of congenic resistant lines is described in Appendix 1.

Klein (36) has published a list of congenic resistant lines, including many not considered here. Other lists of CR and relevant inbred strains (the oldest lists now primarily of historical interest) will be found in references 38, 40, 55, 57, and 58.

A group of congenic resistant lines was started in June 1946 with the assistance of Helen Parker, before the Bar Harbor fire of October 1947 in which the main building of the Jackson Laboratory was destroyed. This group was entirely lost. In the spring of 1948, matings were set up again, using space at the Hamilton Station Laboratory, and mice which had been returned to the Jackson Laboratory from many former beneficiaries of Dr. Little's generous policy in distributing animals. Because it was impossible to tell, in the light of our knowledge at that time, whether some crosses might be substantially better than others, a wide variety of crosses was used. The one invariable restriction was that the parent to which backcrosses were to be made

TABLE 1. Inbred Strains Used in the Production of Congenic Resistant (CR) Strains

Strain[a]	Short symbol	Genotype[b,c]
A/WySn	A	(17) $H-2^a$ Tla^a $Pgk-2^a$ Apl^b
AKR/J	AKR	(17) $H-2^k$ Tla^b $Pgk-2^b$ Apl^b $Ly-2^a$
BALB/c	C	(2) $H-3^c$ $Ir-2^b$ $H-13^{not\ a\ or\ b}$ A (7) c Hbb^d $H-1^b$ (9) $Thy-1^b$ $Fv-2^s$ $H-7^b$ (17) $H-2^d$ Tla^c $Pgk-2^a$ Apl^b $H-9^b$
CBA	CBA	(19) $Ly-1^b$
CE/J	CE	(2) $H-3^b$ $Ir-2^b$ $H-13^b$ A^w (7) $H-4^{not\ a\ or\ b}$ c Hbb^s $H-1^{a,\ b,\ or\ c}$
C3H/HeDiSn	C3H	(2) $H-3^b$ $Ir-2^?$ $H-13^c$ A (7) $H-4^a$ C Hbb^d $H-1^a$ (17) $H-2^k$ Tla^b $Pgk-2^b$ Apl^b
C57BL/6By	B6	Same as C57BL/10Sn except $H-9^{not\ a\ or\ b}$, a difference at an H locus on chromosome 4, and an Ea difference.
C57BL/10Sn	B10	(2) $H-3^a$ $Ir-2^a$ We Un $H-13^a$ a (6) $Ly-2,3^b$ (7) P $H-4^a$ C Hbb^s $H-1^c$ $Ea-2^b$ $Ea-7^b$ (9) $Thy-1^b$ $Fv-2^r$ $H-7^a$ (17) $H-2^b$ Tla^b $Pgk-2^a$ Apl^b $Mlv-1^b$ Hc^1 $H-8^a$ $H-9^a$ $H-10^a$ $H-11^a$ $H-12^a$
DA/HuSn	DA	(17) $H-2^{qp}$ Tla^b $Pgk-2^b$ $Apl^?$
DBA/1	D1	(17) $H-2^q$ Tla^b $Pgk-2^a$ $Apl^?$
DBA/2	D2	(7) Hbb^d $H-1^a$ (17) $H-2^d$ Tla^c $Pgk-2^a$ Apl^b Hc^0 $H-8^b$ $H-11^?$ $Mlv-1^a$
FS/Ei	FS	(7) p $H-4^?$ c^{ch} $sh-1$ Hbb^d $H-1^?$ fr
HTG/SfSn	HTG	(17) $H-2^g$ Tla^b $Pgk^?$ $Apl^?$

125

TABLE 1, continued

Strain[a]	Short symbol	Genotype[b,c]
JK/Sf	JK	(17) $\underline{H\text{-}2^j\ Tla^b\ Pgk\text{-}2^?\ Apl^?}$
KR/Di	KR	(2) $\underline{H\text{-}3^d\ Ir\text{-}2^?\ H\text{-}13^c\ A}$
LG/Ckc	LG/Ckc	(17) $H\text{-}2^{ar}$
LP	LP	(2) $\underline{H\text{-}3^b\ Ir\text{-}2^?\ H\text{-}13^b\ A^w}$ (17) $\underline{H\text{-}2^b(or\ bc)\ Tla^c\ Pgk\text{-}2^c\ Apl^b}$
P/Sn	P	(7) $\underline{p\ H\text{-}4^?\ C\ Hbb^d\ H\text{-}1^?}$ (17) $\underline{H\text{-}2^p\ Tla^?\ Pgk\text{-}2^a\ Apl^b}$
PA/Re	B6-pa	(2) $\underline{pa\ H\text{-}3^?\ H\text{-}13^?}$
PL/J	PL	(6) $\underline{Ly\text{-}2,3^a}$ (9) $\underline{Thy\text{-}1^a}$ (17) $\underline{H\text{-}2^u\ Tla^a\ Pgk\text{-}2^a\ Apl^b}$
RIII	RIII	(17) $\underline{H\text{-}2^r\ Tla^?\ Pgk\text{-}2^b\ Apl^?\ Ea\text{-}2^a}$
SEC	SEC	(7) $\underline{H\text{-}4^?\ c^{ch}\ Hbb^s\ H\text{-}1^{not\ a,\ b,\ or\ c}}$ $Ea\text{-}2^a$
SM/J	SM	(7) $\underline{H\text{-}4^?\ C\ Hbb^s\ H\text{-}1^{not\ a,\ b,\ or\ c}}$ (17) $\underline{H\text{-}2^v\ Tla^b\ Pgk\text{-}2^a\ Apl^a}$
ST/bJ	ST	(17) $\underline{H\text{-}2^k\ Tla^b\ Pgk\text{-}2^b\ Apl^b}$
Uw/Le	UW	(2) $\underline{H\text{-}3^b\ Ir\text{-}2^?\ we\ un\ H\text{-}13^a\ a^t}$
WB/Re	WB	(7) $\underline{H\text{-}4^{not\ a\ or\ b}\ C\ Hbb^d\ H\text{-}1}$ (17) $\underline{p\ H\text{-}4^b\ c\ Hbb^d\ H\text{-}1^b}$
129	129	(2) $\underline{H\text{-}3^b\ Ir\text{-}2^?\ H\text{-}13^b\ A^w}$ (7) $\underline{p\ H\text{-}4^b\ c\ Hbb^d\ H\text{-}1^b}$ (17) $\underline{H\text{-}2^{bc}\ Tla^c\ Pgk\text{-}2^c\ Apl^b\ H\text{-}10^b\ H\text{-}11^b\ H\text{-}12^b\ Ea\text{-}7^a}$

126

Legend to Table 1

^aThe strains listed are those inbred strains which entered, either as first or second parent, into the production of the congenic strains shown in Tables 2-5. In the case of some of the early strains, especially those in Table 2, the substrain used is unknown and is therefore not indicated. Where more than one substrain was used, as in the case of C57BL/6, the currently most relevant is shown.

^bThe loci listed for each inbred strain are, with the exception of a few linked loci, those at which differences occur between one or more congenics and the inbred strain. Chromosome numbers, where known, are indicated in parentheses. Genes closest to the centromere are given at the left of each group of linked genes. Loci whose relative positions are undetermined are bracketed. The references from which information concerning genotype have been taken are: 34 (information re H-1, H-3, H-13, strain KR), 35 (H-3), 39 (Ir-2), 41 (Tla), 43 (Pgk-2), 44 (linkage of Fv-2 and H-7), 58 (non-H-2 alleles), 59 (non-H-2 alleles), 60 (Ea-2, Ea-7, H-2, Ly-2,3, Thy-1, Tla), 61 (a, c, Hbb, Hc, H-2, p), 62 (strains FS and UW), 63 (does BALB/c have fifth Tla allele?), 64 (Apl), 65 (Ea difference between B6 and B10).

^cThe Pgk-2 allele of strain 129 differs in different sublines. The allele in the subline which we have used appears to have been Pgk-2^c.

TABLE 2. CR Strains (Group 1) Produced by the Cross-intercross System; Many of the Second Strains Used in Original Cross Not Inbred[a]

Strain	Short symbol	Generation	N equivalent	Genotype	References
A.BY/Sn	A.BY	G2F17N5F20N5F26	NE11F26	(17) $H-2^b$ Tla^c $Pgk-2^c$ $Apl^?$	7 <u>10</u> 13 <u>14</u> 19 <u>38</u> 41 <u>43</u>[c]
A.CA/Sn	A.CA	G2F21N5F30N5F23	NE11F13	(17) $H-2^f$ Tla^d $Pgk-2^b$ $Apl^?$	7 <u>10</u> 13 <u>14</u> 19 <u>38</u> 41 <u>43</u>[c]
A.SW/Sn	A.SW	N2F14N11F8N1F23	NE14F23	(17) $H-2^s$ Tla^b $Pgk-2^b$ $Apl^?$	3 4 7 9 <u>10</u> 13 <u>14</u> 19 <u>38</u> <u>43</u>[c]
AKR.ALB/Sn	AKR.ALB		Dropped	Non-$H-2$	9 <u>10</u>
AKR.K/Sn	AKR.K		Dropped	(17) $H-2^a$	3 <u>10</u>
AKR.M/oSn	AKR.M/o	G12F70	Dropped	(17) $H-2^m$ Tla^a $Pgk-2^a$ $Apl^?$	<u>10</u> <u>36</u> <u>38</u>
				$Ly-2^b$	
AKR.M/nSn	AKR.M/n	G12F51N6F13	NE12F13	(17) $H-2^m$ Tla^a $Pgk-2^a$ $Apl^?$	39 <u>43</u>
B6.D2/Sn	B6.D2		Dropped	(17) $H-2^d$	1 5
B10.BY/Sn	B10.BY	G13F1G5F?	Dropped[d]	(7) $Hbb^?$ $H-1^b$	<u>10</u> <u>15</u> 20 21 <u>26</u> 34[d]
B10.D2/oSn	B10.D2/o	G12F71	NE6F71	(17) $H-2^d$ Tla^c $Apl^?$	1 3 4 5 6 7 9 <u>10</u> <u>12</u> 15 <u>16</u>
				Hc^0	17 20 29 32 40
B10.D2/o2Sn	B10.D2/o2	G12F54N4F6	NE10F6	(17) $H-2^d$ Tla^c $Apl^?$ Hc^0	33 <u>38</u>
B10.D2/nSn	B10.D2/n	G12F3G6F2G4F53	NE11F53	(17) $H-2^d$ Tla^c $Apl^?$	<u>16</u> <u>17</u> 22 23 25 <u>33</u> 43
B10.D2/n2Sn	B10.D2/n2	G12F3G6F2G4F39N5F11	NE16F11	(17) $H-2^d$ Tla^c $Apl^?$	<u>38</u>
B10.LP/Sn	B10.LP	G12F1G2N2F61	NE9F61	(2) $H-3^b$ $Ir-2^b$ $H-13^b$ A^w	3 4 5 6 7 9 <u>10</u> 11 12 15 18
					20 <u>24</u> 28 29 34 35 39
B10.LP-$H-3^b$/Sn	B10.LP-$H-3^b$	G12F1G2N2F29N4F22	NE13F22	(2) $H-3^b$ $Ir-2^b$	18 <u>21</u> 24 28 34 35 <u>38</u> <u>39</u>[e]

128

TABLE 2, continued

Strain	Short symbol	Generation	N equivalent	Genotype	References
B10.M/Sn	B10.M	G12F13J3F47	NE8F47	(17) $H-2^f$ Tla^d $Pgk-2^b$ $Apl^?$	10 38 41 43[f]
B10.SW/Sn	B10.SW		Dropped	(17) $H-2^d$	10
B10.Y/Sn	B10.Y	G14F10J3F59	NE9F59	(17) $H-2^{pa}$ $Tla^{c?}$ $Apl^?$	10 25 31 38 43[g]
C.B6/Sn	C.B6		Dropped	(17) $H-2^b$	10 20[h]
C3H.K/Sn	C3H.K	G14F53N4F12	NE11F12	(7) c Hbb^d $H-1^b$	2 6 7 9 10 11 15 26 34 38
C3H.LPK/Sn	C3H.LPK		Dropped	(17) $H-2^b$	10
C3H.NB/Sn	C3H.NB	G16F40N4F19	NE12F19	(17) $H-2^p$ $Tla^{c?}$ $Apl^?$	10 38 43
C3H.SW/Sn	C3H.SW	G14F44N4F18	NE11F18	(17) $H-2^b$ $Pgk-2^c$ $Apl^?$	7 10 38 43
D1.C/Sn	D1.C	G14F37N5F14	NE12F14	(17) $H-2^d$ Tla^c $Apl^?$	10 38 43[i]
D1.LP/Sn	D1.LP	G12F43N5F15	NE11F15	(17) $H-2^b$ $Tla^{b?}$ $Apl^?$	10 38 43[j]
D1.ST/Sn	D1.ST		Dropped	(17) $H-2^k$	10
D2.WA/Sn	D2.WA		Dropped	(17) $H-2^b$	10 20[h]
L.WA/Sn	L.WA		Dropped	Non-H-2	10
LP.RIII/Sn	LP.RIII	G12F74	NE6F74	(17) $H-2^r$ Tla^b $Pgk-2^b$ $Apl^?$	10 25 38 43
P.CFW/Sn	P.CFW		Dropped	(17) $H-2^b$	10
RIII.P/Sn	RIII.P		Dropped	(17) $H-2^p$	10
ST.T6/Sn	ST.T6		Dropped	(17) $H-2^d$	10

Legend to Table 2

^{a}In the column headed Generation, G indicates the number of matings by the cross-intercross system, J the number of matings by the cross-backcross-intercross system (appears in Table 3 only), N the number of backcross matings, R the number of irregular matings, and F the number of brother x sister matings. It should be noted that most of the strains in Table 2 which are still maintained have been put through additional matings since they were first established. Their coisogenicity has thus been increased. Insofar as the original strains are maintained in other laboratories, they may show minor genetic differences from the strains maintained at the Jackson Laboratory. In the column headed Genotype, only known gene differences from the inbred partner are given, or in a few cases linked loci where the allele is questionable. Where the chromosome bearing the defining locus is known, this is indicated in parentheses. The References (last column) are arranged chronologically. The references most pertinent to each CR strain are underlined. See Appendix 1 for definition of terms and significance of symbols.

^{b}In the production of the CR strains listed in Table 2, 10 non-inbred animals or stocks were used as the second partners in various initial crosses. These (together with information as to some marker genes which they carried) were: ALB (dd Tt), BY (Aa), CA (Cc CaCa Fufu), CFW (AA BB), K (Aa bb? Cc Kiki), M (aa BB CaCa), SW (Aa BB cc), T6 ($c^{ch}c^{ch}$), WA (Aa Bb ?sh-2), and Y (A^ya) (10).

^{c}The strain originally used as the "inbred partner" in production of strains A.BY, A.CA, and A.SW was A/Lilly (Eli Lilly and Co.), now designated AL. It was found to have been contaminated (13, 14). Later matings (1952 and thereafter) were to A/Sn, and only these matings are included in the

Legend to Table 2, continued

generation numbers. Strain A.CA originally carried <u>Fu</u> linked to H-2^f (10), but this gene was lost in later backcrosses.

^{d}B10.BY was originally classified as H-1^d (15) but is H-1^b (26). It has been dropped at The Jackson Laboratory but may be maintained at the Institute of Molecular Genetics, Prague, Czechoslovakia.

^{e}Strain B10.LP-$\underline{\text{H-3}}^b$/Sn was previously called B10.LP-<u>a</u>/Sn.

^{f}Strain B10.M was originally B6.M (10). Subsequently two matings were made to C57BL/10, so three fourths of the background genes should be from this strain.

^{g}Strain B10.Y was originally B6.Y (10). Subsequently two matings were made to C57BL/10, so three fourths of the background genes should be from this strain. B10.Y was originally classified as H-2^q (10). The test was unambiguous. It was later found to carry the H-2^{pa} haplotype, closely resembling H-2^p. The reason for the discrepancy is unknown; possibly at the time of the original test there were two sublines, the subline used in the original test later being lost.

^{h}See footnote b, Table 3.

^{i}If strain D1.C received an $\underline{\text{H-2}}$ haplotype from the second parent, BALB/c, this should have been H-2^d. By one test, D1.C typed as H-2^d, by a second test it typed as an unknown haplotype (10). It was tentatively assigned the haplotype symbol H-2^c. The existing line of D1.C is, as expected from the original cross, H-2^d.

^{j}There are indications (Snell and Cherry, unpublished) that strain D1.LP carries a variant of the H-2^b haplotype, probably the same as H-2^{bc} found in strains 129 and B10.129 (B6). Possibly the suspected difference between H-2^b and

Legend to Table 2, continued

$\underline{H-2}^{bc}$ is due to differences in linked $\underline{H}$ genes rather than to differences in the MHC itself.

TABLE 3. Non-$\underline{H\text{-}2}$ CR Strains (Group 2), Produced by the Cross-intercross and Cross-backcross-intercross Systems; All Original Matings Between C57BL/10 and Some Other $\underline{H\text{-}2}^b$ Strain[a]

Strain[b]	Short symbol	Generation	N equivalent	Genotype	References
B10.C(41N)/Sn	41N	J3G2J6G2F45	NE8F45	(7) c $\underline{Hbb^d}$ $\underline{H\text{-}1^b}$	23 26 34 38
B10.C(44N)/Sn	44N		Dropped	$\underline{H\text{-}9^b}$	20
B10.C(45N)/Sn	45N	J12G2F52	NE9F52	$\underline{H\text{-}9^b}$	20 21 23 28 34
B10.C(47N)/Sn	47N	J12F55	NE8F55	(9) $\underline{Fv\text{-}2^s}$ $\underline{H\text{-}7^b}$	20 21 22 23 28 34 44
B10.D2(51N)/Sn	51N	J12G2F2	Dropped	(7) $\underline{H\text{-}1^a}$	26
B10.D2(52N)/Sn	52N	J12G2F2	Dropped	(7) $\underline{H\text{-}1^a}$	26
B10.D2(55N)/Sn	55N	J12F8G4F41	NE10F41	$\underline{H\text{-}11^?}$	20 34 38
B10.D2(57N)/Sn	57N	J12F55	NE8F55	$\underline{H\text{-}8^b}$	20 21 28 34
B10.D2(58N)/Sn	58N	J12G2F55	NE9F55	(7) $\underline{Hbb^d}$ $\underline{H\text{-}1^a}$ $\underline{Mlv\text{-}1^a}$	26 28 34 37
B10.129(5M)/oSn	5M/o	G14F1G4F55	NE9F55	(7) c $\underline{Hbb^d}$ $\underline{H\text{-}1^{b1}}$ $\underline{Ea\text{-}7^a}$	15 20 21 22 26 27 28 34 37 38[c] [d]
B10.129(5M)/nSn	5M/n	G14F1G4F22G4F27	NE11F27	(7) c $\underline{Hbb^d}$ $\underline{H\text{-}1^b}$	38
B10.129(6M)/Sn	6M	G14F60	NE7F60	(17) $\underline{H\text{-}2^{bc}}$ $\underline{Tla^c}$ $\underline{Pgk\text{-}2^c}$ $\underline{Apl^?}$	20 32 38 43
B10.129(9M)/Sn	9M			$\underline{H\text{-}12^b}$	
B10.129(10M)/Sn	10M	G16F49	NE8F49	$\underline{H\text{-}10^b}$	20 21 22 27 28 34
B10.129(12M)/Sn	12M	G14F47	NE7F47	$\underline{H\text{-}11^b}$	20 21 22 27 28 34
B10.129(13M)/Sn	13M		Dropped	$\underline{H\text{-}12^b}$ $\underline{H\text{-}3^b}$ $\underline{H\text{-}10^{b?}}$	20 21 22 27 28 32 34 18 22 34 35

Table 3, continued

Strain[b]	Short symbol	Generation	N equivalent	Genotype	References
B10.129(14M)/Sn	14M	G14F10G2R1F	Dropped	(2) H-13^b A^w	18 21 22 24 27 28 34
B10.129(17M)/Sn	17M		Dropped	H-12^b H-$x^{b?}$	20 32
B10.129(20M)/Sn	20M		Dropped	(7) p H-$4^{b'}$ c	15^d
B10.129(21M)/Sn	21M	G16F5G6F5G8F25	NE11F25	(7) p H-$4^{b'}$	15 20 21 22 27 28 34 37 38[d]

[a] See footnote a, Table 2.

[b] In the production of the B10.C and B10.D2 lines, strains C.B6 and D2.WA (also called D2.W) (see Table 2) were used as the second parent. Both, like C57BL/10, the first parent, were H-2^b; hence H-2 did not segregate. Both lines were subsequently dropped. Because of this derivation of the B10.C and B10.D2 lines, it is possible that they carry genes of origin other than B10, BALB/c, and DBA/2. The (M) lines were produced by the cross-intercross system, the (N) lines by the cross-backcross-intercross system.

[c] B10.129(5M)/o was originally classified as H-1^e (15), but is now classified as H-1^b (21, 26).

[d] Strains B10.129(5M), B10.129(20M), and B10.129(21M) were derived from a strain, given the 5M designation, which carried the marker genes p and c as well as H-4^b and H-1^b on the chromosome 7 segment derived from strain 129. Strain B10.129(21M) was derived from a crossover mouse carrying p but not c that appeared in this strain in generation G8. Strain B10.129(5M) was separated from 20M at generation G6 and shown at G14 to be cP/cP, and hence the product of another crossover. The designation 20M was given to the unmodified strain with all four introduced genes.

TABLE 4. CR Strains Produced, in Conjunction with Ralph Graff, by Backcrossing Marker Genes onto the C57BL/10Sn Background[a]

Strain	Short symbol	Generation	N equivalent	Genotype	References
B10.C(28NX)/Sn	28NX	N8F53	NE8F53	(2) $H-3^c$ $Ir-2^?$ $H-13^?$ A	28 34 35
B10.C-H-3^c/Sn	28NX/2	N8F27N3F	NE11F	(2) $H-3^c$	34 35 39 40[b]
B10.C-H-13^a/Sn	28NX/3	N8F27N3F	Dropped	(2) we $H-13^a$ A	Unpublished[b]
B10.CE(30NX)/Sn	30NX	N7G4F46	NE9F	(2) $Ir-2^b$ $H-13^b$ A^w	22 34 38 39
B10.CE(62NX)Grf	62NX	N10F12N1F4	NE11F4	(7) c^e $H-1^{not\ a,\ b,\ or\ c}$	34 36[c]
C10.C3H(29NX)/Sn	29NX	N6G6F10N3R1F26	NE12F26	(2) $H-13^c$ A	24 34
B10.C3H(40NX)/Sn	N6G6F	N6G6F	Dropped[d]	(7) $\overline{Hbb^?\ H-1^a}{}^1\ H-x$	22 26 28 34[d]
B10.FS(63NX)/Grf	63NX	N10F	Dropped	(7) $\overline{p\ H-4^?}{}^1\ c^{ch}\ \overline{sh-1\ Hbb^?\ H-1}{}^b$	36
B10.KR(35NX)/Sn	35NX	N8F27	NE11F	(2) $H-3^d$ $Ir-2^?$ $H-13^c$ A	34 35
B10.KR-H-3^d/Grf	35NX/2	N8F27N3F	Dropped	(2) $H-3^d$	34 35 39[e]
B10.KR-H-13^{c}A/Grf	35NX/3	N8F27N3F27	NE11F	(2) we $H-13^c$ A	34 38[e]
B10.LP(36NX)/Sn	36NX	N8F	Dropped	(2) $H-3^{b?}$ $Ir-2^?$ $H-13^b$ A^w	18 22
B10.LP-H-13^{b}A^w/Sn	36NX/2	N8F17N1R1N2F25	NE11F	(2) $Ir-2^?$ $H-13^b$ A^w	34 38[f]
B10.P(61NX)/Grf	61NX	N10F14N1F3	NE11F3	(7) $\overline{Hbb^?\ H-1}{}^{not\ a,\ b,\ or\ c}$	34 36[c]
B10.PA-H-3^e/Grf	B10.PA	N11F18	NE11F18	(2) $pa\ H-3^e\ a^t$	34 35 39[g]
B10.SEC(64NX)/Grf	64NX	N10F16N1F1	NE11F1	(7) $H-1^{not\ a,\ b,\ or\ c}$	34 36[c]
B10.SM(65NX)/Grf	65NX	N10F13N1F2N1F2	NE12F2	(7) $H-1^{not\ a,\ b,\ or\ c}$	34 36[c]
B10.SWR(38NX)/Sn	38NX	N6F1N3F20	Dropped	(2) $?\ A$	Unpublished

TABLE 4, continued

Strain	Short symbol	Generation	N equivalent	Genotype	References
B10.UW(69NX)/Sn	69NX	N6F4N2F44	NE8F	(2) $H\text{-}3^b$ $\overline{Ir\text{-}2^?\ \overline{we\ un\ a^t}}$	18 $\underline{24}$ 34 35 36
B10.UW-$\underline{a}$/Sn	69NX/2	N6F4N2F	Dropped	(2) $H\text{-}3^b$ $\overline{Ir\text{-}2^?\ \overline{we\ un}}$	18 20 $\underline{24}$ 34 35^h
B10.UW-$H\text{-}3^b\underline{we}$/Sn	69NX/3	N6F4N2F8R3F	Dropped	(2) $\underline{H\text{-}3^b}\ Ir\text{-}2^?\ \overline{we}$	$\underline{24}$ 34 35^h
B10.UW-$H\text{-}3^b$/Sn	69NX/4	N6F4N2F8R3F	Dropped	(2) $\underline{H\text{-}3^b}\ Ir\text{-}2^?$	$\underline{24}$ 34 35^h
B10.WB(66NX)/Grf	66NX	N10F12N1F4	NE11F4	(7) $\underline{Hbb^?\ H\text{-}1^{not\ a,\ b,\ or\ c}}$	34 $\underline{36}^c$

136

Legend to Table 4

^{a}See footnote a, Table 2. The strains marked "Dropped" are not maintained either at the Jackson Laboratory or by Dr. Graff in St. Louis. It is possible that some of them are maintained elsewhere.

^{b}Strains B10.C-$\underline{H\text{-}3}^c$ and B10.C-$\underline{H\text{-}13}^?$ were derived from strains B10.C(28NX) as a result of crossing over following matings of B10.C(28NX) to B10-$\underline{we}^{Bkr}$ (35 and unpublished). B10-$\underline{we}^{Bkr}$ is a C57BL/10Sn congenic carrying a mutation to $\underline{we}$ which occurred in B10.129(21M). B10.129(21M) was at NE11 at the time B10-$\underline{we}^{Bkr}$ was separated.

^{c}These $\underline{H\text{-}1}$ strains were started at the Jackson Laboratory and completed by Dr. Ralph Graff in St. Louis. They were produced by using B10.129(5M)/n ($\underline{c}\ \underline{H\text{-}1}^b$) as the first parent. The presence in this strain of the seventh chromosome gene, $\underline{c}$ (albinism), made it easy to follow the introduction of $\underline{c}$ alleles which served as markers for introducing $\underline{H\text{-}1}$.

^{d}Strain B10.C3H(40NX) was split off from strain B10.C3H (29NX) at N6G6 (unpublished). It has been dropped in Bar Harbor but may be maintained at the Institute of Molecular Genetics in Prague.

^{e}Strains B10.KR-$\underline{H\text{-}3}^d$ and B10.KR-$\underline{H\text{-}13}^c$A were derived from strain B10.KR(35NX) as a result of crossing over following matings of B10.KR(35NX) to B10-$\underline{we}^{Bkr}$ (35 and unpublished; see footnote b).

^{f}Strain B10.LP-$\underline{H\text{-}13}^b A^w$ was derived from B10.LP(36NX) following a mating to C57BL/10Sn at N8F17 and a mating to B10.UW at N8F17N1. It carries the $\underline{H\text{-}3}^a$ allele of B10 instead of the $\underline{H\text{-}3}^b$ allele of LP, but when the loss of $\underline{H\text{-}3}^b$ occurred is unknown. It should be noted that these B10.LP strains in Table 4 are derived from a cross distinct from that which gave rise to the B10.LP strains listed in Table 2.

Legend to Table 4, continued

[g] Strain B10.PA-$\underline{H\text{-}3}^e$ was previously called B10-$\underline{pa}$ $H\text{-}3^e$ $\underline{a}^t$. It was derived by repeated backcrossing of $\underline{pa}$ $H\text{-}3^e$ $\underline{We}$ $\underline{Un}$ from B6-$\underline{pa}$/Re to strain B10.UW (35). The $\underline{a}^t$ allele in B10.PA-$\underline{H\text{-}3}^e$ was derived from B10.UW.

[h] Strain B10.UW-$\underline{H\text{-}3}^b\underline{we}$ was originally called B10.UW-$\underline{we H\text{-}3}^b$. It was derived from B10.UW(69NX).

TABLE 5. CR Strains, Produced in Conjunction with Marianna Cherry, by Backcrossing Serologically Demonstrable Loci onto Inbred Backgrounds[a]

Strain	Short symbol	Generation	N equivalent	Genotype	References
B6.PL-Thy-1^a/Cy	74NS	N10F16		Thy-1^a H-$2^?$	$\underline{40}$
B6.PL-Ly-$2,3^a$/Cy	75NS	N11F15		Ly-2^a Ly-3^a	$\underline{40}$[b]
B6.RIII-Ea-2^a/Cy	76NS	N8F9N3F18	NE11F18	Ea-2^a	$\underline{38}$
B6-Ly-1^a/Cy	B6-Ly-1^a	N12F19	(19)	Ly-1^a	$\underline{40}$[c]
B10.AKM/Sn	B10.AKM	N11F32	(17)	H-2^m Tla^a $Apl^?$	31 $\underline{36}$ 38 $\underline{43}$
B10.DA(80NS)/Sn	80NS	N10F12	(17)	H-2^{qp} Tla^b Pgk-2^b $Apl^?$	$\underline{36}$ 38 $\underline{43}$
B10.PL(73NS)/Sn	73NS	N8F17	(17)	H-2^u Tla^a	$\underline{36}$
B10.RIII-H-2^r/Sn	71NS	N8F26	(17)	H-2^r Pgk-2^b $Apl^?$	31 $\underline{36}$ $\underline{43}$
B10.RIII-Ea-2^a/Sn	72NS	N8F26		Ea-2^a	25 $\underline{38}$
B10.SM(70NS)/Sn	70NS	N10F17	(17)	H-2^v Apl^a	$\underline{36}$ 38 $\underline{42}$
B10.WB(69NS)/Sn	69NS	N10F22	(17)	H-2^{ja}	$\underline{30}$ $\underline{36}$ 38 $\underline{43}$
B10-Ea-2^a Ea-7^a/Sn		N8R1F23	NE8F23	Ea-2^a Ea-7^a	$\underline{38}$[d]
C3H.HTG(82NS)/Cy	82NS	N12F9	(17)	H-2^g Pgk-2^b $Apl^?$	$\underline{36}$ 38 40 $\underline{43}$
C3H.JK/Sn		N1F1N7F22	NE8F22	H-2^j Pgk-2^a $Apl^?$	$\underline{30}$ 31 36 $\underline{38}$ $\underline{43}$
C3H.LG/Ckc/Cy		N12F2N1F2	NE13F2	H-2^{ar}	Unpublished
D1.DA/Sn		N10F8	(17)	H-2^{qp} $Tla^?$ Pgk-$2^?$ $Apl^?$	$\underline{43}$

Legend to Table 5

[a] See footnote a, Table 2.

[b] There have been no crossovers between the specificities defining Ly-2 and Ly-3 and it has been reported that the specificities appear to be on the same molecule (reviewed in 63). However, there is now some evidence that Ly-2 and Ly-3 can be independently precipitated (Gottleib, personal communication). If this is confirmed, the designation Ly-2,3 will be inappropriate.

[c] Strain B6-Ly-1^a was derived from a cross between C57BL/6J and a stock carrying the linkage markers Ca, St, Hm. The stock carrying the markers was of mixed origin and not inbred but had substantial CBA/Ca ancestry. It is likely, therefore, that the Ly-1^a in strain B6-Ly-1^a was of CBA/Ca origin.

[d] The double congenic strain B10-Ea-2^a Ea-7^a was derived from a cross between B6.RIII(72NS) and B10.129(5M)/o. The Ea-2^a allele was thus originally derived from strain RIII and the Ea-7^a allele from strain 129.

TABLE 6. Numbers of Congenic Resistant or Serologically Incompatible Lines Produced by Various Methods

Group	Started (approx.)		Number of lines Tested		Now maintained	
	Original	Derived[a]	Original	Derived[a]	Original	Derived[a]
H-2 and non-H-2 lines produced by cross-intercross (Table 2)	125	5	27	5	14	5
Non-H-2 lines produced by cross-intercross and cross-backcross-intercross (Table 3)	34	3	19	3	10	3
H-1 and H-3 lines produced by backcrossing marker gene (Table 4)	14	9	13	9	9	4
Lines (various) produced by backcrossing and serotyping (Table 5)	16	0	16	0	16	0
Total	189	17	75	17	49	12

[a] Derived lines are lines which were separated from original lines following the removal of a contaminant gene or following the occurrence of a crossover which eliminated the introduced alleles at one or more passenger loci linked to the defining locus.

had to come from an inbred strain, and from a strain for which a transplantable tumor was available. Any mouse, inbred or otherwise, could be used as the second parent and, in fact, partly because all mice were still in short supply, a considerable variety of mice was used. A total of 125 lines were started. Development of congenic resistant lines is a long process and since, in the development of this first group, I was working in unexplored territory, inevitably problems were encountered. It would be five years before I knew whether the effort was paying off.

The most urgent need for the Laboratory at this time was rebuilding, and I was chairman of the Construction Committee to which this undertaking had been assigned. The work of this committee was essentially a full-time job for nearly two years. I had very little time for the congenic line project. This turned out to be one of the least of my problems. My assistant at the time was Sally Lyman, now Dr. Sally Allen, and the project could not have been in better hands.

The major problem in producing the Group 1 lines turned out to be the weakness of some of the histocompatibility barriers. This would have been a minor concern if I had been using skin transplants as the test of susceptibility and resistance, but transplantable tumors, especially as they get older, are less discriminating than skin and may totally override the weaker barriers. We soon found that many of the lines were being lost due to the lack of surviving animals in the inoculated generations. The use of preimmunization reduced but by no means eliminated this problem.

Another and continuing concern was the health of the mice. Here the fire turned out to be a blessing, since the new building, once available, was substantially more sanitary and vermin-free than the original structure. But most colonies at that time were infested with mites and lice, and ours was no exception. Also the transplantable tumors, which were an essential element of the project, were all too effective transmitters of infection. These health problems continued to plague us well beyond the period devoted to producing the first group of CR mice. I well remember the time when Diane Kelton and I broke out with rash on our arms from using a mitricide newly put on the market and widely advertised by a major chemical company. The product was later withdrawn. We also found it necessary at one period to add aureomycin to the drinking water. Gradually, however, animal care and animal health conditions at the Laboratory were improved. Perhaps the single most important development was the introduction of stainless steel mouse boxes. Dr. Green deserves a major share of the credit for these improvements.

Of the 125 lines initially started in Group 1, only 27 were carried through to the point where they could be tested to identify the defining histocompatibility gene which distinguished them from their inbred partner. Here a surprise awaited us. Of these 27 lines, 22 differed from their partner at the H-2 locus. Ten different alleles or, as we would now say, haplotypes were represented (10). This was the first clear evidence of the uniqueness of H-2. There were technical problems in characterizing the locus in the five non-H-2 lines. The lines were on four different inbred backgrounds and the differential loci probably quite polymorphic. This made it difficult to transfer information from one CR line to another. However, it was soon apparent that, with the aid of linked marker genes, we had identified two new loci, H-1 and H-3 (2, 5).

Subsequent tests showed that disparities at H-1 and H-3 impose weaker barriers to transplants of tumor, skin, or ovary than do disparities at H-2 (6, 15). This further emphasized the uniqueness and "strength" of H-2 and foreshadowed the concept of a major histocompatibility complex.

The Group 1 CR lines illustrate a potential feature of such lines which to some extent has been apparent in all the groups. The chromosome segment carrying the defining H locus inevitably carries other linked loci. In due course some of these can be identified, and some of them may turn out to be additional histocompatibility loci. Thus the B10.LP line of Group 1, originally identified as carrying H-3 and the marker locus A or agouti, also turned out to carry another histocompatibility locus, H-13, and an immune response locus, Ir-2 (24, 39). The CR lines, especially if they are derived following only a limited number (6-8) of matings to the inbred parent strain, may also carry contaminant loci from the second parent not borne on the chromosome segment to which selection was applied. It has been shown that this happened a number of times in the Group 1 lines. In some cases the contaminant locus turned out to be of considerable interest, e.g., the complement locus, Hc, by which B10.D2/o was found to differ from B10 and B10.D2/n (16). The contaminants, therefore, have not been an unmixed evil.

When a contaminant or passenger locus is found, it may be of interest by appropriate matings to separate off a subline, lacking the contaminant or passenger but otherwise resembling the original CR line. This produces a new congenic pair differing primarily at the contaminant or passenger locus. Five such derivative sublines have been produced from the original 27 tested lines of Group 1.

Nineteen CR lines and sublines of Group 1 are currently maintained at the Jackson Laboratory. Some of these lines are maintained in many other laboratories also. One of their special values is that they are the principal source of H-2 haplotypes on more than one inbred background. Thus in this group of lines the $H-2^b$ haplotype can be found on the A, C3H, C57BL/10, and DBA/1 backgrounds.

The loss of many lines in Group 1 emphasized the need for refinement in our transplantation techniques. If we were to stick to tumor transplants, which certainly had the advantage of ease and rapidity, they had to be made more discriminating. We had introduced the use of immunization in producing Group 1 mice and had gradually improved our techniques (10). The major change was the shift from immunization with the tumor used for the challenge inoculation, which often resulted in immunization against tumor-specific rather than strain-specific antigens, with disastrous results, to immunization with normal tissues. After the H-1 and H-3 lines became available, we set up systematic tests of different immunization schedules (9). The method finally adopted, when we came to produce the Group 2 lines, involved prior immunization with an intraperitoneal injection of thymus cells and challenge with a radiation-induced leukemia. The leukemia was given as a cell suspension at a controlled dose and preferably was one recently induced. This method proved capable of discriminating relatively weak histocompatibility differences.

Our intent in producing the second group of CR lines was to detect non-H-2 loci. The crosses were chosen accordingly. All lines were on a C57BL/10Sn background. This meant that, once the lines were established, tests to identify the isolated loci would be uniformly feasible. The second parents, like B10, were all $H-2^b$. Three lines were used as second parent, 129, C.B6 and D2.WA. The latter two were congenic lines from Group 1 in which $H-2^b$ had been introduced, respectively, onto the BALB/c and the DBA/2 backgrounds. All mice were preimmunized with thymus and challenged with a B10 x-ray-induced leukemia.

The B10 x 129 crosses were set up first in the spring of 1953, the cross-intercross system being used. The other two crosses were not set up until nearly three years later. The more rapid but more laborious cross-backcross-intercross system was used. Thirty-four lines were originally set up and derivative lines increased this to 38. Nineteen original and four derived lines were carried through to substantial analysis. Six new histocompatibility loci, H-7, H-8, H-9, H-10, H-11, and H-12, were revealed (18, 32). Studies with these lines, using both tumor and skin transplants, emphasized the great range in the histocompatibility barrier imposed by both different non-H-2 loci and different allelic combinations at these loci (21). Ten original and four derivative lines of Group 2 are currently maintained at the Jackson Laboratory (Tables 3 and 6).

It had been shown in the analysis of Group 1 lines that H-1 is in chromosome 7, close to the albino (c) locus, and H-3 and H-13 in chromosome 2 near the agouti (A) locus (2, 5, 25). Hence if c or other appropriate genes in chromosome 7 were introduced by backcrossing from one strain onto the inbred background of another, it would be expected that the H-1 allele of the donor strain also would be introduced. The same reasoning of course applied to A (or A^t or a^t), H-3, and H-13. This provides a method for identifying new alleles at the loci under test.

It was with the intent of exploiting this method that the Group 3 CR lines were established. Dr. Ralph Graff, as part of what has been a long and fruitful collaboration, played a major role in this project.

All Group 3 lines were established on a C57BL/10Sn background. The second parents used as the source of H-3 and H-13 alleles were from strains BALB/c, CE, C3H, KR, PA, SWR, and UW. To the two H-3 alleles already known were added three more, $H-3^c$, $H-3^d$, and $H-3^e$. $H-13^c$ was also added. The second parents used as the source of H-1 alleles were CE, C3H, FS, P, SEC, SM, and WB. The marker genes used were the c (albino) alleles c^{ch} and C. These are both dominant to c. The inbred parent was B10.129(5M)/n (c $H-1^b$). The presence of albinism in this parent made it easy to follow the introduced markers. These lines have not been fully analyzed, but it is clear that they will serve to identify additional H-1 and perhaps H-4 alleles. Details concerning the Group 3 lines will be found in Tables 4 and 6.

Dr. Marianna Cherry has been in large part responsible for the development of Group 4 lines. The lines of this group were produced by backcrossing and serotyping. This method, of course, is applicable only to loci the end product of which is serologically demonstrable. Among histocompatibility loci, in the strict sense of the term, this applies only to H-2. The lines, however, include congenics disparate at blood group (Ea) loci and at loci (Ly, Thy-1) determining lymphocyte membrane alloantigens. The lines are either on a C57BL/6By or a C57BL/10Sn background. Of 16 lines produced, 10 carry H-2 disparities, three Ea disparities, and three lymphocyte antigen disparities (Table 5).

Altogether, 208 original or derivative congenic lines, with disparities either at histocompatibility or membrane alloantigen determining loci, were started. Of these 208, 94 were developed to the point of substantial testing, and 63 are still maintained at the Jackson Laboratory or in the colony of Dr. Ralph Graff. Except for the lines currently maintained, these figures are approximate. Some of the records go back a long way, and I may have overlooked a few lines. The figures include lines primarily produced by Dr. Ralph Graff and Dr. Marianna Cherry, my contribution having been minor. They do not include an important group of H-2 lines (B10.A, B10.BR, and five B10 x A recombinants) produced by Dr. Jack Stimpfling at the Jackson Laboratory by the backcross and serotype method. I had essentially no hand in producing these. They also do not include the very major group of CR lines produced by Dr. Donald Bailey with the use of tail skin grafts, a technique developed by Bailey which renders the tumor transplant method obsolete.

This completes my tale of a 30-year hunt for histocompatibility genes. I hope the hunt has been as useful as it has been enjoyable. It is a satisfaction to know that the histocompatibility genes snared in my congenic resistant lines are thriving in captivity, and have spread beyond the limits of Maine where they were caught, to many other states and foreign countries. It is an even greater satisfaction to know that my wonderful companions in the hunt are still pursuing this fascinating game with new skills and new technologies but with the same lively enthusiasm that has marked the hunt in the past.

APPENDIX 1. THE TERMINOLOGY OF CONGENIC STRAINS

A. Vocabulary Applicable to Congenic Strains

1. A <u>congenic strain</u> is a strain identical or almost identical to a standard inbred strain except for the presence of a chromosome segment introduced by appropriate crosses from a second strain. A strain is usually not regarded as congenic unless there have been at least eight crosses to the inbred strain.

2. The two strains used in the initial cross are referred to as the <u>first parent</u> (the inbred strain to which repeated crosses are made) and the <u>second parent</u>. The first parent must be inbred, the second parent need not be inbred. The first parent, when being contrasted with a congenic strain derived from it, may also be referred to as the <u>inbred partner</u>.

3. A <u>congenic resistant</u> (CR) <u>strain</u> is a congenic strain in which the introduced segment carries a histocompatibility ($\underline{H}$) allele foreign to the inbred partner.

4. The locus by which a congenic strain and its inbred partner differ, and to which selection was applied in producing the congenic strain, is called the <u>differential locus</u>

for this strain pair. The term <u>defining locus</u> has also been employed (63), but current usage seems to favor <u>differential locus</u>. Genes introduced because of linkage with the differential locus are called <u>passenger genes</u>. If any genes from the second parent not linked to the differential locus are still present in the congenic strain, these are called <u>contaminant genes</u>.

5. Usually the differential locus is also the <u>locus of major interest</u>. However, when a CR strain is produced by backcrossing an easily demonstrable marker gene, the passenger H locus may be the locus of major interest. There may, in some cases, be more than one locus of major interest.

6. If a congenic strain carries known contaminant or passenger genes, <u>derivative congenic strains</u> may be produced by eliminating one or more of these. The original congenic strain and its derivative strain may then constitute a congenic pair more closely matched than the original congenic strain and its inbred partner.

7. A <u>double congenic strain</u> is a strain which differs from its inbred partner at two loci, both loci of major interest and both loci determining a similar phenotype. Thus a strain which differs from its inbred partner at both <u>Ea-2</u> and <u>Ea-7</u> (both blood group loci) is a double congenic. The two loci may be linked or unlinked and may be derived from the same second parent or different second parents.

B. Symbols for Congenic Strains

1. The <u>basic symbol</u> for designating a congenic strain consists of the symbols for the first and second parent strains, or of abbreviations for the same, with a period between them. Thus strain B10.D2 came from an initial cross between strains C57BL/10 and DBA/2. Double congenic strains may constitute an exception. If the two loci of major interest came from

different second parents, the basic symbol may be that of the first parent only. Also if the second parent is of mixed origin and has no standard symbol, only the first parent may be designated.

2. Since several sublines may be derived from the same initial cross, and hence have the same basic symbol, an additional <u>subsidiary symbol</u> may be necessary. This may take either of two forms.

a. A number and letter, in parentheses, may be appended to the basic symbol. Thus B10.129(5M) and B10.129(6M) are two strains derived from the initial cross C57BL/10 x 129. In the strains produced by Snell and others at the Jackson Laboratory, the letters used and their meanings are as follows:

M lines produced by cross-intercross system
N lines produced by the cross-backcross-intercross system
NX line produced by backcrossing a marker gene
NS lines produced by backcrossing and serotyping
R lines derived from a recombinant (used by Stimpfling).

b. When the introduced histocompatibility allele, or other allele of major interest, has been identified, a hyphen and the symbol for this allele may be appended to the basic symbol. Thus B10.129(5M) may become B10.129-$\underline{H-1}^b$. Bailey has adopted usage b and it is recommended for newly developed strains.

It is permissible, if for any reason advantageous, to use a symbol combining both symbols a and b, e.g., B10.129(5M)-$\underline{H-1}^b$.

If there are two introduced alleles which, for any reason, are of major interest, both may be indicated, e.g., B10.KR-$\underline{H-13}^c$ A. For double congenics, two appended symbols should regularly be used, e.g., B10-$\underline{Ea-2}^a$ $\underline{Ea-7}^a$.

3. The investigator originating a congenic strain is indicated by appending to the basic and subsidiary symbols the standard abbreviation for the investigator's name (61). Example: B10.129(5M)/Sn or B10.129-$\underline{H-1}^b$/Sn. If two investigators

have collaborated in developing a congenic strain, the name symbol should be that of the investigator currently maintaining the strain (if maintained only by one) or that of the investigator who made the larger contribution.

4. A congenic strain may become split into substrains as a result of mutation or the elimination of contaminant or passenger loci. Thus strain B10.KR(35NX)/Sn, which carried an introduced chromosome segment bearing the foreign alleles $H-3^d$ $Ir-2^?$ $H-13^c$ A, gave rise through recombination to strains bearing only the foreign alleles $\underline{H-3}^d$ and $\underline{H-13}^c$ A (Table 4). Such derivative strains may be designated in either of two fashions.

a. Each strain may be assigned a different subsidiary symbol giving, for each, the locus of major interest (symbol of type 2b). Thus the two lines derived from B10.KR(35NX)/Sn may be designated B10.KR-$\underline{H-3}^d$ and B10.KR-$\underline{H-13}^c$ or B10.KR-$\underline{H-13}^c$ A.

b. A number may be inserted between the slash line and the same symbol. The two lines derived from B10.KR(35NX)/Sn then become B10.KR(35NX)/2Grf (derivative developed by Graff) and B10.KR(35NX)/3Sn. In the past, lower case letters have been used to distinguish derivative strains. Example: B10.D2/o and B10.D2/n (o = old, n = new). This use is not recommended.

C. Short Symbols for Congenic Strains

Since the standard symbols for congenic strains may be awkwardly long, short or abbreviated symbols may be used. Illustrations of several forms that these may take will be found in Tables 2-5. The full symbol should always be given at least once in any published paper.

REFERENCES

1. Snell, G.D. (1953). J. Natl. Cancer Inst. 14:691.

2. Snell, G.D., and Kelton, D. (1953). Proc. Amer. Assoc. Cancer Res. 1:53.
3. Hoecker, G., Counce, S., and Smith, P. (1954). Proc. Natl. Acad. Sci. 40:1040.
4. Snell, G.D. (1955). Transplant. Bull. 2:6.
5. Snell, G.D., Counce, S., Smith, P., and Dube, L. (1955). Proc. Amer. Assoc. Cancer Res. 2:47.
6. Counce, S., Smith, P., Barth, R., and Snell, G.D. (1956). Ann. Surg. 144:198.
7. Snell, G.D. (1957). Ann. N.Y. Acad. Sci. 69:555.
8. Snell, G.D., Wheeler, N., and Aaron, M. (1957). Transplant. Bull. 4:18.
9. Snell, G.D. (1958). J. Natl. Cancer Inst. 20:787.
10. Snell, G.D. (1958). J. Natl. Cancer Inst. 21:843.
11. Winn, H.J., Stevens, L.C., and Snell, G.D. (1958). Transplant. Bull. 5:18.
12. Berrian, J.H., and McKhann, C.F. (1960). Ann. N.Y. Acad. Sci. 87:106.
13. Linder, O., and Klein, E. (1960). J. Natl. Cancer Inst. 24:707.
14. Snell, G.D. (1960). J. Natl. Cancer Inst. 25:1191.
15. Snell, G.D., and Stevens, L.C. (1961). Immunology 4:366.
16. Herzenberg, L.A., Tachibana, T.K., Herzenberg, L.A., and Rosenberg, L.T. (1963). Genetics 48:711.
17. Snell, G.D. (1964). "The 'Old' (Complement Deficient) and 'New' Sublines of Strain B10.D2." Mimeographed, The Production Department, The Jackson Laboratory, Bar Harbor.
18. Snell, G.D., and Bunker, H.P. (1964). Transplantation 2:743.
19. Boyse, E.A., Old, L.J., and Stockert, E. (1965). In "Immunopathology, IVth Internat. Symp." (P. Grabar and P.A. Miescher, eds.), p. 23. Schwabe and Co., Basel.
20. Snell, G.D., and Bunker, H.P. (1965). Transplantation 3:235.

21. Graff, R.J., Hildemann, W.H., and Snell, G.D. (1966). Transplantation 4:425.
22. Graff, R.J., Silvers, W.K., Billingham, R.E., Hildemann, W.H., and Snell, G.D. (1966). Transplantation 4:605.
23. Tennant, J.R., and Snell, G.D. (1966). Natl. Cancer Inst. Monogr. 22:61.
24. Snell, G.D., Cudkowicz, G., and Bunker, H. (1967). Transplantation 5:492.
25. Snell, G.D., Hoecker, G., and Stimpfling, J.H. (1967). Transplantation 5:481.
26. Graff, R.J., and Snell, G.D. (1968). Transplantation 6:598.
27. Stimpfling, J.H., and Snell, G.D. (1968). Transplantation 6:468.
28. Hilgert, I., and Snell, G.D. (1969). Transplantation 7:401.
29. Lengerova, A., and Matousek, V. (1969). Folia Biol. 15:247.
30. Snell, G.D., Cherry, M., and Démant, P. (1971). Transplant. Proc. 3:183.
31. Snell, G.D., Demant, P., and Cherry, M. (1971). Transplantation 11:210.
32. Snell, G.D., Graff, R.J., and Cherry, M. (1971). Transplantation 11:525.
33. Kaliss, N., Bailey, D.W., and Shin, H.S. (1972). Transplantation 14:523.
34. Graff, R.J., and Bailey, D.W. (1973). Transplant. Rev. 15:26.
35. Graff, R.J., Brown, D., and Snell, G.D. (1973). Transplant. Rev. 5:299.
36. Klein, J. (1973). Transplantation 15:137.
37. Taylor, B.A., Meier, H., and Huebner, R.J. (1973). Nature New Biol. 241:184.

38. Snell, G.D., and Bailey, D.W. (1974). "Congenic Resistant and Recombinant Inbred Strains of Mice Available Through the Production Department." Mineographed, The Production Department, The Jackson Laboratory, Bar Harbor.
39. Gasser, D.L. (1976). Immunogenetics 3:271.
40. Cherry, M. (1977). Inbred Strains of Mice 10:28.
41. Flaherty, L., Sullivan, K., and Zimmerman, D. (1977). J. Immunol. 119:571.
42. Womack, J.E., and Eicher, E.M. (1977). Mol. Gen. Genet. 155:315.
43. Eicher, E.M., Cherry, M., and Flaherty, L. (1978). Mol. Gen. Genet. 158:225.
44. Bailey, D.W., and Lilly, F. (1978). Personal communication.
45. Woglom, W.H. (1929). Cancer Rev. 4:129.
46. Gorer, P.A. (1938). J. Path. Bact. 47:231.
47. Irwin, M.R. (1939). Genetics 24:709.
48. Little, C.C. (1941). In "The Biology of the Laboratory Mouse" (G.D. Snell, ed.), p. 279. Blakiston, Philadelphia.
49. Spencer, R.R. (1942). J. Natl. Cancer Inst. 2:317.
50. Gorer, P.A., Lyman, S., and Snell, G.D. (1948). Proc. Roy. Soc. Lond. (Biol.) 135:499.
51. Snell, G.D. (1948). J. Genet. 49:87.
52. Chovnick, A., and Fox, A.S. (1953). Proc. Natl. Acad. Sci. USA 39:1035.
53. Snell, G.D., Smith, P., and Gabrielson, F. (1953). J. Natl. Cancer Inst. 14:457.
54. Allen, S.L. (1955). Genetics 40:627.
55. Snell, G.D. (1964). Meth. Med. Res. 10:8.
56. Green, E.L. (1966). In "Biology of the Laboratory Mouse" 2nd ed. (E.L. Green, ed.), p. 11. McGraw-Hill, New York.
57. Stimpfling, J.H., and Reichert, A.E. (1970). Transplant. Proc. 2:39.

58. Graff, R.J., and Snell, G.D. (1969). Transplantation 8:861.
59. Graff, R.J., Polinsky, S.L., and Snell, G.D. (1971). Transplantation 11:56.
60. Snell, G.D., and Cherry, M. (1972). In "Viruses and Host Genome in Oncogenesis" (P. Emmelot and P. Bentvetzen, eds.), p. 221. North-Holland Publ., Amsterdam.
61. Staats, J. (1972). Cancer Res. 32:1609.
62. Lane, P.W. (1974). "Lists of Mutations and Mutant Stocks of the Mouse." Mimeographed, The Jackson Laboratory, Bar Harbor.
63. Snell, G.D., Dausset, J., and Nathenson, S. (1976). "Histocompatibility." Academic Press, New York.
64. Lalley, P.A., and Shows, F.B. (1977). Genetics 87:305.
65. Cherry, M., Bailey, D.W., and Snell, G.D. (1978). In "Origins of Inbred Mice" (H.C. Morse III, ed.). Academic Press, New York.

George D. Snell

BRIEF AUTOBIOGRAPHY RELATED TO WORK ON INBRED MICE

Margaret C. Green

The Jackson Laboratory

Bar Harbor, Maine

I became a mouse geneticist by marriage. Shortly before marrying Earl Green, I had received my Ph.D. for work on the genetics and radiation cytogenetics of the grasshopper Chorthippus longicornis. At the time there seemed to be little similarity between Chorthippus and Mus musculus, but I discovered many years later that there was, after all, a connection. My Ph.D. adviser was Dr. W.R.B. Robertson, who had noted in a classical paper that the variation in chromosome number between related species of grasshoppers could be accounted for by fusion of two chromosomes with terminal centromeres to make a single chromosome with a median centromere. The species I worked with had three such metacentric chromosomes. These metacentric chromosomes later became known as Robertsonian translocations, and it is now well known that they are characteristic of certain wild subspecies of Mus musculus. They have also occurred in laboratory mice, and are one of the most useful cytogenetic variants available in the mouse.

At the time we were married in 1940, Earl had a postdoctoral fellowship ($300 extra for being married) at the University of Chicago and I had a fresh Ph.D. but no other means of support. We decided that we could live on the fellowship, and that instead of my looking for a job we should join forces and work on Earl's fellowship project, the anatomy and development of short-eared (se/se) mice (1, 2, 3). Since the microscope had been my principal research tool, I became responsible for the embryological and histological studies, a line of endeavor that continued with other mutant genes throughout my research career. This was the easy part of converting from grasshoppers to mice. Some other things were harder.

I was afraid of mice. In our research project, Earl was responsible, among other things, for the mouse husbandry, and he continued this activity after we moved to Ohio State University in 1941. But in 1943 he went into military service and I had no escape from the duty of handling mice. I well remember sweating out those ear-punching sessions, particularly with one strain in which the mice were hyperactive and also had short ears that were extremely hard to punch legibly. Fear of being bitten stayed with me for years. It finally disappeared on a memorable occasion after our move to the Jackson Laboratory in 1956. A mouse bit me while I was showing it to a room full of visitors in the auditorium. Fear of making a spectacle in front of the audience overcame fear of being bitten by the mouse, and instead of throwing the mouse across the room as I would have done under private circumstances, I stood there calmly and let it bite. I was cured.

My work with mice has been mostly with mutant genes, and any contributions to the development of inbred strains has been incidental to the propagation and use of mutant genes. In our first work with the short-ear gene, we compared two different short-eared inbred strains with a different normal-eared inbred strain. We soon realized that this was not a proper procedure for determining the effects of the mutant gene, and set about producing several strains inbred by brother-sister matings with forced heterozygosis for short-ears. This procedure makes the mutant gene available for comparison with its normal allele uncontaminated by differences at other genetic loci. It also allows for selection for a genetic background compatible with viability and fertility of the mutant mice.

After we moved to the Jackson Laboratory, I became responsible for the Mouse Mutant Stocks Center. Some of the stocks in the Center came from George Snell who had also been propagating some of them by inbreeding with forced heterozygosis. We continued the use of this procedure and extended it to many other stocks. We thus produced a number of new inbred strains, some of which have developed characteristics that make them useful for purposes other than propagation of the mutant for which they were originally bred. We used the same breeding system for maintenance of the linkage testing stocks containing multiple dominant genes. These stocks are now highly inbred and some of them have been typed for polymorphic biochemical markers, making them more useful for linkage tests since the additional loci extend the number of chromosomal regions that can be tested in a single linkage cross.

In the operation of the Mouse Mutant Stocks Center, I was lucky enough to become associated with Mrs. Priscilla W. Lane, known as "Skippy," whose enthusiasm and enterprise in developing new stocks and promoting their use have made her a major supplier of mutant mice for research around the world. In addition to all her other activities, Skippy has produced, every two years, a revised list (widely known as the "Lane List") of all the mutant stocks and stocks bearing chromosomal variants at the Jackson Laboratory. An important feature of this list is a section (11 pages in the 1976 list) listing all the congenic and segregating inbred strains at the Laboratory (except for the histocompatibility congenic strains which are listed elsewhere).

Many mutant genes are very deleterious, and it is often difficult to find an inbred background on which the mutant animals are fertile or can survive long enough to be useful for research. This is particularly troublesome for the case of dominant mutations in which the affected mice must be able to reproduce. We devised the procedure of crossing the mutant mice repeatedly to an F_1 hybrid to maximize hybrid vigor. The resulting stocks are not genetically homogeneous, but they bear only the background genes of the two parental strains. Although they are about as variable as an F_2 generation, their tissues can be transplanted into normal F_1 hosts, a procedure that is often useful in studying gene function. This breeding system has greatly increased the ease of maintaining many difficult genes and chromosome variants, including the X/0 stock and several translocations, and has saved considerable time, space, and effort. It is now being used in several other laboratories. The F_1 hybrids we have used are (C57BL/6 X CBA/Ca)F_1 and (C57BL/6 X C3HeB/Fe)F_1.

I began doing linkage studies in the late 1940s with a new mutation that occurred in our stocks at Ohio State University. When we moved to the Jackson Laboratory, I continued linkage tests with many of the unlocated genes in the stocks there. This was before gene mapping became popular and only a handful of mouse geneticists devoted much attention to linkage tests. It was a rather easy routine activity, and I enjoyed watching the loci fall into place on the map. Also, I thought the information was potentially useful, even for genes not intrinsically very interesting in themselves. Any "good" mutant gene, that is, any gene with good viability and fertility, could be a useful marker for studies of other more important loci located in the same vicinity. I always had the feeling that the evidence for the existence of a new locus was not complete until its location was known. Later when the mouse linkage map began to fill up and the advantages of knowing the location of important loci became more obvious, many others entered the field. There is plenty of work to be done, as many known genes are still unlocated, including many very interesting and important biochemical and immunological loci, and new loci are being discovered at an increasing rate.

Perhaps the most useful service I have performed for mouse geneticists has been to summarize linkage information in a linkage map. This effort began in 1958 in preparation for the Tenth International Congress of Genetics held that year in Montreal. Dr. Walter Heston had pointed out that, at the International Congress at Cornell University in 1932, the most recent Genetics Congress in North America, the corn geneticists had constructed a living linkage map by planting mutant corn plants in the form of a map. He suggested that it was now time for a living linkage map of the mouse and that the Jackson Laboratory was the institution to do the job. Members of the staff threw themselves enthusiastically into the task. We designed and constructed a large map with 18 chromosomes made of vertically placed cardboard tubing, each about 4 inches in diameter and 5 feet long, and with small plastic cages holding the mutant mice appropriately arranged at intervals beside the chromosomes. The project was a great success.

To prepare the map I had to assemble all the available linkage data, and I devised a card file system for organizing and storing the information. After that, it was relatively easy to keep the file up to date and to revise the linkage map as often as enough new information accumulated. The version of the map used in the Live Linkage Map was published in the Journal of Heredity in 1959 (4). Revised versions have appeared in several editions of the Handbook of Biochemistry (9), the second edition of Biology of the Laboratory Mouse in 1966 (6), the Handbook of Genetics in 1975 (8), several textbooks of genetics, in Mouse News Letter from 1972 through 1975, and in a number of other publications. Tables of the linkage data on which the map is based have appeared in several editions of the Biology Data Book published by the Federation of American Societies of Experimental Biology (7). Perhaps the most useful form of the map was in the hundreds of Xerox copies distributed to the research staff at the Jackson Laboratory, to visitors and students, and to participants in the short courses in medical and mammalian genetics and the workshops in biochemical genetics held at the Laboratory.

The making of linkage maps requires knowledge of methods of analyzing linkage data and of combining data from different sources. Having acquired some experience in these matters, in 1963 I contributed a chapter on methods of testing for linkage to Burdette's book on Methodology in Mammalian Genetics (5).

Prior to 1972 the linkage map was arranged in order of linkage groups numbered chronologically as they were discovered. A major change in the map occurred in 1972 after it became possible to identify all the chromosomes of the mouse individually by their distinctive banding patterns, and linkage groups had been located on 14 of the 20 chromosome pairs. As chairman of the Committee on Standardized Genetic

Nomenclature for Mice, I arranged for adoption of a standard
system for numbering the chromosomes in order of size. The
Committee decided to recommend that chromosome numbers be
used to replace the old linkage group numbers to bring the
cytological and genetic terminology into a single system.
The standard numbering system and the Committee's recommendations were published that year in the Journal of Heredity
(10). The map is now arranged in order of chromosome number,
and genes have been assigned to all chromosomes.

Construction of the linkage map has little bearing on
the development of inbred strains, the subject of this workshop, but a good linkage map does increase the usefulness of
inbred strains. With the recent explosive increase in the
number of known biochemical and other polymorphic loci, knowledge of the location of these loci on the map makes the
inbred strains more useful as linkage testing stocks.
Strains that differ in the alleles they carry at a number
of loci on different chromosomes become good linkage testing
stocks for new variants that occur in the strains or for differences that are found between the strains. The increasingly complete linkage map will remain a handy tool for users
of inbred strains.

REFERENCES

1. Green, E.L., and Green, M.C. (1942). J. Morphol. 70:1.
2. Green, M.C. (1951). J. Morphol. 88:1.
3. Green, M.C. (1968). J. Exp. Zool. 167:129.
4. Green, M.C., and Dickie, M.M. (1959). J. Hered. 10:1.
5. Green, M.C. (1963). "Methodology in Mammalian Genetics" (W.J. Burdette, ed.), p. 56. Holden-Day, San Francisco.
6. Green, M.C. (1966). In "Biology of the Laboratory Mouse" (E.L. Green, ed.), p. 87. McGraw-Hill, New York.

7. Green, M.C. (1972). In "Biology Data Book," 2nd ed., Vol. 1 (P.L. Altman and D.S. Dittmer, eds.), p. 15. FASEB, Washington.
8. Green, M.C. (1975). In "Handbook of Genetics," Vol. IV. (R.C. King, ed.), p. 203. Plenum Press, New York.
9. Green, M.C. (1976). In "Handbook of Biochemistry and Molecular Biology," 3rd ed., Nucleic Acids, Vol. II (G. D. Fasman, ed.), p. 856. Chemical Rubber Co., Cleveland.
10. Committee on Standardized Genetic Nomenclature for Mice, M.C. Green (Chm). (1972). J. Hered. 63:69.

MARGARET C. GREEN

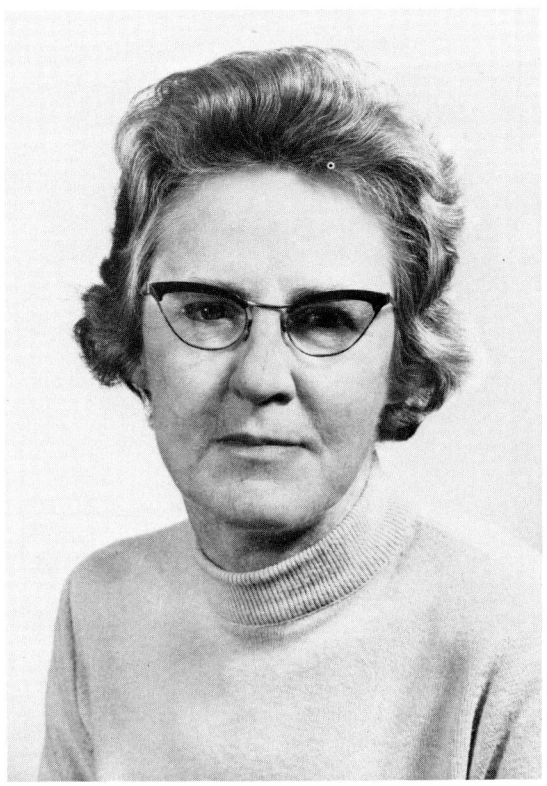

Margaret C. Green

A BRIEF AUTOBIOGRAPHY RELATIVE TO WORK
WITH INBRED AND MUTANT MICE

Earl L. Green

The Jackson Laboratory
Bar Harbor, Maine

I was walking across the campus of Allegheny College in Meadville, Pennsylvania, one day early in June of 1932. Examinations were over. I was almost 19 years old and I had just completed my freshman year of college. Approaching me from the opposite direction was the professor, Dr. Darling, under whom I had taken my first year-long course in biology. Dr. Chester Arthur Darling was a large man with gruff but kindly ways of dealing with students. He had been a professor at Allegheny for nearly 15 years after having completed his study for the Ph.D. degree at Columbia University.

He said to me, "Say, Green, you live in town, don't you? How would you like to take care of the mouse colony this summer? The chap who has been doing the job is going to be away."

I must have said, "Yes," because soon I was engaged in the daily ritual of feeding and watering about 50 pens of mice and the weekly task of trying to clean and repair the wooden pens. I got bored with pouring water into clean dishes, placing them in the mouse pens, and observing soon thereafter that the mice had pushed the bedding into the dishes and, a little later, that they had defecated and urinated into the dishes as well. I'm not sure I knew those four-syllable words at that time. I devised an automatic watering system, based on the hanging-drop principle now in universal use. My system was, however, crude and inefficient compared with later devices. I had trouble getting the aperture, through which the droplets of water were to pass, small enough to suspend the drops and not allow water to run continuously. I was grateful for the advances, made by others, of devising glass "goosenecks" and stainless steel sipper tubes. I continued the husbandry of mice through the rest of my years at Allegheny.

Shortly after I started tending the mice, Dr. Darling referred to one of the stocks as "the strong strain." This made me a bit timid about dealing with them. Even though they looked tame enough, I didn't want to test their strength just yet. The next summer, Dr. Darling introduced me to a visitor, Dr. Leonell C. Strong of the Roscoe B. Jackson Memorial Laboratory in Bar Harbor, Maine. It dawned on me eventually that I had been husbanding "the Strong strain," for Dr. Strong had given some breeding pairs of his Strain A mice to Dr. Darling some years earlier. This was my first contact with an inbred strain of mice.

During my second year at Allegheny, I took a course in heredity. One day, in the mouse room, Dr. Darling asked me if I had thought of any project I might undertake with mice. I said I wanted to cross a black mouse with a brown mouse and to mate the offspring together to see if I got a litter with three black mice and one brown mouse in it.

Dr. Darling said, "What are you going to do if the litter has five mice in it?"

The look of perplexity, commingled with chagrin, that swept over my face must have amused Dr. Darling. The next thing I heard was, "Heh, heh, heh."

It wouldn't have been so bad if he had not laughed. I was playing the role of a carefree trout that had taken a beautiful fly; that laugh sunk the hook in. Damn it, why hadn't I thought of that? I knew very well that mice had litters ranging from one or two up to 12 or 13. Clearly there was only one thing to do: charge ahead and hope for revelation later.

I made up the matings between black and brown mice and between members of the F_1 generation to produce the F_2 generation. I devised a system or record-keeping that, with successive modifications, served me through four decades. With the data in hand, I went to the library -- "Heh, heh, heh," still sounding in my ears -- to see if I could get help in understanding what I had found. Remember the time is 1933, and the statistical resources available to students in a small general college library were primitive compared with those of the present. Actually, I was not able to resolve the question until 1937 when I was in graduate school. That is when Snedecor's first edition of <u>Statistical Methods</u> appeared. Then I learned about attributes (now called discrete variables), probability, the binomial distribution, the computation of expected frequencies, and the testing of deviations of observed from expected numbers by the use of the chi-square distribution.

Having gone to the library for one purpose, I developed the habit of returning for another. The college had just completed a new wing on its library building. In it were individual study carrels near the stacks. The librarian assigned me one of the carrels, and I started to search books and journals for information about mouse genetics. I became familiar with <u>Genetics</u>, <u>The American Naturalist</u>, <u>Journal of Heredity</u>, <u>Science</u>, and many other journals. I also met the names and the works of Morgan, Castle, Little, Wright, Haldane, Darbyshire, Detlefsen, Gates, Snell, Dunn, Keeler, and many others. One day the librarian gave me a copy of a publication she was about to discard. It was a paper-bound monograph of the Carnegie Institution of Washington dated 1914. It contained C.C. Little's doctoral dissertation on mouse breeding experiments, submitted to the Harvard faculty in fulfillment of the requirements for the doctoral degree. It contained several plates, in color, of mice of various coat colors. It would be another 50 years before colored photographs of mice appeared in print for the aid of the novice.

My project as a college student laid the foundation for my lifelong interest in genetics and statistics. What had been an accidental meeting on a college campus led to my decision to major in biology as an undergraduate. That, in turn, led to the decision to pursue graduate work in genetics.

In the fall of 1936, I entered the Graduate School of Brown University in Providence, Rhode Island, as a teaching assistant in comparative anatomy of the vertebrates and as a advisee of Dr. Herbert Eugene Walter and Dr. Paul Baldwin Sawin. The other teaching assistant was a red-haired girl, named Margaret Creighton, from New London, Connecticut.

Dr. Sawin advised me to take a course in statistics offered by Professor A.A. Bennett in the Mathematics Department. The course was a valuable introduction to the concepts and methods of statistics. More valuable was a friendship with Professor Bennett that continued throughout my five years at Brown. He helped a small self-organized group of graduate students to conduct a study of Snedecor's first edition of Statistical Methods.

Dr. Sawin had recently discovered a skeletal variation in his stocks of rabbits. I became interested in seeing if similar variations occurred in existing inbred strains of mice. We procured samples of various strains of mice from the Jackson Laboratory in Bar Harbor and from Dr. Strong who was then at Yale University. It turned out that there were four primary types of axial skeletons with respect to the ratio of thoracic-to-lumbar vertebrae. Some strains had 13/6 as the typical numbers. This was regarded as "normal" or "standard." Some strains, however, had 13/5, some had 13/7, and some had 12/7 as typical numbers. The mice within strains were similar, but not identical. Each strain produced a small percentage of mice that were not typical of the strain. This was my own first-hand encounter with the idea of variation within an inbred strain. Many years earlier, in 1905, Johannsen had established that pure lines of beans exhibit variation and do not respond to selection. The idea of nongenetic variation within inbred strains of mice is still, today, a difficult idea for some people to grasp.

I decided to carry out an intensive analysis of the visibly most variable of the strains, the Bagg albino strain (denoted as BALB/cJ in now-conventional symbols), modeled upon the classical analysis that Sewall Wright had then recently published on the inheritance of three and four toes in guinea pigs. I found that, in the Bagg albino strain, the parent-offspring and the sib-sib correlations were essentially zero, showing that the variation within the strain had primarily nongenetic causes (1).

While I was a graduate student at Brown, I arranged to spend part of two summers, 1938 and 1939, at the Jackson Laboratory, primarily to sample more strains of mice for skeletal variations. Dr. C.C. Little, to whom I first wrote, responded with characteristic enthusiasm inviting me to come to the Laboratory. He arranged for me to be under the tutelage of Dr. W.L. Russell, then on the staff. It was then that I also met, for the first time, Dr. Elizabeth S. Russell, Dr. George D. Snell, and Dr. Walter E. Heston. Aside from the scientific value of those two summers, they were enlightening in another important way. I discovered that to be interested in the genetics of the mouse need not be a solitary endeavor. There were others who were seriously and professionally occupied in advancing knowledge in this domain. At the time I was there, the staff of the Laboratory was deeply occupied in preparing manuscripts for the <u>Biology of the Laboratory Mouse</u>, published in 1941, under the editorship of George Snell.

While I was at Brown, I became interested in the problem of trying to discover the intimate multiple effects of single named mutations of the mouse. A candidate gene for such studies seemed to be the short-ear (<u>se</u>) mutation, discovered in 1921 by Dr. Clara Lynch. I decided to make this the major effort of a year of postdoctoral research at the University of Chicago under the sponsorship of Professor Sewall Wright. Brown University awarded me a Corinna Borden Keen Fellowship to make such a year possible (2).

But first things first. A few days after I arrived in Chicago, Margaret Creighton also, by prearrangement, arrived. Less than a week later, we were married. Margaret had stayed at Brown as a graduate student for two years and then had gone to the State University of Iowa at Iowa City to complete her work for the Ph.D. degree in cytogenetics under the supervision of Dr. William Rees Brebner Robertson. Together, Margaret and I undertook the analysis of the effects of the short-ear gene. Our study was designed as a comparison of mice of two short-eared strains, P and NB, with mice of a normal-eared strain, Bagg albino. It took very little imagination to realize that any difference we might find between the strains could be due to any number of causes other than the alleles of the <u>se</u> locus (3).

I spent much of that year trying to comprehend the details of a paper by Bartlett and Haldane, published in 1935, on the theory of inbreeding with forced heterozygosis. The idea of mating brother and sister mice, deliberately selected to preserve heterozygosity at a specified locus, was easy enough to comprehend. But the method of analysis -- the generation matrix method introduced by Bartlett and Haldane -- required a bit more study. In any case, we decided to create some new inbred strains, each segregating for the two alleles at the se locus. Of the five strains we started at that time, two have survived to the present: SEA/GnJ and SEC/1Gn. To the best of our knowledge, these were the first strains of mice deliberately bred as segregating inbred strains. Now there are dozens of such strains. They are among the strains of choice when one wishes to discover the effects of a named mutation on any aspect of the biology of the mouse.

We moved to Columbus, Ohio, -- mice and all -- in 1941 upon my appointment to the faculty of the Ohio State University. The pursuit of the multiple effects of the alleles of the se locus was carried forth after that time largely by Margaret Creighton Green (4). In the time available, which was not much because of military service and, after the war, because of heavy teaching loads, I continued the analysis of skeletal variations by carrying out classical crosses between several pairs of strains differing in skeletal types (5-11).

During our period at Ohio State University, I discovered a marked difference in skeletal types between sublines of the C3H strain. I had procured samples of C3H mice from eight sources and found that they fell into two groups: one group with five lumbar vertebrae, the other with six. I recommended that C3H mice be carefully denoted as C3H/St or C3H/He (12). During this same period, I also carried out the genetic analysis of a new condition, called furless (fs) (13).

In 1956, Margaret and I moved our household goods to Bar Harbor and our mouse colony to the Jackson Laboratory. I had just been appointed to succeed Dr. C.C. Little as director of the Laboratory, and we had both been appointed to the research staff.

My major research effort, started at this time and continuing for about 18 years, was an attempt to detect the effects of various levels of ionizing radiation on the genetic makeup of small populations of mice. I started the experiments with two kinds of mice: 1) a founding population, called "inbred," that traced ancestry to a single pair in the C57BL/10Gn strain, and 2) a founding population, called "hybrid," that traced ancestry to four inbred strains, C57BL/6J, DBA/2J, C3HeB/FeJ, and BALB/cJ. I deliberately varied the sizes of the experimental populations to provide four levels of inbreeding in the populations in each experiment. The idea was that lower levels of inbreeding favor the accumulation of recessive mutations, both viable and lethal, and that the "hybrid" populations would be able to withstand the effects of accumulated mutations more easily than the "inbred." The surprising outcome of these experiments was that essentially nothing affecting fitness traits seemed to happen in the populations. They continued for 20 generations with essentially no change (14-19).

Several other investigators had designed somewhat comparable experiments about the same time. At a symposium held at the Jackson Laboratory in 1964 and in subsequent publications, they reported essentially negative results as well (20, 21).

Another project required less of my time. In the early 1960s, the Jackson Laboratory was producing about a million mice per year in 18 different inbred strains and six F_1 hybrid generations. I organized a large-scale study to estimate the natural mutation rates of the mouse. The project extended over seven years and entailed examining seven million mice; it became a major endeavor of Dr. Margaret M. Dickie and Dr. Gunther Schlager (22).

Throughout my period at the Jackson Laboratory, I tried to make various aspects of the mouse-breeders art comprehensible to research workers in other fields, so they could make better choices of the kinds of mice to use in their research. For instance, when should one use inbred vs. random-bred mice? How much genetic uniformity can one expect after a given number of generations of brother-sister inbreeding? How efficient are the methods of producing segregating inbred strains and congenic inbred strains? What is the special value of coisogenic inbred strains? I had two opportunities to write expository papers on these questions for publication in the scientific literature (23, 24).

Another major endeavor was the organizing and editing of a second edition of the <u>Biology of the Laboratory Mouse</u>, which came out in 1966 and was republished in 1975 (25).

I had the good fortune to discover four useful mutations in the mice in my research stocks: opossum ($\underline{Ra}^{op}$), pale ears ($\underline{ep}$), shambling ($\underline{shm}$), and a remutation to albinism ($\underline{c}^{2J}$) (26-29).

Between the summer of 1932, when I started working with mice, and the summer of 1975, when I retired, I think I have learned a fair amount about the genetics and biology of the mouse. I might even be able to handle a question such as, "What will you do if there are five mice in the litter?"

REFERENCES

1. Green, E.L. (1941). Genetics 26:192.
2. Green, E.L., and McNutt, C.W. (1941). J. Hered. 32:94.
3. Green, E.L., and Green, M.C. (1942). J. Morphol. 70:1.
4. Green, E.L., and Green, M.C. (1946). Am. Naturalist 80:619.
5. Green, E.L., and Green, M.C. (1946). J. Morphol. 78:105.
6. Green, E.L., and Green, M.C. (1946). J. Morphol. 78:113.
7. Green, E.L. (1951). Genetics 36:391.
8. Green, E.L., and Russell, W.L. (1951). Genetics 36:641.
9. Green, E.L. (1954). J. Natl. Cancer Inst. 15:609.
10. Green, E.L., and Green, M.C. (1959). J. Hered. 50:109.
11. Green, E.L. (1962). Genetics 47:1085.
12. Green, E.L. (1953). Science 117:81.
13. Green, E.L. (1954). J. Hered. 45:115.
14. Green, E.L. (1964). Genetics 50:417.
15. Green, E.L. (1964). Genetics 50:423.
16. Green, E.L., and Les, E.P. (1964). Genetics 50:497.
17. Green, E.L., Roderick, T.H., and Schlager, G. (1964). Genetics 50:1053.
18. Green, E.L. (1968). Radiat. Res. 35:263.
19. Green, E.L. (1968). Mutat. Res. 6:437.
20. Roderick, T.H. (ed.). (1964). Genetics 50:1023.
21. Green, E.L. (1968). Ann. Rev. Genetics 2:87.
22. Green, E.L., Schlager, G., and Dickie, M.M. (1965). Mutat. Res. 2:457.
23. Green, E.L., and Doolittle, D.P. (1963). In "Methodology in Mammalian Genetics" (W.J. Burdette, ed.), p. 3. Holden-Day, Inc., San Francisco.

24. Green, E.L. (1966). In "Biology of the Laboratory Mouse," 2nd edition (E.L. Green, ed.) p. 11. McGraw-Hill Book Co., New York.
25. Green, E.L. (ed.). (1966). "Biology of the Laboratory Mouse," 2nd edition. McGraw-Hill Book Co., New York; 1975, Dover Publications, Inc., New York.
26. Green, E.L., and Mann, S.J. (1961). J. Hered. 52:223.
27. Lane, P.W., and Green, E.L. (1967). J. Hered. 58:17.
28. Green, E.L. (1967). J. Hered. 58:65.
29. Green, E.L. (1968). J. Hered. 59:59.

Earl L. Green

HOW IT BEGAN

Clyde Keeler

Central State Hospital
Milledgeville, Georgia

First of all, let me say that I am happy to have a part in honoring some of my genetic colleagues who have made the greatest contributions to mouse genetics with their development and in-depth investigation of inbred strains of mice. These research activities have far-reaching implications for human and animal medicine, pharmacology, animal husbandry, and other biological fields.

It will be my role to say something about the environment and facilities under which the earliest mouse researches were carried out. I shall mention some of the individuals who served in those early days as supporting personnel.

The drama begins at the Bussey Institution of Harvard, spreading to the Station for Experimental Evolution of the Carnegie Institution of Washington at Cold Spring Harbor, Long Island, then to the Jackson Memorial Laboratory, and finally to laboratories in other parts of the world.

Benjamin Bussey answered the call of the sea in his teens and returned to Boston in middle life, fabulously wealthy. He bought an immense tract of land from the Weld family near Forest Hills and settled down to become a gentleman farmer, building his mansion on high ground. A few feet from his back door he shrewdly constructed a one-room school house which he gave to the county. This meant that the schoolmaster would live at the Bussey mansion where he would be available for reading and writing letters as well as answering questions about the complexities of terrestrial life with which Bussey was not completely familiar.

Benjamin Bussey imported pedigreed sheep and cattle from England. He raised wheat among other crops, employing the 100 male servants that he brought ashore with him. They were all sworn to absolute secrecy as to their former activities on the high seas.

The Bussey will envisaged an agricultural school in great detail that eventually became the Bussey Institution for Research in Applied Biology. The will stipulated the design of the Bussey Building, a Gothic revival in style, that gave it a medieval, monastic appearance. It bore flying butresses and its ve-ri-tas escutcheon was boldly carved upon the pediment of the portico. A tall flag-like iron weathervane was raised proudly aloft on the roof.

A single tree and a small barn wagonshed and one horse stall hid behind the Bussey Building, that otherwise stood alone. A single tree graced the long driveway and under it on weekdays for many years Dr. Castle parked his Model T with its brass radiator that he polished with xylol from time to time. On one occasion he absentmindedly tossed the xylol soaked rag onto the driver's seat and sat down on it.

The spacious interior of the Bussey Building was finished with uninspiring white plaster walls and extremely high ceilings. The long white hall on the second floor bore the only decorations: an unappreciated row of 21 famous Japanese prints strung up too high for temptation and therefore too high for artistic contemplation. They represented the complete series of "The Street Cries of Tokyo" and were the gift of Professor Uichanco of Manila, a former Bussey student.

The will described the precise spot on which the building should be erected, exactly so many feet back from the road. Apparently derived from Benjamin Bussey's former use of the word "shipmates," his will stated that students attending the Bussey Institution would be called "inmates." And so, we are honoring three "Bussey inmates" today.

The impressive building in Gothic revival appeared so ecclesiastical that several times lovelorn couples came holding hands and requested a priest to marry them. Once, while Prof. Castle was mating rabbits, such a couple approached and asked for the marriage rite, mistaking the mild Harvard professor for a Catholic clergyman.

West of the Bussey Building there was a dog yard in which Clarence Little raised canines for his study of coat colors and Dr. Castle raised pigeons. In the early days the Bussey was bursting: a veritable menagerie of rats, rabbits, and guinea pigs. At one time it maintained a screech owl, a sparrow hawk, a tortoise shell male cat, and even a skunk which I deodorized alone because all prospective assistants suddenly found that they had pressing appointments elsewhere.

Aside from Dr. Castle, the Bussey faculty consisted of Drs. Wheeler and Brues in entomology, Dr. East in plant genetics, and Prof. Bailey in plant anatomy.

Dean William Morton Wheeler passionately hated genetics that had become so popular compared with his static entomology. He wrote a paper, "The Dry Rot of Our Academic Biology," from which I quote: "Genetics, what a promising bud, but oh how constricted at the base."

And again he fulminated that Nobel prize winner Thomas H. Morgan, the _Drosophila_ geneticist, had "strained at a gnat and swallowed an hypothesis!"

This did not go down very well with Professors Castle and East, the Bussey geneticists.

Mr. Patch was the caretaker of the Bussey grounds. He was over 70 when I first knew him, having worked for the Bussey estate ever since he was 14. He had even known the last of the original 100 silent Bussey sailor men.

Mr. Patch plowed the ground and raised the cabbages and carrots that fed Dr. Castle's rabbits and, when the Bussey boys ate the discard rabbits from Dr. Castle's experiments, they snitched cabbage and carrots to cook with them. Mr. Patch had a key to Dr. Castle's office where the alcohol was kept and his Sunday morning visits appeared therapeutic. However, Dr. Castle had extreme confidence in Mr. Patch. Once Dr. Castle told me if their opportunities had been exchanged, probably Patch would be a better professor and Castle a poorer caretaker.

The second member of the supporting team was Mrs. Kelley who hastened over to the Bussey each morning to greet the professors cheerily, and could be found flicking up dust from the furniture with an enormous instrument made of turkey tail feathers. She never quite mastered the vocabulary of the intelligencia, but she tried. On Monday mornings she distributed what she called the "immaculate towels," and when on one occasion she came upon a box with a light in it, she soliloquized: "What's this? Why, it's an insinuator -- no, it's not an insinuator -- it's an inclabator."

Jovial Billy Reardon, caretaker of the rat room, was fresh off a freighter from Ireland. One day he came into the rat room and found AMA gold medalist Gregory Pincus with a pan in one hand and with the other he was dropping pinches of something from the pan onto a row of cages one after another. Billy dropped the heavy bag of feed that he was carrying, straightened up, and folded his arms.

"An' for Gawd sake, Pincus, an' what <u>are</u> ye doin'?"

"Oh, I'm just giving them a little clover."

"Well, I'm shore glad to know it. For a moment I thought ye was administhering the hooly sacramint!"

The Bussey Building had a distinctive aroma, a none too delicate blend of mouse, rat, and rabbit, and one allergic medical student became so faint that he could not pass the threshold. One Bussey inmate wearing Billy Reardon's unwashed overalls was refused entrance to a fasionable Boston New Year's Ball.

A gray frame house served as the Bussey dormitory. There lived Mrs. Kelley, her husband, and her in-laws. There at one time also lived thirteen assorted doctoral candidates and postdoctoral researchers on fellowships.

After supper Bulgarian Doncho Kostoff might defend communism, or Dr. Hu might sit on the table with his legs folded under him and his hands held appropriately to expound on the nihilism of Buddhist nirvana. Doncho felt that the Soviet government was justified in shooting some 150 Moscow homeless waifs found to have hoof and mouth disease to prevent an epidemic. Doncho had published scientific papers in English, French, German, Bulgarian, and Esperanto.

Though Dr. Hu declared that all in this world is unreal and imagination, he carefully locked a typed draft of his two-volume thesis in a bank vault until the printers had issued his printed edition of "The Phanerogams of China." We thought that this was an extreme precaution for a thesis made of "nothingness" as its author declared.

Each student took his turn at cooking for a week, and there was an eternal soup on the stove to be quickly warmed up for between meals. It might start out with a ham bone and navy beans, mutate through many constitutions, as cans of this and that were added, to end up several weeks later as chicken and noodles.

One evening Dr. Myers came down from the Bussey mansion at dusk and inadvertently kicked into a pair of amorous skunks. He buried his clothes, of course.

One night a thief reached through a window with a rake and stole Prof. Brues' pants containing his wallet. The police found and identified the pants the next day. We were all afraid that the Harvard Lampoon would hear of the incident. Just imagine the possible headline, and then shudder:

"Harvard professor's pants found in arboretum. Says they were stolen."

Aside from the ancient brass-framed microscopes, the Bussey equipment included an imposing cylindrical slide rule five feet long. There was a box calculator, an all-brass contraption of encased cogwheels, buttons, slides, and stamped numbers. There was a handle that, when turned, suggested audibly some relationship to a McCormick thrashing machine. There was a versatile portrait camera and a carbon arc for photography.

Here mouse genetics, as we know it, was born with the research on coat colors in mice by phenomenal Clarence Cook Little, Phi Beta Kappa, captain of the track team, collegiate title holder in shot put and discus, assistant dean of Harvard College in his senior year, and later founder of the Cancer Society and the Jackson Memorial Laboratory.

I have here a copy of Dr. Little's thesis study of coat color in mice, bearing witness to the author's connection with both the Bussey Institution and the Carnegie Institution of Washington Station for Experimental Evolution at Cold Spring Harbor.

Dr. Little exhibited a complete collection of stuffed mouse skins demonstrating coat color inheritance in mice at the International Congress of Genetics at Ithaca.

Mouse researchers always cooperated with the Mouse Fancy and so when the Boston Cat Club invited the Mouse Fanciers to hold a joint Cat and Mouse Show, a number of Bussey types of mice were exhibited. And it was there that a well known Mouse Fancier of the cloth was entrapped. It was really a conspiracy. An innocent looking young lady stepped forward and asked the minister:

"Aren't you afraid to handle mice?"

"Why, of course not, they can run all over your hands, like this, and are perfectly harmless."

"But aren't you afraid to put them on your shoulder?"

"Now, here is one on my shoulder, see!"

"Well, would you be afraid to have one on your head?"

The minister placed a mouse on his exceptionally bald head to demonstrate his bravery, and a light bulb flashed behind the decoy. The picture hit the press wires. Within a matter of several hours the papers hit the streets of the nation and the clergyman was no longer an employed Episcopal priest.

The Bussey was well represented at the Harvard tercentenary celebration with charts and models on exhibition in Cambridge.

There was a large chart depicting the known mutations arranged according to their linkages. There was a chart with ceramic mouse models showing the size effects produced by each of six recessive gene mutations in mice, studied by Castle, Gates, Reed, and Law.

There were large plaster models demonstrating degenerate retina and absent corpus callosum. Dr. Castle's rabbit skins were displayed and ceramic rabbit models showed their genetic relationships. Cat genetics was featured.

If mouse genetics was born at the Bussey, it was cradled at Cold Spring Harbor in the Carnegie Institution of Washington Station for Experimental Evolution. There we again meet C.C. Little, Leonell Strong, and E. Carlton MacDowell, a former Bussey inmate noted especially for his early studies of virus associated with leukemia in mice. There were Bea Johnson and Amelia Vicari. There was also from time to time Halsey J. Bagg, a student of behavior, who developed the famous Bagg strain of inbred albino mice. He published with Little on defects in the Bagg strain.

It was also with this strain of albinos that I was introduced in 1923 to mammalian genetics through my finding of what is now known as "degenerate retina," still our best animal model for retinal degeneration in man.

The Carnegie Station for Experimental Evolution was much better equipped than the Bussey with an office building separate from the mouse laboratories.

As a college student I had hitchhiked from Marion, Ohio, in the summer of 1920 to Cold Spring Harbor where I served as headwaiter, collected lobsters and starfish, and retrieved pig embryos in a slaughter house to pay for expenses and a ride back home.

There, in 1920, I met many outstanding geneticists. Charles B. Davenport ran the Experiment Station. Herbert Eugene Walter, Sidney Isaac Kornhouser, and H.M. Parshley were the mainstays of the Summer Field Course in Biology.

On the hill stood the Eugenics Record Office operated by Dr. Harry Laughlin and Dr. Muncie. Leslie Peckam was their factotum, who constructed family pedigrees for publications of the staff.

At the Station for Evolution in Botanical Genetics was Albert F. Blakeslee, president of the American Association for the Advancement of Science, and his assistant, John Belling, inventor of the famous "Belling smear technique" for microscopy. These men and their summer assistants were doing spectacular research with the pig weed <u>Datura</u> <u>stramonium</u>, identifying various types characterized by three copies of particular chromosomes instead of two.

Hooper Hall contained the great dining room plus some dormitory space upstairs. The dining facilities served all personnel of the Record Office, the resident geneticists, and the summer course students.

Mrs. Davenport was dietician in 1920. That year she tried valiantly to stay within the budget by serving up Oscar Riddle's discard pigeons from his experiments on modification of the sex ratio. The menu said "squab." She dessicated these old birds. The diners merrily tapped out a tattoo chorus with their knives on the solid rib cages of these mummified atrocities, and the diners to a man, woman, and child returned the birds to the kitchen untasted.

Once again the dietician attempted to economize by having the carpenter armed with a sledge hammer do battle with Tetraceratops. He was the old four-horned Algerian ram with which Dr. Davenport kept going his study on the "Inheritance of Supernumerary Horns in Sheep."

The carpenter eventually won. However, this was in the days before papaya tenderizers and, besides, Tetraceratops had a strong bodily odor. So his tough inedible chunks all went back to the kitchen. His life had been sacrificed in vain.

I leave to your imagination what happened when in the kitchen maggots were found in the cheese.

Because of my doctorate studies on degenerate retina in mice, I was asked by Dr. Howe in 1927 to join the Howe Laboratory of Ophthalmology at the Harvard Medical School to teach and research in genetics, but after twelve years I resigned because I could not convince my conservative medical colleagues that genetics would ever produce anything of value to medical science. A few were sympathetic, but most were skeptical.

The head of my department, crusty old Prof. Frederick Verhoeff, declared: "Heredity always runs out. You never see a defect in a human pedigree in more than two or three generations!" Once he even induced Dr. Howe to drop me from the payroll because "medical genetics has no future." The final report that I turned in must have been convincing because he promptly hired me back again with a $500 raise in salary.

Some medics maintained that medical school research should be confined to the human subject, and one colleague declared: "You can never convince me that immunity and susceptibility in man can be distributed according to a 3-1 ratio."

During my tenure at the medical school I kept my genetic experiments going at the Bussey. In mice I discovered the first genetic heart defect, studied coloboma iridis, corneal opacities, rosetted fur, discovered $wavy_2$, as well as spine and tail anomalies. I examined hypo- and hyperglycemia, cystic kidneys, lethal anemia, harelip, hydrocephalus, and other defects.

I spread out to research glucose tolerance in Yale rats, blood groups in rabbits, sex deficiency in the tortoise shell male cat, and kidney stones in Dalmatian dogs.

Eventually my genetic interests led me into behavior genetics associated with coat color genes in rats, mink, and foxes where tameness may be synthesized by an appropriate combination of certain coat color genes. I extended these principles to man with my studies of albinism and red hair.

Shortly after I took my doctorate, Dr. Castle told me seriously that there was no future in studying mice because there were no more gene mutations in mice. They had all been studied.

Dr. East tried to get me to become a botanical geneticist! Nevertheless, my attitude was that it was merely a matter of many researchers being on the lookout for mouse mutations because there must be a whole spectrum of mouse mutations paralleling known mutations in man. So I wrote a small book to popularize mouse mutations. The accomplishments of our honorees show how wrong Dr. Castle could be at times.

In search for possible new mutations I went to China, Japan, and Turkey. I spent three days examining the thousands of stuffed mouse skins in the British Museum. I found blues, chocolates, pink eyed buffs, and albino types from all parts of the world.

When I finished my survey I thanked the curator for his kindness in allowing me to examine the British Museum mouse skins and told him that I had at the Bussey three varieties that he did not have, and that I would gladly send him specimens.

That little man became highly insulted and angered. He stretched several inches taller in his striped trousers and gray spats, looked down his nose at me, and screamed:

"Sir, if they are not in our collections, they do not exist!"

The ancient literature recorded that albino mice were used for auguries in Babylon, Egypt, and Troy long before the time of Christ. It was the same in China where all albino mice caught in the wild had to be turned over to the magistrates, and 26 such albinos are listed in Chinese history as found between 307 and 1641 A.D.

I wondered about the tiny Island of Tenedos at the mouth of the Dardanelles where stood the temple of Apollo God of Mice (Apollo Smintheus) since long before the Trojan War. Aristotle and other ancient writers told of the white mice raised under Apollo's altar. I went to Tenedos in 1930 and learned that in 1929 an albino house mouse was living in a garden shed a stone's throw from the site of the ancient temple of Apollo. Barring possible mutation, that could mean as much as 3,000 years of population inbreeding and with three generations a year the number of generations is staggering.

During this trip to catch mice in Asia Minor, I was captured and tried for espionage by the Turkish military. When it was learned that I was sponsored by a fellowship from Harvard University, the prosecution was incredulous.

The colonel who served as prosecuting attorney bluntly declared: "The government of Turkey maintains that there is no university on earth crazy enough to send a man half way 'round the world to catch mice!"

My interpreter bravely defended me with: "Yes, but there is one university crazy enough to send a man half way around the world to catch mice, and that is Harvard University in Cambridge, Massachusetts."

From the biomedical standpoint, our lowly housemouse must be awarded the title of "man's best friend."

Among the many uses of mice in pharmaceutical laboratories is drug research: first infecting and then trying to find a cure. There is testing of drugs for carcinogenic and teratogenic effects as well as toxicity of their products. Drugs are tried out on mice before going to other laboratory animals. Specialized strains of mice are employed for detecting diabetic agents, epileptic agents, and others. Mice are used to reveal behavior patterns and pattern effects of therapeutic drugs. Mice are involved in the standardization of sera.

A center for disease control employs mice to isolate virus, four types of fungi, for producing rabies antibodies, and for botulism titrations. They are used for the identification of the Hymenolepis tapeworm, <u>Trypanosoma cruzi</u>, and Echinococcus as well as the Toxoplasmosis parasite. Mice can separate pneumococcus from streptococcus viridans. There is a mouse test for Staphylococcus virulence, one for rat mite fever, and one for Leptospira of kidney and bladder. There are mouse tests for identifying rabies and equine encephalitis.

Elsewhere, mice are used to study weightlessness, radiation, toxic substances in chemical plants, and problems of the psychology laboratory. Man-mouse leukocyte hybrids are used in mapping human chromosomes. Mice are used in a pregnancy test. Mouse organ tissues are employed as substrates in the study of PKU and other human enzyme deficiencies.

Sminthophiles with curiosity might like to carry out modest but valuable experiments of their own on mice, employing simple apparatus. Perhaps some already do. In any case, old hands in the mouse genetics business will be glad to advise and assist the Mouse Fanciers.

Let me conclude with the observation that the ways of rodent geneticists are still often misunderstood by the laity. For example, one day a little old lady met me on the street and bubbled:

"Oh Doctah Keelah, I am so glad to see you. You know, this morning I saw the most horrible, big, old dead rat lying in the guttah, and, of course, I thought of you."

Then she became frustrated, blushed, and added:

"Oh, oh, oh, I didn't mean it <u>quite</u> that way. I meant that I thought you could use it in one of your experiments."

I judged that the conversation had gone far enough, so I did not stop to explain to the lady that I was really a mouse geneticist at heart and that in my opinion dead rats do not reproduce very well.

Clyde Keeler

Inbred Strains: Mutations, Biochemistry

Donald W. Bailey, Chairman

SOURCES OF SUBLINE DIVERGENCE AND THEIR
RELATIVE IMPORTANCE FOR SUBLINES
OF SIX MAJOR INBRED STRAINS OF MICE[1]

Donald W. Bailey

The Jackson Laboratory
Bar Harbor, Maine

A number of genetic differences among sublines of various highly inbred strains of mice have been reported (1-9). Such differences should come as no surprise, for they were predicted by the early population geneticists. Nevertheless, with the number of sublines increasing, especially through development of new congenic lines and specific-pathogen-free colonies, the number of encounters with subline differences will be growing and their vitiating effects on research no doubt will be frequently felt.

Subline differences arise by the gradual differential fixation of genes at loci that were heterozygous from three possible causes: 1) contamination from outcrossing, 2) incomplete inbreeding, and 3) mutation. In this paper we shall

---

[1] This work was supported by NIH Research Grants No. GM22878 from the National Institute of General Medical Sciences and No. AI13130 from the National Institute of Allergy and Infectious Diseases.

consider the relative importance of these three sources of subline differences, and we shall see how existing sublines of six major inbred strains of mice may be expected to have diverged due to these sources.

Contamination

Genetic contamination often will become obvious if a full-sib mating system is strictly adhered to, for the presence of new recessive as well as dominant coat color genes will soon make themselves evident. However, in a strain in which inbreeding has been relaxed, the presence of contaminant recessive genes may not be so apparent. (This points out one of the hazards of not maintaining a strict mating regimen.) The number of contaminant genes that eventually contribute to fixed gene differences between subsequently derived inbred sublines will be inversely proportional to the effective number of breeders maintained during the generations of random mating between the time of the outcross and the resumption of inbreeding.

Contamination can be identified quite effectively by routine skin graft exchanges or allozyme typing for genetic quality control. Moreover, contamination can be identified as the likely source of subline divergence if the number of gene differences is significantly greater than that possible either from mutation or from residual heterozygosity, or if the new allelic differences specifically match those of the suspected contaminating strain. Contamination may indeed be the source of most existing major subline differences, but because it is based on human error we shall not consider it further here.

Residual Heterozygosity

For present considerations, we define residual heterozygosity as that proportion of loci with genotypic mating combinations that are not yet in the genetically fixed state (both parents homozygous for the same allele) at a specified generation of inbreeding. This definition is more inclusive and more appropriate for the present purpose than the commonly used definition, namely, that proportion of individuals at a given generation of inbreeding that are still heterozygous at a specified locus (10).

Fisher (11), using the concept of "junctions" and the generation matrix method, was able to estimate the number and lengths of heterogenic tracts (chromosomal segments that are still of heterogeneous origin and thus potential carriers of residual heterozygosity) at specified generations of inbreeding. From these estimates he was also able to calculate the probability of no heterogenic tract remaining (complete genetic fixation, barring mutation) after progressive amounts of inbreeding (Fig. 1). We have used Fisher's method with the chromosome number of 20 and an estimated mouse genome length of 1500 cM (12); Fisher had used the estimate of 2500 cM.

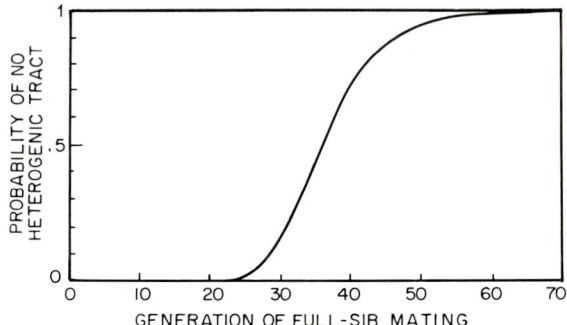

FIGURE 1. The probability of no heterogenic tract remaining after the indicated number of generations of full-sib mating in mice.

From the curve in Figure 1, we see that the probability of being completely free of heterogenic tracts does not attain even the 0.5 level until F36 nor the 0.99 level until F60. Purity is not easily attained.

The probability of complete purity takes into account only the heterogeneity of the genomes of the original pair of mice but no subsequently arising mutations. The question that more realistically should be asked is: At what generation does residual heterozygosity become no more important than mutation as a source of subline differentiation? We can answer this question by comparing the numbers of fixed gene differences arising from the two sources.

By Fisher's method we can estimate the number and lengths of heterogenic tracts after a specified number of generations of full-sib mating as in columns 2 and 3 of Table 1. Then, if the number of structural genes in the mouse genome is 30,000 (12), the proportionate number of structural genes in the heterogenic tracts after different amounts of inbreeding can be estimated as in column 5 of Table 1. If only 35% of these loci were heterozygous at the time of the strain's origin (13) and 43% of these in turn contribute to subline differences (as based on proportions of the genotypic mating combinations and the probabilities of the alleles in these combinations becoming differentially fixed in two sublines), then the number of fixed gene differences between two sublines that branched at a given generation of inbreeding and continually maintained by full-sib mating for at least 20 more generations can be estimated as in column 6 of Table 1. [The estimates in column 4 can be calculated directly from the generation matrix without regard to chromosome number or genome length. In these calculations and all others in this paper, the effects of selection are ignored.]

TABLE 1. Theoretical Effects of a Full-sib Mating Regimen on the Mouse Genome

Generation	Heterogenic tracts				Number of structural genes	
	Number	Length (cM)	Proportion of total genome	In heterogenic tracts		Differentially fixed between sublines that branched at this generation
10	42.7	323.2	0.215	6500		975
15	22.9	112.0	0.075	2240		336
20	10.7	38.8	0.026	780		117
25	4.7	13.5	0.009	270		40
30	2.0	4.7	0.003	93		14
35	0.8	1.6	0.001	32		5
40	0.3	0.6	0.000	11		2
45	0.1	0.2	0.000	4		1
50	0.0	0.1	0.000	1		0

The estimated 35% heterozygosity that we have employed is derived from observations on electrophoretic variants of enzymes and soluble proteins in Mus musculus (13). On one hand, it can be argued that heterozygosity may not have been as high as this in the fanciers' mouse populations from which the strains of laboratory mice arose, because the effective breeding population from time to time was probably quite small. On the other hand, many fanciers although maintaining small populations probably continually exchanged breeders and occasionally introduced variant mice from wild populations of different subspecies and in this way probably would have maintained much of the original genetic heterogeneity. However, even if the heterozygosity were reduced, the estimate of 35% seems within reason and perhaps still on the conservative side considering the arguments Lewontin (14) gives for such estimates being too low.

Mutation

The expected number of differentially fixed genes ($\underline{n}$) arising from new mutation is expressed by the equation:

$$\underline{n} = (G_1 + G_2 - 7)\mu,$$

where G_1 and G_2 are the numbers of generations that sublines 1 and 2, respectively, have been inbred since branching, μ is the mutation frequency per gamete per specified block of loci, and the number 7 corrects for the effect of mutations of recent generations having not yet been fixed. If one wishes to include unfixed loci, instead of subtracting 7 in the above equation, subtract 4 for recessives, and add 20 for dominants. This subject is treated in more detail in a recent paper (9).

If there are 30,000 structural genes (12) and the average mutation frequency per gamete per locus is 10^{-5}, then the mutation rate for all structural genes would be about 0.3 per gamete, and the number eventually becoming differentially fixed in two sublines would be:

$$(G_1 + G_2 - 7)(0.3).$$

Mutation rate as used here is not a true mutation rate but rather a practicable one that results from a confounding of the probabilities of origination, viability, observability, and reproducibility of the mutant.

The value 10^{-5} is generally used as a rough estimate of spontaneous mutation rate in mammals. However, published estimates of mutation rate in mice vary greatly and often are lower than 10^{-5} (15). Still it seems reasonable to round off the estimate to this higher value for the present considerations. Most mutation rates in mice are based on loci that determine visible traits such as coat color. The number of viable mutations that actually occur at these loci is probably much greater, because surely not all mutations result in observable phenotypic changes. On the other hand, it can be argued that it is the observable changes that we are truly concerned about in subline divergence because these are what affect experiments, so why consider undetected mutations? However, as experiments deal more and more with traits at the molecular level of gene expression, more mutations with subtle effects will become evident and the apparent rate will rise. Thus, with our present state of knowledge, the rate we have chosen seems reasonable and indeed may prove to be too low.

It is of interest to note that the spontaneous mutation rate for viability polygenes in <u>Drosophila melanogaster</u> has been estimated to be 0.3 mutations per gamete per genome and a mutation rate of 5×10^{-5} per cistron (16). This would suggest that the spontaneous mutation rate we have chosen, if anything, is too low.

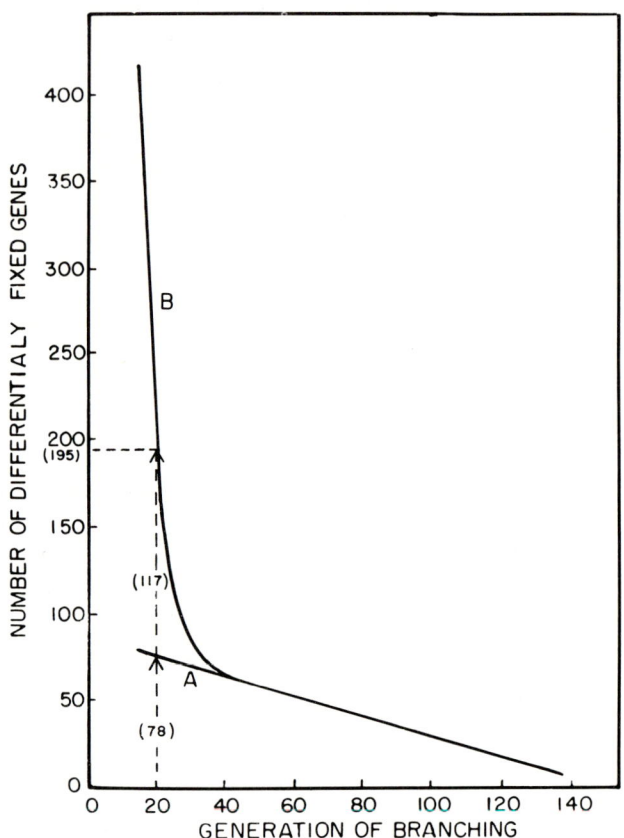

FIGURE 2. The number of fixed gene differences expected between hypothetical sublines relative to the generation at which they branched. Both sublines are considered to have been equally inbred for 154 (an arbitrarily chosen number) generations of full-sib mating. To correct for additional (or fewer) generations of inbreeding, add (or subtract) (0.3) times the sum of the generations that both sublines have been inbred beyond (or before) F154. Curve A shows the contribution from new mutations alone, while curve B shows the contribution of residual heterozygosity in addition to that of mutation. The example presented in the text is represented by dashed lines and numbers in parentheses.

A Comparison of Sources

From the equation for $\underline{n}$ and from values in column 6 of Table 1, we have constructed Figure 2 to show the contributions of residual heterozygosity and mutation to subline divergence. The expected number of genes differentially fixed between two sublines, each inbred for 154 generations (an arbitrarily chosen number), is shown in relation to the generation at which the branching occurred. Curve A shows the contribution from new mutations alone, while curve B shows the contribution of residual heterozygosity in addition to that of mutation. As an example, two sublines branching at generation F20, and both continuing under full sib mating for 134 additional generations to F154, will differ at 117 loci due to residual heterozygosity that was present at the time of branching (Table 1) and will differ at $[2(154-20)-7](0.3) = 78$ more loci due to mutations, most of which have occurred since branching, for a total of 195 loci with fixed gene differences.

The chance of any particular researcher encountering a genetic difference between sublines depends of course on the number of genes that determine the trait he is studying. For example, if there are 330 histocompatibility (H) genes in the mouse, then the expected number of H-gene differences between two sublines would be $330/30,000 = 0.01$ of that indicated in Figure 2. In the example given above, two sublines inbred for 154 generations but having branched at generation 20 would be expected to differ at $(0.01)(195) = 2$ H loci. One of these differences would have originated by mutation.

The above estimate of 330 H loci is an average of estimates obtained by three different approaches: 1) The mutation rate of the block of H loci averaged from studies conducted at two different laboratories is 3×10^{-3} (8). If we assume the mutation rate per H locus is the same as that for other loci, i.e.,, 10^{-5} (15), then $3 \times 10^{-3}/10^{-5}$ = 300 H loci. 2) One corollary of Snell's laws of transplantation is that mutations showing a loss of antigen can only be detected at heterozygous loci (in the present case, these would be those loci at which the parental strains, C57BL/6By and BALB/cBy, of the F_1 mice being assayed for mutations carry different alleles). So the number of loss-type mutations out of the total number found should be proportionate to the number of H loci by which the two parental strains differ out of the total number of all H loci in the species. This argument has been set forth in more detail (17), where the estimate obtained was 430 H loci. 3) An individual wild mouse has been estimated to be heterozygous on the average at about 11% of its loci (13). Since the BALB/cBy and C57BL/6By strains have been estimated to differ at 29 or more H loci (18), and since each inbred strain theoretically is genetically equivalent to a single gamete drawn at random from a wild population, then the F_1 derived from these two strains would be equivalent to a representative individual from a wild population, and because it is heterozygous at 29 or more H loci the total number of H loci in the species would be estimated at $(100/11)(29) = 264$.

We have constructed an equivalence curve in Figure 3 so that we can compare the relative importance of the two sources of subline differences. Points in the area above and to the right of the curve indicate mutation to be the most important contributor, and below and to the left, residual heterozygosity to be the more important contributor. As an example, if the nomenclature committee were to recommend (as discussed later) that each subline that branched prior to F24 should be given a distinctive symbol, then to be reasonable, the committee should also recommend, after examining Figure 3, that a subline inbred for $200/2 = 100$

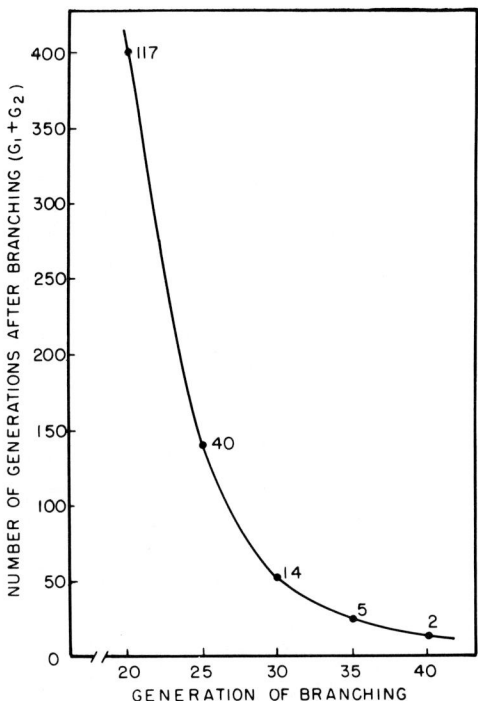

FIGURE 3. An equivalence curve that allows comparison of the relative importance of mutation and residual heterozygosity as sources of subline divergence. The fixed-gene differences derived from heterozygosity at the branching generation (X axis) and those derived from mutations during the generations of separation (Y axis) have equal contributions where perpendiculars intersect at the curve. The estimated number of fixed-gene differences contributed by either source alone is indicated by values entered at various points along the curve. For example, two sublines that branched at generation F25 would be expected to have gained from residual heterozygosity 40 fixed-gene differences, and if subsequently maintained separately for 144 generations would be expected to have gained from accumulated mutations another 40 fixed-gene differences for a total of 80. Contributions from the two sources would have been equal.

generations after branching also should be given a distinctive symbol. We assume here that any subline to which this one is compared will also be inbred 100 generations after branching so that $G_1 + G_2 = 200$.

Minimally Attainable Genetic Uniformity

Sublines will differ not only by <u>fixed</u> mutations but also by <u>unfixed</u> mutations arising in recent generations and not yet having had a chance to become fixed by inbreeding. The number of unfixed gene differences between two hypothetical sublines is estimated at about 27 μ(9), and if μ ~ 0.3, then (27)(0.3) ~ 8. On the average, four of these will be segregating in each subline, and because they are segregating, only about 60% of the mice in each subline will be carrying any one of the four mutant genes. For perspective, it would take, on the average, 40 generations of full-sib mating before the number of unfixed genes from residual heterozygosity would equal the number (four) from mutation and 10 additional generations before <u>only</u> the unfixed genes from mutation would be of consequence.

These unfixed gene differences from mutation become the limit of attainable genetic uniformity within a subline, outside of the routine screening of individuals. We can continually assay for fixed-gene differences between sublines, as discussed below in regard to maintaining parallel lines to assure that no differences exist in the trait of interest, but unfixed gene differences are always potentially present and may be found between members of the same subline. However, the probability that any of these mutations will be affecting the trait of interest depends on what proportion of the genome affects that trait and in most cases this will be extremely small.

Importance to Sublines of Existing Strains

These calculations can be used to evaluate the differences expected between sublines in various existing inbred strains of mice. To do this we have constructed pedigree charts in Figures 4A to F of six major strains based on information gleaned from past issues of <u>Mouse News Letter</u> and from its companion issue, <u>Inbred Strains of Mice</u>. We have used the recorded year of branching instead of the inbreeding generation of branching, because often only the year was given in the entries and generation counts often were begun anew in each laboratory. We have found the number of generations progressed each year to vary among laboratories, but it averages close to 2.5, the value we have used in interpreting these charts.

We have then placed the graph of curve B in Figure 2 (contribution of both residual heterozygosity and mutation) adjacent to each chart so that differences between sublines can be estimated directly from the point of the branching. The graph adjacent to the DBA chart (Fig. 4F) includes a roughly estimated effect of the reported intercrossing of two sublines in 1929. The C57BL/Ks subline in Figure 4E probably is a product of genetic contamination, a conclusion reached because that subline differs from other C57BL/6 sublines at all loci in the H-2 complex as well as at least three other histocompatibility loci (19).

The branching in most cases is likely to have occurred even earlier than that indicated in the chart, perhaps three to five generations earlier, because the sender may not always have taken into consideration the branches in his own colony when selecting breeding pairs for shipment.

It will be seen in Figures 4A through F that a number of early branching sublines will probably be found to differ at many loci due to both residual heterozygosity from the early

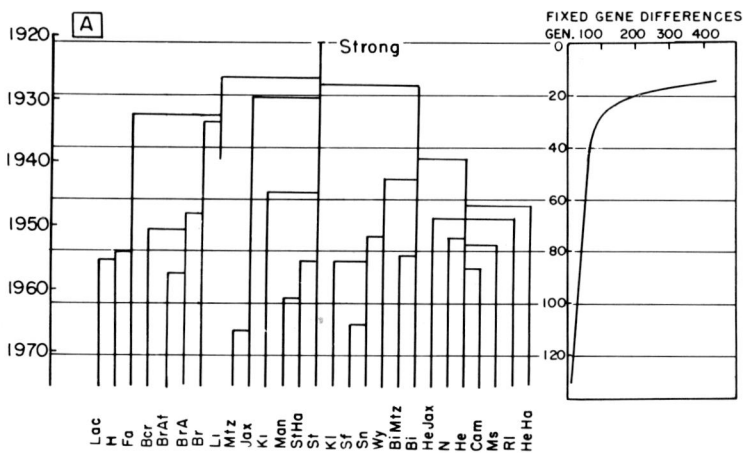

FIGURE 4A

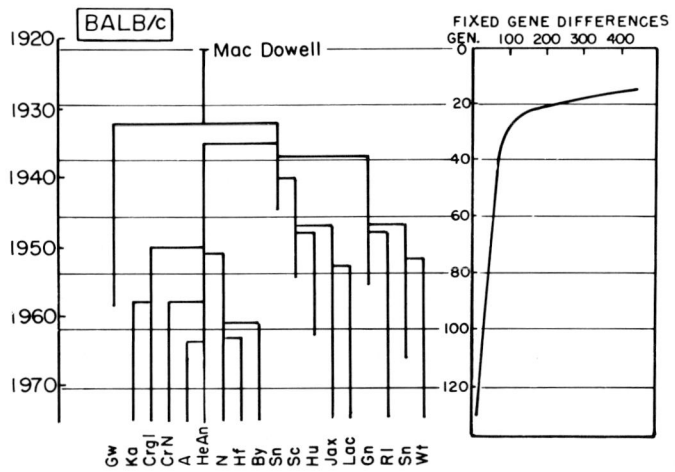

FIGURE 4B

FIGURE 4A-F. Diagrams of subline derivations in six major inbred strains of mice. Graphs on right of each figure indicate number of fixed-gene differences estimated to exist today between sublines that branched at the indicated generation of inbreeding.

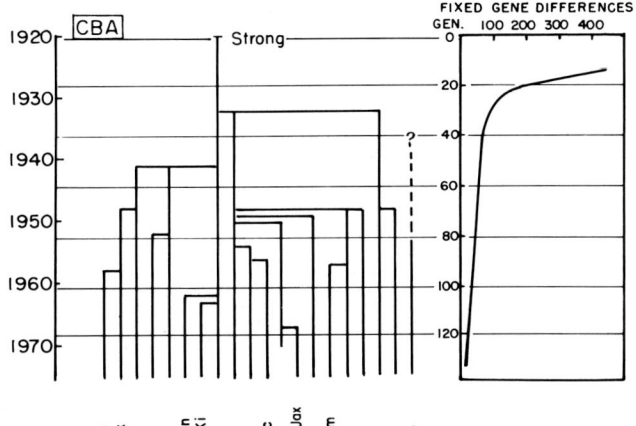

FIGURE 4C

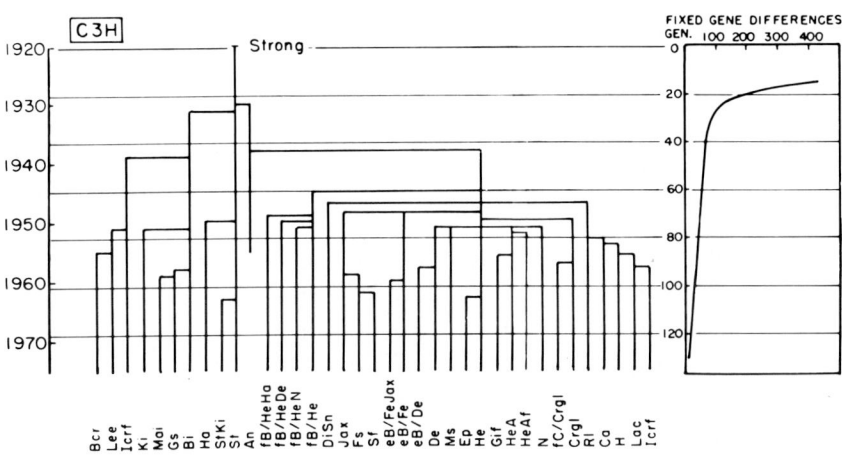

FIGURE 4D

212　　　　　　　　　　　　　　　　　　INBRED STRAINS: MUTATIONS, BIOCHEMISTRY

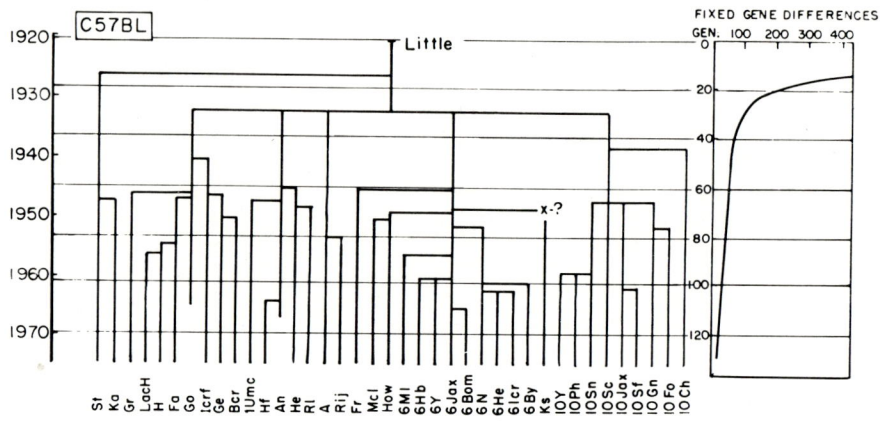

FIGURE 4E

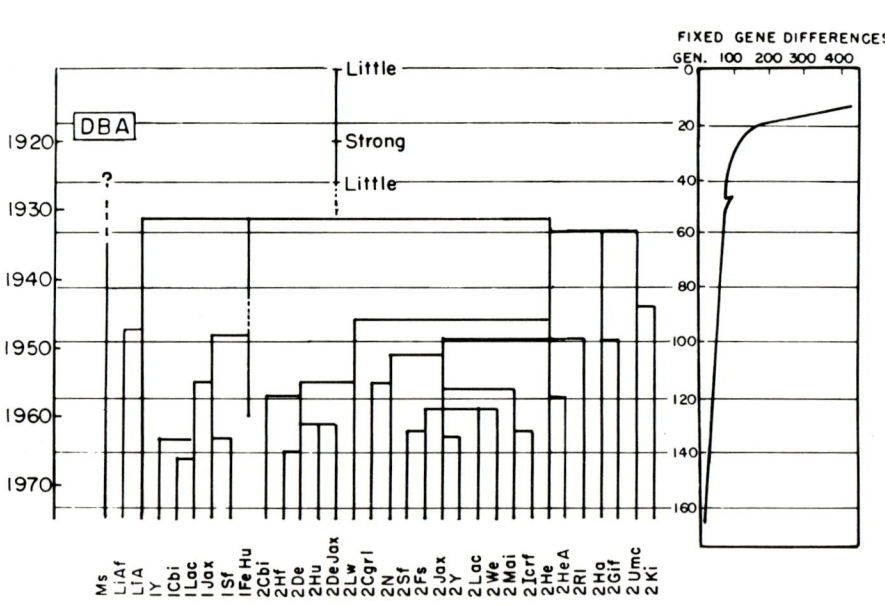

FIGURE 4F

branching and to mutation through many generations of separate maintenance. There is especially early branching of sublines in the C57BL and A strains.

Ways of Controlling Divergence

The problem of subline divergence can be kept under control by several procedures, some more immediately practical than others.

1) Divergence can be avoided by restocking from a common source colony, but this would have to be done by all relevant laboratories. Historically this is a practice not easily established.

2) Divergence can also be slowed down by placing embryos of each strain in frozen storage (20), and restocking the experimental colony from time to time from the frozen source. This could be a common international frozen source from which all laboratories restocked their colonies.

3) Divergence can be detected as it occurs by the investigator maintaining two parallel lines of each strain in his colony (Mobraaten, personal communication). Mice from the two lines can be compared in all pertinent experiments.

4) The research worker could be alerted to potentially large differences between sublines by a revised nomenclature system. A current nomenclature rule (21) allows sublines arising in any generation after F8 to bear the same strain name. It would help the research worker considerably if such sublines arising before, say, F24 be given distinctive symbols. In the same vein, it would be helpful for any subline, as well as its branches (sub-sublines), that has been separately maintained for 100 generations or more (see Figure 3) also to be given a distinctive symbol.

5) The latter procedure unfortunately would have to await the action of an international committee. In the meantime, each experimenter can and should publish a key reference (e.g., in <u>Mouse News Letter</u>) which provides the full ancestral information on his sublines for others to evaluate relationships of their sublines with his and which he should always cite in his pertinent papers.

REFERENCES

1. Acton, R.T., Blankenhorn, E.P., Douglas, T.C., Owen, R.X., Hilgers, J., Hoffman, H.A., and Boyse, E.A. (1973). Nature New Biol. 245:8.
2. Ciaranello, R.D., Lipsky, A., and Axelrod, J. (1974). Proc. Natl. Acad. Sci. 71:3006.
3. Glode, L.M., and Rosenstreich, D.L. (1976). J. Immunol. 117:2061.
4. Maurer, P.H., Merryman, C.F., and Jones, J. (1974). Immunogenetics 1:398.
5. Olsson, M., Lindahl, G., and Ruoslaht, E. (1977). J. Exp. Med. 145:819.
6. Rechicigl, R., Jr., and Heston, W.E. (1963). J. Natl. Cancer Inst. 30:855.
7. Taniguchi, M., Tada, T., and Tokuhisa, T. (1976). J. Exp. Med. 144:20.
8. Bailey, D.W. (1976). In "Basic Aspects of Freeze Preservation of Mouse Strains" (O. Muhlbock, ed.), p. 67. Gustav-Fischer Verlag, Stuttgart.
9. Bailey, D.W. (1977). Ciba Found. Symp. 52:291.
10. Wright, S. (1921). Genetics 6:167.
11. Fisher, R.A. (1949). "The Theory of Inbreeding." Oliver and Boyd, Edinburgh.
12. McKusick, V.A., and Ruddle, F.H. (1977). (Footnote 12) Science 196:390.

13. Selander, R.K., and Yang, S.Y. (1969). Genetics 63:653.
14. Lewontin, R.C. (1973). Ann. Rev. Genet. 7:1.
15. Schlager, G., and Dickie, M.M. (1967). Genetics 57:319.
16. Mukai, T., Chigusa, S.I., Mettler, L.E., and Crow, J.F. (1972). Genetics 72:335.
17. Bailey, D.W. (1968). In "Advance in Transplantation" (J. Dausset, J. Hamberger, and G. Mathe, eds.), p. 317. Munksgaard, Copenhagen.
18. Bailey, D.W., and Mobraaten, L.E. (1969). Transplantation 7:394.
19. Graff, R.J. (1970). Transplant. Proc. 2:15.
20. Whittingham, D.G. (1976). In "Basic Aspects of Freeze Preservation of Mouse Strains" (O. Mühlbock, ed.), p. 45. Gustav-Fischer Verlag, Stuttgart.
21. Staats, J. (1976). Cancer Res. 36:4333.

GENETIC QUALITY CONTROL OF THE LABORATORY MOUSE
(Mus musculus)

Harold A. Hoffman

Comparative Pathology Section
Veterinary Resources Branch
Division of Research Services
National Institutes of Health
Bethesda, Maryland

The genetic background of laboratory animals can be a critical factor in the success of an experiment. There is no greater testimonial to this statement than our present knowledge of tissue transplantation acquired through the understanding of histocompatibility loci in inbred mice (1). One only has to review the historical development of our understanding of mammary carcinoma in mice to appreciate the role of genetically defined animals (2).

The National Institutes of Health recently has been designated by the World Health Organization as one of the three international genetic resource centers that will provide well-defined and characterized breeding pairs of laboratory animals to any institution involved in biomedical research. The Division of Research Services, which maintains over 100 inbred strains and substrains of mice, therefore, has an obligation to the scientific community to continue the propagation of laboratory animals by the standards of scientific integrity set forth by the distinguished group of scientists we are honoring at this conference.

The Division of Research Services recently established a genetic monitoring laboratory to ensure the genetic integrity of the inbred strains and substrains of mice produced by the Veterinary Resources Branch. The genetic monitoring laboratory provides quality assurance through routine monitoring of the inbred colonies of mice for gene definition. The monitoring service provides biochemical and immunologic information for studying and controlling the heredity of laboratory inbred mice. At the present time 33 specific genetic loci located on 13 of the 19 autosomal chromosomes of the mouse are used to define the genetic constitution (genetic profile) of each individual inbred strain and substrain. The genetic markers which constitute the genetic profile fall into three categories: morphological markers, biochemical markers, and immunological markers. Each inbred strain and substrain has a unique distribution of these marker genes which is used to differentiate one strain from another.

Another feature of the genetic monitoring program is the computer-assisted utilization of the genetic profiles in the form of a genetic data bank. The computer can be used to compare several inbred strains of mice for gene similarities and differences, to identify an inbred strain or group of inbred strains by a specific gene distribution, and to aid the research scientist in choosing the appropriate inbred strain for his research project.

Genetic Monitoring Program

Colony Management. The genetic monitoring program is primarily concerned with the genetic quality assurance of the various inbred strains of mice (Table 1). The program

TABLE 1. Genetic Monitoring Program

1. Colony management
2. Training and consultation
3. Genetic profiles
4. Genetic surveillance

is also involved in all areas of the breeding and maintenance of the colonies which affect the genetic integrity of each inbred strain. Colony management involves the proper execution of the mating system to avoid creating sublines within an inbred strain. In a brother by sister system of mating, the number of generations needed to trace back to a common ancestor should be kept to a minimum. The optimum situation occurs when the breeders at any generation trace back to a common ancestor in two generations. The production demands upon a colony regulate the number of breeders which in turn determines the number of generations back to the common ancestor. A second factor that plays a critical role in determining colony size is the reproductive performance of the inbred strain. The number of successful or fertile matings plus litter size influences the breeder count or size of the inbred colony. Good colony management is the first line of defense in protecting the genetic integrity of an inbred strain and is not substituted by sophisticated monitoring or surveillance of the inbred colony.

<u>Training and Consultation</u>. The training of biological laboratory technicians in the principles of simple Mendelian genetics, systems of mating, and genetic monitoring provides a technical staff with an overall understanding of the types of problems that can occur in an inbred animal colony. For example, by knowing the distribution of three coat color genes -- agouti, brown, and albino -- in the inbred strains,

the biological laboratory technician can detect mating errors and thus prevent the loss of an entire inbred line. If the biological laboratory technician understands that F_1 hybrids from an interstrain cross are all identical, then he can recognize problems of F_1-hybrid coat color segregation which could be caused either by a mating error or by a mutation at one of the coat color loci.

The genetic monitoring laboratory can assist the biomedical research scientist to choose the animal strain that best fits the needs of his research program. Attention can be directed to specific genetic markers which are unique to an inbred strain that could influence or interact with the characteristic under investigation.

Genetic Profiles. A genetic profile (Table 2) of an inbred strain is defined by three types of genetic markers: morphological traits, biochemical characteristics, and immunological markers. The location of the genes controlling these genetic markers on specific chromosomes (chromosome linkage map) and the computerization of this information (genetic data bank) are different forms of a genetic profile and offer different means of using the information.

The concept of a genetic profile is based on the fact that each inbred strain of mice has located on its chromosomes a unique set of genes which distinguishes it from

TABLE 2. Genetic Profiles

1. Genetic characteristics:
    Morphological
    Biochemical
    Immunological
2. Chromosome linkage map
3. Genetic data bank

other inbred strains of mice. The development and subsequent maintenance of an inbred strain by a strict system of inbreeding will produce a strain of mice which possesses identical, or nearly identical, sets of pairs of genes. Inbred strains of mice can be grouped into two classes according to their origin. The first category is called inbred strains of independent origin. Inbreeding was started in a group of mice which was not related to and independent of other known colonies of mice. The greatest number of genetic differences would be expected to exist between strains of mice of

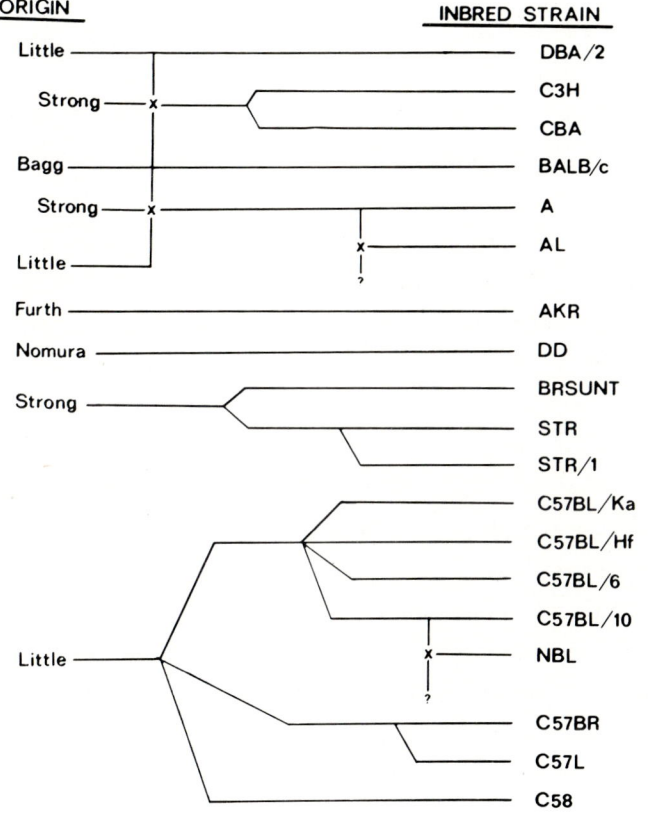

FIGURE 1. Origin of some of the inbred strains and substrains of mice maintained at the National Institutes of Health.

independent origin. The more widely used strains of independent origin are BALB/c, C57BL, and DBA/2 (Fig. 1).

The second category of inbred mice is called derived inbred, strains. Inbred strains in this category were started by first crossing two strains of mice, usually of independent origin, then by inbreeding genetic characteristics from both parental strains are fixed in the new line. Popular derived inbred strains in use today are A, CBA, and C3H (Fig. 1). Inbred strains which were derived from a non-inbred heterogeneous stock are classified as substrains. Gene similarities and differences between the sublines would depend upon the degree of heterozygosity in the original common stock and upon chance fixation of genes during the inbreeding process. The most widely used groups of inbred strains in this category are the C57BL sublines, C58 and C57BR (Fig. 1).

There are more than 320 mapped genes on the 20 pairs of chromosomes in the laboratory mouse (3). For the 19 autosomal pairs of chromosomes our goal is to have at least three genetic markers on each autosome; one locus near each end of the chromosome and one or more loci near the center of the chromosome. At the present time we are using 33 markers on 13 autosomes (Fig. 2 and Table 3). The 33 genetic markers that constitute a genetic profile of an inbred strain have been classified into four functional groups: 1) common morphological characteristics such as coat color and coat texture; 2) non-enzyme proteins found in the serum and red blood cells; 3) enzyme proteins which can be assayed in the liver, kidney, and red blood cell; and 4) immunologic markers such as cell surface antigens of thymocytes and lymphocytes.

These genetic markers represent both qualitative and quantitative gene expression. For example, the isozymes, realized by elecophoretic techniques, of the enzyme dipep-

TABLE 3. Thirty-three Loci Which Constitute the Genetic Profiles

Chromosome	Locus	Locus description
1	Id-1	Isocitrate dehydrogenase
1	Dip-1	Dipeptidase
2	Hc	Hemolytic complement
2	Cs-1	Catalase
2	a	Non-agouti
3	Car-2	Carbonic anhydrase
4	Lv	Delta-aminolevulinate dehydratase
4	Mup-1	Major urinary protein
4	b	Brown
4	Gpd-1	Glucose-6-phosphate dehydrogenase
5	Pgm-1	Phosphoglucomutase
5	Gus	Glucuronidase
6	Ly-2	Lymphocyte antigen
6	Ly-3	Lymphocyte antigen
6	wa-1	Waved
6	Ldr-1	Lactate dehydrogenase regulator
7	Gpi-1	Glucose phosphate isomerase
7	c	Albino
7	Hbb	Hemoglobin beta-chain
8	Es-1	Esterase
9	Thy-1	Thymus cell antigen
9	d	Dilute
9	Mod-1	Malic enzyme
9	Trf	Transferrin
11	Hba	Hemoglobin alpha-chain
11	Es-3	Esterase
14	Es-10	Esterase
14	s	Piebald
17	H-2K	Histocompatibility
17	H-2D	Histocompatibility
17	Tla	Thymus leukemia antigen
17	Ce-2	Catalase
19	Ly-1	Lymphocyte antigen

tidase (4) (Dip-1, chromosome 1) and kidney esterase (5) (Es-3, chromosome 11) are qualitative expressions of different alleles of structural genes at their respective loci. The electrophoretic isozymes of lactate dehydrogenase regu-

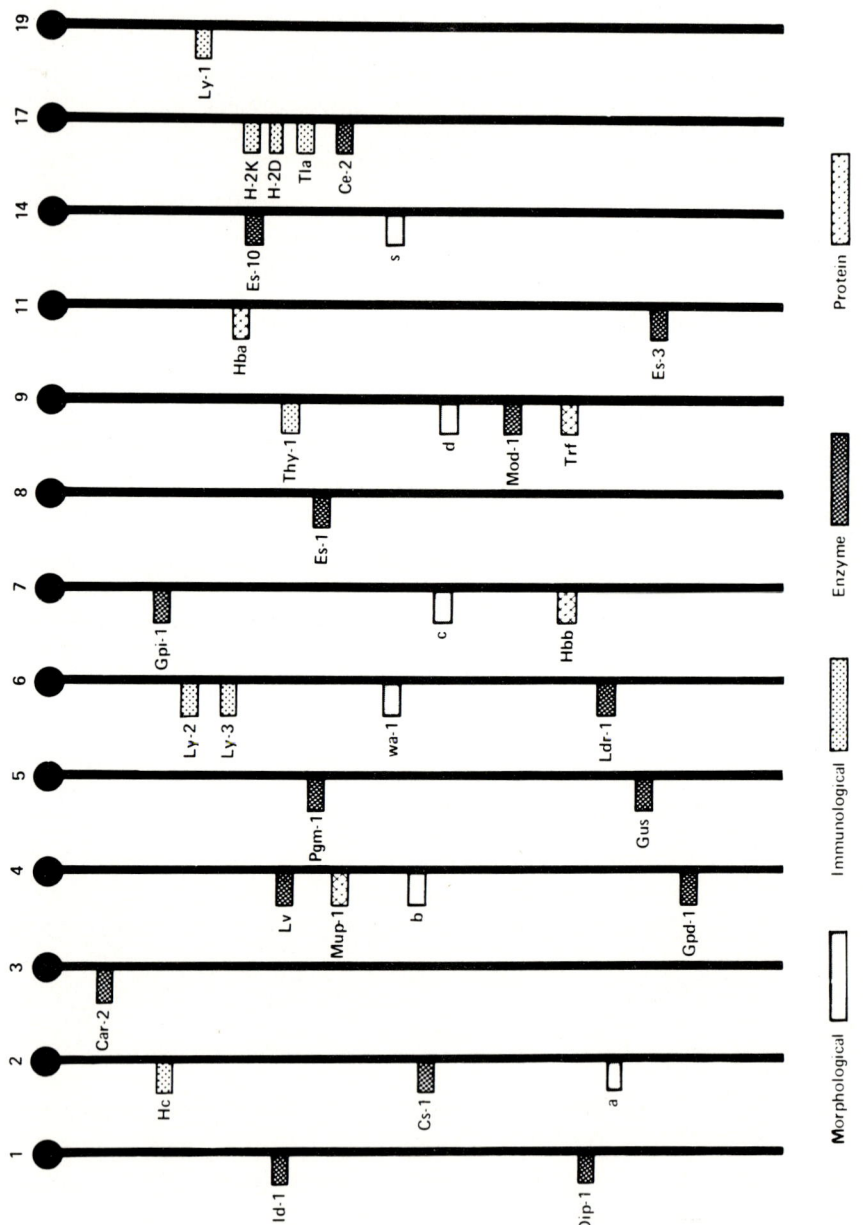

FIGURE 2. Linkage map of the laboratory mouse of genetic markers used for the genetic profile.

lator (6) (Ldr-1, chromosome 6) and kidney catalase (7) (Ce-2, chromosome 17) represent, respectively, a quantitative expression of a regulator gene, the amount of LDH b-protein present, and a post-translational modification of the enzyme catalase which alters its electrophoretic mobility. Additional examples of quantitative gene expression are erythrocyte catalase (8) (Cs-1, chromosome 2) and delta-aminolevulinate dehydratase (9) (Lv, chromosome 4) which are characterized by conventional spectrophotometric analyses measuring total enzyme activity.

Table 4 is a computerized representation of the classical linkage map (Fig. 2). The loci are listed according to the chromosome on which they are located and are listed in linear order starting from the centromere. The body of the table contains the different alleles carried by each inbred strain. The number of different loci and their order of presentation by the computer can be specified to conveniently compare the genetic profiles of two or more inbred strains. One can look at one locus for a specified inbred strain or look at the entire genotype for all the inbred strains listed in the data bank.

The computer can determine allelic similarities and differences at any number of specified loci between selected inbred strains. Table 5 presents a comparison made by the computer for strains C57BL/10, NBL, and DBA/2. The NBL strain -- prior designation C57BL/10-H-2^d or B10B -- was thought to originate as a coisogenic strain with C57BL/10 from a supposed mutation from H-2^b to H-2^d. Mitchison (10) in 1955 and Snell (11) in 1958 presented evidence based on tumor transplantation studies that NBL was not coisogenic with C57BL/10. Later in 1963, Finlayson (12) showed that the urinary protein (Mup-1, chromosome 4) of NBL was of the same type found in strains BALB/c and DBA/2 -- both H-2^d

TABLE 4. Computer Listing of the Strain Profiles for the 45 Inbred Strains of Mice Maintained at the National Institutes of Health

Genetic Profile
Strain Profile

Chromosome Locus	1 Id-1	1 Dip-1	2 Hc	2 Cs-1	2 a	3 Car-2	4 Lyb-2	4 Lv	4 Mup-1	4 b	4 Gpd-1
A/He	a	b	0	?[1]	a	b	b	a	a	b	b
AKR	b	b	0	?	a	a	c	a	a	B	b
AL	a	b	0	?	a	b	?	?	a	b	b
BALB/cAn	a	a	1	a	A	b	b	a	a	b	b
BDL	a	?	0	?	a	a	?	?	a	b	b
BRSUNT	b	b	0	?	a	a	?	?	a	b	b
CBA	b	b	1	?	A	b	b	a	a	B	b
CBA/Ca	b	b	1	?	A	a	b	?	a	B	b
CBA/T6	b	b	1	?	A	a	b	?	a	B	b
C3H/He	a	b	1	?	A	b	b	a	a	B	b
C3HfB/He	a	b	?	?	A	b	?	?	a	B	b
C3H/StWi	b	b	?	?	A	b	?	?	?	B	b

Chromosome Locus	5 Pgm-1	5 Gus	6 Lyt-2	6 Lyt-3	7 Gpi-1	7 c	7 Hbb	8 Es-1	9 Thy-1	9 Mod-1	9 Trf
A/He	a	a	b	b	a	c	d	b	b	a	b
AKR	a	h	a	a	a	c	d	b	a	b	b
AL	b	a	?	?	a	c	d	b	?	a	b
BALB/cAn	a	a	b	b	a	c	d	b	b	a	b
BDL	a	?	?	?	b	c	?	b	?	b	a
BRSUNT	b	a	?	?	b	C	d	b	?	b	b
CBA	b	h	a	b	b	C	d	b	b	b	a
CBA/Ca	b	b	?	?	b	C	d	b	?	b	a
CBA/T6	b	?	a	b	b	C	d	b	b	b	a
C3H/He	b	h	a	b	b	C	d	b	b	a	b
C3HfB/He	b	?	?	?	b	C	d	b	?	a	?
C3H/StWi	a	?	?	?	b	C	d	b	?	b	?

Chromosome Locus	11 Hba	11 Es-3	14 Es-10	17 H-2K	17 H-2D	17 Tla	17 Ce-2	19 Lyt-1
A/He	a	c	?	k	d	a	a	b
AKR	a	c	?	K	k	b	b	b
AL	?	c	a	?	?	?	a	?
BALB/cAn	b	a	a	d	d	c	a	b
BDL	?	?	a	?	?	?	?	?
BRSUNT	?	a	b	?	?	?	a	?
CBA	?	c	?	k	k	b	b	a
CBA/Ca	?	c	?	k	k	b	b	?
CBA/T6	?	c	?	k	k	b	b	a
C3H/He	c	c	b	k	k	b	b	a
C3HfB/He	?	c	b	?	?	?	b	?
C3H/StWi	?	c	b	?	?	?	b	?

TABLE 4 Continued

Chromosome Locus	1 Id-1	1 Dip-1	2 Hc	2 Cs-1	2 a	3 Car-2	4 Lyb-2	4 Lv	4 Mup-1	4 b	4 Gpd-1
C57BL/Hf	a	c	1	g	a	a	?	?	b	B	a
C57BL/Ka	a	a	?	?	a	?	?	?	b	B	a
C57BL/6	a	a	1	g	a	a	b	b	b	B	a
C57BL/10	a	a	1	g	a	a	?	c	b	B	a
C57L	b	a	1	?	a	b	a	a	b	b	a
C58	a	a	1	?	a	a	a	c	b	B	a
DBA/1	b	b	1	?	a	a	a	a	a	b	a
DBA/2	b	b	0	?	a	b	a	a	a	b	b
DD	a	c	0	?	A	?	?	?	a	?	b
FVB	a	b	1	?	?	b	?	?	x	?	?
GR	a	b	0	?	A	a	c	?	b	B	b
HSFR	a	b	1	?	A	?	?	?	a	B	b

Chromosome Locus	5 Pgm-1	5 Gus	6 Lyt-2	6 Lyt-3	7 Gpi-1	7 c	7 Hbb	8 Es-1	9 Thy-1	9 Mod-1	9 Trf
C57BL/Hf	a	?	?	?	b	C	s	a	?	b	b
C57BL/Ka	a	?	?	?	b	C	s	a	?	b	b
C57BL/6	a	b	b	b	b	C	s	a	b	b	b
C57BL/10	a	b	b	b	b	C	s	a	b	b	b
C57L	a	?	b	b	a	C	s	a	b	b	b
C58	a	a	a	a	a	C	s	b	b	b	b
DBA/1	b	b	a	b	a	C	d	b	b	a	b
DBA/2	b	b	a	b	a	C	d	b	b	a	b
DD	b	?	?	?	a	c	d	b	?	a	b
FVB	a	?	?	?	b	c	d	?	?	a	?
GR	b	?	a	b	b	c	s	b	b	a	?
HSFR	a	?	?	?	b	c	s	b	?	b	b

Chromosome Locus	11 Hba	11 Es-3	14 Es-10	17 H-2K	17 H-2D	17 Tla	17 Ce-2	19 Lyt-2
C57BL/Hf	?	a	a	b	b	?	a	?
C57BL/Ka	?	a	?	?	?	?	a	?
C57BL/6	a	a	a	b	b	b	a	b
C57BL/10	a	a	?	b	b	b	a	b
C57L	a	a	?	b	b	c	a	b
C58	b	c	?	k	k	a	a	b
DBA/1	?	c	?	q	q	b	?	a
DBA/2	a	c	b	d	d	c	a	a
DD	?	c	?	?	?	?	a	?
FVB	?	c	a	?	?	?	a	?
GR	?	b	a	d	x	c	a	b
HSFR	?	c	?	?	?	?	a	?

TABLE 4 Continued

Chromosome Locus	1 Id-1	1 Dip-1	2 Hc	2 Cs-1	2 a	3 Car-2	4 Lyb-2	4 Lv	4 Mup-1	4 b	4 Gpd-1
HSFS	a	b	?	?	A	?	a	?	a	B	b
I/St	a	b	0	?	a	?	a	?	a	b	b
LAN	a	a	1	?	A	?	?	?	a	?	b
NBL	b	b	0	h	a	a	?	?	a	B	c
NFR	a	b	1	?	A	b	?	?	a	B	b
NFS	a	b	1	?	A	?	?	?	a	B	b
NGP	b	b	0	?	a	b	?	?	x	?	a
NH	b	b	0	?	?	?	?	?	a	?	b
NZB	a	c	1	?	a	a	?	?	a	B	b
NZW	b	?	1	?	A	a	?	?	a	b	b
P	b	b	1	?	a	a	?	?	b	b	a
RIII	a	b	1	?	?	b	?	?	b	?	b

Chromosome Locus	5 Pgm-1	5 Gus	6 Lyt-2	6 Lyt-3	7 Gpi-1	7 c	7 Hbb	8 Es-1	9 Thy-1	9 Mod-1	9 Trf
HSFS	a	?	a	a	b	c	d	b	b	a	b
I/St	b	a	a	b	a	C	s	b	b	b	b
LAN	a	?	?	?	a	c	d	b	?	a	b
NBL	a	?	a	b	b	C	d	b	?	a	b
NFR	b	?	?	?	b	c	s	b	?	b	b
NFS	a	?	?	?	b	c	d	b	?	a	b
NGP	b	?	?	?	b	c	d	b	?	b	b
NH	b	?	?	?	a	C	d	b	?	b	b
NZB	b	b	b	b	a	C	d	b	b	b	b
NZW	b	?	?	?	a	c	d	b	?	a	b
P	b	a	?	?	a	C	d	b	?	b	b
RIII	a	?	?	?	a	c	s	b	?	b	b

Chromosome Locus	11 Hba	11 Es-3	14 Es-10	17 H-2K	17 H-2D	17 Tla	17 Ce-2	19 Lyt-1
HSFS	?	c	?	?	?	a	a	b
I/St	?	c	?	j	j	b	a	b
LAN	?	a	?	?	?	?	a	?
NBL	?	a	a	d	d	c	a	b
NFR	?	c	a	?	?	?	a	?
NFS	?	a	?	?	?	?	a	?
NGP	?	?	a	?	?	?	a	?
NH	?	a	?	?	?	?	a	?
NZB	?	?	?	d	d	a	b	b
NZW	?	?	?	?	?	?	?	?
P	c	a	?	p	p	?	a	?
RIII	?	c	?	r	r	?	a	b

TABLE 4 Continued

Chromosome Locus	1 Id-1	1 Dip-1	2 Hc	2 Cs-1	2 a	3 Car-2	4 Lyb-2	4 Lv	4 Mup-1	4 b	4 Gdp-1
SJL	b	b	1	?	A	b	c	a	a	?	b
SPM	a	a	1	?	?	?	?	?	b	B	b
ST/b	a	b	0	?	a	b	?	a	a	b	b
STAR	b	b	?	?	?	b	?	?	b	?	?
STR	b	b	1	?	?	?	?	?	a	b	?
STR/1	b	b	1	?	?	?	?	?	a	b	b
STS	a	b	?	?	?	b	?	?	b	?	b
WB/Re	a	b	0	?	a	?	?	?	a	B	a
129	a	b	1	?	A	a	b	c	a	?	a

Chromosome Locus	5 Pgm-1	5 Gus	6 Lyt-2	6 Lyt-3	7 Gpi-1	7 c	7 Hbb	8 Es-1	9 Thy-1	9 Mod-1	9 Trf
SJL	b	b	b	b	a	c	s	b	b	a	a
SPM	a	?	?	?	b	C	s	b	?	b	b
St/b	a	a	b	b	a	c	d	b	b	b	b
STAR	a	?	?	?	b	c	s	x	?	a	?
STR	b	?	?	?	b	C	s	b	?	b	a
STR/1	b	?	?	?	b	C	?	b	?	b	?
STS	b	?	b	b	b	c	d	b	b	b	?
WB/Re	b	?	b	b	b	C	d	b	b	b	b
129	a	a	b	b	a	?	d	b	b	a	?

Chromosome Locus	11 Hba	11 Es-3	14 Es-10	17 H-2K	17 H-2D	17 Tla	17 Ce-2	19 Lyt-1
SJL	c	c	b	s	s	a	a	b
SPM	?	c	?	?	?	?	a	?
ST/b	c	b	?	k	k	b	b	b
STAR	?	a	b	?	?	?	a	?
STR	?	b	?	?	?	?	a	?
STR/1	?	a	?	?	?	?	a	?
STS	c	c	a	?	?	c	a	b
WB/Re	?	a	a	j	d	c	a	b
129	d	c	b	b	c	c	a	b

[1]Allele not determined for National Institutes of Health subline.

strains -- and different from that found in C57BL/10. These facts prompted the name change from C57BL/10-H-2^d to NBL and placed NBL in the category of strains of uncertain origin (Fig. 2). In Table 5 C57BL/10 was specified as the reference strain, and NBL and DBA/2 are compared with C57BL/10 for gene similarities and differences. When a strain has the same allele at a locus as the reference strain, the computer does not print its gene symbol; if the strains differ at a locus, then the computer prints the gene symbol to indicate the difference. For the 19 loci used in the comparison, which represents 11 chromosomes, C57BL/10 and DBA/2 differ by 17 loci, C57BL/10 and NBL differ by 12 loci, and DBA/2 and NBL differ by six loci. Although BALB/c and NBL only differ by seven of the 19 loci used for comparison, BALB/c can be eliminated as a possible outcross strain on the basis of four loci. NBL and DBA/2 have the same alleles at Id-1 and Dip-1 (chromosome 1), Hc (chromosome 2), and the Ly-2,3 complex (chromosome 6), whereas BALB/c and C57BL/10 have the alternate genes at these loci. The strain comparison supports the conclusions from the transplantation studies that NBL was the result of a C57BL/10 outcross and also indicates that DBA/2 is the most likely candidate as the outcross strain.

Another way to utilize the genetic data bank is to have the computer search the strain profiles for inbred strains with a specific gene or set of genes. In this way, possible linkage associations can be detected for an unlinked gene with a gene located on a specific chromosome. For example, after classifying the inbred strains for kidney catalase (Ce-2) and entering the data into the data bank, a search was made for the strains carrying the allele Ce-2^b. A computer listing (Table 6) indicated that seven inbred strains were classified as Ce-2^b. Strain profiles for the seven strains showed that H-$2K^k$ (chromosome 17) was present in all seven

TABLE 5. Computer Listing of Gene Similarities and Differences for Inbred Strains C57BL/10, NBL, and DBA/2

Genetic Profile
Strain Differences

	C57BL/10	NBL	DBA/2
Id-1	a	b	b
Dip-1	a	b	b
Hc	1	0	0
Mup-1	b	a	a
b	B		b
Gpd-1	a	c	b
Pgm-1	a		b
Ly-2	b	a	a
Ly-3	b		
Gpi-1	b		a
Hbb	s	d	d
Es-1	a	b	b
Mod-1	b	a	a
Es-3	a		c
H-2K	b	d	d
H-2D	b	d	d
Tla	b	c	c
Ce-2	a		
Ly-1	b		a

strains. A search for the strains carrying $H\text{-}2K^k$ listed nine inbred strains with this allele (Table 7). Seven of the nine strains are $Ce\text{-}2^b$, suggesting that Ce-2 is located on chromosome 17. A linkage analysis of first backcross mice confirmed the linkage of Ce-2 and H-2K and placed Ce-2 two map units from H-2K.

TABLE 6. Computer Listing of a Gene Search for Kidney Catalase Gene $Ce\text{-}2^b$

Genetic Profile
Gene Search
There are 7 strains with allele b at locus Ce-2

| AKR | CBA | CBA/Ca | CBA/T6 | C3H/He | C3HfB/He | ST/b |

TABLE 7. Computer Listing of a Gene Search for Histocompatibility Gene $H\text{-}2K^k$

Genetic Profile
Gene Search
There are 9 strains with allele k at locus H-2K

| AKR | CBA | CBA/Ca | CBA/T6 | C3H/He | C3HfB/He | C57BR/cd |
| C58 | ST/b | | | | | |

It is noteworthy that in light of this close linkage that the exceptions to the $Ce\text{-}2^b$ and $H\text{-}2K^k$ association are C58 and C57BR. These two strains have a common origin (Fig. 1) and are genetically independent of the other $H\text{-}2^k$ inbred strains. This could indicate a possible subtle difference between $H\text{-}2^k$ haplotypes from different gene pools.

Genetic Surveillance. Progeny from foundation colony breeders which contribute to the main line of descent of an inbred strain are monitored for their conformity with their strain profile. Four to ten chromosomes are routinely examined every generation, and every third generation a complete genetic profile is performed on each inbred strain. Tissues from the breeding pairs destined to become the common ancestors are frozen and stored in liquid nitrogen. In this way genetic profiles from mice of the current generation of inbreeding can be compared with their ancestors several generations removed.

CONCLUSION

Considerable time and effort have been invested in the genetic monitoring program because we believe that these profiles will vastly improve the validity and reproducibility of research investigations. Not only can we control the genetic integrity of the inbred strains of mice through our genetic profile system, but we can selectively breed mice that have a specific set of biochemical and immunologic characteristics. The genetic profiles have been computerized for ready availability and we are encouraging the development of new animal models. For specific research projects, anyone desiring to attempt such a program is encouraged to consult with the NIH genetic monitoring laboratory for collaborative work along these lines.

REFERENCES

1. Klein, J. (1975). "Biology of the Mouse Histocompatibility-2 Complex." Springer-Verlag, New York.
2. Heston, W. (1974). J. Hered. 65:262
3. Mouse News Letter. (1977). No. 57, page 6.
4. Lewis, W., and Truslove, G. (1969). Biochem. Genet. 3:493.
5. Ruddle, F., and Roderick, T. (1965). Genetics 51:445.
6. Hutton, J., and Roderick, T. (1970). Biochem. Genet. 4:339.
7. Hoffman, H., and Grieshaber, C. (1976). Biochem. Genet. 14:59.
8. Feinstein, R., Howard, J., Braun, J., and Seaholm, J. (1967). Genetics 56:559.

9. Russell, R., and Coleman, D. (1963). Genetics 48:1033.
10. Mitchison, N. (1955). J. Exp. Med. 102:157.
11. Snell, G. (1958). J. Natl. Cancer Inst. 21:843.
12. Finlayson, J., Potter, M., and Runner, C.R. (1963). J. Natl. Cancer Inst. 31:91.

BIOCHEMICAL POLYMORPHISMS -- DETECTION,
DISTRIBUTION, CHROMOSOMAL LOCATION, AND APPLICATIONS

John J. Hutton

Department of Medicine

VA Hospital and University of Kentucky

Lexington, Kentucky

The application of starch gel electrophoresis and specific staining procedures has led to the detection of genetically controlled structural variation in enzymes from different inbred strains of mice. Polymorphism at loci controlling the structure of proteins is extensive so that it is frequently easy to find a variant of a specific enzyme. The average mouse in the wild is heterozygous at 10% of loci and the percentage of polymorphic loci in the species is about 40% (63). Genetic variability among independently derived inbred strains of laboratory mice seems similar to variability observed in the wild. Electrophoretic techniques alone cannot detect all variants, since some amino acid differences resulting from mutation do not cause a change in the electrical charge on the protein (55). Other techniques such as comparative measurements of thermal lability of enzymatic activities in different strains (14, 30, 40, 45), reactivity of molecules with alkylating agents (34), solubility of proteins in phosphate buffers (68), enzymatic activity with different substrates (43, 82), activity per unit weight of tissue (1, 14, 30, 35, 36, 44, 50, 55, 56, 81), and distribution of activity among subcellular organelles (1, 44, 56) have been helpful in finding "variants." It is difficult to prove that the variation is caused by changes in the primary amino acid sequence of the protein and not caused by post-translational modifications such as glycosylation (2, 17), or by changes in binding sites for the protein in a membrane of a subcellular organelle (44), changes in the concentration of an intracellular stabilizing factor (23, 35), or mutation in a regulatory element affecting the rate of synthesis of the protein without affecting structure (55, 56). Only in the case of hemoglobin have electrophoretic and solubility variants been proved to result from genetic mutation in structural genes with changes in the amino acid sequences of the α- and β-chains (28, 60, 68).

Biochemical variants constitute a distinct class of mutations with internationally recognized rules for choosing appropriate names and symbols of new loci affecting enzymes and other biochemical characteristics of mice (13). Rules for identifying chromosomes and their associated genetic linkage groups have also been defined (12, 48).

Table 1 lists loci which probably determine the structure of proteins and have been assigned to chromosomes. Prior to 1972 genes that segregated together in breeding experiments were said to be members of the same "linkage group." Linkage groups were identified by number and most could not be assigned to specific chromosomes. Because of recent advances in cytogenetics, genetic linkage groups have now been assigned to chromosomes and chromosome numbers have replaced linkage groups. Loci in Table 1 are listed by chromosomal location. Methods and tissues generally used to detect variants at each locus are given. The list of references is lengthy, but not exhaustive,

TABLE 1. Chromosomal Assignments of Genes That Determine Protein Structure

Chromosome (linkage group)	Locus	Protein	Detection	Reference
1 (XIII)	Dip-1	dipeptidase	E (K)	6,8,43,52,74
	Id-1	isocitrate dehydrogenase	E (K)	6,31,37,49,52,74
2 (V)	Hao-1	α-hydroxyacid oxidase	E (L)	6,18
	Svp-1	seminal vesicle protein	E	6,59
3	Car-1	carbonic anhydrase	E (B)	6,19,74
	Car-2	carbonic anhydrase	E (B)	6,19,74
4 (VIII)	Gpd-1	hexose-6-P dehydrogenase	E (K)	3,6,36,37,66,74
	Mup-1	urinary protein	E	6,25,74
	Pgd-1	6-P-gluconate dehydrogenase	E (K)	3,6,52
	Pgm-2	phosphoglucomutase	E (B,K,L)	6,8,52,72,74
5 (XVII)	Alb-1	albumin	E (B)	6,52,54
	Gus	β-glucuronidase	E,T (K,L)	6,55,56,74
	Map-1	α-mannosidase	E (L)	6,17
	Mor-1	mitochondrial malate dehydrogenase	E (K,L)	6,52,70,80
	Pgm-1	phosphoglucomutase	E (B,K,L)	6,8,37,52,72,74
6 (XI)	Ldr-1	lactic dehydrogenase	E (B)	6,37,52,74
7 (I)	Es-8	esterase	E (B)	5,6,52,74
	Gpi-1	glucose-P isomerase	E (B,K,L)	4,6,10,15,26,27,34,37,52,74
	Hbb	hemoglobin β-chain	E (B)	6,28,34,60,68,69,74
	Hby	hemoglobin embryonic chain	E (B)	6,68,75
	Mod-2	mitochondrial malic enzyme	E (K,L)	6,8,24,52,70,74
8 (XVIII)	Es-1	esterase	E (B,K,L)	6,52,74,77,78,82
	Es-2	esterase	E (B,K,L)	6,52,57,67,74,77,78,82
	Es-5	esterase	E (B)	6,57,74,77
	Es-6	esterase	E (K)	58,74,77,78,82
	Es-7	esterase	E (B)	6
	Es-9	esterase	E (K)	6
	Es-11	esterase	E (K,L)	6,57
	Got-2	mitochondrial glutamic oxaloacetic transaminase	E (K,L)	6,16,52,74,77
	Gr-1	glutathione reductase	E (K)	6,53,74
	Prt-2	pancreatic proteinase	E	6,77
9 (II)	Bge	β-galactosidase	E (K,L)	2,6
	Lap-1	leucine arylaminopeptidase	E	6,81,74
	Mod-1	soluble malic enzyme	E (K,L)	6,8,32,36,37,52,70,71,74
	Mpi-1	mannose-P isomerase	E (K,L)	6,51,52,74
	Trf	transferrin	E (B)	6,74
10 (X)	Apk	acid phosphatase	E (K)	6
11 (VII)	Es-3	esterase	E (B,K)	6,52,64,65,74,82
	Hba	hemoglobin α-chain	S (B)	6,34,68,69,74
12 (XVI)	Amy-1	salivary amylase	E	6,38,73
	Amy-2	pancreatic amylase	E	6,38,73
14 (III)	Es-10	esterase	E (B,K)	6,74,79
	Np-1	nucleoside phosphorylase	E (B,K,L)	6,52,79
15 (VI)	Gdc-1	α-glycerol-P dehydrogenase	E,T (K)	6,40,74
	Gpt-1	glutamic pyruvate transaminase	E (B,K,L)	6,11,20,74
17 (IX)	Glo-1	glyoxalase	E (B)	47
	Map-2	α-mannosidase	E (L)	6,17
	Pgk-2	testicular P-glycerate kinase	E,T	6,76
19 (XII)	Got-1	soluble glutamic oxaloacetic transaminase	E (K,L)	6,7,16,52
X (XX)	Ags	α-galactosidase	T (K,L)	6,45
	Pgk-1	P-glycerate kinase	E (B,K,L)	6,41,52
	Phk	muscle phosphorylase kinase	T	6,30,33

Methods of detecting variants are abbreviated: E = electrophoresis, T = thermal inactivation, S = solubility of protein in phosphate buffer. Tissues commonly used to prepare extracts for detecting variants are indicated in parentheses in the column labeled "Detection." B = blood, K = kidney, L = liver.

and makes no attempt to assign priority of discovery. References describe the variants, their distribution among strains, segregation with other markers in breeding experiments, and applications to solution of several fundamental biological problems.

The distribution of alleles at 28 loci on 17 chromosomes is summarized in Table 2. These data can be collated from original articles listed in Table 1 or, with minor additions, from recent reviews (6, 74). All strains are inbred except M. m. molossinus. Accumulating representative animals for typing is laborious and the distribution of alleles at loci such as Svp-1, Gdc-1, Glo-1, and Phk is not well described. In the case of Phk the b allele is associated with muscle disease so standard inbred strains except I/LnJ are probably normal with the a allele. Each of the 20 strains has a unique pattern of alleles. It is easy to see that biochemical typing of strains can serve both to detect genetic contamination in production colonies (42) and to estimate whether strains are genetically related. Alleles such as $Pgm-2^b$, Gus^h, $Map-1^h$, Hbb^p, $Mod-2^a$, $Es-1^a$, $Got-2^a$, Apk^m, $Es-3^a$, $Es-3^b$, Hba^d, $Amy-1^b$, $Es-10^c$, and $Gpt-1^c$ are only found in one or two of the strains. Variant alleles at Got-1 and Pgk-1 are present in partially inbred mice recently derived from wild populations (6, 7).

In searching for a new variant it is important to examine as many inbred strains as possible, since rare alleles may not be found immediately. A group of inbred strains with no known overlapping pedigrees should be studied first. One "best group" that has been recommended for initial screening includes C57BL/6J, SM/J, SWR/J, AU/SsJ, and PL/J (63). To these should be added strains recently derived from wild mice or substrains of M. musculus such as M. m. molossinus and castaneus.

The positions of loci on chromosomes are illustrated in Figure 1. This map should be compared to the human gene map

TABLE 2. Distribution of Alleles at 28 Loci on 17 Chromosomes

Chromosome	1	1	2	3	4	4	5	5	5	6	7	7	7	8	8	9	9	10	11	11	12	14	15	15	17	19	X	X
Locus	Dip-1	Id-1	Svp-1	Car-1	Gpd-1	Pgm-2	Gus	Map-1	Pgm-1	Ldr-1	Gpi-1	Hbb	Mod-2	Es-1	Got-2	Bgе	Mod-1	Apk	Es-3	Hba	Amy-1	Es-10	Gdc-1	Gpi-1	Glo-1	Got-1	Pgk-1	Phk
A/HeJ	b	a	–	a	b	a	a	b	a	a	a	d	a	b	b	a	a	a	c	a	a	a	b	a	–	a	b	–
AKR/J	b	a	–	a	b	a	b	b	a	a	a	d	–	b	b	a	b	a	c	a	a	b	b	b	a	a	b	–
AU/ssJ	b	b	a	a	b	a	b	b	b	a	b	b	b	b	b	a	b	a	c	e	a	b	a	a	–	–	b	–
BALB/cJ	a	a	b	a	b	a	a	a	b	a	b	d	b	b	b	a	b	a	d	d	a	a	c	a	a	a	b	–
C3H/HeJ	b	a	a	a	b	a	b	b	b	a	b	d	b	b	a	a	b	a	c	c	a	b	b	a	–	a	b	–
C57BL/6J	a	a	b	a	b	a	b	a	b	–	a	d	b	b	b	b	a	a	a	a	a	a	a	a	a	a	b	a
C57L/J	–	–	–	a	a	–	a	b	a	a	a	d	–	a	–	a	–	a	–	a	–	a	–	a	–	–	a	–
DBA/2J	b	b	–	–	b	–	b	b	b	b	b	d	b	b	b	b	a	a	c	c	a	b	b	a	a	a	b	b
DE/J	b	a	–	a	b	a	a	b	a	b	–	d	–	b	–	b	b	a	c	e	–	–	–	a	–	a	b	–
I/LnJ	b	b	–	a	b	a	b	b	b	b	a	d	b	b	b	b	b	a	c	–	b	b	–	b	–	a	b	b
IS/Cam	–	a	–	a	b	a	–	–	–	–	–	d	–	a	a	–	a	m	c	e	–	a	–	–	–	–	–	–
LP/J	a	a	–	a	b	a	b	b	b	a	a	d	b	b	b	b	a	a	c	a	a	b	b	a	a	a	b	b
MA/J	b	a	–	a	b	a	–	–	–	a	–	s	–	b	b	–	b	a	c	–	a	a	–	a	b	–	b	–
Molossinus	b	–	–	a	b	b	b	–	b	a	b	s	b	b	b	b	b	a	c	b	b	c	–	a	–	a	b	–
PL/J	b	b	–	a	b	a	b	b	b	b	a	d	b	b	b	b	b	a	c	e	a	b	–	a	–	a	b	–
RIII/2J	b	a	–	a	b	a	b	b	b	a	a	s	–	b	b	b	b	a	b	–	–	b	b	–	b	–	b	–
RF/J	b	b	–	a	b	b	b	b	b	a	b	d	b	b	b	b	a	a	c	a	a	b	b	a	–	a	b	–
SM/J	b	a	–	a	b	a	–	–	–	b	a	s	–	b	b	b	b	a	c	–	–	b	–	a	–	–	b	–
ST/bJ	b	b	–	a	b	a	b	–	a	–	a	d	a	b	b	b	b	a	c	c	–	b	–	a	–	a	b	–
SWR/J	b	a	–	a	b	a	b	b	b	b	b	s	b	b	a	b	b	a	c	c	a	a	b	a	–	a	b	–

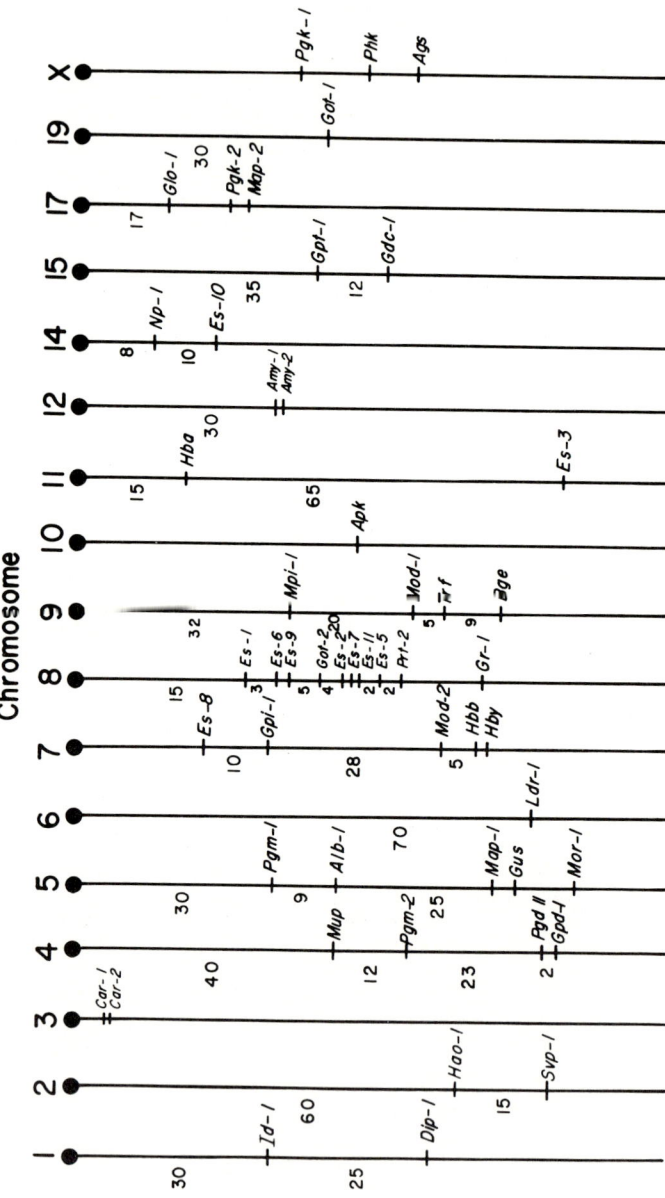

FIGURE 1. Chromosomal location of genes specifying the structure of proteins. Symbols are defined in Table 1 and distribution of alleles among strains is given in Table 2. Genetic variants have been described at each locus and assignment to chromosomes is based upon demonstration of genetic linkage in breeding experiments. There are no convenient biochemical markers on chromosomes 13, 16, and 18.

(46). All chromosomes except 13, 16, and 18 have at least one polymorphic biochemical locus. Additional loci have been assigned to chromosomes by somatic cell hybridization or studies of gene dosage effects associated with the X chromosome. These include glucose-6-phosphate dehydrogenase (9, 21, 41), phosphoglycerate kinase (9, 41), and hypoxanthine guanine phosphoribosyl transferase (9, 22) on the X chromosome, and thymidine kinase and galactokinase on chromosome 11 (9, 39). Centromeres are located at the top of Figure 1 and genetic distances in cM are given where estimates are available. If nongenic DNA is fairly evenly distributed among chromosomes, then the average length of a mouse chromosome should be 60-90 cM (48). This means that markers at two loci spaced around 50 cM apart with one 10 to 30 cM from either end of the chromosome would serve in breeding experiments to detect whether a segregating gene was on that chromosome. These conditions are met reasonably well by markers on chromosomes 1, 4, 5, 7, 8, 9, and 11. Other chromosomes are less well marked. It is remarkably easy to design crosses in which alleles at 10 or more biochemical loci are segregating, thereby testing genetic linkages on multiple chromosomes and in some cases meeting conditions for a 3-point cross to determine the sequential order of loci (5, 37, 79). Polymorphic biochemical variants have many practical advantages when used as markers in linkage analysis. Generally, alleles determining biochemical polymorphisms are codominant and fully penetrant. None has deleterious effects on viability so that segregation ratios are not distorted in breeding experiments. Unfortunately, animals must sometimes be killed to determine their genotype and many of the testing procedures are laborious and expensive. The marker to be tested for linkage may not be found in an inbred strain which can be paired with another to achieve segregation of large numbers of markers.

Four examples of tight genetic linkage of enzymes with similar functions are shown in Figure 1. Each of these may represent the tandem duplication of one or more loci. The structural loci of carbonic anhydrase (Car-1 and Car-2 on chromosome 3) are within 1.5 cM of one another (19). The two isozymes of carbonic anhydrase in the mouse are structurally related to the two human isozymes which share 60% sequence homology. The Amy-1 and Amy-2 loci on chromosome 12 may represent a similar duplication, but there is no information about the amino acid sequence of the two proteins. On chromosome 7 there is the interesting example of very tight linkage of the locus controlling the structure of the y-chain of mouse embryonic hemoglobin with the locus controlling the structure of adult hemoglobin β-chain. The y- and β-chains of hemoglobin are structurally similar (28, 68), so the Hbb and Hhy loci probably represent tandem duplication of a structural gene. The genes at the two loci have evolved so that one functions in the embryo and the other in the adult. The most extensive area of duplication of structural genes is on chromosome 8. Seven esterases (Es-1, Es-2, Es-5, Es-6, Es-7, Es-9, and Es-11) are clustered within a 12 cM segment (57, 78). Three of these loci (Es-2, Es-5, and Es-11) are within 0.58 cM of one another and probably represent tandem duplication of a single ancestral gene. A similar cluster of four esterase loci is present in the rat, suggesting that an autosomal segment comprising at least 15 cM of the rat and mouse genomes has remained relatively intact with respect to genetic content during rodent speciation (82).

Both chromosomal and mitochondrial DNA are present in animal cells and it is not clear which mitochondrial proteins are coded by nuclear as opposed to mitochondrial genes. Through studies of the distribution of electrophoretically distinct isozymes among subcellular fractions from mice of different inbred strains, structural loci of mitochondrial and soluble forms have been identified and assigned to chromosomes. Figure 2 illustrates variants of mitochondrial glutamate oxaloacetate transaminase. The slow phenotype in the figure is designated GOT-2B, the fast GOT-2A, and the heterozygote GOT-2AB (16). GOT-2AB animals possess three distinct bands with the intermediate AB band having stronger histochemical staining activity than either the A or B band alone. The approximate 1A:2AB:1B

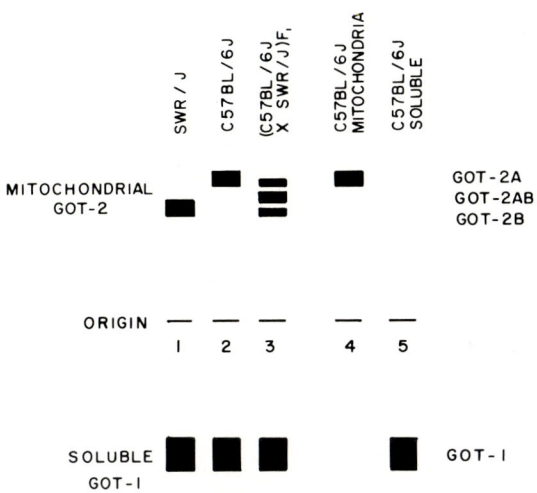

FIGURE 2. Electrophoretic forms of glutamate oxaloacetate transaminase (16). Electrophoretic variants of the mitochondrial isozyme are illustrated. These are apparent when enzymes from inbred strains SWR/J and C57BL/6J are compared. The mutation affecting the electrophoretic mobility of GOT-2 does not affect GOT-1, so the mitochondrial and soluble forms of GOT are structurally different proteins controlled by different genetic loci.

heterozygous phenotype can best be explained if the A and B subunits hybridize at random as dimers (16). Mutant genes for at least four mitochondrial enzymes have been described in the mouse. Mitochondrial glutamate oxaloacetate transaminase is coded by a nuclear locus (Got-2) on chromosome 8, mitochondrial malic enzyme (Mod-2) is on chromosome 7, mitochondrial malate dehydrogenase (Mor-1) is on chromosome 5, and mitochondrial proline dehydrogenase (Pro-1) is on an autosome and is not linked to the loci of the other mitochondrial enzymes. There is no apparent pattern of association of these four nuclear genes coding for mitochondrial enzymes (70, 80).

Variants of isozymes provide powerful tools for analysis of the genetic effects of radiation and chemical mutagens on gametes. It is possible to design experiments measuring mutations affecting the amino acid sequence of specific proteins as well as changes in gene dosage brought about by duplication or deletion of segments of chromosomes. To do this it is necessary to treat a mouse with mutagen and then to mate it to a mouse of another inbred strain which differs in genotype from the treated animal at as many loci as possible. Offspring should be heterozygous at multiple known loci and this heterozygosity can be detected by techniques listed in Table 1. Genetic deletion at a locus is indicated by recovery of only one parental allele, that of the untreated parent. Structural change in an enzyme because of genetic mutation may be detected by changes in electrophoretic pattern or some other property of specific proteins in the hybrid offspring. Duplication of a locus may cause changes in the relative activity of isozymes in the hybrid. The value of testing mice heterozygous at multiple biochemical loci is that genetic deletion and duplication can be detected as well as simple changes in amino acid sequence. Searches for specific biochemical variants can be facilitated by use of chromosomal inversions to recover specific segments of chromosomes of treated animals (62). The map of biochemical loci (Fig. 1) assigns positions of genes and can be used to choose the appropriate inversion.

Using biochemical techniques and screening of heterozygous F_1 mice, new radiation-induced mutations were sought at the hemoglobin Hba and Hbb loci (69). Five hemoglobin variants were found among 8621 F_1 progeny of irradiated animals. In three offspring the genetic contribution from the irradiated father was not expressed with regard to the α-chain. These mice are thought to have deletion of genes controlling the structure of the α-chain and may represent mouse alpha-thalassemics. One of the progeny carried a tandem duplication involving Hbb and one probably carried a double nondisjunction of chromosome 7 (Hbb and Hby). The finding that major chromosomal aberrations can mimic hemoglobin mutations indicates the need to follow F_1 screening with thorough cytogenetic analysis (69). Clearly, the availability of maps assigning specific biochemical loci to specific locations on chromosomes makes possible the interpretation of combined biochemical and cytogenetic studies.

Figure 3 illustrates the use of biochemical, coat color, and neurological mutants in the analysis of radiation-induced lethal albino alleles (24, 29). The c^{14Cos} and c^{6H} mutations were induced by radiation and both are lethal when homozygous. Homozygotes of c^{14Cos} become hypoglycemic and die perinatally, whereas homozygous c^{6H} is an embryonic lethal. Heterozygous c^{14Cos}/c^{6H} mice are albino but do not die so c^{14Cos} and c^{6H} are assigned to different complementation groups and are thought to represent different, but overlapping deletions. The Mod-2 locus of mitochondrial malic enzyme is on chromosome 7 one cM from the albino locus (c) between c and sh-1. The MOD-2 phenotypes of normal, $c^{14Cos}/+$, $c^{6H}/+$, and c^{14Cos}/c^{6H} mice

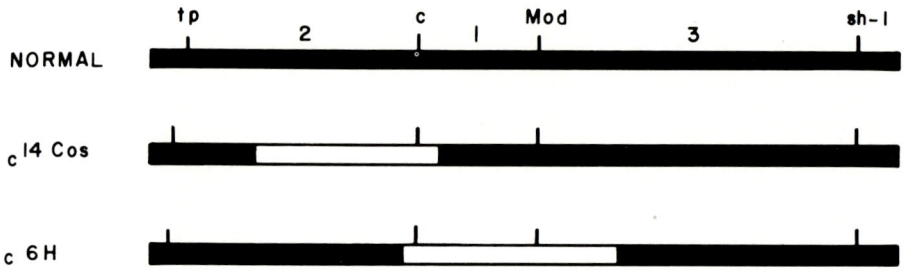

FIGURE 3. Genetic maps of a segment of mouse chromosome 7. The alleles c^{14Cos} and c^{6H} are radiation-induced albino mutations which are lethal when homozygous. It has been proposed that these represent genetic deletions of the extent indicated (24, 29).

were determined. The Mod-2 allele of c^{6H} was not expressed in the hybrids, whereas the allele of c^{14Cos} was expressed. The c^{6H} deletion appears to extend at least one cM from the c locus and involves the structural gene of a known enzyme. The availability of a genetic map (Fig. 1), tables of distributions of biochemical alleles (Table 2), and techniques for distinguishing homozygosity from heterozygosity (Table 1) at loci greatly facilitates analysis of complex mutations.

Genetic variants of polymeric isozymes have been used to analyze a variety of interesting problems in developmental biology. Figure 4 illustrates the use of variants of glucose phosphate isomerase (Gpi-1). Homozygous $Gpi-1^b$ mice are composed of cells of identical genotype (Fig. 4, cell type 1). Homogenates of these cells give a single banded electrophoretic pattern of GPI-1B activity (Fig. 4, electrophoretic pattern 1). Homozygous $Gpi-1^a$ mice are composed of cells of identical genotype (Fig. 4, cell type 2). Homogenates of these cells give a

JOHN J. HUTTON

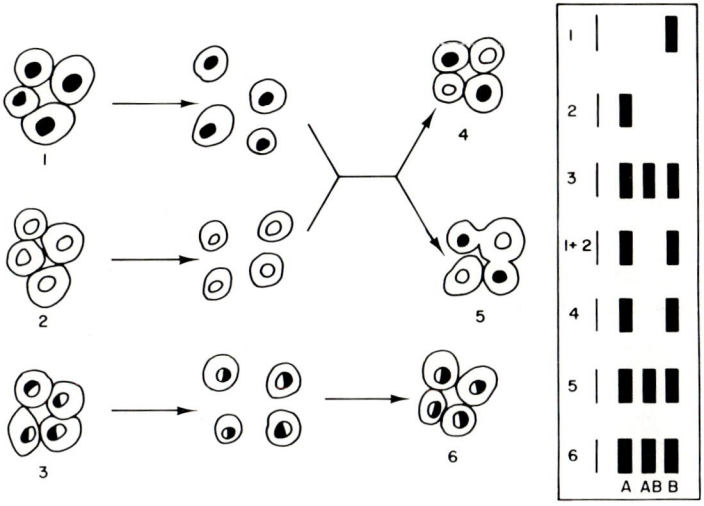

FIGURE 4. Electrophoretic patterns of glucose phosphate isomerase (GPI-1) from cells and mixtures of cells of differing genotype at Gpi-1. Cells of type 1 are homozygous Gpi-1^b/Gpi-1^b and give electrophoretic phenotype GPI-1B. Cells of type 2 and 3 are of genotypes Gpi-1^a/Gpi-1^a and Gpi-1^a/Gpi-1^b, respectively, and give electrophoretic phenotypes GPI-1A and GPI-1AB, respectively. The GPI-1AB phenotype is characterized by the presence of the heteropolymeric AB enzyme. Mixtures of homozygous cells of types 1 and 2 or mixtures of homogenates from these cells have electrophoretic phenotype 1 plus 2 (GPI-1B plus GPI-1A, not GPI-1AB). If cells of type 1 and 2 are mixed in chimeric embryos or in tissue culture, electrophoretic pattern 4 (GPI-1A plus GPI-1B) will result if cell fusion does not occur (cell type 4), but electrophoretic pattern 5 (GPI-1AB) will result if cell fusion does occur (cell type 5).

single banded electrophoretic pattern of GPI-1A activity (Fig. 4, electrophoretic pattern 2). Heterozygous Gpi-1^a/Gpi-1^b mice express both the Gpi-1^a and Gpi-1^b alleles in all cells and since glucose phosphate isomerase is a dimer, the heteropolymer GPI-1AB comprises the major electrophoretic component (Fig. 4, electrophoretic pattern 3). Bands of GPI-1A and GPI-1B activity are also present. Mixtures of homogenates from homozygous $Gpi-1^a$ mice with homogenates from $Gpi-1^b$ homozygotes give rise to two electrophoretic bands (GPI-1A and GPI-1B, Fig. 4, electrophoretic pattern 1 plus 2) without formation of heteropolymer. Formation of heteropolymer requires that the $Gpi-1^a$ and $Gpi-1^b$ alleles both be active within the same cytoplasm. If somatic cells of genotype $Gpi-1^a/Gpi-1^a$ and $Gpi-1^b/Gpi-1^b$ are mixed, two results can occur. The cells can form mixtures without cell fusion. Homogenates of mixtures of cells will consist of GPI-1A and GPI-1B enzyme without heteropolymer (Fig. 4, cell type 4, electrophoretic pattern 4). Alternatively, the cells can fuse to form heterokaryons. Both types of Gpi-1 alleles are active within the same cytoplasm. Homogenates of heterokaryons will consist of GPI-1A, GPI-1AB, and GPI-1B enzyme (Fig. 4, cell type 5, electrophoretic pattern 5).

Whether cell fusion occurs has been examined in several mouse tissues using the principles illustrated in Figure 4 (4, 27). Trophoblast differentiation in the mouse embryo is characterized by the formation of giant cells whose nuclei contain as much as 500-1,000 times the haploid amount of DNA. Among the mechanisms proposed are endomitosis, fusion of fetal cells, or fusion of maternal and fetal cells. GPI-1A embryos were transferred to GPI-1B mothers, and chimeric GPI-1A ⟷ GPI-1B embryos were also made. Fetal tissues were examined for the presence of heteropolymer, GPI-1AB, which would indicate cell fusion. No heteropolymer was found in electrophoretic analysis of trophoblast from either the transferred or chimeric embryos. Trophoblast cells do not functionally incorporate maternal DNA nor do they form syncytial heterokaryons by cell fusion. The multinucleated cells must result from endomitosis. Using identical principles and chimeric mice, individual somites from 8 to 9 day embryos have been found to arise from more than one precursor cell (26). Myogenesis occurs by actual fusion of uninucleated myoblasts so that the heteropolymer GPI-AB is present in skeletal muscle (26), but not in cardiac muscle or other tissues.

Electrophoretic variants of glucose phosphate isomerase have been used to study the time of paternal gene activation during early embryogenesis of the mouse (10). GPI-1A females were mated to GPI-1B males and the hybrid embryos were examined for GPI isozymes during preimplantation stages (Fig. 5). The heteropolymer GPI-1AB, indicating activity of the paternal gene, was first detected in the late blastocyst, day 5. No increase in total GPI activity occurred on day 5 so that quantitative measurement of GPI activity does not detect paternal gene activation.

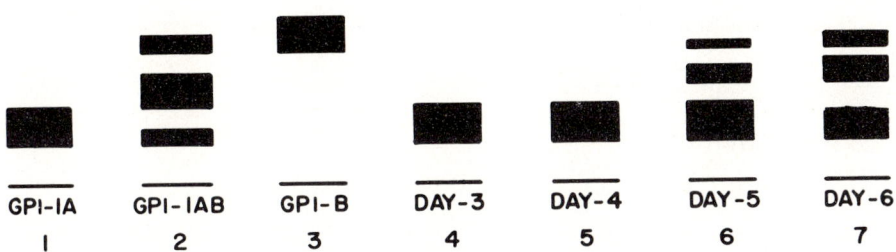

FIGURE 5. Isozymes of glucose phosphate isomerase found in embryos of various ages (10). The mother is $\underline{Gpi\text{-}1^a/Gpi\text{-}1^a}$, the father is $\underline{Gpi\text{-}1^b/Gpi\text{-}1^b}$, the embryo $\underline{Gpi\text{-}1^a/Gpi\text{-}1^b}$. Phenotype GPI-1A is present in the mother, GPI-1B in the father. GPI-1AB is the pattern expected in a heterozygous adult. Blastocysts were removed from the mother on days 3, 4, 5, and 6 after fertilization and their isozymic patterns were examined by starch gel electrophoresis. GPI-1AB and GPI-1B which represent paternal gene products were first observed on day 5.

This review of biochemical polymorphisms has of necessity been brief. Discussion has been limited to structural variants of proteins. Extensive references have been provided in Table 1 with emphasis on the recent literature. The space devoted to uses of variants may seem disproportionate, but this information is not easily collated and the potential importance of protein variants in solution of biological problems is not widely appreciated.

ACKNOWLEDGMENTS

The author acknowledges the assistance of Drs. V.M. Chapman and J.E. Womack in preparing Table 2 and Figure 1. Reference 6 was provided to me prior to publication.

REFERENCES

1. Blake, R.L., Hall, J.G., and Russell, E.S. (1976). Biochem. Genet. 14:739.
2. Breen, G.A.M., Lusis, A.J., and Paigen, K. (1977). Genetics 85:73.
3. Chapman, V.M. (1975). Biochem. Genet. 13:849.
4. Chapman, V.M., Ansell, J.D., and McLaren, A. (1972). Develop. Biol. 29:48.
5. Chapman, V.M., Nichols, E.A., and Ruddle, F.H. (1974). Biochem. Genet. 11:347.
6. Chapman, V.M., Paigen, K., Siracusa, L., and Womack, J. (1978). Biochemical Variation: Mouse Inbred and Genetically Defined Strains of Laboratory Animals. Handbook. Federation of American Societies of Experimental Biology.
7. Chapman, V.M., and Ruddle, F.H. (1972). Genetics 70:299.
8. Chapman, V.M., Ruddle, F.H., and Roderick, T.H. (1971). Biochem. Genet. 5:101.
9. Chapman, V.M., and Shows, T.B. (1976). Nature 259:665.

10. Chapman, V.M., Whitten, W.K., and Ruddle, F.H. (1971). Develop. Biol. 26:153.
11. Chen, S.H., Donahue, R.P., and Scott, C.R. (1973). Biochem. Genet. 10:23.
12. Committee on Standardized Genetic Nomenclature for Mice (1972). J. Hered. 63:69.
13. Committee on Standardized Genetic Nomenclature for Mice (1973). Biochem. Genet. 9:369.
14. Daniel, W.L. (1976). Biochem. Genet. 14:1003.
15. DeLorenzo, R.J., and Ruddle, F.H. (1969). Biochem. Genet. 3:151.
16. DeLorenzo, R.J., and Ruddle, F.H. (1970). Biochem. Genet. 4:259.
17. Dizik, M., and Elliott, R.W. (1977). Biochem. Genet. 15:31.
18. Duley, J., and Holmes, R.S. (1974). Genetics 76:93.
19. Eicher, E.M., Stern, R.H., Womack, J.E., Davisson, M.T., Roderick, T.H., and Reynolds, S.C. (1976). Biochem. Genet. 14:651.
20. Eicher, E.M., and Womack, J.E. (1977). Biochem. Genet. 15:1.
21. Epstein, C.J. (1969). Science 163:1078.
22. Epstein, C.J. (1972). Science 175:1467.
23. Erickson, R.P. (1974). Biochem. Genet. 11:33.
24. Erickson, R.P., Eicher, E.M., and Gluecksohn-Waelsch, S. (1974) Nature 248:416.
25. Finlayson, J.S., Potter, M., Shinnick, C.S., and Smithies, O. (1974). Biochem. Genet. 11:325.
26. Gearhardt, J.D., and Mintz, B. (1972). Develop. Biol. 29:27.
27. Gearhardt, J.D., and Mintz, B. (1972). Develop. Biol. 29:55.
28. Gilman, J.G. (1972). Science 178:873.

29. Gluecksohn-Waelsch, S., Schiffman, M.B., Thorndike, J., and Cori, C.F. (1974). Proc. Natl. Acad. Sci. 71:825.
30. Gross, S.R., Longshore, M.A., and Pangburn, S. (1975). Biochem. Genet. 13:567.
31. Henderson, N.S. (1965). J. Exp. Zool. 158:263.
32. Henderson, N.W. (1966). Arch. Biochem. Biophys. 117:28.
33. Huijing, F., Eicher, E.M., and Coleman, D.L. (1973). Biochem. Genet. 9:193.
34. Hutton, J.J. (1969). Biochem. Genet. 3:507.
35. Hutton, J.J. (1971). Biochem. Genet. 5:315.
36. Hutton, J.J., and Coleman, D.L. (1969). Biochem. Genet. 3:517.
37. Hutton, J.J., and Roderick, T.H. (1970). Biochem. Genet. 4:339.
38. Kaplan, R.D., Chapman, V., and Ruddle, F.H. (1973). J. Hered. 64:155.
39. Kozak, C.A., and Ruddle, F.H. (1977). Somatic Cell Genet. 3:121.
40. Kozak, L.P. (1974). Biochem. Genet. 12:69.
41. Kozak, L.P., McLean, G.K., and Eicher, E.M. (1974). Biochem. Genet. 11:41.
42. Krog, H.H. (1976). Biochem. Genet. 14:319.
43. Lewis, W.H.P., and Truslove, G.M. (1969). Biochem. Genet. 3:493.
44. Lusis, A.J., Tomino, S., and Paigen, K. (1977). Biochem. Genet. 15:115.
45. Lusis, A.J., and West, J.D. (1976). Biochem. Genet. 14:849.
46. McKusick, V.A., and Ruddle, F.H. (1977). Science 196:390.
47. Meo, T., Douglas, T., and Rijnbeek, A. (1977). Science 198:311.
48. Miller, O.J., and Miller, D.A. (1975). Ann. Rev. Genet. 9:285.
49. Mintz, B., and Baker, W.W. (1967). Proc. Natl. Acad. Sci. 58:592.

50. Mishkin, J.D., Taylor, B.A., and Mellman, W.J. (1976). Biochem. Genet. 14:635.
51. Nichols, E.A., Chapman, V.M., and Ruddle, F.H. (1973). Biochem. Genet. 8:47.
52. Nichols, E.A., and Ruddle, F.H. (1973). J. Histochem. Cytochem. 21:1066.
53. Nichols, E.A., and Ruddle, F.H. (1975). Biochem. Genet. 13:323.
54. Nichols, E.A., Ruddle, F.H., and Petras, M.L. (1975). Biochem. Genet. 13:551.
55. Paigen, K. (1971). In "Enzyme Synthesis and Degradation in Mammalian Systems" (M. Rechcigl, ed.), p. 1. Karger, Basel.
56. Paigen, K. (1975). J. Cell Physiol. 85:379.
57. Peters, J., and Nash, H.R. (1977). Biochem. Genet. 15:217.
58. Petras, M.L., and Sinclair, P. (1969). Can. J. Genet. Cytol. 11:97.
59. Platz, R.D., and Wolfe, H.G. (1969). J. Hered. 60:187.
60. Popp, R.A. (1973). Biochim. Biophys. Acta 303:52.
61. Roderick, T.H., and Hawes, N.L. (1970). Proc. Natl. Acad. Sci. 67:961.
62. Roderick, T.H., and Hawes, N.L. (1974). Genetics 76:109.
63. Roderick, T.H., Ruddle, F.H., Chapman, V.M., and Shows, T.B. (1971). Biochem. Genet. 5:457.
64. Ruddle, F.H., and Roderick, T.H. (1965). Genetics 51:445.
65. Ruddle, F.H., and Roderick, T.H. (1966). Genetics 54:191.
66. Ruddle, F.H., Shows, T.B., and Roderick, T.H. (1968). Genetics 58:599.
67. Ruddle, F.H., Shows, T.B., and Roderick, T.H. (1969). Genetics 62:393.
68. Russell, E.S., and McFarland, E.C. (1974). Ann. N.Y. Acad. Sci. 241:25.

69. Russell, L.B., Russell, W.L., Popp, R.A., Vaughan, C., and Jacobson, K.B. (1976). Proc. Natl. Acad. Sci. 73:2843.
70. Shows, T.B., Chapman, V.M., and Ruddle, F.H. (1970). Biochem. Genet. 4:707.
71. Shows, T.B., and Ruddle, F.H. (1968). Science 160:1356.
72. Shows, T.B., Ruddle, F.H., and Roderick, T.H. (1969). Biochem. Genet. 3:25.
73. Sick, K., and Nielsen, J.T. (1964). Hereditas 51:291.
74. Staats, J. (1976). Cancer Res. 36:4333.
75. Stern, R.H., Russell, E.S., and Taylor, B.A. (1976). Biochem. Genet. 14:373.
76. Vandeberg, J.L., Cooper, D.W., and Close, P.J. (1973). Nature New Biol. 243:48.
77. Watanabe, T., Ogasawara, N., and Goto, H. (1976). Biochem. Genet. 14:999.
78. Womack, J.E. (1975). Biochem. Genet. 13:311.
79. Womack, J.E., Davisson, M.T., Eicher, F.M., and Kendall, D.A. (1977). Biochem. Genet. 15:347.
80. Womack, J.E., Hawes, N.L., Soares, E.R., and Roderick, T.H. (1975). Biochem. Genet. 13:519.
81. Womack, J.E., Lynes, M.A., and Taylor, B.A. (1975). Biochem. Genet. 13:511.
82. Womack, J.E., and Sharp, M. (1976). Genetics 82:665.

GENETIC CONTROL OF ENZYME ACTIVITY

Kenneth Paigen

Molecular Biology Department
Roswell Park Memorial Institute
New York, New York

We are here to honor a group of pioneers in mammalian genetics for their achievements in developing the inbred mouse as a genetic tool. In doing so we honor them not only for the significance their achievement has for today, but also for the even greater promise it has for the future. That promise comes from the application of the inbred mouse and its genetics to new questions, questions whose answers could only be dreamed about in the early days of mammalian genetics. Perhaps the most graphic tribute we can give is to describe these applications. One of them is the study of gene regulation that grew out of a marriage between mammalian genetics and biochemistry. It is interesting that both fields had their origins at very nearly the same time. The study of intermediary metabolism began with the discovery of the first sugar phosphate in 1905, almost simultaneous with the beginnings of mouse genetics. However, it was not until 1941 that Dr. Strong first reported an enzyme difference between mouse strains (1), in what is to my knowledge the first report in any organism of an enzyme variant, and it was in 1952 that Dr. Law carried out the first genetic analysis of an enzyme difference (2), describing the inheritance pattern of the low glucuronidase phenotype that has occupied so much attention over subsequent years.

My task is to illustrate the types of genetic variation that influence enzyme activity and the experimental approaches, both genetic and biochemical, used to analyze them. Because the subject is too large to encompass in one brief talk, the illustrations are necessarily selected. I shall put some emphasis on β-glucuronidase because it is the most extensively studied system and also the one with which I am most familiar. Various aspects of the genetics (3), biochemistry and intracellular localization (4), and hormonal induction (5) of this enzyme have been reviewed.

Over the years it has proved both conceptually and operationally useful to divide enzyme variants into four groups. The composition of these groups has varied slightly from time to time as additional insights have been gained. Table 1 lists these categories and the genes that are now known to fall into one or another category with the exception that I have not listed the many structural mutants known. There are simply too many of these and a separate catalogue of them is available (6). Of course, to some measure the assignment of a locus to one or another category depends on our frame of reference. For example, the Eg gene is almost certainly the structural gene for the anchor protein egasyn, but is listed here among the processing genes for its role in the intracellular localization of β-glucuronidase. Similarly, the Tfm locus, that is probably the structural gene for the androgen receptor protein, is listed here among regulatory genes for its role in regulating hormonal induction of other proteins.

One group of interesting, but not easily categorized mutants is the series of radiation-induced deletions overlapping the albino locus. Their biochemical phenotypes have

TABLE 1. Gene Classes in Enzyme Control

Type	Examples
Structural	Many, see (6)
Processing	
Modification	Map-1, Map-2, Apl
Turnover	Ce
Localization	Eg
Organelle	bg, le, ep, pe, pa, mr, ru
Regulatory	
Systemic	Lv, Bgs
Effector response	Gur, (rennin)
Receptors	Tfm, Ah
Temporal	
Proximate	Gut, Bgt, Int, Asr-1
Distant	Tag, Tem, Rec
Unknown	Pd

been extensively studied by Gluecksohn-Waelsch and co-workers (7-10). Using complementation testing in heterozygotes, it has been possible to define a number of subregions with distinct biochemical functions within one short distance.

Structural Genes

Dr. Hutton has just discussed structural genes in considerable detail. I shall only point out one additional aspect in relation to determination of enzyme activity. A great deal of intracellular processing of polypeptide chains occurs post-translationally, and this processing is quite significant in the eventual realization of enzyme function (see below). Appropriate processing requires that each polypeptide sequence be recognized correctly, so that amino acid substitutions within the recognition region may alter processing and eventually enzyme function. One such case has been described. A structural mutation in β-glucuronidase influences its partition between two intracellular compartments.

β-Glucuronidase is present in lysosomes in virtually all cells except erythrocytes, and in some tissues is also present in endoplasmic reticulum (11, 12). In liver, for example, nearly half of the enzyme is membrane bound. Genetic, biochemical, and immunological evidence indicate that the enzyme from both sites is derived from the same structural gene (11, 13-15), located near the distal end of chromosome 5 (11, 16-19). The product of this structural gene is an enzyme tetramer of four identical subunits that is glycosylated and has been purified to homogeneity (21). Three alleles of the structural gene have been described in the literature, Gus^b, the standard form of the enzyme, Gus^a, an electrophoretic variant with more rapid mobility, and Gus^h, a variant with decreased thermostability. Several additional alleles are known, including one with increased thermostability and another with a new electrophoretic mobility (20). With respect to enzyme localization it has turned out that not all allelic forms of the enzyme respond equally to the intracellular mechanisms involved, and that the GUS-H allozyme is not distributed between the two sites in the same manner that other enzyme forms are (11, 22), presumably because of a change in its recognition properties.

Processing Genes

Among the second category of genes in earlier versions of Table 1 (22, 23) only loci affecting intracellular localization were included; these were considered architectural. Subsequent experience suggests that the second category should be enlarged to include a variety of mutations affecting the post-translational life of an enzyme molecule.

Modification. The final completion of enzyme chains by post-translational modification includes a variety of processes including proteolytic cleavages and addition of a wide variety of conjugant groups. Several mutants have been reported that are altered in enzyme conjugation.

The acid α-mannosidase present in lysosomes carries sialic acid in liver, but not in kidney. All mouse strains appear to have enzyme molecules with the same electrophoretic mobility in kidney, but vary in extent of sialylation and hence electrophoretic mobility of the liver enzyme; at least two loci are involved (24, 25), the Map-1 gene located on chromosome 17 near the H-2 complex and the Map-2 gene on chromosome 5 near Gus. The Map-1 locus may be identical with the Apl locus described by Lalley and Shows (26) which affects the electrophoretic mobility of acid phosphatase in liver but not in kidney. These genes probably determine the presence of specific sialyl transferases. Their existence makes the interesting point that a family of transferases probably exists, each with a spectrum of enzymes it is capable of sialylating, since other acid hydrolases that are sialylated in liver but not in kidney are not affected by variation at these loci (27-29).

Turnover. Rates of enzyme synthesis and degradation have an equal influence on the final concentration of enzyme activity in a tissue (30). However, it is not clear that there is much specificity in the system used for intracellular protein turnover, or that variations in protein turnover are used as a regulatory mechanism for individual proteins. Although the turnover of lactate dehydrogenase varies considerably from one tissue to another, the ratio of lactate dehydrogenase turnover to the turnover of average cellular proteins remains nearly the same from one tissue to another. Thus, modulations in relative concentrations of the enzyme must be achieved by altering its synthesis relative to that of other proteins (31). It is not surprising, then, that only one variant has so far been described with a specific alteration in the intracellular degradation of a protein. This is the Ce locus determining the rate of catalase degradation in liver (32, 33). Several substrains of the C57 family have twice the catalase activity of other strains as a result of decreased enzyme turnover in liver but not in kidney. Fast versus slow turnover segregates as a single Mendelian factor with fast turnover dominant. The locus involved is not linked to the structural gene.

Architectural. Approximately half of all cellular protein exists incorporated into one cellular organelle or another, but relatively little is known as to how this is achieved beyond the recent discovery that N-terminal peptide sequences act as signals deciding whether nascent chains attach to membranes (34) and the genetic evidence for a special class of anchor proteins involved in localization. This evidence comes from the only mutation affecting intracellular localization that has been described and involves the Eg locus and β-glucuronidase localization (14). This lack of mutants may reflect the difficulty of recognizing mutants with localization defects since failure to successfully localize an enzyme may not cause an obviously different phenotype from failure to synthesize the protein in the first place. It was possible to detect the Eg mutant because β-glucuronidase is unusual in being present in both lysosomes and endoplasmic reticulum. The mutant was detected as a loss of enzyme from the endoplasmic reticulum site without loss of enzyme from the lysosomal site. The properties of this mutant and the mechanisms of intracellular localization of glucuronidase have been reviewed recently (4).

Briefly, the polypeptide product of the single glucuronidase structural gene, Gus, is processed and assembled into two distinct tetrameric forms of the enzyme, L and X, that differ slightly in both charge and molecular weight (15, 35, 36). The L form is almost entirely located in lysosomes where it is the only enzyme form present. The X form is entirely located in endoplasmic reticulum, where it occurs complexed with from one to four molecules of the protein egasyn (15, 36). Formation of these complexes, or M forms, is apparently required to anchor or stabilize the binding of X to the membrane (37). Mutation in the Eg gene results in loss of egasyn (38, 39) and absence of enzyme bound to membranes (14, 40). Eg is located on chromosome 8 unlinked to the glucuronidase structural gene (14, 41). Heterozygotes have reduced levels of egasyn but normal binding of enzyme to membrane (39). Only tissues containing egasyn possess membrane bound glucuronidase (4, 42).

Organellar Function. Once a protein has become integrated into a cellular organelle its fate, at least in part, is tied to that of the organelle itself. A case in point is the series of mutants described by Swank and co-workers (43) that appear to be deficient in the transport of lysosomes across cell membranes. The initial observation was that the beige mutant, which possesses enlarged lysosomes in many cells and is an analog of Chediak-Higashi syndrome in the human, accumulates excessive amounts of β-glucuronidase during induction of this enzyme in kidney proximal tubule epithelial cells (44, 45). Normally these cells excrete most of their induced enzyme by a process that appears to involve extrusion of entire lysosomal contents into the lumen of the collecting tubule. The beige mutant is deficient in carrying out this process and consequently accumulates large amounts of enzyme. Following the observation by Meisler (46) that another pigmentation mutant, light ears, shares a similar defect, four additional pigmentation mutants were found defective in lysosome transport (43). Lysosomes and melanosomes apparently share common mechanisms of transport across membranes. The various mutations are unlinked and recessive, and presumably affect different functional steps in the process of transport.

Regulation

Regulatory genes are distinguished by their effect on protein synthesis rather than merely altering apparent enzyme activity. Activity changes can result from a variety of causes that do not involve true regulation, including alterations in processing and structure. At the present time it is useful to distinguish between regulatory sites controlling systemic levels of enzyme, those determining the response of an enzyme to an effector molecule such as a hormone, and those affecting the receptor molecules involved in effector responses.

<u>Systemic</u>. There are now two examples of genetic variants in which enzyme activity is altered in a more or less constant manner among all tissues and at different stages of development. The first discovered was the <u>Lv</u> locus determining levels of aminolevulinate dehydratase in several tissues during both fetal and adult stages of development (47). High strains have several times the activity of low strains and heterozygotes are intermediate. A third allele with intermediate levels has been reported (48). The locus is linked to the enzyme structural gene (49). Enzyme purified from both high and low strains has identical kinetic and physical properties and the same catalytic activity per molecule (50, 51). Labeling studies have shown that the kinetics of enzyme turnover are the same in high and low strains and that the difference lies in the rate of enzyme synthesis.

A systemic regulatory locus for β-galactosidase has also been found that in many respects resembles the Lv locus. It acts uniformly in different tissues and stages of development, is closely linked to the structural gene, and controls enzyme activity by regulating enzyme synthesis. The structural gene for murine β-galactosidase is defined by an electrophoretic polymorphism, Bge, and is located on chromosome 9 (27). Although multiple forms of β-galactosidase are present in tissues, these are all derived from this single structural gene, and represent charge differences arising from differential sialylation of the enzyme among tissues as well as several molecular weight forms resulting from pH dependent aggregation of the enzyme (27). The systemic regulator, Bgs, is very closely linked to Bge, and no recombination was observed between the two (27). Strains carrying the Bgs^h allele have approximately twice the enzyme level of Bgs^d strains in a variety of tissues throughout development (52, 53). Heterozygotes show additive inheritance, possessing intermediate enzyme levels. Antibody titration shows that high enzyme activity reflects an increase in the number of enzyme molecules present (52) and recent labeling experiments show this results from increased enzyme synthesis (54). The difference in activity seen between Bgs^d and Bgs^h strains has been used as a histochemical marker for the recognition of genetically marked cells in chimeric mice (55).

Induction. Mutants with altered response to enzyme induction have received considerable attention for the insights they offer into mechanisms of regulation of gene transcription and the control of physiological responses in tissues. Three systems have been examined so far, all involving induction by androgens. These include glucuronidase induction in epithelial proximal tubule cells of kidney, induction of rennin activity in submandibular gland, and induction of the major urinary proteins (Mups) in liver.

A massive increase in the glucuronidase content of kidney proximal tubule epithelial cells occurs following administration of androgens in the mouse (19, 56, 57). Several other enzymes, notably D-amino acid oxidase and alcohol dehydrogenase, are also induced, but to lesser extents. There is a very small induction of other lysosomal hydrolases. Much of the induced glucuronidase is excreted into urine. Electron micrographs (3) show the accumulation of membranous material inside lysosomes and large amounts of similar material appearing in the proximal tubule lumen.

Inbred strains fall into two major classes of high and low inducibility (19). Additional rare phenotypes may exist (20). The system has been characterized in some detail using A/J and C57BL/6J as the prototype high and low strains (19). Among the androgen-inducible proteins only glucuronidase is affected, and the difference in inducibility is independent of the specific hormone used and the dosage.

Examination of recombinant inbred lines between BALB/c and C57BL/6 and crosses between A/J and C57BL/6 showed that inducibility is controlled by a single locus, Gur, closely linked to the structural gene (19, 58). Despite differences in reported properties, the O locus described by Ohno and coworkers (59) is probably identical with Gur. Originally several ostensible recombinants were observed between the new regulatory site and the structural gene (19), but true recombinants were not found in a more extensive analysis of this cross using newer techniques allowing the survival of test animals and subsequent breeding (5). Ostensible recombinants did arise at very low frequency but proved to be errors of diagnosis.

Pulse labeling studies have shown that Gur locus control is a cis acting regulation of enzyme synthesis (19). Heterozygotes carrying Gur^a/Gur^b have intermediate levels of enzyme because one chromosome is producing enzyme rapidly and the other slowly.

The analysis of recombinant inbred lines showed the presence of genes that affect the final level of induced enzyme activity, but not the early kinetics of induction (58). Labeling studies showed that unlike Gur these genes do not act at the level of enzyme synthesis and very likely affect enzyme excretion (60).

Molecular studies now give some indication as to the nature of Gur gene function. A very sensitive enzymatic assay for glucuronidase mRNA has been developed based on the formation of catalytically active mouse enzyme in Xenopus oocytes following injection of mouse mRNA (61). Measurements of mRNA levels were made in fully induced A/J and C57BL/6J mice as well as in a congenic strain carrying the Gur^a allele of strain A backcrossed onto the C57BL/6 background. They showed that induction is an accumulation of mRNA activity and that the Gur locus determines the amount of mRNA activity present (62). Thus, the Gur locus is a regulatory site controlling mRNA activity. Detailed kinetic studies of the changes in rates of enzyme synthesis and levels of mRNA activity during induction and deinduction have been carried out (63). They suggest that Gur controls mRNA activity by determining the hormonal response of a rate limiting factor in the synthesis of messenger activity and that this factor turns over very slowly. The most obvious candidate for such a factor is the activation of chromatin, but this is still a hypothesis.

Several additional genes are known to affect glucuronidase induction. One is the Tfm mutant that is deficient in androgen binding protein (64-66) and hence loses induction (67). Several others act through their effects on the pituitary. Hypophysectomized animals are only inducible to about 5% of the level of intact animals (68), and hormone replacement experiments suggest that the missing factor is growth hormone (69). In accord with this, induction is lost in the dwarf mutant that lacks all pituitary function, as well as in the little mutant that is a relatively specific deficiency in growth hormone (69). These mutations all affect enzyme synthesis.

Because enzyme excretion is the major mechanism for removal of induced enzyme, genetic factors affecting excretion influence enzyme accumulation. This is apparent in the segregation among recombinant inbred strains of genes unlinked to the structural gene (58), and the elevated enzyme accumulation seen in beige and other pigmentation mutants defective in lysosome excretion (see Processing above).

Rennin induction in submaxillary gland is also controlled by a single locus with additive inheritance that acts at the level of enzyme synthesis (70). There are multiple rennin isozymes, and both basal and induced activities are genetically affected, making it uncertain where in the induction process the gene acts.

The major urinary proteins of the mouse (Mups) are a family of three closely related proteins that are synthesized in the liver and excreted in the urine. Their interest lies in the very large synthesis achieved after androgen induction; depending upon the genetic background, this is in the range of 10-30 mg of mup protein per day, and represents the highest percentage of any single protein synthesized in liver. Originally, Hudson et al. (71) described a locus, Mup-a, that was thought to represent a structural polymorphism in which mup-1 and mup-2 were allelic forms derived from the same structural gene. More recently Szoka and Paigen (72) have found that all mouse strains synthesize all three mups, and that the Mup-a locus is, in fact, regulatory. Analysis of recombinant inbred lines between BALB/c and C57BL/6, which carry the $Mup-a^1$ and $Mup-a^2$ alleles, respectively, has shown that the Mup-a locus determines the relative proportions of the three Mups synthesized after induction, but that other loci are involved in other aspects of the phenotype, including determination of the relative proportions before induction, the kinetics of induction, and the total level of Mup production reached at the final induced state.

Receptors. The testicular feminization mutant, Tfm, is deficient in androgen receptor protein (64-66) and has lost all androgen-inducible functions in the mouse. Its existence raises the question as to why, if the same receptor protein functions in all tissues, a different set of proteins is induced in each tissue. An additional specificity factor must exist.

The other receptor mutant known involves the receptor protein for planar hydrocarbons that functions in the induction of the aryl hydrocarbon hydroxylase system and some other proteins (73, 74). Segregation at the Ah locus determines the binding affinity of this receptor (75) and has important effects on the relative susceptibility of various sites to hydrocarbon induction of tumors (76).

Temporal Genes

Genetic factors controlling development have fascinated biologists for many years. Historically, the primary emphasis has been on morphogenetic mutants. Now, with the advent of biochemical genetics it is possible to ask questions about the genetic systems determining developmental programs for the products of individual structural genes. This approach allows us to focus on questions of programming itself and thereby avoid the confusion attending phenotypic changes that may turn out to be metabolic consequences of altered protein function. Taking this approach does not assume any particular model for genetic regulation of differentiation; rather, we appreciate that the nature of our findings will eventually help us choose between models. In this case, the mouse has led the work in other organisms. The mouse systems most extensively examined have been those controlling H-2, α-galactosidase, β-galactosidase, and β-glucuronidase. The properties of these systems and the general concepts of temporal genes have been reviewed (77).

The first temporal gene element identified in mice proved to be in close proximity to the glucuronidase structural gene. The developmental program for β-glucuronidase has been defined for a number of mouse tissues (78), and is coordinate with that of β-galactosidase (79), and α-galactosidase (80). In some strains of mice there is a time in development at which enzyme activity begins an abrupt decline relative to wild type levels. Each tissue does this at a different time; in liver it begins at 12-15 days of age. This change in enzyme programming is determined by a single locus, Gut, that maps in very close proximity to the structural gene (78). The Gut allele producing low activity is invariably associated with the Gus^h structural allele. The combination results in low heat sensitive activity and has been used as a histochemical marker for genetically distinct cells in chimeras (81, 82). Heterozygotes show additive inheritance, but there is contradictory evidence as to whether Gut acts cis or trans (83, 84). The Gut site determines developmental patterns prenatally as well as postnatally. Within a few days after fertilization there is a dramatic increase in the β-glucuronidase activity of individual embryos, and the extent and timing of this increase differ in mice carrying various alleles of Gut (81).

Recent studies by Ganschow (86) indicate that the Gut site, like Gur, controls the rate of enzyme synthesis. Pulse labeling studies have shown that abrupt changes in rates of enzyme synthesis determined by Gut are the cause of the developmental changes in enzyme activity.

A proximate temporal element has also been postulated for the regulation of aryl sulfatase development (87). Another, whose location is not known, appears to affect three enzymes in the pyrimidine degradation pathway (88).

A temporal locus has been found for α-galactosidase, but in this case it is not linked to the structural gene. Nevertheless, it shows additive inheritance. The α-galactosidase structural gene is on the X chromosome of the mouse (89), and the temporal gene specifically determining the liver developmental pattern of α-galactosidase is autosomal (80).

Both proximate and distant temporal elements participate in programming the developmental pattern of β-galactosidase levels in liver. C57BL/6 mice have an enzyme development curve that parallels that of other strains in most tissues. However, in liver it begins to deviate at 25 days of age and steadily increases its enzyme activity until in adult life it has twice the level that would otherwise be expected. Crosses between strains carrying alleles of the systemic regulator Bgs showed that a single locus determines this difference and is closely linked to the β-galactosidase complex on chromosome 9 that contains Bgs and Bge (90). However, examination of other crosses and recombinant inbred lines indicates that in some cases the failure to undergo this timed elevation in liver is determined by genetic factors unlinked to the β-galactosidase complex on chromosome 9 (91). These facts are illustrated by examining the developmental phenotypes of C57BL/6 and C57BL/10 congenic strains carrying various substitutions of the β-galactosidase complex on chromosome 9 (Table 2). The rise in liver is absent in the congenic carrying a β-galactosidase complex derived from CBA, which lacks the elevation seen in C57BL/6. On the contrary,

TABLE 2. β-Galactosidase Activity of Inbred and Congenic Strains

	β-Galactosidase Activity (units/g)				
Strain	Liver	Brain	Kidney	Spleen	Heart
C57BL/6	18.1	6.3	32.5	35.7	2.16
BALB/c	10.2	6.1	30.8	34.6	2.09
B10. BALB/c congenic	19.2	7.2	36.1	32.1	2.54
CBA	5.90	2.8	13.4	13.2	0.95
B6. CBA congenic	4.90	2.5	14.5	13.9	0.78

the rise is present in the congenic carrying an insertion of the β-galactosidase complex from BALB/c, which also lacks the C57BL/6 liver elevation. Thus, the genetic factors preventing developmental elevation of liver enzyme were carried along with the β-galactosidase complex when it came from CBA, but not when it came from BALB/c.

The H-2 system was defined by Boubelik et al. (92) and contains an additional complexity in its genetic structure. H-2 antigen is present on all cells during fetal development except erythrocytes. H-2 antigen appears on erythrocytes after birth; this may occur immediately after birth (the early phenotype) or it may come approximately four days later (the late phenotype). The timing of appearance is not related to the act of birth, but rather to the process of erythrocyte differentiation itself, since the same differences in timing are seen in the erythrocytes that appear after marrow transplants are made into irradiated hosts. In this case the timing of appearance is determined by the donor genotype.

A genetic model was developed from examination of various recombinants and congenic lines and from the segregation of phenotypes during the establishment of five recombinant inbred lines derived from a cross between strains with early and late phenotypes. The model suggests a temporal element, Int, that is in close proximity to the D end of the H-2 region, and acts cis in determining timing control. Int has been separated from H-2, and in recombinants selected between H-2D and the nearby marker Tla segregated with Tla. A second timing system is present that contains two genes Rec and Tem. Tem carries early and late timing alleles, and segregation at Rec determines whether these can override Int. If Rec and Int alleles match then override occurs, if they do not match override does not occur. In order to account for the results it is necessary to assume either that Rec and Tem interact in a cis fashion, although they are 20 centimorgans apart, or more simply, that Rec and Tem are on the X chromosome and thus show haploid expression.

The first major finding in the study of temporal genes is the existence of both proximate and distant sites determining developmental patterns of specific proteins. Second, it is probable that the proximate sites are cis-acting; the evidence for this is good in the H-2 system and for proximate temporal genes seen in other organisms (77). Third, the distant sites show additive inheritance when examined quantitatively. Fourth, there appear to be specific interactions between the proximate and distant sites, these are seen in the H-2 and β-galactosidase systems in mice and also in the amylase system in Drosophila. An obvious possibility is that proximate sites are the receptors for signals transmitted by the distant sites. If there are signals, the fact that regulation is dosage dependent has strong implications regarding possible mechanisms. Fifth, the relative simplicity of these genetic systems is suggested by the high proportions of all mutants discovered that are proximate sites. Sixth, the systems are relatively specific; the temporal genes determining programs for each enzyme segregate independently of each other even when the programs determine the coordinate regulation of enzymes in development.

CONCLUSIONS

Perhaps the major conclusion we can draw is that inbred mouse strains and their derivatives, recombinant inbred lines and congenic strains, have proved to be a very powerful tool for examining questions of gene regulation and cell biology in higher eukaryotes, and especially mammals. The pioneers we honor today built very well indeed. Without the strains they developed it would have been impossible to attack these questions with any great measure of success. Success has also depended on the extensive natural polymorphism for regulatory mechanisms that must exist in the wild, and that was frozen in the inbred lines when they were first established, providing the basic material for research.

This extensive variation must have evolutionary significance. It appears that regulatory elements may be as polymorphic as structural genes have been shown to be in recent years. In this context it is significant that so much regulatory information for an enzyme proves to be linked to the structural gene. Indeed, what appears to be segregating in natural populations are gene complexes, chromosome regions whose DNA sequences contain both a structural gene and much associated regulatory information. The existence of such complexes raises the important question of how these two kinds of information are organized in relation to each other.

From the frequency with which regulatory information turns out to be closely linked to the structural gene it is reasonable to assume that a very high proportion of all regulatory information is so linked. We assume that selective forces must exist to maintain this close linkage. The most obvious would be the advantage derived from co-selection of compatible combinations of regulatory elements, including some that are necessarily linked to the structural gene for the regulation of transcription. Linkage disequilibrium may well be a reflection of the mechanisms developed to maintain this close association. From this standpoint it is interesting to consider that to whatever extent the extensive natural polymorphism for structural variants is maintained by natural selection, selection is probably operating on the linked regulatory information rather than on the enzyme structure per se. This would be true whether the same base sequences determine both regulation and structure or whether they are determined by separate sequences.

Looking at the total genetic variation in the realization of enzyme activity, it is surprising how much involves post-translational processing mechanisms rather than regulation of enzyme synthesis. Processing genes are not linked to the enzyme structural gene and generally show recessive/dominant inheritance, in contrast to the regulatory information influencing enzyme synthesis, much of which is closely linked and all of which shows additive inheritance. An interesting point is that additive inheritance can only be seen because enzyme levels are not self-regulated. Mammals, unlike microorganisms, lack a sensing mechanism to determine what enzyme levels are actually present and adjust their regulatory mechanisms accordingly. Thus, heterozygotes for enzyme deficiencies typically have intermediate levels of enzyme. Taken together, the absence of sensing mechanisms in regulating enzyme levels and the presence of additive regulatory information strongly imply that the mechanisms utilized in mammals are not similar to the repressor-like mechanisms seen in microorganisms. It appears that we are, in fact, looking at novel modes of regulation whose analysis is our future challenge.

REFERENCES

1. Figge, F.H.J., and Strong, L.C. (1941). Cancer Res. 1:779.
2. Law, L.W., Morrow, A.G., and Greenspan, E.M. (1952). J. Natl. Cancer Inst. 12:909.
3. Paigen, K., Swank, R.T., Tomino, S., and Ganschow, R.E. (1975). J. Cell. Physiol. 85:379.
4. Lusis, A.J., and Paigen, K. (1977). In "Isozymes," vol. II (M.C. Rattazzi, J.G. Scandalios, and G.S. Whitt, eds.), p. 64. A.R. Liss, New York.
5. Swank, R.T., Paigen, K., Davey, R., Chapman, V., Labarca, C., Watson, G., Ganschow, R.E., Brandt, E.J., and Novak, E. (1978). In "Recent Progress in Hormone Research" vol. 34 (R.O. Greep, ed.). Academic Press, New York.
6. Chapman, V.M., Paigen, K., Siracusa, L., and Womack, J. In "FASEB Biological Handbook on Inbred Laboratory Animals," in press.
7. Erickson, R.P., Gluecksohn-Waelsch, S., and Cori, C.F. (1968). Proc. Natl. Acad. Sci. USA 59:437.
8. Gluecksohn-Waelsch, S., and Cori, C.F. (1970). Biochem. Genet. 4:195.
9. Thornkide, J., Trigg, M.J., Stockert, R., Gluecksohn-Waelsch, S., and Cori, C.F. (1973). Biochem. Genet. 9:25.
10. Gluecksohn-Waelsch, S., Schiffman, M.B., Thorndike, J., and Cori, C.F. (1974). Proc. Natl. Acad. Sci. USA 71:825.
11. Paigen, K. (1961). Exp. Cell Res. 25:286.
12. Fishman, W.H., Goldman, S.S., and Delellis, R. (1967). Nature 213:457.
13. Lalley, P.A. and Shows, T.B. (1974). Science 185:442.
14. Ganschow, R.E., and Paigen, K. (1967). Proc. Natl. Acad. Sci. USA 58:938.

15. Tomino, S., and Paigen, K. (1975). J. Biol. Chem. 250: 1146.
16. Law, L.W., Morrow, A.G., and Greenspan, E.M. (1952). J. Natl. Cancer Inst. 12:909.
17. Paigen, K., and Noell, W.K. (1961). Nature 190:148.
18. Sidman, R.L., and Green, M.C. (1965). J. Hered. 56:23.
19. Swank, R.T., Paigen, K., and Ganschow, R.E. (1973). J. Mol. Biol. 81:225.
20. Chapman, V.M. Personal communication.
21. Tomino, S., Paigen, K., Tulsiani, D., and Touster, O. (1975). J. Biol. Chem. 250:8503.
22. Paigen, K., and Ganschow, R. (1965). Brookhaven Symp. Biol. 18:99.
23. Paigen, K. (1971). In "Enzyme Synthesis and Degradation in Mammalian Systems" (M. Rechcigl, ed.). Karger Press, Basel, Switzerland.
24. Dizik, M., and Elliott, R.W. (1977). Biochem. Genet. 15:31.
25. Dizik, M., and Elliott, R.W. (1978). Biochem. Genet., in press.
26. Lalley, P.A., and Shows, T.B. (1977). Genetics 87:305.
27. Green, G., Lusis, A.J., and Paigen, K. (1977). Genetics 85:73.
28. Lusis, A.J., Breen, G.A.M., and Paigen, K. (1977). J. Biol. Chem. 252:4613.
29. Lusis, A.J., and Paigen, K. (1976). Biochem. Biophys. Acta 437:487.
30. Schimke, R.T. (1969). Current Topics Cell Regulation 1:77.
31. Nadal-Ginard, B. (1978). J. Biol. Chem. 253:170.
32. Rechcigl, M., Jr., and Heston, W.E. (1967). Biochem. Biophys. Res. Comm. 27:119.
33. Ganschow, R.E., and Schimke, R.T. (1969). J. Biol. Chem. 244:4649.

34. Devillers-Thiery, A., Kindt, T., Scheela, G., and Blobel, G. (1975). Proc. Natl. Acad. Sci. USA 72:5016.
35. Ganschow, R.E., and Bunker, B.G. (1970). Biochem. Genet. 4:127.
36. Swank, R.T., and Paigen, K. (1973). J. Mol. Biol. 77:371.
37. Smith, K. (1976). Ph.D. Thesis, The Turnover of β-Glucuronidase in Normal and Mutant Mice.
38. Lusis, A.J., Tomino, S., and Paigen, K. (1976). J. Biol. Chem. 251:7753.
39. Lusis, A.J., Tomino, S., and Paigen, K. (1977). Biochem. Genet. 15:115.
40. Ganschow, R.E., and Paigen, K. (1968). Genetics 59:335.
41. Karl, T.R., and Chapman, V.M. (1974). Biochem. Genet. 11:367.
42. Lusis, A.J., and Paigen, K. (1977). J. Cell Biol. 73:728.
43. Swank, R.T., Novak, E., Brandt, E.J., and Skudlarek, M. (1978). In "Protein Turnover and Lysosomal Function" (D. Doyle and H. Segal, eds.). Academic Press, New York.
44. Brandt, E.J., Elliott, R.W., and Swank, R.T. (1975). J. Cell Biol. 67:774.
45. Brandt, E.J., and Swank, R.T. (1976). Am. J. Pathol. 82:573.
46. Meisler, M. (1978). J. Biol. Chem., in press.
47. Russell, R.L., and Coleman, D.L. (1963). Genetics 48:1033.
48. Hutton, J.J., and Coleman, D.L. (1969). Biochem. Genet. 3:517.
49. Coleman, D.L. (1971). Science 173:1245.
50. Coleman, D.L. (1966). J. Biol. Chem. 241:5511.
51. Doyle, D., and Schimke, R.T. (1969). J. Biol. Chem. 244:5449.

52. Felton, J., Meisler, M., and Paigen, K. (1974). J. Biol. Chem. 249:3267.
53. Lundin, L., and Seyedyazdani, R. (1973). Biochem. Genet. 10:351.
54. Berger, F., Meisler, M., and Paigen, K. Unpublished.
55. Dewey, M., Gervais, A., and Mintz, B. (1976). Develop. Biol. 50:68.
56. Fishman, W.H. (1951). Ann. N.Y. Acad. Sci. 54:548.
57. Fishman, W.H. (1965). In "Methods in Hormone Research," vol. 4 (R.I. Portman, ed.). Academic Press, New York.
58. Swank, R.T., and Bailey, D.W. (1973). Science 181:1249.
59. Dofuku, R., Tettenborn, U., and Ohno, S. (1971). Nature New Biol. 234:259.
60. Watson, G., and Paigen, K. (1978). Biochem. Genet., in press.
61. Labarca, C., and Paigen, I. (1977). Proc. Natl. Acad. Sci. 74:4462.
62. Paigen, K., Labarca, C., and Watson, G. Unpublished.
63. Watson, G., Davey, R., and Paigen, K. Unpublished.
64. Bullock, L.P., and Bardin, C.W. (1972). J. Clin. Endocrinol. Metab. 35:935.
65. Attardi, B., and Ohno, S. (1974). Cell 2:205.
66. Gehring, U., and Tomkins, G.M. (1974). Cell 3:59.
67. Ohno, S., and Lyon, M.F. (1970). Clin. Genet. 1:121.
68. Swank, R.T., Davey, R., Joyce, L., Reid, P., and Macey, M.R. (1977). Endocrinol. 100:473.
69. Swank, R.T. Personal communication.
70. Wilson, C.M., Erdös, E.G., Dunn, J.F., and Wilson, J.D. (1977). Proc. Natl. Acad. Sci. USA 74:1185.
71. Hudson, D.M., Finlayson, J.S., and Potter, M. (1967). Genet. Res. 10:195.
72. Szoka, P., and Paigen, K. (1978). Genetics, in press.
73. Gielen, J., Goujan, F.M., and Nebert, D.W. (1972). J. Biol. Chem. 247:1125.

74. Thomas, P.E., Kouri, R.G., and Hutton, J.J. (1972). Biochem. Genet. 6:157.
75. Poland, A., and Glover, E. (1974). Molec. Pharm. 11:389.
76. Thomas, P.E., Hutton, J.J., and Taylor, B.A. (1973). Genetics 74:655.
77. Paigen, K. (1977). In "Proceedings Fifth International Congress of Human Genetics, Mexico City," p. 33. Excerpta Medica, Amsterdam.
78. Paigen, K. (1961). Proc. Natl. Acad. Sci. USA 47:1641.
79. Meisler, M., and Paigen, K. (1972). Science 177:894.
80. Lusis, A.J., and Paigen, K. (1975). Cell 6:371.
81. Condamine, H., Custer, R.P., and Mintz, B. (1971). Proc. Natl. Acad. Sci. USA 68:2032.
82. Feder, N. (1967). Nature 263:67.
83. Herrup, K., and Mullen, R.J. (1977). Biochem. Genet. 15:641.
84. Lusis, A.J. Personal communication.
85. Wudl, L., and Chapman, V.M. (1976). Develop. Biol. 48:104.
86. Ganschow, R.E. (1975). In "Third Int. Conf. Isozymes" (C.L. Markert, ed.). Academic Press, New York.
87. Daniel, W.L. (1976). Biochem. Genet. 14:1003.
88. Dagg, C.P., Coleman, D.L., and Fraser, G.M. (1964). Genetics 49:979.
89. Lusis, A.J., and West, J.D. (1978). Genetics, in press.
90. Paigen, K., Meisler, M., Felton, J., and Chapman, V.M. (1976). Cell 9:533.
91. Berger, F., Breen, G.A.M., and Paigen, K. Unpublished.
92. Boubelik, M., Lengerova, A., Bailey, D.W., and Matousek, V. (1975). Develop. Biol. 47:206.

Inbred Strains: Viruses

Lloyd Law, Chairman

GENETIC REGULATION OF TYPE C VIRUSES IN MOUSE LEUKEMIA[1]

Frank Lilly

Department of Genetics
Albert Einstein College of Medicine
New York, New York

Leukemia is a disease which occurs spontaneously among mice of most -- perhaps all -- laboratory strains, with an incidence considerably in excess of that seen in humans. Aside from the curious and exceedingly useful strains (e.g., AKR and C58) which show a near 100% incidence of the disease by one year of age, mice of most other strains, if observed throughout their life spans, show an incidence of at least a few percent, occurring usually well after one year of age. The bulk of these leukemias are lymphatic, frequently T-cell leukemias or lymphomas, although a few strains show a significant incidence of other types, such as the Hodgkin's-like disease of SJL mice.

In addition to spontaneously occurring leukemia, the disease may be readily induced in mice by leukemogenic viruses of the C-type, RNA oncovirus family and also by treatment with x-rays and chemical carcinogens of several different types. Susceptibility to the induction of the disease varies widely

---

[1]The author's work is supported by a contract from the Virus Cancer Program and grants, all from the National Cancer Institute, Bethesda, Maryland.

among mice of different inbred strains, however, and high susceptibility to one inducing agent is not necessarily accompanied by a similar level of susceptibility to another agent.

A considerable body of evidence points to the involvement of endogenous murine leukemia viruses (MuLVs) in the leukemias occurring in the high-incidence strains. It is less certain, however, that the same process is necessarily involved in all cases of spontaneous or induced leukemia occurring in low-incidence strains. As with cancer in general, mouse leukemia may well represent a cellular phenotype which may be attained by several different pathways, some associated with viruses and some not. In any case, viral leukemias in mice are of special interest because the pathways followed in their genesis seem likely to be the first or among the first such pathways to be elucidated. Nevertheless, our present understanding of the details of viral leukemogenesis is fragmentary at best.

Genetic analyses of strain differences in the occurrence of spontaneous or induced leukemias in mice have indicated some of the major subdivisions of the pathways in viral leukemogenesis (1). A number of single genes have been identified which are polymorphic in the laboratory mouse and which govern events and/or conditions that appear to be favorable to the course of the disease. Specifically, three broad categories of genes have been revealed: (a) chromosomally located viral genomes, either complete or defective, representing at least two different varieties of murine leukemia viruses (MuLV) which are transmitted vertically as Mendelian traits; (b) genes which govern the expression of MuLV genomes by, for example, influencing the horizontal spread between nearly cells of the viral infection; and (c) genes which influence the capacity of the host to respond immunologically to virus-infected and/or transformed cells.

Inherited Viral Genomes. Different classes of nondefective MuLVs share a common mechanism of replication (2). They depend upon their endogenous RNA-dependent DNA polymerases (reverse transcriptases) for the formation of a DNA copy of the viral RNA immediately following penetration of the host cell and uncoating; this DNA copy is inserted into a chromosomal location and serves as the source of viral information for all succeeding events in the replication cycle. If this chromosomal insertion occurs in a somatic cell, then the viral genome will be transmitted to all its daughter cells, but other cells of the individual can become similarly infected only by means of a new infectious event mediated by a mature virus particle produced by the original infected cell. If, however, the chromosomal insertion occurs in a germline cell, then it may be transmitted to an individual of the next generation. In this case the viral genome will be present in all cells of the individual and will be transmitted by it to future generations as a Mendelian character.

Mice infected with MuLV of exogenous origin do not routinely transmit their infection in this manner, suggesting that germline cells are protected by special mechanisms from infection with these viruses. But Mendelian transmission by mice infected in vitro as early embryos has been demonstrated (3).

Mice of the inbred AKR strain show high levels of ecotropic MuLV, infectious for murine cells but noninfectious in cells of most other species from soon after birth (4). Studies by Rowe and his collaborators (5) have indicated that this virus derives from MuLV genomes situated at two different chromosomal sites -- one of which is very precisely mapped with respect to closely linked markers. That these genes, designated Akv-1 and Akv-2, do in fact consist of viral genomes, rather than regulators of viral genomes located elsewhere, has been proven by experiments showing that nucleic acid sequences specific to this particular MuLV map uniquely at these two chromosomal sites in AKR mice (6).

Ecotropic MuLV genomes are present chromosomally in mice of strains other than AKR as well, both high- and low-leukemic. From what is presently known of their chromosomal locations, it appears that the sites of insertion of these viral genomes are rarely, if ever, the same in unrelated mouse strains. This observation clearly implies that possible insertion sites may be very numerous in the host DNA, although they may not all be of equally ready access for the proviral DNA. In a parallel manner, it is clear that <u>xenotropic</u> MuLVs, closely related to ecotropic MuLVs but showing a markedly different host range pattern, are also present in the genomes of most if not all mouse strains (7), again showing different strain-specific integration sites into the host chromosomal DNA.

<u>Control of Virus Expression</u>. Studies in crosses of high-leukemic AKR mice with mice of low-leukemic strains indicate that the introduction of genes capable of suppressing the expression of endogenous MuLV genomes results in suppression of the high-leukemic phenotype. Apparently it is not sufficient that the viral genome be present in a mouse for leukemia to occur; it must also be expressed in the form of infectious viral particles.

The <u>Fv-1</u> gene is the best studied example of a gene which suppresses MuLV expression and also leukemogenesis. The $\underline{Fv-1}^b$ allele, present in BALB/c and other inbred strains, possesses the dominant property of suppressing the replication of certain MuLVs, including the ecotropic MuLV of AKR mice, in the tissues of young mice (8), and as a consequence the pattern of segregation of this gene in AKR x BALB/c crosses coincides closely with the pattern of leukemia suppression (9). The molecular events associated with this virus suppression are not yet elucidated, but it is clear that restriction of cell-to-cell spread of virus from rare spontaneously producing cells is the basis of the low-MuLV phenotype associated with the $Fv-1^b$ allele in these crosses (10).

Recent studies by Allen Mayer in my laboratory have indicated that another allele at the Fv-1 locus, that present in the RF strain, also suppresses leukemia in crosses with AKR mice, but it does so without showing the severe suppressive effect of $Fv-1^b$ on ecotropic MuLV expression. Rather, this allele exerts its suppressive effect on the xenotropic MuLV which is detected in AKR mice at around six months of age. Thus suppression of either ecotropic or xenotropic MuLV expression appears sufficient to suppress leukemogenesis, a finding that lends support to the current idea that the occurrence of the disease results from a collaborative effort on the part of both types of MuLV (11).

Control of Tumor Cell Outgrowth. Following virus replication and spread, one or more cells of the thymic lymphoid series may be transformed, acquiring malignant growth potential. Failure of the host to suppress the outgrowth of such cells then results in clinically recognizable disease and death. Suppression of tumor growth appears to result from an immunologic response to virus-induced, tumor-specific antigens expressed on the surfaces of leukemia cells. The occurrence of this immune response is strongly influenced by the host's genotype at the complex H-2 region (12) of chromosome 17.

The I subregion of the H-2 complex includes the determinants of immune response genes, which markedly influence the capacity of mice to respond to specific antigenic determinants (13). Thus it is readily conceivable that certain tumor antigens might induce strong responses leading to tumor rejection in some hosts but only weak responses with little effect on tumor growth in others. Such a system is exemplified by the X.1 antigen of x-ray-induced leukemias of BALB/c mice (14). Hosts bearing the $\underline{I}$ region of the $\underline{H-2}^b$ haplotype respond strongly to X.1, whereas hosts possessing only the $\underline{H-2}^d$ or $\underline{H-2}^k$ haplotypes are nonresponders with little capacity to reject X.1-positive tumor cells.

Recent evidence suggests another different mechanism whereby the $\underline{H-2}$ type of the host may influence rejection of virus-induced tumor cells. Cytolytic T lymphocytes specific for viral antigens show a marked preference $\underline{in}$ $\underline{vitro}$ for target cells of the same $\underline{H-2}$ type as that of the tumor cell which induced the response (15). It appears that this $\underline{H-2}$ restriction phenomenon is attributable to the fact that the antigen recognized consists of a viral molecule complexed on the cell surface with a molecule governed by the $\underline{K}$ or $\underline{D}$ loci of the $\underline{H-2}$ region (16). Not all products of the polymorphic $\underline{K}$ and $\underline{D}$ loci show the capacity to form an appropriate complex with viral molecules, and mice possessing only non-complexing $\underline{K}$ and $\underline{D}$ molecules do not exhibit a cytotoxic T cell response to their leukemia cells because the cells lack the proper antigenicity.

The genetic evidence summarized above makes a strong case for the involvement of C-type viruses in the leukemias occurring spontaneously in AKR mice, adding thereby further weight to the observations of Gross (17) that the disease is transferable under certain conditions with cell-free extracts. Comparable viruses have also been isolated from leukemias induced with x-irradiation or with hydrocarbon carcinogens, suggesting that induction of endogenous virus expression was also the critical factor in these leukemias (18, 19).

It has long been known that skin painting of mice of some strains with the carcinogen, 3-methylcholanthrene (MCA), results in a high incidence of lymphatic leukemia. Recent findings indicate that this response is specific for mice homozygous for the recessive Ah^d allele (20). Mice possessing a dominant Ah^b allele, and who therefore respond to the treatment by augmented production of the inducible enzyme, aryl hydrocarbon hydroxylase, show a strong skin tumor response rather than leukemia. In this case it is clear that the Ah^d genotype of the mice does not cause the leukemia but merely represents a prerequisite for its occurrence in response to MCA. It is quite possible that the detailed mechanism of this response involves endogenous MuLV as an intermediate, and if so the marked influence of the Ah locus on this disease amply suggests the complexity of the interactions between the host, the virus, and the environment in the genesis of leukemia. The study of genetic factors in mouse leukemia has contributed enormously to our understanding of the disease, but it is clear that we have merely scratched the surface of this complex problem.

REFERENCES

1. Steeves, R.A., and Lilly, F. (1977). Ann. Rev. Genet. 11:277.
2. Tooze, J. (1973). "The Molecular Biology of Tumor Viruses." Cold Spring Harbor Laboratory, Cold Spring Harbor.
3. Jaenisch, R. (1976). Proc. Natl. Acad. Sci. USA 73:1260.
4. Rowe, W.P., and Pincus, T. (1972). J. Exp. Med. 135:429.
5. Rowe, W.P. (1972). J. Exp. Med. 136:1272.
6. Chattopadhyay, S.K., Rowe, W.P., Teich, N.M., and Lowy, D.R. (1975). Proc. Natl. Acad. Sci. USA 72:906.
7. Kozak, C., and Rowe, W.P. (1978). Science 199:1448.
8. Rowe, W.P., and Hartley, J.W. (1972). J. Exp. Med. 136:1286.
9. Lilly, F., Duran-Reynals, M.L., and Rowe, W.P. (1975). J. Exp. Med. 141:882.

10. Rowe, W.P. (1973). Cancer Res. 33:3061.
11. Hartley, J.W., Wolford, N.K., Old, L.J., and Rowe, W.P. (1977). Proc. Natl. Acad. Sci. USA 74:789.
12. McDevitt, H.O., and Benacerraf, B. (1969). Adv. Immunol. 11:31.
13. Klein, J. (1975). "Biology of the Mouse Histocompatibility-2 Complex." Springer-Verlag, New York.
14. Sato, H., Boyse, E.A., Aoki, T., Iritani, C., and Old, L.J. (1973). J. Exp. Med. 138:593.
15. Doherty, P.C., and Zinkernagel, R.M. (1975). J. Exp. Med. 141:502.
16. Blank, K.J., and Lilly, F. (1977). Nature (London) 269:808.
17. Gross, L. (1978). Cancer Res. 38:485.
18. Lieberman, M., and Kaplan, H.S. (1959). Science 130:387.
19. Ball, J.K., and McCarter, J.A. (1971). J. Natl. Cancer Inst. 46:751.
20. Duran-Reynals, M.L., Lilly, F., Bosch, A., and Blank, K.J. (1978). J. Exp. Med. 147:459.

CHROMOSOMAL LOCATION OF C-TYPE VIRUS GENOMES
IN THE MOUSE

Wallace P. Rowe
Janet W. Hartley

Laboratory of Viral Diseases
National Institute of Allergy and Infectious Diseases
National Institutes of Health
Bethesda, Maryland

A unique and important set of recently recognized genes of the mouse are those that are the genomes of endogenous retroviruses (i.e., the RNA viruses that replicate by transcribing the genome into a DNA copy that integrates into cellular DNA). During the course of evolution of many vertebrates, DNA copies of retrovirus genomes have become established as germ line chromosomal elements. In the mouse there are at least two distinct classes of retrovirus in the germ line, i.e., the B-type, or mammary tumor viruses, and the C-type, or leukemia viruses; it is the latter that will be discussed here.

C-type viruses of mice, or murine leukemia viruses (MuLV), fall into three major categories: the ecotropic, xenotropic, and amphotropic classes. Ecotropic viruses are capable of infecting mouse cells but not heterologous species cells; xenotropic viruses infect only heterologous cells; while amphotropic viruses infect both categories. These viruses are closely related to one another in that the virion structural proteins p30 and reverse transcriptase are the same. The viruses differ, however, in the envelope glycoproteins, which confer the type-specific interference, neutralization, and host range phenotypes.

The ecotropic viruses are typified by the classical AKR MuLV; comparable strains are found in other high lymphoma strains of mice, such as C58, PL/J, and C3H/Fg. The most familiar example of a xenotropic MuLV is the endogenous virus of NZB mice, long thought to be a noninfectious MuLV, but shown by Levy and Pincus to be infectious for a variety of non-murine cells. Amphotropic viruses to date are known to occur only in certain populations of wild Mus musculus.

The three classes of MuLV show quite distinct patterns in their natural occurrence. Ecotropic viruses are carried by some strains of mice, but are not found in others. Among those strains that do carry the ecotropic viruses there are marked differences in the level of expression. The high lymphoma strains show high titers of ecotropic MuLV throughout life, with virus appearing in the tissues at almost the time of birth. The virus can be induced from embryonic cells of such mice in tissue culture by treatment with the thymidine analogues 5-IUdR and 5-BUdR, and tissue culture cloning studies show that all cells possess the inducible viral genome.

Other strains of mice, referred to as low-ecotropic-virus yielders, on occasion yield the same MuLV as found in high virus mice, but it only appears late in life and in low titer. Strains with this pattern include BALB/c, C57BL, C57BR, DBA/2, and many others. Other strains, including C57L, NZB, 129, and NIH Swiss and its inbred derivative, NFS/N, have never yielded ecotropic MuLV, either in vivo or in vitro. Nucleic acid hybridization studies with viral cDNA probes have shown that the complete set of genome sequences of ecotropic MuLV is present in the cellular DNA of high- and low-virus mice, but a portion of the sequences, amounting to about 20% of the genome, is not present in DNA of the virus-negative strains.

The high-ecotropic virus phenotype shows classic Mendelian heredity patterns, and a number of the loci for ecotropic virus inducibility have been mapped. The mapped loci include two on chromosome 7, the Akv-1 and Fgv-1 loci, present in AKR/J and C3H/FgLw, respectively; they occupy non-allelic sites, Akv-1 being 12 map units from Gpi-1, and Fgv-1 being very close to Hbb. Two as yet unnamed loci in C57BL/10 and its H-2 congenic derivative B10.BR have also been mapped; one is on chromosome 8, linked to Es-1, and the other is on chromosome 11, linked to Es-3.

Strains of mice that are high for expression of ecotropic MuLV generally carry multiple loci for virus induction. AKR carries two such loci, C3H/Fg carries three, and C58 carries three or possibly four. The multiple loci are probably the result rather than the cause of the high virus phenotypes, since hybrids carrying only one virus-inducing locus per cell usually show as high virus expression as do the inbred strains with homozygosity at multiple loci. Indeed, we have seen on several occasions appearance of new loci for virus induction in congenic mice carrying the Akv-1 locus.

While much mapping of the ecotropic genomes in the various strains of mice remains to be done, it seems very likely that they are not at allelic sites in different strains of laboratory mice. Rather the loci must be thought of as being strain specific.

That these virus-inducing loci indeed represent viral genomes has been shown clearly by the nucleic acid hybridization studies of Chattopadhyay et al. (1, 2). When cellular DNA of various mouse strains is tested for the ability to hybridize with single stranded radiolabeled DNA probes prepared for ecotropic MuLV, and the number of copies of the virus-related sequences is determined from the kinetics of the reaction, a generally excellent correlation is found between the biologic and biochemical observations. The quantitative hybridization analyses show that there are two major populations of viral sequences in the cell DNA. One set, which corresponds to the genes for the MuLV group-specific proteins, is present at a frequency of roughly 50 copies per haploid genome. A second set of sequences, corresponding to the type-specific (presumably envelope) genes, is present in a small number of copies, ranging from 1 to 3 or 4 copies per haploid genome in the ecotropic virus-positive strains, and being absent in the ecotropic virus-negative strains. The strains of mice that show multiple loci on Mendelian segregation analysis show multiple copies by hybridization kinetics, while those that are low in virus or are high-virus strains with a single locus, such as PL and RF, show a lower copy number by hybridization. In segregating crosses of hybrids carrying the Akv-1 locus on the ecotropic virus-negative NIH Swiss background, the presence of the type-specific nucleic acid sequences segregated with virus inducibility and with chromosome 7 markers, establishing that viral DNA sequences are indeed present at this locus.

The pattern of xenotropic virus distribution among mouse strains is somewhat different. While there are marked differences in expression of xenotropic virus between mice, it seems probable that all strains of mice have the capacity to produce the virus. NZB mice regularly show high levels of spontaneous expression, while some other strains manifest lower levels of spontaneous virus production or are IUdR-inducible in tissue culture. In still other strains, xenotropic virus has been detected only on rare occasions, and it is not inducible from their cells in culture; strain 129 has not yielded virus, but it expresses MuLV glycoproteins with xenotropic antigenic specificity. As with the ecotropic virus, in the mouse strain with high level expression of xenotropic virus (NZB), Mendelian segregation analysis shows that there are multiple loci for virus induction, while in the strains with lower inducibility levels there is only a single locus. Nucleic acid hybridization studies have shown that the complete genome of xenotropic viruses is present in the DNA of all mouse strains tested, and in multiple copies, ranging from 2 to as many as 8 per haploid genome. In this case, the number of copies does not correlate with the biologic activity. Whether the multiple copies are clustered or scattered throughout the genome is not yet known. The mapping studies to date have succeeded in locating loci for induction of xenotropic virus in three cases, C57BL/10, BALB/c and the Lp (loop tail) linkage testing stock; in all three instances the locus is on chromosome 1, 10 to 20 map units from <u>Dip-1</u>. Whether the three loci are allelic has yet to be established.

The amphotropic MuLV strains show a third pattern of natural history. As mentioned, these viruses have been detected only in certain colonies of feral mice, and have never been detected in laboratory strains. In their occurrence in wild mice, transmission seems to be primarily maternal, with infection by transuterine or milk-borne routes; genetic transmission has not yet been clearly established. However, the complete set of viral genome sequences is present in the cellular DNA of all mice, both feral and laboratory; and, again, the subset of viral sequences that corresponds to type-specific virion components is present in multiple copies, differing in number in the inbred strains. It seems quite possible that the amphotropic genomes are not present in the chromosomal DNA as complete linear copies, but they may require genetic recombination between products of separated loci before virus of this type is formed.

Indeed, there is much mounting evidence that new viruses are being generated constantly within the lifetime of high-virus mice. A new class of viruses called MCF (for "mink cell focus inducing") strains, since they induce foci of cell damage or transformation in a tissue culture line of mink lung cells, emerges in high lymphoma strains prior to the time of onset of lymphomas. These viruses represent unique genetic recombinants between ecotropic virus and some as yet unidentified class of endogenous MuLV sequences, perhaps representing xenotropic genomes. Each of these viruses appears to be a unique recombinant in the envelope gene region and as such does not represent the product of a single chromosomal locus, but is rather the outcome of an interaction between endogenous viruses coded by two (or more?) separate loci.

REFERENCES

1. Chattopadhyay, S.K., Lowy, D.R., Teich, N.M., Levine, A.S., and Rowe, W.P. (1974). Cold Spring Harbor Symposium on Quantitative Biology 39:1085.
2. Chattopadhyay, S.K., Hartley, J.W., Lander, M.R., Kramer, B.S., and Rowe, W.P. J. Virol., in press.
3. Datta, S.K., and Schwarz, R.S. (1977). Virology 83:449.
4. Gardner, M.B. Current Topics in Microbiology and Immunology, in press.
5. Hartley, J.W., Wolford, N.K., Old, L.J., and Rowe, W.P. (1977). Proc. Natl. Acad. Sci. USA 74:789.

6. Kozak, C., and Rowe, W.P. (1978). Science 199:1448.
7. Lilly, F., and Pincus, T. (1977). Adv. Cancer Res. 17:231.
8. Rowe, W.P. (1973). Cancer Res. 33:3061.
9. Rowe, W.P. (1977). The Harvey Lectures Series 71:173.

EXPRESSION OF XenCSA, A CELL SURFACE ANTIGEN
RELATED TO THE MAJOR GLYCOPROTEIN (gp70) OF XENOTROPIC
MURINE LEUKEMIA VIRUS, BY LYMPHOCYTES OF INBRED MOUSE STRAINS

Thomas M. Chused

Laboratory of Microbiology and Immunology
National Institute of Dental Research

Herbert C. Morse III

Laboratory of Microbial Immunity
National Institute of Allergy and Infectious Diseases
National Institutes of Health
Bethesda, Maryland

There are several classes of endogenous murine leukemia viruses (MuLV) which can be defined by host range and interference patterns (1). Ecotropic MuLV, such as the Gross-AKR type, can only infect mouse cells (2). Xenotropic MuLV are able to infect a wide range of xenogeneic cells but are almost totally incapable of infecting mouse cells (3). Amphotropic MuLV, recovered from wild mice, are able to infect both murine and non-murine cells (1, 4). The genetic information coding for xenotropic and amphotropic viruses is integrated into the chromosomal DNA of all inbred mouse strains (5) while that coding for ecotropic virus is present in the DNA of most but not all strains (6, 7). The physical structure of all three classes of MuLV is similar (Fig. 1). There are three structural genes: *gag* for "group specific

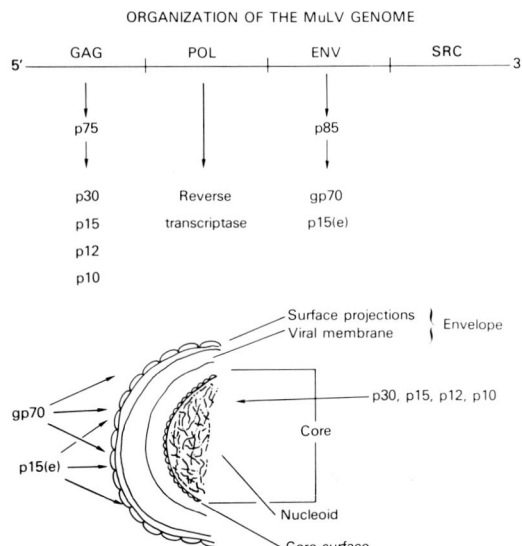

FIGURE 1. Genome and structural components of type-C murine leukemia viruses.

antigen" codes for a 70,000 dalton polyprotein precursor which is cleaved to produce the internal proteins p30, p15, p12 and p10 ("p" for protein, the number referring to its size in 1,000's of daltons). Pol produces reverse transcriptase. Env codes for a glycosylated 90,000 dalton precursor which is split into gp70 and p15e, the "gp" indicating a glycoprotein and "e" standing for "envelope" to distinguish it from the core-associated p15 (8). Of the 70,000 daltons of gp70, 30,000 is carbohydrate. Gp70 is the external coat protein of the virion (9). It determines, in large part, the host range of the virus, presumably by mediating attachment to the target cell.

Three recent observations have further focused our attention on MuLV gp70. First, Jensen, Lerner and co-workers developed a method of fingerprint analysis of gp70 by chromatography of tryptic peptides of purified and iodinated gp70 (10). This technique has revealed a surprisingly extensive polymorphism of gp70 that may be as extensive as that of H-2. While the fingerprints of gp70 from different MuLV isolates fall into families which correspond to the three host range classes, the only identical patterns to date are those of the viruses coded by the Akv-1 and Akv-2 genes (11). The second finding concerns Hartley and Rowe's isolation of novel viruses from spontaneous lymphomas in several strains of mice which are infectious for both murine and xenogeneic cells and form foci on mink lung cells (12). They are termed "MCF" for "mink cell focus-inducing." Isolates of this class of MuLV possess gp70 molecules which, by fingerprint analysis and neutralization testing, appear to be genetic recombinants between ecotropic and xenotropic viruses (11). The third observation is that gp70's which resemble MuLV proteins antigenically and by fingerprint analysis are found in the serum and on the cell surface of sperm and certain epithelial tissues in the absence of virus production (13, 14). Because the products of the many genes that must be present to code for various gp70's may function as differentiation antigens or be involved in cellular regulation and would be expected to mediate such functions at the cell surface, our group has undertaken the long-term study of the types of gp70 present on cell membranes and their genetic control.

The studies to be described in this chapter concern the expression on mouse lymphocytes of gp70 closely related to that of xenotropic MuLV. We have produced an antiserum which reacts with xenotropic MuLV by immunizing New Zealand White rabbits with SIRC (Serum Institute Rabbit Corneal) cells infected with xenotropic MuLV from NZB mice (15). By virus neutralization, which depends on antibody to gp70, the antiserum is quite specific for xenotropic virus (Table 1). Not

TABLE 1. Neutralization of Various Strains of MuLV by
Antiserum to NZB Xenotropic MuLV Infected SIRC Cells

Virus Group	Strain	Serum neutralizing titer
Xenotropic	BALB-IU-1	640
	NZB-IU-1	320
	NZB-6	160
	NZB-IU-6	320
	NZB-S-1	640
	AKR-T-6	320
Ecotropic	M-MLV	<20
	AKR-L1	<20
Amphotropic	4070-A	<20
Mink cell focus	247	80

surprisingly, it also neutralizes the recombinant MCF virus. To determine exactly what antibody specificities are present in the serum we radioiodinated disrupted xenotropic MuLV, exposed it to the antiserum, and analyzed the precipitate on polyacrylamide gels (Fig. 2). With saturating amounts of antiserum, gp70, p30 and two smaller components are precipitated (Fig. 2a). When limiting amounts of reagent are used, only the gp70 is brought down, indicating that the highest titer is against gp70 (Fig. 2b).

To detect gp70 on the cell surface we have used the technique of flow microfluorometry as embodied by the fluorescence-activated cell sorter (Fig. 3) (16). The cells to be examined are entrained in the center of a rapidly flowing jet of liquid which is traversed by a laser beam. Molecules of fluorochrome present on the cells are excited by the laser beam and their

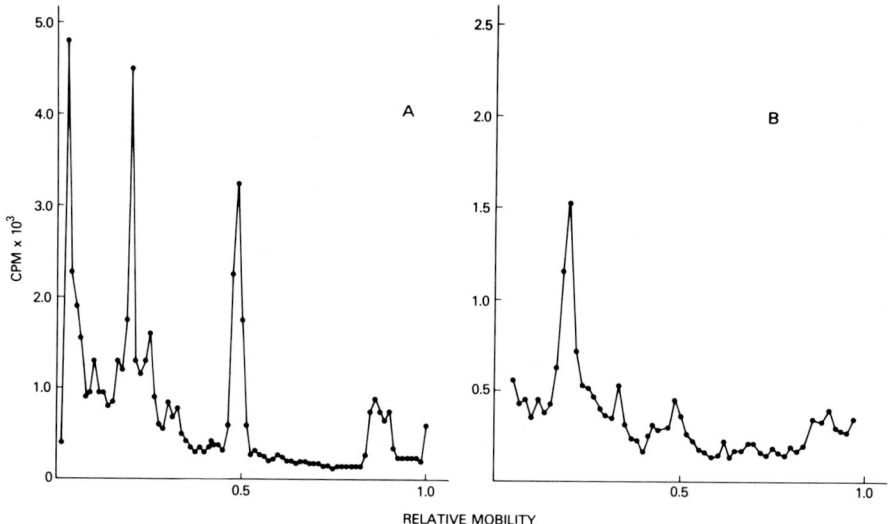

FIGURE 2. SDS-polyacrylamide gel electrophoresis analysis of iodinated xenotropic MuLV precipitated by rabbit antiserum to xenotropic MuLV-infected SIRC cells at saturating level of antiserum (5 µl - panel a) and limiting level (0.1 µl - panel b).

fluorescence is detected by a photomultiplier tube perpendicular to both the liquid jet and the laser beam. The output is presented as a histogram of the number of cells (ordinate) in each of the 1,000 channels into which the fluorescence distribution (abscissa) is divided.

By utilizing the computer associated with the system to determine the mean channel number of the fluorescence distribution after staining with the fluoresceinated antiserum, we obtain a highly reproducible measure of the amount of gp70 present on the cell. This is given in arbitrary, but constant fluorescence units. We have done this for non-SIRC

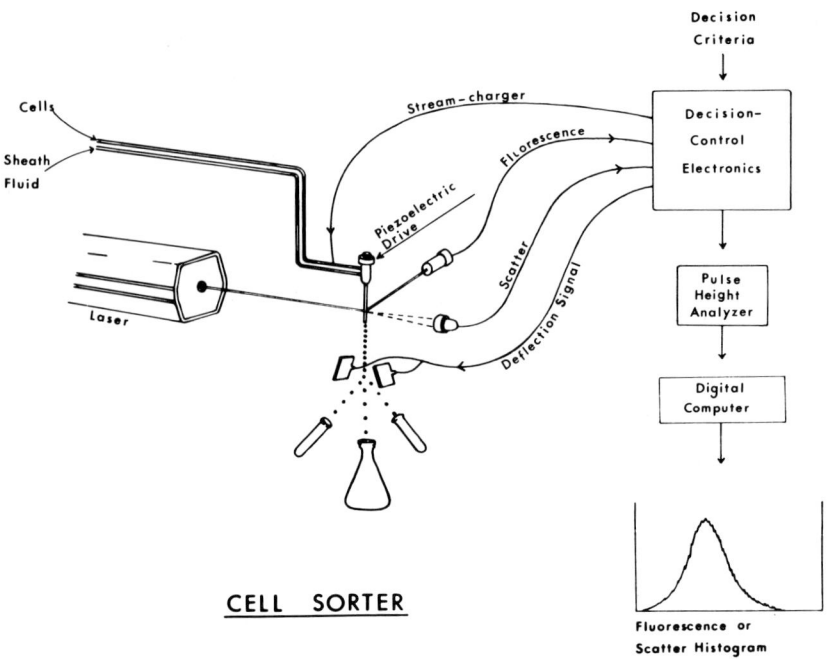

FIGURE 3. Schematic diagram of the fluorescence activated cell sorter (courtesy of Dr. John Weinstein).

cell lines infected with different viruses after staining with the antisera to NZB xenotropic MuLV-infected SIRC cells (Table 2). There is 10 to 15 times more staining of cells infected with xenotropic virus than of cells infected with amphotropic or ecotropic virus, implying relative specificity for xenotropic MuLV-coded gp70. Despite the differences in fingerprint patterns of gp70 obtained from different xenotropic viruses, mink cells infected with xenotropic MuLV isolates from NZB, NZW, DBA/2, AKR, and BALB/c mice all stain brightly with the antiserum to the NZB virus, and antisera to the BALB xenotrope give results that are virtually identical to those obtained with the anti-NZB xenotropic MuLV serum.

Table 2. Staining of Cell Lines Infected with Various MuLV by Antiserum to NZB Xenotropic MuLV Infected SIRC Cells

Cells	Virus		Mean fluorescence
	Class	Strain	
Mink lung	-	-	(218)
	Xenotropic	NZB-6	65,100
	Xenotropic	BALB-IU-2	42,800
	Amphotropic	4070-A	8,440
NIH-3T3	-	-	(270)
	Ecotropic	AKR-2T	4,000

The immunoglobulin was isolated from this antiserum, digested with pepsin to produce $F(ab')_2$ fragments, fluoresceinated, and absorbed with C57BL/6 spleen cells to remove heteroantibody. We have termed the gp70 antigen detected by this type of antiserum XenCSA, for Xentropic envelope-related cell surface antigen.

The staining of NZB lymphocytes by this reagent is shown in Figure 4. The left panel shows a unimodal distribution of XenCSA on NZB thymocytes. In contrast, the right panel shows that NZB spleen cells display a bimodal distribution of XenCSA with a narrow peak of dull cells and a broad shoulder of more brightly staining cells. The dull peak is the T cells and the bright peak the B cells. This was shown by staining NZB spleen cells with anti-mouse immunoglobulin and separating the non-staining T cells from the surface immunoglobulin-bearing B cells by the cell sorter. The separated T cells were stained with anti-XenCSA and rerun on the sorter (Fig. 5). They gave the dotted profile with a peak which is superimposable on the dull peak of the unseparated spleen cells. The brighter cells are missing and hence must be the surface

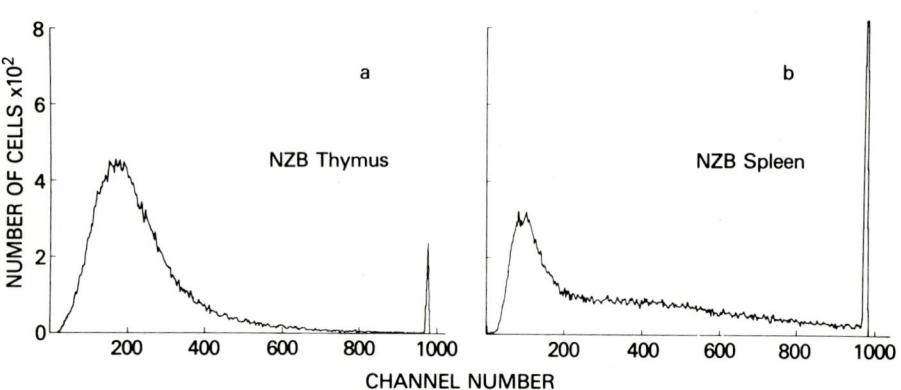

FIGURE 4. XenCSA expression by NZB thymocytes (a) and spleen cells (b) determined by flow microfluorometry (FMF) with the fluorescence activated cell sorter after staining with fluorescein-conjugated $F(ab')_2$ from the antiserum to xenotropic MuLV-SIRC.

immunoglobulin-bearing B cells which were removed. Thus the level of XenCSA expression by NZB lymphocytes depends on their differentiation pathway. We have obtained the same result with NZW and DBA/2 spleen cells.

Although the antiserum to xenotropic MuLV is primarily directed against gp70, it was necessary to determine if this was the only cell surface molecule recognized by the antiserum. To this end, spleen cells from several strains were surface labeled with ^{125}I, and detergent extracts of them were precipitated with the antisera and analyzed on polyacrylamide gels (Fig. 6). In each case, the only antigenic species recognized by the antisera had a molecular weight of 70,000 daltons. This reaction was abolished by absorption

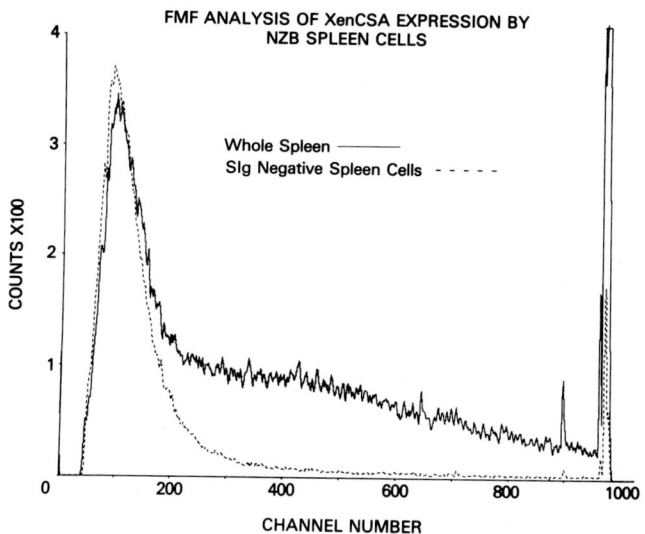

FIGURE 5. XenCSA expression by NZB spleen cells (solid line) is bimodal. If the surface immunoglobulin-bearing B cells are removed and the remaining surface immunoglobulin negative T cells are stained for XenCSA (dotted line), only the lower peak remains, indicating that the brighter cells in the original population are B cells.

of the antiserum with xenotropic MuLV-infected mink cells but not uninfected mink cells. Similarly, only a 70,000 dalton protein was precipitated by the antiserum from iodinated cell surface extracts of the exogenously infected cell lines. The uptake of this molecule by a lentil lectin column indicates that it is a glycoprotein. The lack of precipitation by antibody to p30 indicates it is not the gag polyprotein.

Thus prepared, we set out on a tour through the subject of this book--the inbred strains of mice. We found great variation in XenCSA expression (17). Examples of the three

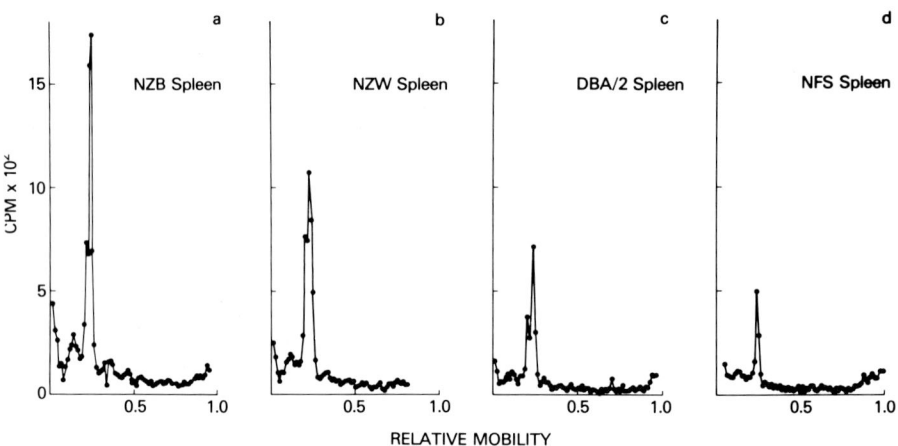

FIGURE 6. SDS-polyacrylamide gel electrophoresis analysis of molecules precipitated by anti-xenotropic MuLV antiserum from detergent lysates of surface iodinated spleen cells from four strains of mice. Only a molecule with a molecular weight of approximately 70,000 daltons was precipitated.

observed phenotypes are shown in Figure 7. DBA/2, NZB, and NZW are the most positive strains with thymocytes and both T and B cells of the spleen staining. A large group of strains resemble RIII with lower levels of staining of only a portion of spleen cells, presumably B cells, and no thymocyte staining. Another large group, represented by strain MA, had low levels of XenCSA expression on both thymocytes and spleen cells. The distributions of the averages of the computer-derived means of XenCSA staining for four or more mice of each of 99 inbred strains are shown in Figure 8. The triangles are the spleen cell means, the dots the thymocyte means, and the squares the sum of both. Both distributions show a break at 60 so we have arbitrarily scored means above 60 as

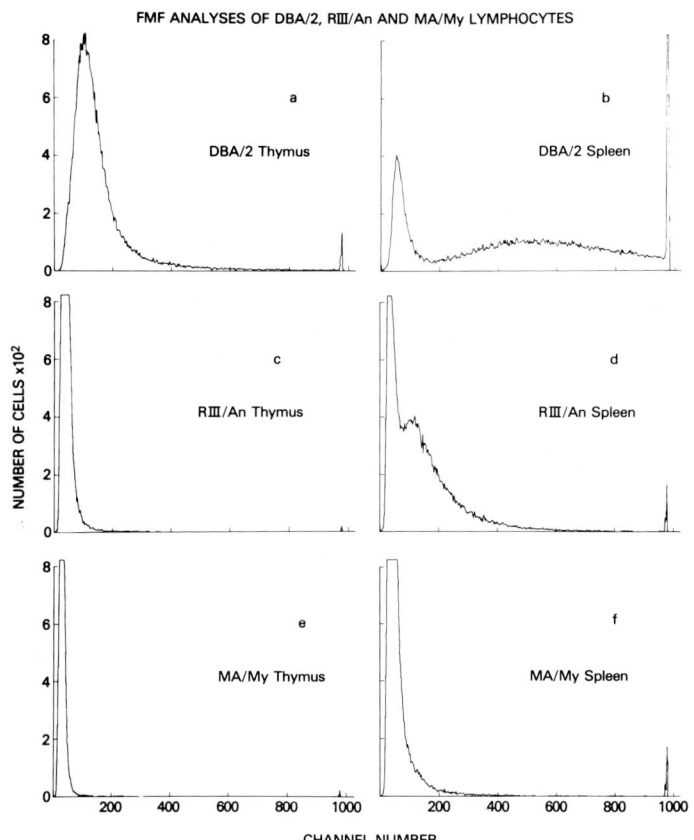

FIGURE 7. FMF analysis of thymocytes (a, c, e) and spleen cells (b, d, f) from strains representative of the three basic patterns of XenCSA: 1) DBA.2J (high thymus, high spleen); 2) RIII/AnN (low thymus, high spleen); and 3) MA/MyJ (low thymus, low spleen).

high and below as low for XenCSA expression. Strains expressing high XenCSA levels on both thymocyte and spleen cells are listed in Table 3.

In addition to NZB, NZW, and DBA/2, the C57BL/6 congenic for G_{IX} and the C57BR-G_{IX}^+ mutant are XenCSA positive. Spleen

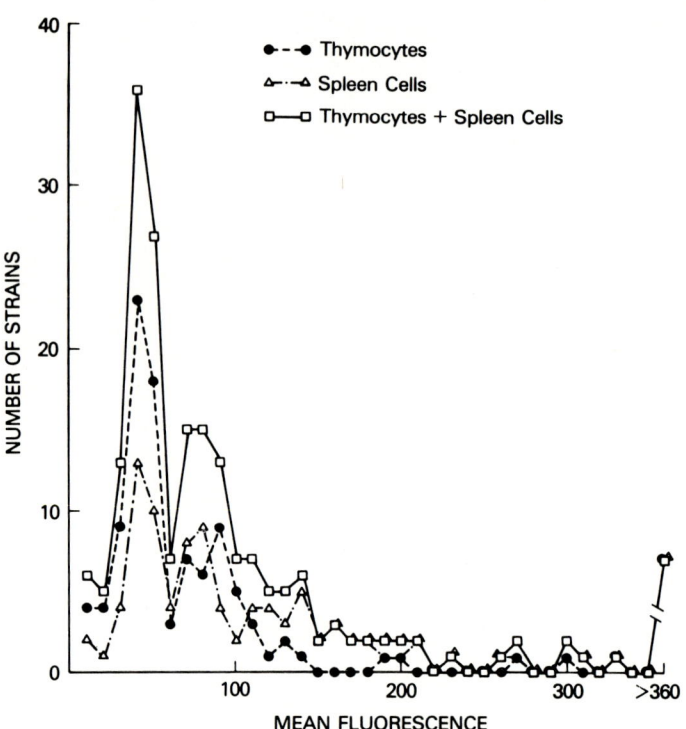

FIGURE 8. Mean fluorescence values of spleen cells and thymocytes from 99 strains of inbred mice. Each point represents the number of strains with mean fluorescence values falling in that decile.

high, thymus low strains are given in Table 4. These include AKR and C57BL/6. Finally, the strains with low XenCSA levels on both populations are shown in Table 5. The prototype G_{IX} positive strain, 129, and the non-mutant C57BR/cd are in this group. These data show that expression of XenCSA varies markedly between strains. XenCSA levels do not correlate with alleles of the Tla, Pca-1 or H-2 loci or with Gross cell surface antigen. XenCSA expression does show partial correlation with G_{IX}. An illustration of the lack of H-2 effect

TABLE 3. Inbred Strains Expressing High Levels
of XenCSA on Thymocytes and Spleen Cells

	Mean fluorescence	
Strain	Thymus	Spleen
NZB/BlN	270	393
NZW/BlN	134	410
DBA/2J	130	424
C57BL6-G_{IX}^{+}	108	127
CE/J	99	137
C58/J	96	184
LG/J	94	146
C3H/AnfCum	88	114
SM/J	87	330
C57BR-G_{IX}^{+}	79	121
ST/bJ	78	121
A/J	67	76

and an indication of the reproducibility of the method is shown in Table 6. A/SnGrf and three H-2 congenics derived from this strain have very similar XenCSA levels. The same is true of C3H/DiSnGrf and the congenic C3H.SW. Two of the B10 congenics resemble the parent strain. B10.D2/(n) which was derived from the XenCSA negative B10.D2/(o) by further backcrossing to B10 is XenCSA positive, strongly suggesting that a mutation occurred during the derivation of this strain.

We have begun to investigate the genetic control of XenCSA expression. One portion of these studies utilized recombinant inbred strains, derived by brother-sister matings of the F_2 cross of two inbred strains. We have examined 24 such strains derived from the strongly XenCSA positive DBA/2 (D2) and the low XenCSA C57BL/6 (B6) strains (Morse, H.C. III,

TABLE 4. Inbred Strains Expressing High Levels of XenCSA on Spleen Cells But Not on Thymocytes

	Mean fluorescence	
Strain	Thymus	Spleen
C3Hf/FgLw	56	136
AU/SsJ	42	136
C3H/BiMai	48	128
DW	43	116
HRS/J	36	108
RIII/AnN	17	100
C3H/DiSnGrf	46	97
SJL/J	44	89
PL/J	37	88
C57BL/6J	50	80
AKR/J	46	75
CBA/N	48	73
LP/J	52	71
AL/N	37	68
DBA/1J	43	66

Taylor, B., et al., in preparation). XenCSA expression by the (B6D2)F_1 was intermediate between the parental levels. The XenCSA levels of the recombinant inbred lines ranged from BXD-2, which resembles B6, to BXD-14, which matches the D2 profile (Fig. 9). Many of the lines, despite being homozygous at 99% of their genetic loci, were intermediate between the parental profiles. Averages of the log mean fluorescence for 6 to 11 samples of thymus and spleen for

TABLE 5. Inbred Strains Expressing Low Levels of XenCSA on Thymocytes and Spleen Cells

Strain	Mean fluorescence	
	Thymus	Spleen
129/J	45	41
101/Cum	44	38
C57L/J	42	47
C57BR/cdJ	39	42
SWR/J	37	41
P/J	32	51
BALB/cAnN	31	38
RF/J	28	50
SEC/1ReJ	24	40
C57BL/10J	15	27
MA/MyJ	5	33
BDP/J	4	12

TABLE 6. Expression of XenCSA on Lymphocytes of Mice Congenic for H-2

Background strain	H-2 congenic	Mean fluorescence	
		Thymus	Spleen
A/SnGrf		67	69
	A.SW/J	86	109
	A.CA/J	69	85
	A.By/J	68	72
C3H/DiSnGrf		46	97
	C3H.SW/SnJ	41	67
C57BL/10ScSnJ		26	25
	B10.A/SgSnJ	23	21
	B10.D2/(o)J	33	34
	B10.D2/(n)J	103	181

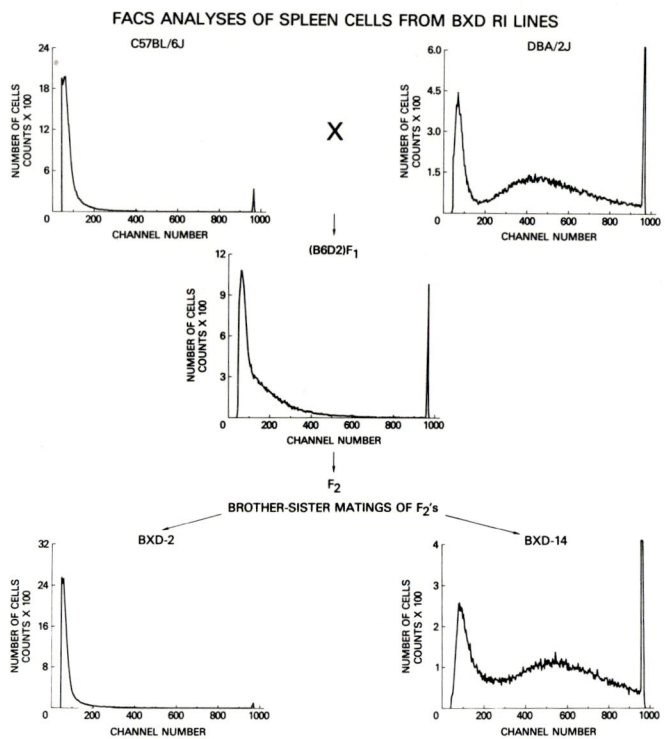

FIGURE 9. XenCSA expression by spleen cells from C57BL/6J by DBA/2 (BXD) recombinant inbred (RI) lines. The parental, F_1 and two RI lines (low XenCSA BXD-2 and high XenCSA BXD-14) are illustrated.

each line are shown in Figure 10. The stippled areas include the mean and 95% confidence limits of the parental strain XenCSA levels. Spleen values, in the lower panel, show partial correlation with thymus levels. For the purpose of genetic analyses, lines with both spleen and thymus values within the range of one of the parents were classified as being of that type. The remainder were assigned to the parental type whose range contained the majority of the individual mean fluorescence values. By this latter criterion, BXD-1 and 12 are considered D2-like and BXD-23, 8, 21, 15 and 6 are

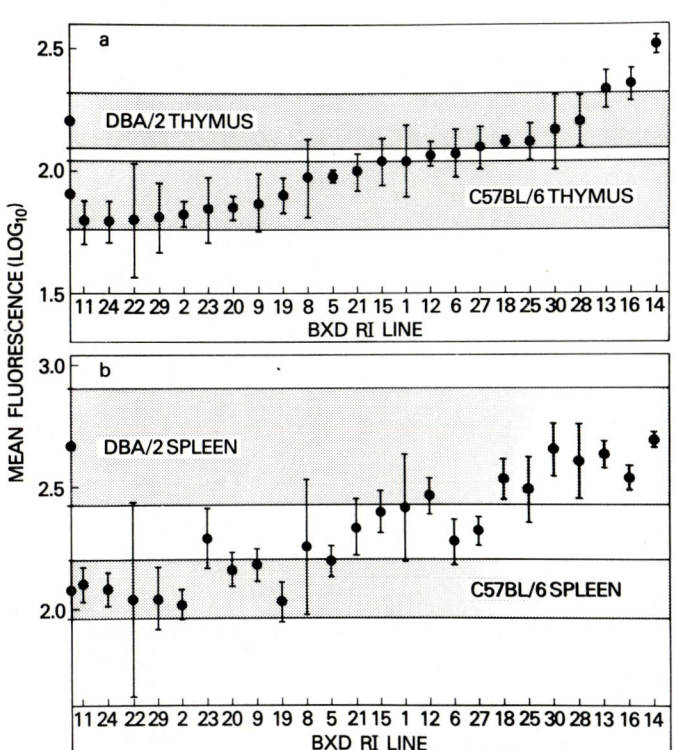

FIGURE 10. XenCSA expression by thymocytes and spleen cells of BXD RI lines. See text for details.

scored as B6-like. Comparison of these somewhat arbitrary XenCSA assignments with the alleles at the 60 loci for which these lines have been characterized shows a close association with the Fv-1 type (Table 7). The D2 allele, $Fv-1^n$, is associated with high XenCSA levels and the B6 type, $Fv-1^b$, with low XenCSA expression. The correlation with Gpd-1, which is 1 map unit from Fv-1, is similar. Three of the four discordant lines, 5, 8 and 21, have intermediate XenCSA levels which makes their classification uncertain. BXD-25 appears

TABLE 7. Inheritance of Chromosome 4 Markers in C57BL/6J x DBA/2J (BXD) Recombinant Inbred Lines

	$Fv\text{-}1^n$ (D2)	$Fv\text{-}1^b$ (B6)
XenCSA high (D2)	9	1
XenCSA low (B6)	3	11

P = .001 (Fisher)
Recombination frequency = .055

	$Gpd\text{-}1^b$ (D2)	$Gpd\text{-}1^a$ (B6)
XenCSA high (D2)	8	2
XenCSA low (B6)	3	11

P = .006 (Fisher)
Recombination frequency = .077

clearly D2-like despite being $Fv\text{-}1^b$ and is being studied further to see if it is a true recombinant. Taking the data at face value gives an apparent recombination frequency of 5% between Fv-1 and the gene controlling XenCSA. To confirm this linkage we have examined 54 (B6D2)F_1 x D2 backcross mice. Again, a spectrum of XenCSA expression is observed. Twenty-six of the 54 were in the D2 range, 5 in the B6 range, and 23 were intermediate between the parental strains. The result of Gpd-1 typing all and Fv-1 typing 28 of these mice was the same as with the recombinant inbred lines: an apparent interval of 7 map units between Fv-1 and the gene influencing XenCSA.

There were three apparent recombinants between Fv-1 and the XenCSA controlling gene in this group. One, which typed as $Fv\text{-}1^n$ but had low XenCSA expression, was progeny tested by mating with D2. It did not breed true since all six offspring scored high for XenCSA. With respect to the DBA/2 system, we conclude that the effect of this locus is regulatory rather than structural since XenCSA can still be expressed in the presence of the low allele. The data at hand do not allow us to distinguish between a pleiotropic effect of Fv-1 or the effects of a closely linked gene which regulates XenCSA independently of Fv-1. We feel the closeness of the linkage to Fv-1 makes quasi-linkage, as described for Fv-1-incompatible crosses in the G_{IX} system (18), unlikely.

Finally, we would like to mention several observations made during the genetic analysis of infectious xenotropic MuLV and XenCSA in NZB mice which illustrate the problems and complexities of XenCSA typing (Chused, T.M., et al., in preparation). As our background strain we have used NFS, an inbred Swiss, $Fv-1^n$, mouse from which we have not recovered infectious ecotropic or xenotropic virus. Seventy-five (NZB x NFS)F_1 x NFS backcross mice were tested for virus production and XenCSA expression. In agreement with the studies of Datta and Schwartz (19), we also find that NZB contains two independently segregating xenotropic viruses. The first produces large numbers of foci when spleen cells are cocultivated with mink lung cells carrying the mouse sarcoma genome but not the leukemia virus genome (20), which we term the S^+L^- virus. Sixty-two percent of the backcross mice each produced about 40% of the number of foci given by the parental NZB spleen cells in a clear-cut, all-or-none fashion indicating semi-dominant inheritance. In subsequent backcross generations 51.5% of the progeny have been S^+L^- positive. The second virus does not give foci in the mink S^+L^- cells but grows in uninfected mink cells and is detected by cytoplasmic immunofluorescence with antibody to p30. It can only be tested for in the absence of the S^+L^- virus which also grows in mink cells. Fourteen of the 29 S^+L^- negative backcross mice produced the second virus which we call "FA."

TABLE 8. Effect on XenCSA of Selecting for Virus Production in Successive Backcrosses of NZB Genes into NFS

Selected for	Virus production	XenCSA (Mean fluorescence)			
		Backcross generation			
		2	3	4	5
Mink S^+L^- virus	+	200	167	196	117
	−	113	65	100	56
Mink S^+L^- virus	+	301	222	199	
& high XenCSA	−	170	143	142	
Mink S^+L^- virus	+	102	152		
& low XenCSA	−	47	56		
Mink FA virus	+	146	84	118	
	−	83	81	55	

The most interesting result is that 84% of the backcross mice were XenCSA positive.

We are now deriving several lines from the first backcross group by successive backcrosses of males with females of the virus negative NFS strain. Table 8 shows the mean XenCSA levels in the virus positive and virus negative members of each generation of backcross progeny. For reference, the NFS background strain has a mean XenCSA level of 40 with the 95% confidence limits being 4 to 77. There are several points to be made: First, although it is not indicated by the data in the table, the variation in XenCSA levels among both virus positive and virus negative segregants of these crosses is much greater than that observed in the inbred lines we have examined. Second, production of virus is associated with higher XenCSA levels. Third, there are mating-to-mating variations in mean XenCSA levels. The

S^+L^- line being selected for average XenCSA level has shown cyclic variation over four generations. This is suggestive of maternal antibody or some other suppressive factor. Fourth, the S^+L^- line being selected for low XenCSA has so far not bred true, again suggesting an influence of factor(s) other than the viral genome on XenCSA expression. Fifth, in the line which carries both viruses and was derived from one of the five first backcross mice which expressed XenCSA levels as high as NZB, all 63 progeny through N4 are XenCSA positive, implying a very large number of XenCSA producing genes or an unusual mechanism. The same phenomenon was observed in N4 of the S^+L^- line where the mean XenCSA level of the virus-negative segregants was 100, even though the virus-negative mice of the preceding generation did not have significantly elevated XenCSA expression. We have not been able to demonstrate infectious spread of XenCSA when tested by several different methods and do not have an explanation for this phenomenon. Sixth, there is a line at N3 which expresses XenCSA in the absence of both viruses. We have not yet tested enough progeny to determine the segregation ratio. Thus NZB mice contain at least three phenotypically distinct semi-dominant genes that affect XenCSA expression by lymphocytes.

We anticipate that future studies will determine the genetic control of XenCSA expression and its relationship to cell differentiation. We are also investigating its role in the autoimmune disease of NZB mice.

REFERENCES

1. Hartley, J.W., and Rowe, W.P. (1976). J. Virol. 19:19.
2. Gross, L. (1951). Proc. Soc. Exp. Med. 78:342.
3. Levy, J.A., and Pincus, T. (1970). Science 170:326.

4. Rasheed, S., Gardner, M.B., and Chan, E. (1976). J. Virol. 19:13.
5. Callahan, R., Benveniste, R.E., Lieber, M.M., and Todaro, G.J. (1974). J. Virol. 14:1394.
6. Chattopadhyay, S.K., Rowe, W.P., Teich, N.M., and Lowy, D.R. (1975). Proc. Natl. Acad. Sci. USA 72:906.
7. Chattopadhyay, S.K., Lowy, D.R., Teich, N.M., and Rowe, W.P. (1975). Cold Spring Harbor Symp. Quant. Biol. 39:1085.
8. Bolognesi, D.P., Montelaro, R.C., Frank, H., and Schafer, W. (1978). Science 199:183.
9. Witte, O.N., Weissman, I.L., and Kaplan, H.S. (1973). Proc. Natl. Acad. Sci. USA 70:36.
10. Elder, J.H., Jensen, F.C., Bryant, M.L., and Lerner, R.A. (1977). Nature 267:23.
11. Elder, J.H., Gautsch, J.W., Jensen, F.C., Lerner, R.A., Hartley, J.W., and Rowe, W.P. (1977). Proc. Natl. Acad. Sci. USA 74:4676.
12. Hartley, J.W., Wolford, N.K., Old, L.J., and Rowe, W.P. (1977). Proc. Natl. Acad. Sci. USA 74:789.
13. DelVillano, B., Nave, B., Croker, B.P., Lerner, R.A., and Dixon, F.J. (1975). J. Exp. Med. 141:172.
14. McClintock, P.R., Ihle, J.N., and Joseph, D.R. (1977). J. Exp. Med. 146:422.
15. Morse, H.C., Chused, T.M., Boehm-Truitt, M., Mathieson, B.J., Sharrow, S.S., and Hartley, J.W. Submitted for publication.
16. Loken, M.R., and Herzenberg, L.A. (1975). Ann. N.Y. Acad. Sci. 254:163.
17. Morse, H.C., Chused, T.M., Mathieson, B.J., Sharrow, S.O., and Hartley, J.W. Submitted for publication.

18. Stockert, E., Boyse, E.A., Sato, H., and Itakura, K. (1976). Proc. Natl. Acad. Sci. 73:2077.
19. Datta, S.K., and Schwartz, R.S. (1976). Nature 263:412.
20. Peebles, P.T. (1975). Virol. 67:288.

BIOLOGY OF MAMMARY TUMOR VIRUSES[1]

Robert D. Cardiff
Bruce W. Altrock

Department of Pathology
School of Medicine
University of California
Davis, California

Mouse mammary tumors are the result of interactions between 1) a tumor virus, 2) a genetically susceptible host, 3) hormones, 4) the hosts' immunological responsiveness, and 5) other factors. The understanding of this complex system has been enriched by contributions from many disciplines including virology, genetics, endocrinology, and immunology. Advances in these fields have complemented one another when applied to the mouse mammary tumor model. As a result, the mouse mammary tumor system has become a prototype for research in solid tumors. This review is designed to alert the readers to the current concepts of the relationship of one factor, the mouse mammary tumor virus (MMTV), to the others involved in mouse mammary tumor biology. It ends on a speculative theme anticipating future developments in the field.

---

[1]This work was supported in part by NIH-NCI contract NO1 CP61013 within the Virus Cancer Program, by NIH grant 21454, and PHS training grant 05245.

Early studies of mouse mammary tumors emphasized hereditary factors. Inbred strains of mice were, in part, developed to study the genetics of this and other neoplasms. As a result of the development of these strains, investigators in Europe and the United States were able to study the genetics of low and high breast cancer strains of mice. These experiments culminated in the recognition by the staff of the Jackson Laboratories and by Korteweg in the Netherlands of an extrachromosomal factor causing mammary adenocarcinomas in mice (1, 2). Shortly thereafter, Bittner discovered that this factor was transmitted via the milk of high tumor incidence mothers to their offspring (3). This factor became known as the Bittner agent, milk agent, mouse mammary tumor agent, and, following morphological and biophysical characterization (4, 5), the mouse mammary tumor virus.

The mouse mammary tumor virus is similar in many respects to other RNA tumor viruses (6). As with the other members of the Retraviridae family, MMTV has a high molecular weight polyploid RNA genome (7, 8) which is associated with the characteristic Retraviridae enzyme RNA-dependent DNA polymerase and is located within the viral core (9). The core is surrounded by a lipid envelope which has external projections or spikes. The distinctive eccentric location of the core within this virus led Bernhard to call the mature mouse mammary tumor virion "B-particle" to distinguish its structural arrangement from the C-type retrovirus particles (10). The MMTV core apparently arises from a precursor, the intracytoplasmic A-particle (11, 12), which contains RNA-dependent DNA polymerase and the internal viral polypeptides (p28, p14 and p10) in precursor form (12, 13). The virion buds from the surface of infected cells where it acquires its lipid envelope and surface projections (6, 14). These projections are thought to be composed of two major glycoproteins with molecular weights of 52,000 and 36,000 daltons (15, 16). The mouse mammary tumor virus is the prototype virus for the type B oncornaviruses because of its biophysical, biochemical, and morphological characteristics and because of its tumorigenic capabilities (17).

The genetic information of MMTV can be found in two phases, RNA and DNA (18). The DNA phase is integrated into the chromosomal DNA of all mice thus far studied and can be transmitted genetically in this form (an endogenous viral sequence) (19). In most strains of mice, these endogenous or germinal DNA sequences are not usually expressed and reports of their recovery as mature virions are rare (6). The GR mouse strain has been an interesting exception in which the germinal DNA sequences are expressed (20, 21).

The B-particle (exogenous virus) contains the RNA phase and is transmitted extracellularly via the milk in most mouse strains (22). Although B-particles, infectious agents and/or viral antigens have been reported in a variety of organs, the exogenous particle appears to primarily infect the mammary gland (23). The exogenous virus usually becomes integrated into the DNA of the infected mammary cells and can be passed in this manner to subsequent generations of daughter cells (24). These events are associated with mammary tumorigenesis. Mammary tissue thus has been the focus of MMTV research.

Although the emphasis has been on the virus in mammary adenocarcinoma, the reader is reminded that the traditional point of view envisions mammary cancer as a multifactorial disease. This point of view, most effectively espoused by Dr. Walter Heston (25), emphasizes genetic, hormone, and viral cofactors. The theoretical modes of action of these cofactors were expanded by the description of an intermediate preneoplastic tissue (the hyperplastic alveolar nodule) (26, 27).

Normal Mammary Epithelium → Preneoplastic Epithelium → Neoplastic Epithelium (28)

This introduced two possible steps at which these cofactors could interact: either at the step of conversion of normal epithelium to preneoplastic epithelium (nodulogenesis) or at the step of conversion of preneoplastic epithelium to neoplastic epithelium (tumorigenesis). Since not all of the cells participate in these events, it is proper to say that nodule cells <u>emerge</u> from normal epithelium and that tumors <u>emerge</u> from nodule cells (29). This clearly implies a selective proliferation of a subpopulation of cells in each of the pre-existing populations. The critical virological question is how the virus affects these populations and how genetic, hormonal, immunological, and other cofactors intervene to modify the virological events.

<u>Genetics</u>. Although early mouse mammary tumor studies concerned the hereditary aspect of cancer, the discovery of the viral milk agent tended to reduce the emphasis on genetic factors (25). Fortunately, studies of genetic susceptibility to mammary cancer continued and later complemented the virological work (6, 25).

In 1965, Mühlbock reaffirmed male transmission of mammary cancer in the relatively new European mouse strain, GR (30). In 1968, Bentvelzen, through a study of the GR strain, came to the then startling conclusion that the mouse mammary tumor virus was transmitted by germ cells of the GR mouse (21). Thus, passage of the high-risk agent to offspring was not limited to transmission through the mother's milk.

Subsequently the Dutch investigators found, using GRxC57BL crosses, that tumor development followed a simple Mendelian ratio (21, 31). They proposed that mouse mammary tumorigenesis was controlled by a single gene (32). This concept was extended to define a single gene locus (MTV-2) controlling the expression of MMTV in the milk of GR mice. A congenic GR strain lacking the MTV-2 gene has been described which has a low tumor incidence (33).

The single gene hypothesis has been disputed by others. Nandi and Helmich interpreted their data to mean that at least two genes were involved in mammary tumorigenesis (34). In more extensive experiments using C57BLxGR backcrosses, Heston and Parks concluded that the tumor incidence in the second backcross generation was inconsistent with a single gene hypothesis (25, 36).

The correct interpretation of these various experiments has been hotly debated (36). All appear to agree, however, that a limited number of genes influence viral expression and tumorigenesis. Implicit in most interpretations has been the idea that genetic factors and virus expression were interrelated in the control of tumor incidence and latency. These factors are further complicated by the numerous host factors which exert control on the various phases of the viral infection. For example, it has been observed that susceptibility to a given virus stock varies with the test mouse strain used (37-39). Bentvelzen has reviewed a number of potential host factors which may influence viral infection and neoplastic transformation (6, 30).

Immunology. One prominent host factor capable of influencing MMTV infection and tumorigenesis is the host's immunological responsiveness to MMTV-related antigens. Our understanding of the immunology of MMTV has been considerably extended in the last few years but is still far from complete. The current data indicate that the quality as well as quantity of an animal's response to MMTV may be important. An inappropriate immune response to MMTV may be ineffective or, worse yet, may actually favor the genesis and/or growth of an MMTV-induced tumor.

Early immunological studies used tissue transplantation techniques to investigate mammary tumor antigenicity. The predominant antigenicity of MMTV-induced tumors showed little variation from tumor to tumor and was attributable to the presence of exogenous viral antigens (40, 46). Non-viral non-cross-reacting mammary tumor antigens were reported (47-51), but these were clearly much less immunogenic than the viral antigens.

In initial studies, mice infected with exogenous MMTV as neonates appeared to be tolerant to viral-associated antigens (43). These neonatally infected animals did not exhibit a detectable anti-viral immunity of their own and apparently could not be immunized against challenge with transplanted tumor cells. Blair et al. (52) later reported the detection of serum antibodies against MMTV-associated antigens following immunization of infected mice. This report was confirmed and extended by other investigators (53-55).

With the development of sensitive, reproducible means of assessing cell-mediated immunity in vitro came the observation that mice infected with exogenous MMTV exhibit a naturally occurring cellular response to MMTV-induced tumor cells (47, 56). Serological factors were later described in these mice which inhibited (blocking factors) or otherwise affected the ability of lymphocytes (from infected mice) to kill mammary tumor cells in vitro (48, 57, 58).

Serum from exogenously infected mice has been shown to contain antibodies directed against MMTV antigens (59-62). In addition, MMTV antigens have also been found in these sera (63). Serum MMTV antigens may combine with circulating anti-MMTV antibodies and neutralize their effectiveness by forming immune complexes which then accumulate in the glomeruli of the kidney (60).

Strains of mice which carry only the endogenous form of MMTV can be successfully immunized against MMTV and protected against challenge with transplanted mammary tumor cells (40-46). These mice respond to MMTV-associated antigens with the production of humoral and cellular immune responses which, unlike responses seen in exogenously infected mice, are effective against MMTV or mammary tumor tissue.

Despite the antigenicity of MMTV-infected tissues, and despite the presence of both cellular and humoral responses to viral antigens, mice infected with exogenous MMTV develop mammary tumors. Attempts to augment the immune responsiveness of exogenously infected mice to viral antigens specifically (53, 64-69) or to generally stimulate their immune systems (43) have proved ineffective in protecting them and frequently have accelerated mammary tumor development. Immunosuppressive measures, on the other hand, have been shown to improve an infected host's ability to resist MMTV-induced tumorigenesis (70-74). Uninfected mice can be successfully immunized in such a way as to resist subsequent challenge with tumor cells (75). A host's response to MMTV and MMTV-induced tumor cells may or may not be effective in protecting the animal from mammary tumors, but clearly influences the course of events in MMTV infection and tumorigenesis. The quality, quantity, and timing of an animal's anti-MMTV response appear to be the critical factors in determining its effectiveness.

Hormones. The endocrine status of the mouse has long been recognized to play an important role in mouse mammary tumorigenesis (76-79). Classical endocrine organ ablation and hormonal replacement experiments have demonstrated hormone control of this system and have emphasized the role of lactogenic hormones (i.e., prolactin) and estrogen (77). Increasing the number of pregnancies also increases the risk of mouse mammary tumorigenesis (14, 78, 79). Nulliparous mice of some strains show a lower tumor incidence than their multiparous counterparts (14, 80-83). Many early workers thought that the increased tumor incidence associated with hormonal stimulation was related to increased levels of viral expression (14).

With the development of in vitro systems, the role of hormones could be systematically studied. McGrath first demonstrated that glucocorticoids (e.g., cortisol, hydrocortisone) induce virus expression in primary cultures of mammary tumor cells (84). Subsequently, glucocorticoid induction of viral expression has been demonstrated in long-term primary cell cultures (85), in several established mouse mammary tumor cell lines (86-88), and in heterologous MMTV-infected cells (88, 89). MMTV has been found to be inducible by a number of glucocorticoids, but the most effective steroid has been the synthetic hormone dexamethasone (86, 87, 90). Dexamethasone induction of MMTV requires the appropriate hormone receptors (14, 87) and has been shown to induce an almost instantaneously increased rate of MMTV RNA synthesis (91). This increased RNA synthesis did not require DNA or protein synthesis (91). The hormone effects have been shown to be dose dependent (91).

Synergistic interactions have been described between insulin and hydrocortisone (84), insulin, hydrocortisone and estradiol (14, 87), and insulin and prolactin (92). These effects have not had the full benefit of molecular analysis. As a result, the exact mechanisms and full significance of hormonal action and interaction in this system remain unclear. It does appear, however, that the same hormones which produce mammary tumors in the animal also stimulate virus production in vitro (14, 87). It remains to be seen whether this is a primary effect on viral synthesis or merely reflects optimal tissue culture conditions.

Recent demonstrations of infection of heterologous (feline kidney, rat hepatoma, and mink lung) cells in vitro with MMTV have created important new avenues of research (89). Heterologous cell infection has already provided new insight into the hormonal regulation of virus expression, into the antigenic structure of the virus, and into the molecular biology of the virus. It will make the study of virus infection and replication possible in previously uninfected cells and lead the way to genetic experiments.

Virus. The basic, but often unstated, assumption in many studies in the mouse mammary tumor system is that virus expression is related to tumorigenesis. This assumption is consistent with the widely accepted oncogene hypothesis which was developed for the endogenous C-type viruses (93) and has been modified to fit the murine mammary tumor virus system (6, 94, 95).

Investigators have long related titers of virus in the milk of various strains to the mammary tumor incidence of those strains (14). Numerous studies involving virus titration curves have attempted to relate the amount of virus which a group of mice receive to the ultimate tumor or nodule incidence in that group (14, 96-101). More recently, the development of more sensitive quantitative assays has reinforced these impressions. For example, nucleic acid hybridization studies showed that the highest levels of MMTV RNA resided in the organs of those strains of mice having the highest tumor incidence (102). Others, using radioimmunoassays, have related the high levels of viral antigens in the milk of a given strain to the tendency of that strain to develop cancer (103, 104).

An example of the quantitative relationship between virus and tumorigenesis has come from our studies of the inbred mouse strain BALB/c (tumor incidence less than 1%) and the virus-infected subline BALB/cfC3H (tumor incidence of 87% at

16 months of age). BALB/c females have no detectable MMTV antigens in their milk and their BALB/c lactating mammary glands contain no detectable MMTV by electron microscopy, by immunoperoxidase, or by radioimmunoassay (62, 105). The BALB/c normal organs and lactating mammary gland contain approximately seven endogenous MMTV gene copies per diploid BALB/c cell. MMTV RNA can be detected in the BALB/c cytoplasmic extracts (106).

Our BALB/cfC3H colony was established at the Simonsen Laboratories, Gilroy, California (through the auspices of the Drug Evaluation Branch, National Cancer Institute), by foster nursing BALB/c pups on milk agent-bearing C3H/He mothers. Once infected, the subsequent generations of BALB/cfC3H females were nursed on their own mothers. With a tumor incidence of 87%, the colony contains a relatively high proportion of animals which can be expected to develop tumors (high risk) and a few which will not (low risk).

As one possible parameter to define tumor risk before tumor development, MMTV concentrations in the milk of individual BALB/cfC3H animals were determined by radioimmunoassay (107) and the animals were held for tumor development. The animals segregated cleanly into two groups. All animals which developed tumors by 16 months of age had high levels of MMTV in their milk by the second lactation. The animals with low levels of virus in their milk did not develop tumors by the time the experiment was terminated at 16 months. The 95% confidence intervals for viral expression levels in these two groups of mice did not overlap.

The same pattern of expression was observed in a large group of BALB/c animals which were experimentally injected with one of seven log dilutions of MMTV at 4-5 weeks of age (107). Those injected animals which developed tumors by 16 months of age previously expressed high levels of MMTV in their milk by the third lactation while those which did not develop tumors expressed, at best, very low levels of viral antigens (107). In general, more mice in high dosage groups showed the high viral expression levels indicative of high risk of tumor development. The development of high virus levels was, however, an all or none phenomenon in that those animals in the high dosage groups which did not produce high levels of virus did not develop tumors.

The relationship between input dose of virus, subsequent infection and tumorigenesis was observed by calculating the 50% infectious dose (ID_{50}) and 50% tumor dose (TD_{50}) of MMTV (107, 108). The major effect of injecting mice with lower doses of MMTV was to increase the latency period prior to detectable milk virus expression. That is, the calculated ID_{50} for infection by the first lactation was approximately 19 mµg/mouse (50% of mice given this dose would be expected to have MMTV positive milk at the first lactation), but by the seventh pregnancy, most of the animals, regardless of the dose of MMTV received, expressed MMTV milk antigens and were scored as infected. The calculated ID_{50}, therefore, dropped to 0.01 mµg/mouse or about 20,000 particles per mouse when milk from later lactations was examined.

Tumorigenesis was also dose dependent. The calculated TD_{50} at 16 months of age with all but one of the "high risk" high expressor animals developing tumors was 24 mµg/mouse (107). A comparison of the infectious dose and tumor dose data reveals that a four log difference exists between the final ID_{50} and the TD_{50}. Thus, while MMTV in this context is necessary for tumorigenesis, there are many host-virus interactions which do not result in tumors.

The role of high MMTV input and high MMTV expression in the high-risk animal is not understood at a cellular level, but the immunoperoxidase technique provides one means of examining the cellular events involved. Almost all BALB/cfC3H tumor cells express high levels of MMTV antigens when examined by immunoperoxidase (105). In contrast, the BALB/cfC3H lactating mammary gland consists of small clusters of mammary epithelial cells which are positive for MMTV antigens and many cells which do not express any viral antigens. If one assumes (an unproven assumption) that only those cells expressing antigen are infected with exogenous virus, then the mammary epithelium is heterogenous, consisting of infected and uninfected cells. Since the hyperplastic alveolar nodules all express viral antigen, it is also reasonable to assume that the nodule cells are derived from those mammary epithelial cells which express MMTV antigens. The viral antigen expressing cells could be the only cells at risk of undergoing nodulogenesis and/or tumorigenesis.

Molecular hybridization studies also suggest that tumors emerge from a select subpopulation. Morris et al. (109) demonstrated that non-mammary organs in most mouse strains contain five to nine copies of MMTV DNA per diploid genome as calculated from nucleic acid hybridization kinetic analysis. Mammary tumors, however, contain as many as 30 copies of MMTV DNA per diploid genome. The additional MMTV copies in tumors were interpreted to imply infection with exogenous MMTV.

Using the same techniques, V. Morris, H. Varmus and C. Cohen in collaboration with our laboratory found that BALB/c and BALB/cfC3H normal non-mammary organs from our colony have approximately seven copies of MMTV DNA per diploid genome. The BALB/cfC3H tumors have a variable MMTV copy number up to 44 copies per cell. The lactating mammary gland might be expected to contain an intermediate MMTV copy number but primiparous and multiparous normal lactating mammary glands had the same number of MMTV copies as the non-mammary organs, i.e., seven copies per diploid genome.

Since tumors contain more MMTV DNA than the normal lactating mammary gland, the viral DNA copies may have increased during the tumorigenic process, or the tumor cells may arise from a subpopulation of cells with a high MMTV copy number. Drs. Schlom and Drohan at the National Cancer Institute, using their "tumor-specific" MMTV probe, found high levels of tumor-specific sequences in BALB/cfC3H tumors but lower levels in the BALB/cfC3H lactating mammary gland (110). As a result of reconstruction experiments, the lactating mammary gland was estimated to have less than one tumor-specific MMTV DNA per cell. Slightly higher estimates were obtained from reconstruction restriction endonuclease experiments (111).

Simple arithmetic tells us that an average of one copy per cell could also be ten copies in one out of every ten cells. This arrangement would then be consistent with the clusters of cells observed in the immunoperoxidase studies. The data are all consistent with the notion that BALB/cfC3H mammary tumors arise from a subpopulation of cells. The ultimate form of such a subpopulation is a monoclonal proliferation.

The clonal origin of a number of other spontaneous and experimental tumors has been well documented (112). Polyclonal tumors, on the other hand, occur less frequently. Previous studies by Mintz using allophenic chimeras (113, 114) and by Hilgers on hormone-sensitive tumors (115) have suggested a clonal origin for mouse mammary tumors. Recent experiments using restriction endonuclease mapping have also implied that mouse mammary tumors are monoclonal (116, 117).

Support for the clonal origin of tumors in other systems has been based on three types of observations: 1) sex-linked polymorphic isoenzymes such as glucose-6-phosphate dehydrogenase in heterozygous females indicate that all tumor cells originate from a single cell (112), 2) immunoglobulin produced by myelomas or B cell lymphomas tend to be of a single type, indicating clonal origin (119), and 3) cytogenetic studies indicate that some tumors have a uniform karyotypic aberration, such as the Philadelphia chromosome in acute myelogenous leukemia (118, 119). Unfortunately, such marker systems which would be useful in the study of mouse mammary tumors have not been developed in inbred strains of mice. The clonal hypothesis in mouse mammary tumors awaits a technology which can test its validity.

If independently verified, a monoclonal origin for mouse mammary tumors would mean that they originate from a single somatic cell. This would have a number of important biological implications which can simply be paraphrased from Boveri (120): 1) the initial transformation event is rare, 2) the critical event is far removed from the final phenotypic expression of the cancer cell, and 3) virus infection is not in and of itself sufficient for tumorigenesis. As such, MMTV can be seen in its proper perspective as a cofactor in neoplasia since mammary neoplasia requires a series of events (121). As these events are unraveled with the powerful tools now available, a new chapter in the biology of the mouse mammary virus will be written.

ACKNOWLEDGMENTS

The authors are indebted to Dr. Harold Varmus, Dr. Craig Cohen, and Dr. Vincent Morris of the University of California, San Francisco, and Dr. Jeffrey Schlom and Dr. William Drohan of the National Cancer Institute for sharing preprints of their work, permitting discussion of our unpublished data, and for their interest in our joint ventures and their many hours of discussion. The authors wish to acknowledge the work of Dr. L.R. Ayyagari and Ms. Judith St. George which has contributed to the concepts discussed herein and to express their appreciation to Ms. J.K. Lund for her editorial assistance.

REFERENCES

1. Staff of the Roscoe B. Jackson Memorial Laboratory. (1933). Science 78:465.
2. Korteweg, R. (1934). Ned. Tijdschr. Geneeskd. 78:240.
3. Bittner, J.J. (1936). Science 84:162.
4. Visscher, M.B., Green, R.G., and Bittner, J.J. (1942). Proc. Soc. Exp. Biol. Med. 49:94.
5. Bryan, W.R., Kahler, H., Shimkin, M.B., and Andervont, H.B. (1942). J. Natl. Cancer Inst. 2:451.
6. Bentvelzen, P. (1974). Biochim. Biophys. Acta 355:236.
7. Duesberg, P.H., and Cardiff, R.D. (1968). Virology 36:696.
8. Duesberg, P.H. (1968). Proc. Natl. Acad. Sci. 60:1511.
9. Teramoto, Y.A., Cardiff, R.D., and Lund, J.K. (1977). Virology 77:135.
10. Bernhard, W. (1958). Cancer Res. 18:491.
11. Smith, G. (1967). Cancer Res. 27:2179.
12. Tanaka, H. (1977) Virology 76:835.
13. Cardiff, R.D., Puentes, M.J., Young, L.J.T., Smith, G.H., Teramoto, Y.A., Altrock, B.A., and Pratt, T.S. (1978). Virology 85:157.
14. Nandi, S., and McGrath, C.M. (1973). Adv. Cancer Res. 17:353.
15. Cardiff, R.D., Puentes, M.J., Teramoto, Y.A., and Lund, J.K. (1974). J. Virol. 14:1293.
16. Sarkar, N.H., Taraschi, N.E., Pomenti, A.A., and Dion, A.S. (1976). Virology 69:677.

17. Dalton, A.J., Melnick, J.L., Bauer, H., Braudreau, G., Bentvelzen, P., Bolognesi, D., Gallo, R., Graffi, A., Haguenau, R., Heston, W., Huebner, R., Todaro, G., and Heine, V.I. (1974). Intervirology 4:201.
18. Varmus, H.E., Bishop, J.M., Nowinski, R.C., and Sarkar, N.H. (1972). Nature New Biol. 238:189.
19. Mühlbock, O., and Bentvelzen, P. (1968). Perspect. in Virol. 6:75.
20. Schlom, J., Drohan, W., Teramoto, Y., Hand, P., Colcher, D., Callahan, R., Todaro, G., Kufe, D., Howard, D., Gautsch, J., Lerner, R., and Schidlovsky, G. (1978). In "Origins of Inbred Mice" (H.C. Morse III, ed.). Academic Press, New York.
21. Bentvelzen, P. (1968). "Genetic Controls of the Vertical Transmission of the Mühlbock Mammary Tumor Virus in the GR Mouse Strain," p. 1. Hollandia Publ. Co., Amsterdam.
22. Moore, D.H., Charney, J., and Holben, J.A. (1974). J. Natl. Cancer Inst. 52:1757.
23. Bentvelzen, P. (1972). Int. Rev. Exp. Pathol. 11:259.
24. Drohan, W., Kettmann, R., Colcher, D.D., and Schlom, J. (1977). J. Virol. 21:986.
25. Heston, W.E., and Parks, W.P. (1975). J. Genet. Cytol. 17:493.
26. DeOme, K.B., Faulkin, L.J., Jr., Bern, H.A., and Blair, P.B. (1959). Cancer Res. 19:515.
27. DeOme, K.B., Nandi, S., Bern, H.A., Blair, P.B., and Pitelka, D. (1962). In "Proceedings of the International Conference on the Morphologic Precursors of Cancer" (L.M. LeCam and J. Neyman, eds.), p. 349. Division of Cancer Research, Perugia.

28. DeOme, K.B. (1968). In "Proceedings of the Fifth Berkeley Symposium on Mathematical Statistics and Probability" (J. Neyman, ed.), p. 649. University of California Press, Berkeley.
29. Cardiff, R.D., Wellings, S.R., and Faulkin, L.J., Jr. (1977). Cancer 39:2734.
30. Mühlbock, O. (1965). Europ. J. Cancer 1:123.
31. Bentvelzen, P. (1972). In "RNA Viruses and Host Genome in Oncogenesis" (P. Emmelot and P. Bentvelzen, eds.), p. 309. North-Holland Pub. Co., Amsterdam.
32. Van Nie, R., Verstraeten, A.A., and de Moes, J. (1977). Int. J. Cancer 19:383.
33. Van Nie, R., and de Moes, J. (1977). Int. J. Cancer 20:588.
34. Nandi, S., and Helmich, C. (1974). J. Natl. Cancer Inst. 52:1567.
35. Heston, W.E., Smith, B., and Parks, W.P. (1976). J. Exp. Med. 144:1022.
36. Heston, W.E., and Parks, W.P. (1977). J. Exp. Med. 146:1206.
37. Andervont, H.B. (1940). J. Natl. Cancer Inst. 1:147.
38. Moore, D.H., Charney, J., and Holben, J.A. (1974). J. Natl. Cancer Inst. 52:1757.
39. Moore, D.H., Holben, J.A., and Charney, J. (1976). J. Natl. Cancer Inst. 57:889.
40. Lavrin, D.H., Blair, P.B., and Weiss, D.W. (1966). Cancer Res. 26:293.
41. Lavrin, D.H., Blair, P.B., and Weiss, D.W. (1966) Cancer Res. 26:929.
42. Dezfulian, M., Lavrin, D.H., Shen, A., Blair, P.B., and Weiss, D.W. (1967). In "Carcinogenesis: A Broad Critique," p. 365. Williams and Wilkins Co., Baltimore.

43. Weiss, D.W., Lavrin, D.H., Dezfulian, M., Vaage, J., and Blair, P.B. (1966). In "Viruses Inducing Cancer -- Implications for Therapy" (W.F. Burdette, ed.), p. 138. University of Utah Press, Salt Lake City.
44. Weiss, D.W. (1967). In "Proceedings of the Fifth Berkeley Symposium on Mathematical Statistics and Probability" (L.M. LeCam and J. Neyman, eds.), p. 657. University of California Press, Berkeley.
45. Morton, D.L., Goldman, L., and Wood, D.A. (1969). J. Natl. Cancer Inst. 42:321.
46. Morton, D.L. (1969). J. Natl. Cancer Inst. 42:311.
47. Heppner, G.H., and Pierce, G. (1969). Int. J. Cancer 4:212.
48. Heppner, G.H. (1969). Int. J. Cancer 4:608.
49. Vaage, J. (1968). Nature (London) 218:101.
50. Vaage, J. (1968). Cancer Res. 28:2477.
51. Morton, D.L., Miller, G.F., and Wood, D.A. (1969). J. Natl. Cancer Inst. 42:289.
52. Blair, P.B., Lavrin, D.H., Dezfulian, M., and Weiss, D.W. (1966). Cancer Res. 26:647.
53. Blair, P.B. (1968). Cancer Res. 28:148.
54. Hilgers, J., Daams, J.H., and Bentvelzen, P. (1971). Isr. J. Med. Sci. 7:154.
55. Fink, M.A., Feller, W.F., and Sibal, L.R. (1968). J. Natl. Cancer Inst. 41:1395.
56. Blair, P.B., Lane, M., and Yagi, M. (1964). J. Immunol. 112:693.
57. Blair, P.B., and Lane, M. (1974). J. Immunol. 112:439.
58. Blair, P.B., and Lane, M. (1974). J. Immunol. 113:1446.
59. Muller, M., Hageman, P.C., and Daams, J.H. (1971). J. Natl. Cancer Inst. 47:801.
60. Pascal, R.R., Rollwagen, F.M., Harding, T.A., and Schiavone, W.A. (1975). Cancer Res. 35:302.

61. Ihle, J.N., Arthur, L.O., and Fine, D.L. (1976). Cancer Res. 36:2840.
62. Altrock, B.W., and Cardiff, R.D. (1977). Proc. Am. Soc. Micro. #E148.
63. Ritzi, E., Martin, D.S., Stolfi, R.L., and Spiegelman, S. (1976). Proc. Natl. Acad. Sci. 73:4190.
64. Burton, D.S., Blair, P.B., and Weiss, D.W. (1969). Cancer Res. 29:971.
65. Nishioka, K., Irie, R.F., Inove, M., Chang, S., and Takeuchi, S. (1969). Int. J. Cancer 4:121.
66. Irie, R.F., Nishioka, K., Tachibana, T., and Takeuchi, S. (1969). Int. J. Cancer 4:150.
67. Charney, J., and Moore, D.H. (1972). J. Natl. Cancer Inst. 48:1125.
68. Charney, J., Holben, J.A., Cody, C.M., and Moore, D.H. (1976). Cancer Res. 36:777.
69. Stutman, O. (1976). Cancer Res. 36:739.
70. Martinez, C. (1964). Nature 203:1188.
71. Law, L.W. (1966). Cancer Res. 26:551.
72. Heppner, G.H., Wood, P.C., and Weiss, D.W. (1968). Isr. J. Med. Sci. 4:1195.
73. Squartini, F., Olivi, M., and Bolis, G. (1970). Cancer Res. 30:2069.
74. Lappe, M.A., and Blair, P.B. (1970). Proc. Am. Assoc. Cancer Res. 11:47.
75. Blair, P.B. (1971). Isr. J. Med. Sci. 7:161.
76. Bern, H.A. (1960). Science 131:1039.
77. Furth, J. (1957). Cancer Res. 17:454.
78. Blair, P.B. (1968). Curr. Top. Micro. Immunol. 45:1.
79. Mühlbock, O. (1956). Adv. Cancer Res. 4:371.
80. Bagg, H.J. (1926). Amer. Naturalist 60:234.
81. Bittner, J.J. (1946-47). Harvey Lecture Series 42:221.
82. Mühlbock, O. (1950). J. Natl. Cancer Inst. 10:1259.

83. Mühlbock, O., and van Rijssel, T.G. (1954). J. Natl. Cancer Inst. 15:73.
84. McGrath, C.M. (1971). J. Natl. Cancer Inst. 47:455.
85. Young, L.J.T., Cardiff, R.D., and Ashley, R.L. (1975). J. Natl. Cancer Inst. 54:1215.
86. Fine, D.L., Plowman, J.K., Kelley, S.P., Arthur, L.O., and Hellman, E.M. (1974). J. Natl. Cancer Inst. 52:1881.
87. Parks, W.P., Scolnick, E.M., and Kozikowski, E.H. (1974) Science 184:158.
88. Ringold, G.M., Cardiff, R.D., Varmus, H.E., and Yamamoto, K. (1977). Cell 10:11.
89. Vaidya, A.B., Lasfargues, E.Y., Heubel, E.Y., Lasfargues, J.C., and Moore, D.H. (1976). J. Virol. 18:911.
90. Ringold, G., Lasfargues, E.Y., Bishop, J.M., and Varmus, H.E. (1975). Virology 65:135.
91. Ringold, G.M., Yamamoto, K.R., Tomkins, G.M., and Bishop, J.M. (1975). Cell 6:299.
92. Yang, J., Enami, J., and Nandi, S. (1977). Cancer Res. 37:3644.
93. Huebner, R.J., and Todaro, G.J. (1969). Proc. Natl. Acad. Sci. 64:1087.
94. Bentvelzen, P. (1972). In "RNA Viruses and Host Genome in Oncogenesis" (P. Emmelot and P. Bentvelzen, eds.), p. 309. North-Holland Pub. Co., Amsterdam.
95. Hilgers, J., and Bentvelzen, P. (1978). Adv. Cancer Res. 17:143.
96. Nandi, S., Haslam, S., Helmich, C., and Ritter, R. (1971). J. Natl. Cancer Inst. 46:1309.
97. Moore, D.H., Pillsbury, N., and Pullinger, B.D. (1969). J. Natl. Cancer Inst. 43:1263.

98. Charney, J., Pullinger, B.D., and Moore, D.H. (1969). J. Natl. Cancer Inst. 43:1289.
99. Moore, D.H., Charney, J., and Pullinger, B.D. (1970). J. Natl. Cancer Inst. 45:561.
100. Moore, D.H., Charney, J., and Holben, J.A. (1974). J. Natl. Cancer Inst. 52:1757.
101. Moore, D.H., Holben, J.A., and Charney, J. (1976). J. Natl. Cancer Inst. 57:889.
102. Varmus, H.E., Quintrell, N., Medeiros, E., Bishop, J.M., Nowinski, R.C., and Sarkar, N.H. (1973). J. Mol. Biol. 79:663.
103. Noon, M.C., Wolford, R.G., and Parks, W.P. (1975). J. Immunol. 115:653.
104. LoGerfo, P., Silverstein, G., and Charney, J. (1974). Surgery 76:16.
105. St. George, J.A., Cardiff, R.D., Young, L.J.T., and Faulkin, L.J. (1978). Proc. Am. Assoc. Cancer Res. #204.
106. Ayyagari, L.R., Cardiff, R.D., Young, L.J.T., Puma, J.P., and Faulkin, L.J. (1978). Proc. Am. Soc. Micro. #S272.
107. Altrock, B.W., and Cardiff, R.D. (1978). Proc. Am. Soc. Micro. #S197.
108. Remington, R.D., and Schork, M.A. (1970). "Statistics with Application to the Biological and Health Sciences." Prentice-Hall, Inc., New Jersey.
109. Morris, V.L., Medeiros, E., Ringold, G.M., Bishop, J.M., and Varmus, H.E. (1977). J. Mol. Biol. 114:73.
110. Drohan, W., Kettman, R., Colcher, D., and Schlom, J. (1977). J. Virol. 21:986.
111. Shank, P.R., Cohen, J.C., Varmus, H.E., Yamamoto, K.R., and Ringold, G.M. Proc. Natl. Acad. Sci. U.S., in press.
112. Failkow, P.J. (1974). New Engl. J. Med. 291:26.
113. Mintz, B., Custer, R.P., and Donnelley, A.J. (1971). Int. Rev. Exp. Pathol. 10:143.

114. Mintz, B., and Slemmer, G. (1969). J. Natl. Cancer Inst. 43:87.
115. Sluyser, M., Nouwen, T., Hilgers, J., and Calafat, J. (1977). Cancer Res. 37:1986.
116. Varmus, H.E., Cohen, J.C., Ringold, G.M., Shank, P.R., Morris, V.L., Cardiff, R.D., and Yamamoto, K.R. (1978). UCLA-ICN Abstr.
117. Varmus, H.E., Cohen, J.C., Ringold, G.M., Shank, P.R., Morris, V.L., Cardiff, R.D., and Yamamoto, K.R. In preparation.
118. Nowell, P.C. (1977). Am. J. Pathol. 89:459.
119. Nowell, P.C. (1976). Science 194:23.
120. Bovari, J., and German, I. (1974). "Chromosomes and Cancer." Wiley and Co., New York.
121. Rapp, R., and Reed, C. (1977). Cancer 40:419.

DIVERSITY OF MOUSE MAMMARY TUMOR VIRUS GENETIC INFORMATION AND GENE PRODUCTS IN RODENTS

J. Schlom, W. Drohan, Y. Teramoto,
P. Hand, D. Colcher, R. Callahan, and G. Todaro

Laboratory of Viral Carcinogenesis
National Cancer Institute
National Institutes of Health
Bethesda, Maryland

D. Kufe

Sidney Farber Cancer Institute
Boston, Massachusetts

D. Howard

Meloy Laboratories, Inc.
Springfield, Virginia

J. Gautsch and R. Lerner

Scripps Clinic and Research Foundation
La Jolla, California

G. Schidlovsky

Brookhaven National Laboratory
Upton, New York

The murine model is widely used to study factors involved in the etiology of mammary carcinoma. It became evident as early as 1936 (1) that a filterable agent, i.e., a virus, is involved in at least some mouse strains in the causation of mammary cancer. Over the past four decades, experimental systems have been developed in numerous mouse strains, and in almost every strain studied, a mouse mammary tumor virus (MMTV) has been revealed (2, 3). Table 1 surveys the variations in incidence of spontaneously occurring mammary tumors, latent period to tumor, types of tumors produced, and whether mouse mammary tumor virions are observed in mammary tumors by electron microscopy. It should be noted that mammary tumors appear "early" (before one year) in the high incidence mouse strains C3H, RIII, and GR, and "late" (after one year) in low and moderate incidence strains C3HfC57BL and BALB/c.

A question that has remained unanswered concerning the origin of mammary oncogenesis in the mouse is: How many mouse mammary tumor viruses are there? There are at least two considerations involved in this question: (a) each mouse strain contains its own mouse mammary tumor virus or viruses, which are partially related or distinct from other viruses of other mouse strains, and (b) there is only one MMTV and each mouse strain exerts its own control over various properties of this virus, such as its expression or its virulence. The answer to this question is essential to our understanding the etiology of the disease in each system studied.

A recent achievement has been the ability to productively infect heterologous cells with various MMTV isolates (4, 5). The MMTVs grown in feline cells share nucleic acid and immunologic properties with their mouse grown counterparts. These reagents have now made possible the delineation between viral-coded vs. host-derived entities. In the nucleic acid hybridization and immunologic studies presented here, both

TABLE 1. Mammary Tumor Incidence in Various Mouse Strains

Mouse Strain	Mammary tumor incidence (%)	Latent period (months)	Type of mammary tumor	Virions in tumor
C3H	100	7	Fast growing, hormone-independent	++
RIII	96	9	Fast growing, hormone-independent	++
GR	100	3	Hormone-dependent plaques, progress to hormone independent	++
C3HfC57BL	35	19	Slow growing, hormone-independent	+
BALB/c	10	14	Slow growing, also acanthomas	neg.
C57BL	<1	24	Hormone-independent	neg.

murine and feline grown MMTVs were employed with identical results, thus demonstrating the viral-coded nature of the nucleic acids and proteins being studied.

HETEROGENEITY IN MMTV-CODED GENE PRODUCTS

Type- and Group-Specific Reactivities of the MMTV Major Envelope Glycoprotein

The major surface component of the MMTV virion is a 52,000d glycoprotein (gp52). The other major protein components of the MMTV virion are a 36,000d glycoprotein (gp36), and the 28,000 (p28), 14,000 (p14), and 10,000 (p10) dalton polypeptides. With the development of sensitive radioimmunoassays for the whole MMTV virion (6), or for purified MMTV polypeptides (7-10), it has become possible to precisely analyze similarities or differences among MMTVs from different mouse strains. Several investigators have previously shown that the gp52 of MMTVs share group-specific antigenic determinants (7-10). Experiments have recently been described (11, 12) that also demonstrate type-specific antigenic determinants on the gp52 molecule.

MMTVs from RIII and C3H mice, i.e., MMTV(RIII) and MMTV(C3H), were used in a competitive radioimmunoassay (RIA) to compete for the binding of anti-MMTV(C3H) to [^{125}I]-labeled MMTV(C3H) virions. In this system, increasing amounts of MMTV(RIII) competitor gave a shallower slope than that given by the isologous MMTV(C3H) (Fig. 1A). At the highest input of competing MMTV(RIII) protein employed, i.e., 100 µg, only 75% inhibition was obtained while less than 1 µg of MMTV(C3H) resulted in the same competition. In a "group-specific" assay using anti-MMTV(C3H) vs. [^{125}I]MMTV(RIII), both viruses competed identically, i.e., with comparable inputs and with the same slope.

The addition of increasing amounts of MMTV(GR) into the anti-MMTV(C3H) vs. [^{125}I]MMTV(C3H) system did not cause complete inhibition of the precipitation of [^{125}I]MMTV(C3H) (Fig. 1B). Even at high inputs of protein, MMTV(GR) was incapable of competing for all the antibodies binding to MMTV(C3H). The anti-MMTV(C3H) serum, therefore, appears to contain an antibody population that is directed toward antigenic determinants that are present in MMTV(C3H) but not in MMTV(GR). To further amplify the type-specific reactions observed, anti-MMTV(C3H) serum was absorbed with MMTV(GR) and the immune precipitate was removed by centrifugation. The resulting absorbed serum retained its ability to bind [^{125}]MMTV(C3H). This binding could be completely inhibited by the addition of MMTV(C3H) or MMTV(C3H) gp52, but was not inhibited by the addition of up to 10,000 ng MMTV(GR) as competitor (Fig. 1C). Additional type-specific reactivities among the various MMTVs also exist. These include differ-

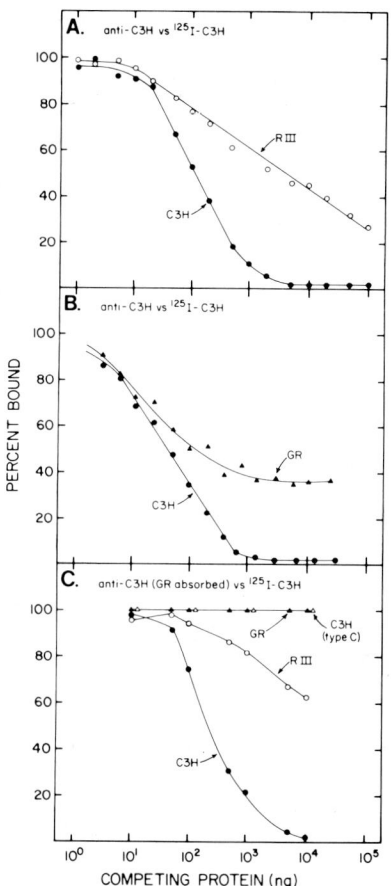

FIGURE 1. Competition radioimmunoassay for MMTV gp52. (A) Anti-MMTV (C3H) serum at a 1:10,000 input dilution was used to precipitate 10,000 cpm of [^{125}I]MMTV(C3H) intact virions. MMTV(RIII) (o) and MMTV(C3H) (●) were used as competitors. (B) Anti-MMTV(C3H) serum at a 1:10,000 input dilution was used to precipitate [^{125}I]MMTV(C3H) whole virions. MMTV(C3H) (●) and MMTV(GR) (▲) were used as competitors. (C) Anti-MMTV(C3H) serum, absorbed with MMTV(GR), was used at a 1:1,000 dilution to precipitate [^{125}I]MMTV (C3H) whole virions. MMTV(C3H) (●), C3H type-C virus (Δ), MMTV(RIII) (o), and MMTV(GR) (▲) were used as competitors.

ences between MMTV(C3H) and the endogenous MMTV of C3H obtained from C3HfC57BL mice (11, 12).

The type and group specificities of MMTVs grown in feline cells (5) were indistinguishable from the reactivities observed with murine grown MMTVs, thus providing strong evidence that the MMTV gp52 antigens are viral coded. The analysis of feline grown MMTV further excludes the possibilities that the observed antigenic differences were due to either differences in murine antigenic determinants of the different mouse strains producing the virus, or to host-coded differences in glycosylation of virions.

The identification of the type-specific differences for different MMTVs is now being used in several laboratories to monitor the host's immune response to mammary tumorigenesis, as well as in studies seeking trans-species reactivities with MMTVs. These studies further delineate the molecular diversity of viruses that can be involved in the etiology of mammary carcinoma within a given species.

Type- and Group-Specific Reactivities of the Major Internal Protein of MMTV

We have recently developed a competition RIA for the 28,000d major internal protein (p28) of MMTV (13). When this assay was conducted with high antibody dilutions for maximum sensitivity, no differences were observed among MMTVs from RIII, GR, C3H, or C3HfC57BL mice or with mammary tumor extracts from those mice. To demonstrate type-specific reactivities associated with the MMTV p28 polypeptide, assay conditions of low antibody dilution were used.

The binding of antisera prepared against the p28 of MMTV from RIII mice to [^{125}I]MMTV(RIII) p28 could be completely inhibited by the addition of 1 μg of purified MMTV(RIII) p28 (Fig. 2A). The addition of increasing amounts of

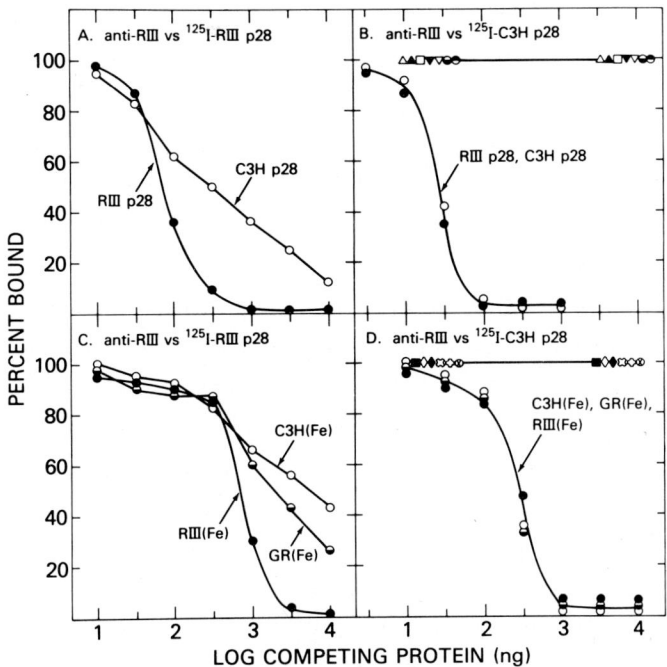

FIGURE 2. Group-specific and type-specific reactivities of MMTV p28. Anti-MMTV(RIII) p28 was used at a final dilution of 1:1,000 to precipitate [^{125}I]MMTV(RIII) p28 in A and C (type-specific assays) and [^{125}I]MMTV(C3H) p28 in B and D (group-specific assays). Purified MMTV p28s and MMTV virions grown in feline cells (Fe) were used as competitors as indicated. The following were also used as competitors: Δ, langur type-D retrovirus; ▲, Mason-Pfizer virus; ☐, squirrel monkey retrovirus; ▼, RD-114 virus; ∇, avian myeloblastosis virus; ⊖, bovine leukemia virus; ⊙, guinea pig virus; ■, M432 Mus cervicolor virus; ◇, endogenous C3H/T10 virus; ◆, Rauscher murine leukemia virus; X, extract of CrFK feline cells; +, fetal calf serum proteins; ⊗, purified MMTV gp52.

MMTV(C3H) p28 also competed for this binding, but with a shallower slope characteristic of a cross-reacting (14) but not identical antigen (Fig. 2A). Changing the radioactive antigen from [^{125}I]MMTV(RIII) p28 to [^{125}I]MMTV(C3H) p28 and maintaining the same antibody dilution revealed that both MMTV(RIII) p28 and MMTV(C3H) p28 competed identically in the radioimmunoassay (Fig. 2B). Thus, the MMTV p28 appears to contain both indistinguishable, i.e., group-specific, antigenic determinants as well as distinguishable, i.e., type-specific, antigenic determinants (13).

Both type-specific and group-specific reactivities were retained when MMTVs were used that were grown in the same feline CrFK cell line (Fig. 2C). All three of the MMTVs grown in feline cells, however, competed identically in the anti-MMTV(RIII) p28 vs. [^{125}I]MMTV(C3H) "group-specific" p28 radioimmunoassay (Fig. 2D).

Using the combination of gp52 and p28 RIAs, levels of MMTV gene expression can now be analyzed in terms of both viral glycoproteins and nonglycoproteins. This type of comparison is important in light of recent evidence of noncoordinate polypeptide chain initiation of glycosylated vs. nonglycosylated MMTV proteins in infected cells (15). These combined radioimmunoassays should further elucidate the mechanisms of MMTV replication and gene expression in murine mammary tumors.

Tryptic Peptide Analyses of MMTV Gene Products

Tryptic peptide analyses (16) have been performed on the 52,000 and 36,000 dalton glycoproteins and the nonglycosylated 28,000, 14,000 and 10,000 dalton proteins of the highly oncogenic MMTVs of C3H, RIII, and GR mice (17). Each virus was grown in both murine and feline cells to ensure the viral-coded nature of each peptide analyzed. The gp36-38 and p14 peptide maps of all three MMTVs were indistinguishable. Both the gp52 (Fig. 3) and the p28 (Fig. 4) of MMTV(C3H), however, could be clearly distinguished from the corresponding proteins of MMTV(RIII) and MMTV(GR), regardless of whether the viruses were grown in feline or murine cells. The p10 of MMTV(RIII), on the other hand, was clearly different from that of MMTV(C3H) and MMTV(GR) (17). Therefore, tryptic peptide analysis of three MMTV proteins, gp52, p28, and p10, can serve to distinguish these three viruses from one another. These studies represent further strain-specific markers for several MMTV gene products. Thus, as for type-C retroviruses of the mouse (16), MMTVs form a multigene family, the final extent of which is not yet known.

Figures 3 and 4 are on pages 352 and 353.

FIGURE 3. Tryptic peptide maps of MMTV gp52s. Two-dimensional fingerprints of the gp52s from the MMTVs were performed as described (16, 17). The peptide maps of gp52s from mouse and feline grown viruses are shown in the top and bottom rows, respectively. The arrows in the MMTV(C3H) gp52 maps show an additional peptide which distinguishes this protein from the gp52s of the RIII and GR viruses.

FIGURE 4. Tryptic peptide maps of MMTV p28s. Two-dimensional fingerprints of tyrosine-containing tryptic peptides of the p28s of various MMTVs grown in murine and feline cells. The arrows indicate the differences between the p28 of MMTV(C3H) grown in either murine or feline cells and the MMTVs of RIII and GR mice.

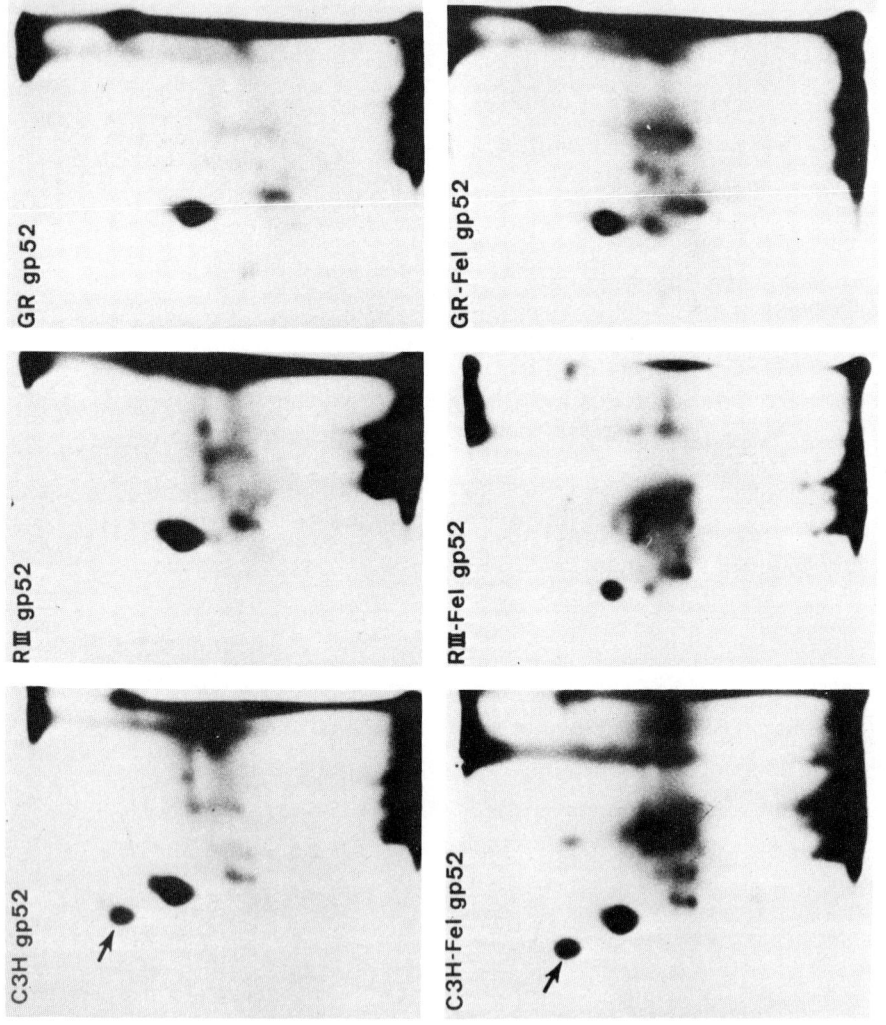

FIGURE 3. See legend on page 351.

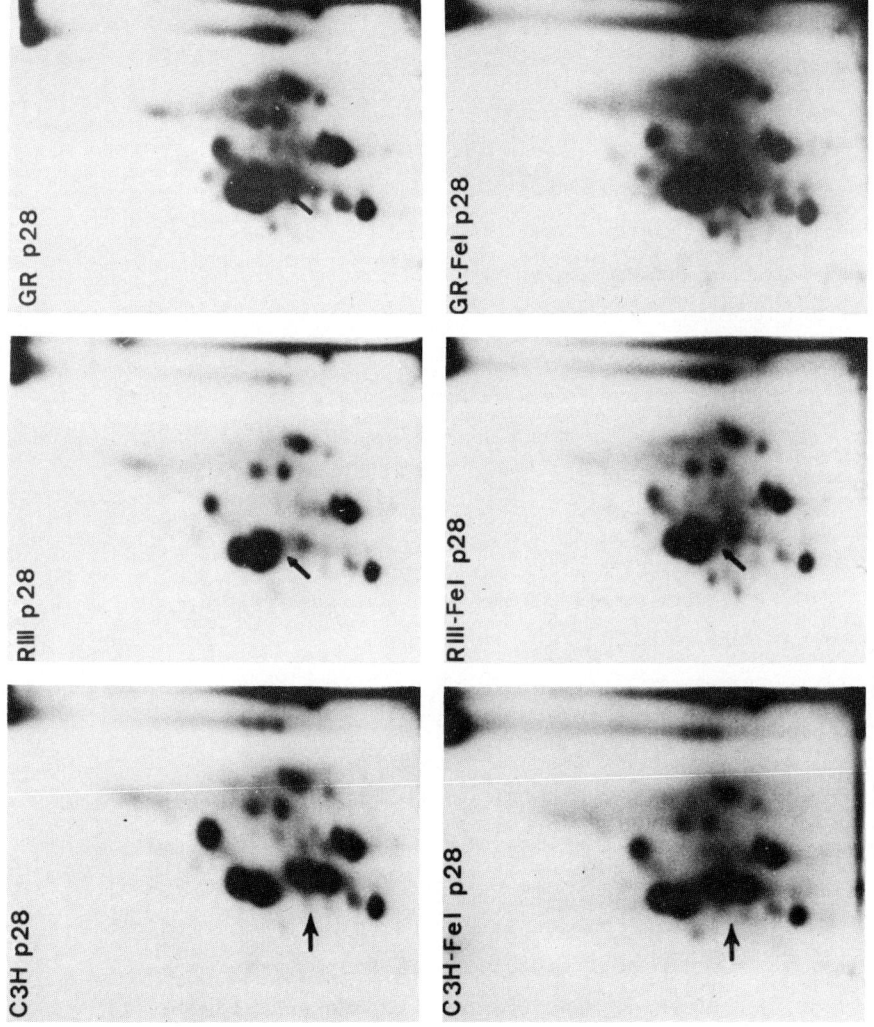

FIGURE 4. See legend on page 351.

HETEROGENEITY IN MMTV GENOMES AND MODE OF TRANSMISSION

Studies employing both MMTV radioactive 60-70S RNA and MMTV cDNA probes have demonstrated that MMTV proviral sequences are present in the DNA of all strains of laboratory mice. Copy numbers are in the low repetitive range for normal tissues such as liver and are higher in mammary tumors (18-23).

Biological studies have shown that MMTVs can be transmitted in different mouse strains either by the milk or via the germline (1-3). Occasionally, MMTV may also be transmitted by male seminal fluids to females, which in turn can transfer the virus to their progeny via the milk (2, 3). Other modes of transmission are, of course, possible. The question that arises is: Can one distinguish if an MMTV has been introduced into a given mouse via the germline (i.e., as a germinal provirus or virogene) or via some non-germline mechanism, such as via the placenta, milk, seminal fluid, or as a plasmid? The term "horizontal transmission" is not used here due to the confusion that would arise from such modes of viral transmission as via the placenta or as a plasmid in a germ cell. If an MMTV was introduced into a mouse via the germline, one would expect to find MMTV proviral sequences equally distributed in the DNA of all tissues of that mouse. On the other hand, if an MMTV was introduced into a mouse by some other mechanism, one would expect to see an uneven distribution of MMTV sequences in the DNA of different tissues of that mouse. To address these points, we used the technique of molecular hybridization.

MMTV(C3H) was isolated from supernatant fluids of a C3H mammary tumor cell line. The 60-70S RNA from these virions was purified and iodinated to a specific activity of approximately 2×10^7 cpm per µg as described previously

(20). This RNA was then hybridized at various C_ot values to DNA from C3H mammary tumor cells and DNA from an apparently normal C3H liver; hybridization to sheep DNA was used as a control. C_ot is defined as the product of the DNA concentration in moles of deoxyribonucleotide per liter and time in seconds. As assayed by resistance to ribonuclease (RNase A and T_1) digestion, hybridization to sheep and other non-rodent DNAs remained consistently at less than 6% up to a C_ot of 35,000 and was thus scored as nonspecific background. Hybridization between the iodinated MMTV(C3H) 60-70S RNA and DNA from C3H mammary tumors was approximately 60% (Fig. 5A). This value was consistently higher than the maximum extent of hybridization between this MMTV(C3H) RNA and DNA from livers or other normal organs of C3H mice (Fig. 5A).

The C_ot 1/2 value was approximately 380 for the hybridization between MMTV(C3H) [^{125}I]RNA and C3H mammary tumor DNA, and approximately 440 for C3H liver DNA. For comparison, poly A enriched C3H cellular RNA (selected by poly U-Sepharose chromatography), representing messenger RNA, was also iodinated and hybridized to C3H liver DNA. The C_ot 1/2 value obtained using this RNA was approximately 3,100, thus representing the value obtained with "unique" DNA. The results depicted in Figure 5A demonstrate that both the C3H mammary tumor cell line and C3H liver contain MMTV proviral sequences in the low repetitive range (20). The lower $C_ot_{1/2}$ value obtained with the C3H mammary tumor cell line DNA indicates that there are more MMTV proviral sequences in this DNA than in the DNA of the C3H liver. The differences in final percent hybridization, i.e., approximately 60% for the C3H tumor cell line DNA, and approximately 50% for the C3H liver DNA, however, may be indicative of one or a combination of two phenomena: (a) there are quantitatively more MMTV proviral sequences in the mammary tumor DNA than the liver DNA; and (b) the DNA of the C3H mammary tumor cells

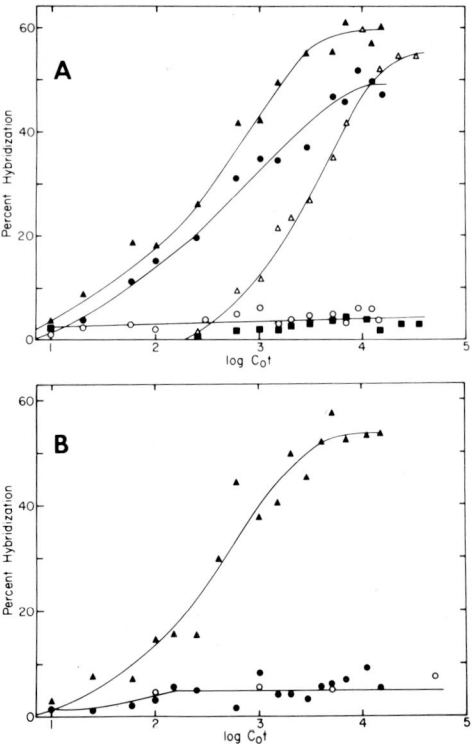

FIGURE 5. (A) Molecular hybridization of [^{125}I]-labeled MMTV(C3H) 60-70S RNA to DNAs as described (20). DNA from the C3H mammary tumor cell line (▲); normal C3H liver (●); and sheep lung (o). For comparison, hybridizations were performed between [^{125}I]-labeled poly(A)-enriched mouse RNA and C3H liver DNA (Δ) and calf thymus DNA (■). (B) Hybridization of recycled [^{125}I]-labeled MMTV(C3H) RNA to cellular DNAs. Iodinated MMTV(C3H) 60-70S RNA was extensively hybridized to normal C3H liver, and the unhybridized fraction was recovered by hydroxylapatite column chromatography as described (20). The recycled RNA was then hybridized to the following cellular DNAs: C3H mammary tumor cell line (▲); C3H liver (●); and sheep lung (o).

contain sequences of the MMTV genome that are not found in the DNA of C3H liver cells. To answer this question, recycling experiments were performed.

Recycling of MMTV 60-70S [^{125}I]RNA

To determine if there are any MMTV sequences in mammary tumors that are not present in the DNA of an apparently normal organ, i.e., the liver of a C3H mouse, iodinated MMTV(C3H) 60-70S RNA was first hybridized to a vast excess of C3H liver DNA. Liver DNA was chosen because murine livers have been shown to be negative for most MMTV markers (2, 3). MMTV(C3H) 60-70S RNA (300,000 cpm) was first annealed to 30 mg of normal C3H liver DNA at 68°C to a $C_o t$ of 20,000 as described previously (20). The unhybridized single stranded [^{125}I]RNA eluting from the hydroxylapatite column at 0.14 M sodium phosphate was termed "recycled RNA." This RNA was then concentrated and reannealed to C3H mammary tumor DNA and to C3H liver DNA to a $C_o t$ of 20,000. As can be seen in Figure 5B, the recycled MMTV(C3H) RNA failed to hybridize above background levels to the DNA of normal C3H livers. This demonstrates that the recycling procedure effectively removed all portions of the MMTV(C3H) [^{125}I]RNA that are complementary to the DNA of normal C3H liver. The recycled [^{125}I]RNA, however, hybridizes extensively with DNA from C3H mammary tumor cells (Fig. 5B), thus demonstrating the existence of MMTV sequences in C3H mammary tumor DNA that are absent in liver DNA of that same mouse (20).

Distribution of Non-Germline Transmitted Murine Mammary Tumor Virus Sequences

Studies were conducted to determine the natural distribution of the recycled MMTV sequences in various murine DNAs. The recycled MMTV(C3H) [^{125}I]RNA was first hybridized to DNA from RIII livers and RIII mammary tumors that arise early in life. Complementary sequences were found in the DNA of all "early" mammary tumors tested but not in the DNA of livers (Table 2).

The GR strain of mice is of great interest as a result of genetic studies that have demonstrated the transmission of MMTV in this strain as a one-gene dominant characteristic (2). The recycled MMTV(C3H) [^{125}]RNA was hybridized to GR mammary tumor and liver DNAs. Complementary sequences are present in the DNAs of both GR mammary tumors and GR livers (Table 2) as well as other GR organs tested. DNA of "late" occurring mammary tumors of the low and moderate incidence strains BALB/c and C3HfC57BL (Table 1) were also analyzed for the presence of the recycled sequences in their DNA and were consistently found negative. Since the C3HfC57BL strain was originated by C3H mice foster nursed on C57BL mothers, a strain of mice devoid of overt MMTV in its milk (2, 3), this is further evidence that the recycled sequences are part of the milk transmitted MMTV(C3H) and are not germline transmitted.

Livers of BALB/c, C57BL/6N, and C57BL/10SCN mice were shown to contain some MMTV proviral information (18, 19, 21-23); they do not contain, however, sequences homologous to the MMTV(C3H) recycled RNA (Table 2).

It thus appears that the virus or viruses responsible for causing early mammary tumors in C3H, RIII, and GR mice are easily distinguishable from the virus or viruses causing late mammary tumors in C3HfC57BL mice and BALB/c mice. Furthermore, a virus similar to the highly oncogenic non-germline transmitted C3H virus appears to be a germinal provirus in the DNA of the GR strain, the only strain in which genetic evidence has been presented for a one-gene dominant characteristic for MMTV. One possible explanation, therefore, is that a virus similar to the non-germline transmitted viruses

TABLE 2. Hybridization of Recycled MMTV(C3H) [^{125}I]RNA to DNAs of Mammary Tumors and Apparently Normal Tissues of Different Strains of Mice

Source of DNA	% Hybridization
C3H mammary tumor	44
C3H liver	7
RIII mammary tumor	47
RIII liver	7
GR mammary tumor	40
GR liver	38
BALB/c mammary tumor	7
BALB/c liver	7
C3HfC57BL mammary tumor	7
C57BL/6N liver	4
C57BL/10SCN liver	6
Ovine lung	6

Six hundred cpm of MMTV recycled [^{125}I]RNA were hybridized to 500 µg of cellular DNA to a C_ot of 15,000 (in 0.4 M NaPB, pH 6.8, and 0.05% SDS) and assayed using RNase as previously described (20).

of C3H and RIII has become integrated as an endogenous virus of GR.

MMTV-RELATED GENETIC INFORMATION IN THE DNAs OF OTHER RODENTS

In view of the varied distribution of MMTV proviral sequences in the mouse, we set out to determine if MMTV proviral sequences are present in other rodents. Furthermore, to determine if sequences related to, but not identical to, MMTV could be detected in the DNAs of other rodents, conditions of hybridization were relaxed by lowering the temperature at which hybridizations were carried out from 68°C to 54°C, and the assay of resulting RNA-DNA duplexes was accomplished by raising the salt concentration from the standard conditions of 2 x SSC (1 x SSC is 0.15 M NaCl and 0.015 Na citrate) to 8 x SSC. Specificity is still maintained using relaxed conditions since no hybridization above 6% background is observed to DNA obtained from bovine and canine tissue or E. coli.

Molecular hybridization experiments were carried out to determine the presence of "MMTV-specific" sequences (using standard conditions) and "MMTV-related" sequences (using relaxed conditions of hybridization and an RNase assay) in other Mus species. Results are given in Table 3 where values have been normalized by subtracting the 6% background hybridization level and using as 100% the hybridization observed between MMTV [^{125}I]60-70S RNA and the homologous C3H mammary tumor cell line DNA. As can be seen in Table 3, when standard conditions of hybridization and RNase assay are used, the subspecies of Mus musculus, i.e., Mus musculus molossinus and Mus musculus Peru Atteck, appear to contain less MMTV sequences as endogenous provirus. When standard conditions of hybridization are used, however, no MMTV-specific sequences are detected in the DNAs of other Mus species, i.e., Mus cervicolor or Mus caroli (24).

When relaxed conditions of hybridization and RNase assay are used, the degree of hybridization detected using DNAs of C3H or RIII Mus musculus tissues remains approximately the same as does the degree of hybridization to Mus musculus molossinus and Mus musculus Peru Atteck DNAs. Using these relaxed conditions, however, MMTV-related sequences are now detected in the DNAs of Mus cervicolor and Mus caroli mice (Table 3). Kinetic analysis of this hybridization reveals

TABLE 3. Hybridization of [^{3}H]MMTV (Fel) 60-70S RNA to Rodent DNAs: Normalized Values

Source of DNA	Standard[a]	Relaxed
Mouse		
Mus musculus - mammary tumor (C3H)	100[b]	100
- liver (C3H)	56	56
- liver (RIII)	60	56
Mus musculus molossinus	22	22
Mus musculus Peru Atteck	18	22
Mus cervicolor	0	20
Mus caroli	0	18
Rat (Rattus norvegicus)		
Wistar	0	20
Lewis	0	17
Osborn-Mendel	0	17
Sprague-Dawley	2	19
AXC	0	22
feral	0	17
Hamster	0	0
Guinea pig	0	0
Mink	0	0
Bovine	0	0
Feline	0	0

[a] Standard and relaxed conditions of hybridization and RNase assay are described (20).

[b] Values were normalized as follows: the 6% (average) background hybridization observed to non-rodent DNAs was subtracted from all values. The hybridization value to the DNA of the cell line from which the MMTV RNA was produced was given a value of 100%, and all other values were normalized to that number accordingly.

$C_o t_{1/2}$ values of 150 and 600 for Mus cervicolor and Mus caroli DNAs, respectively, as compared to 500 for Mus musculus and 3,100 for murine "unique" sequences (24). Thermal analysis studies revealed ΔT_m values 3.2 and 3.4°C lower for hybrids formed between MMTV RNA and DNA from Mus cervicolor and Mus caroli, respectively, than observed with Mus musculus DNAs. These results are in general agreement with a recent report (21) using MMTV cDNA probes.

MMTV-related DNA in Rats. As seen in Table 3, five different laboratory strains of rats, as well as feral rats, all contained approximately the same degree of MMTV-related information in their DNA when using relaxed conditions of hybridization and RNase assay. DNAs from various organs of a feral rat were also tested for the presence of MMTV-related information and they all contained the same degree of MMTV-related information (24). The sequences detected thus appear to be endogenous, i.e., germline transmitted in rats.

The MMTV 60-70S [^{3}H]RNA, obtained from virions from the supernatant fluids of MMTV-infected feline (Fel) cells (5), was also used in these experiments and gave the same results as the MMTV RNA obtained from virus grown in murine cells. This RNA was used to rule out the possibility that normal murine or feline cellular RNA or DNA was contaminating the MMTV 60-70S RNA preparation. Thus, the hybridization observed to rat DNA is interpreted as the result of the presence of nucleic acid sequences related to the mouse mammary tumor virus genome.

Fisher, or F344, rats from several colonies were also examined and found to contain the same degree of MMTV-related information in their DNA as other rat strains (Table 3). However, F344 rats obtained from certain colonies appeared to contain additional MMTV-related information in their DNA, indicating a possible infection and integration by MMTV or

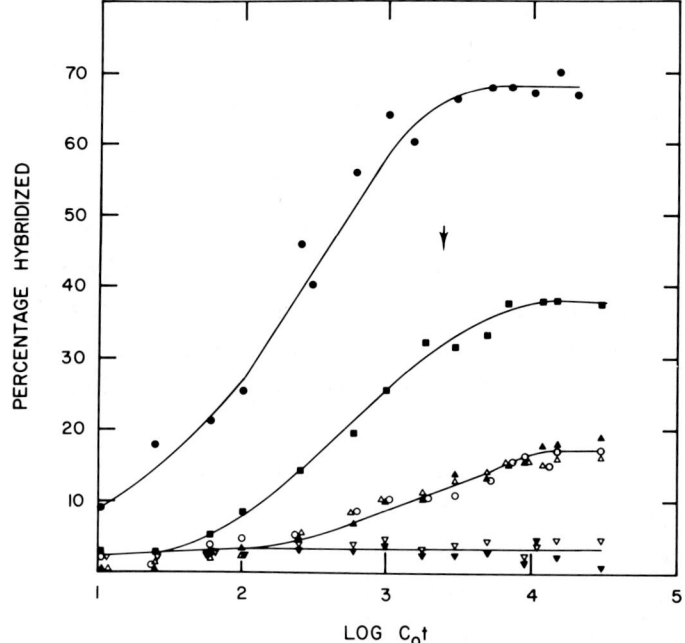

FIGURE 6. Kinetic analysis of hybridization between MMTV 60-70S [^{125}I]RNA and rat cellular DNAs using relaxed conditions of hybridization and RNase assay (see text). Each point represents the hybridization between 1,000 cpm of [^{125}I]MMTV 60-70S RNA and 500 μg of DNA from the following: C3H mammary tumor cell line (●); C3H liver (■); AXC rat (▲); Lewis rat (o); Wistar rat (Δ); hamster (▼); and guinea pig (▽). The arrow indicates the $C_0t_{1/2}$ for the hybridization between [^{125}I]-labeled polyadenosine selected normal rat RNA [selected on Poly U-Sepharose 4B (20)], and rat DNA under relaxed conditions of hybridization.

an MMTV-related virus. The nature and distribution of these additional sequences are currently being investigated.

To determine at what frequency the MMTV-related sequences are present in rat DNA, MMTV radioactive 60-70S RNA was hybridized to the DNAs of five different laboratory strains of rats as well as C3H mouse liver and mammary tumor DNA. As seen in Figure 6, the kinetics of hybridization to the DNAs of several strains were extremely similar. The $C_o t_{1/2}$ values obtained for all five strains were approximately 800 (24). The hybridization of poly A selected radioactive rat RNA to "unique" rat DNA gave a $C_o t_{1/2}$ value of approximately 2,500, indicating that the endogenous MMTV-related sequences in rat DNAs are present in the "low repetitive" range.

To determine the fidelity of hybrids formed between MMTV radioactive RNA and rat DNA, the thermal stability of the RNA-DNA duplexes formed was analyzed by hydroxylapatite chromatography. The Tm values observed were 5°C lower than those obtained with homologous Mus musculus DNAs.

CHARACTERIZATION OF A NEW B-TYPE VIRUS, RELATED TO MMTVs, FROM MUS CERVICOLOR

A virus, morphologically indistinguishable from the type-B MMTVs of the laboratory mouse Mus musculus, has been identified in the milk of Mus cervicolor popaeus mice (25, 26). Group-specific radioimmunoassays for the gp52 (Fig. 7A) and the p28 (Fig. 7B) of MMTV demonstrate that this new virus shares some antigenic determinants with both of these MMTV proteins (25). This reactivity is clearly different, however, from that observed with all MMTVs tested from M. musculus. The Mus cervicolor B-type virus has a density of 1.16 g/ml in sucrose and a virion-associated DNA polymerase with a divalent cation preference for Mg^{2+} over Mn^{2+}. Competitive molecular hybridization experiments showed little if any

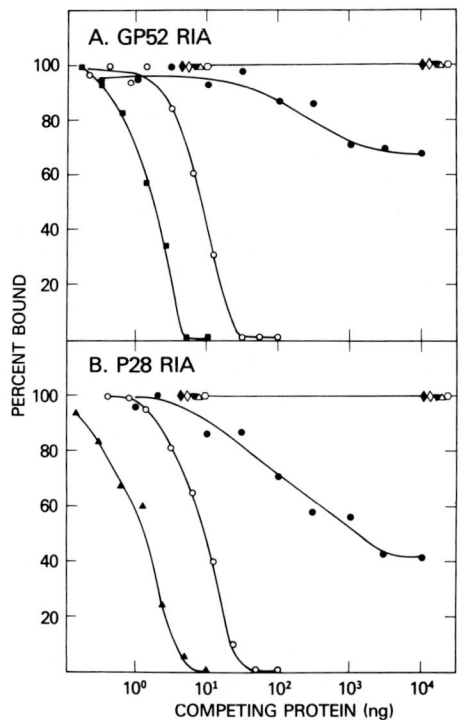

FIGURE 7. Reactivity of <u>Mus cervicolor</u> type-B virus in MMTV gp52 and p28 radioimmunoassays. (A) Anti-MMTV(C3H) gp52 serum was used to precipitate [^{125}I]-labeled MMTV(RIII) gp52. Increasing amounts of the following competitors were used: purified gp52 of MMTV(RIII) (■); MMTV(RIII) (o); <u>Mus cervicolor</u> type-B virus (●). Other competitors were tested at multiple protein inputs, but only two points are shown for clarity. Symbols: CERV-CI (◇); CERV-CII (◆); M432 virus (▼); C57BL milk (Δ); and MuLV from C3H10T1/2 cells (◯). (B) Anti-MMTV(RIII) p28 serum was used to precipitate [^{125}I]-labeled MMTV(C3H) p28. Competitors are as in (A), plus MMTV(C3H) p28 (▲).

nucleic acid homology with MMTVs. Radioimmunoassays also clearly differentiate this virus from the other viruses previously identified from M. cervicolor: M432, CERV-CI, and CERV-II (26, 27). These studies thus identify the first virus from another species that is immunologically related to the MMTVs of M. musculus. Similar particles were also observed in a spontaneous M. cervicolor mammary tumor (25).

Milk of some feral and various inbred strains of M. musculus has previously been shown to be a source of MMTV, regardless of the mode of transmission of the virus (1-3, 28). Similarly, M. cervicolor milk appears to be a good source of B-type virus, particularly those mice from the Tak province of Thailand. Experiments are now in progress, involving use of this resource, to develop "interspecies" RIAs for the p28 and gp52 viral proteins to use as probes for the detection of antigen expression related to mammary tumor viruses in other rodent and more distantly related species.

REFERENCES

1. Bittner, J.J. (1936). Science 84:162.
2. Bentvelzen, P. (1972). Int. Rev. Exp. Path. 11:259.
3. Nandi, S., and McGrath, C.M. (1973). Adv. Cancer Res. 17:353.
4. Lasfargues, E.Y., Lasfargues, J.C., Dion, A.S., Greene, A.E., and Moore, D.H. (1976). Cancer Res. 36:67.
5. Howard, D.K., Colcher, D., Teramoto, Y.A., Young, J.M., and Schlom, J. (1977). Cancer Res. 37:2969.
6. Cardiff, R.D., Puentes, M.J., Teramoto, Y.A., and Lund, J.K. (1974). J. Virol. 14:1293.
7. Parks, W.P., Howk, R.S., Scolnick, E.M., Oroszlan, S., and Gilden, R.V. (1974). J. Virol. 13:1200.
8. Verstraeten, A.A., vanNie, R., Kwa, H.G., and Hageman, Ph.C. (1975). Int. J. Cancer 15:270.
9. Ritzi, E., Baldi, A., and Spiegelman, S. (1976). Virology 75:188.

10. Sheffield, J.B., Daly, T., Dion, A.S., and Taraschi, N. (1977). Cancer Res. 37:1480.
11. Teramoto, Y.A., Kufe, D., and Schlom, J. (1977). Proc. Natl. Acad. Sci. USA 74:3564.
12. Teramoto, Y.A., Kufe, D., and Schlom, J. (1977). J. Virol. 24:525.
13. Teramoto, Y.A., and Schlom, J. (1978). Cancer Res. 38:1990.
14. Hunter, W.M. (1973). In "Handbook of Experimental Immunology" (D.M. Weir, ed.), p. 17.1. Blackwell Scientific Publications, Oxford.
15. Schochetman, G., and Schlom, J. (1976). Virology 73:431.
16. Elder, J.H., Jensen, F.C., Bryant, M.L., and Lerner, R.A. (1977). Nature 267:23.
17. Gautsch, J.W., Lerner, R., Howard, D., Teramoto, Y.A., and Schlom, J. J. Virol., in press.
18. Varmus, H.E., Bishop, J.M., Nowinski, R.C., and Sarkar, N. (1972). Nature New Biol. 238:191.
19. Varmus, H.E., Stavnezer, J., Medeiros, E., and Bishop, J.M. (1975). In "Comparative Leukemia Research 1973, Leukemogenesis" (Y. Ito and R.M. Dutcher, eds.), p. 451. University of Tokyo Press, Tokyo.
20. Drohan, W., Kettmann, R., Colcher, D., and Schlom, J. (1977). J. Virol. 21:986.
21. Morris, V.L., Medeiros, E., Ringold, G.M., Bishop, J.M., and Varmus, H.E. (1977). J. Mol. Biol. 114:73.
22. Schlom, J., Colcher, D., Drohan, W., and Kettmann, R. (1977). In "Progress in Experimental Tumor Research" (F. Homberger, ed.), 21:1. S. Karger, Basel.
23. Cardiff, R. (1978). In "Origins of Inbred Mice" (H.C. Morse III, ed.). Academic Press, New York.
24. Drohan, W., and Schlom, J. Submitted for publication.
25. Schlom, J., Hand, P., Teramoto, Y.A., Callahan, R., Todaro, G., and Schidlovsky, G. J. Natl. Cancer Inst., in press.

26. Callahan, R., and Todaro, G. (1978). In "Origins of Inbred Mice" (H.C. Morse III, ed.). Academic Press, New York.
27. Callahan, R., Sherr, C.J., and Todaro, G.J. (1977). Virology 80:401.
28. Rongey, R.W., Hiavacova, A., Lara, S., Estes, J., and Gardner, M.B. (1973). J. Natl. Cancer Inst. 50:1581.

Inbred Strains: Cell Surface Antigens, Mapping

Bernard Amos, Chairman

MINOR HISTOCOMPATIBILITY GENES
AND THEIR ANTIGENS[1]

Ralph J. Graff

Waldheim Department of Surgery

Jewish Hospital of St. Louis

Washington University School of Medicine

St. Louis, Missouri

Progress in transplantation immunogenetics has depended heavily on the development of inbred mouse strains. The earliest observations indicating that acceptance of allografts was at least partially heritable were made in closed mouse colonies. With the production of inbred mouse strains, the basic principles of transplantation immunogenetics developed more readily: outstanding examples were the finding that the acceptance of allografts is a polygenic trait, and that histocompatibility genes can be isolated in congenic strain pairs.

Since the present report will deal with histocompatibility genes and their antigens, particularly the minor histocompatibility genes and antigens, it would be appropriate to define these terms. Histocompatibility antigens (H-antigens) are found on a broad distribution of tissues and immunological reaction against them results in the rejection of antigen-

---

[1] Supported by U.S. Public Health Service Research Grant AI 07437 from the National Institute of Allergy and Infectious Diseases and Veterans Administration Grant MRIS #3177.

bearing tissues. This definition excludes tissue-specific antigens such as the blood group antigens (the Ea series in the mouse) and the lymphocyte antigens (the Ly series in the mouse). In addition, it excludes modulating antigens -- antigens that evoke immunological reactions not resulting in tissue rejection (TL in the mouse). The term minor H-antigen is complementary to the term major H-antigen. The major H-antigens are under the genetic control of the H-2 complex, a group of closely linked genes on chromosome 17. The minor H-antigens are controlled by genes scattered throughout the remainder of the chromosomes in addition to four genes located on chromosome 17.

Estimates of the Number of H-Loci

The role of genetics in transplantation was first suggested by Loeb (1) and Tyzzer (2), but it remained for Little (3) in his classical publication, "A Possible Mendelian Explanation for a Type of Inheritance Apparently Non-Mendelian in Nature," to hypothesize that the susceptibility or resistance to the growth of an allograft was under the control of multiple genes, subsequently named histocompatibility genes by Snell (4). In a later publication, Little and Tyzzer (5) produced F_2 mice from the initial cross of a Japanese waltzing mouse and a common house mouse. They used the acceptance or rejection of a Japanese waltzing mouse tumor by the F_2 mice to estimate that the Japanese waltzing mouse and common house mouse differed by 14-15 H-genes. The technique suffered from two shortcomings: 1) because of the large number of H-genes by which the parental strains differed, very few grafts were accepted and the resultant error was great, and 2) the technique allowed only one opportunity for crossover; therefore, multiple closely linked H-genes behaved as one gene. Bailey and Mobraaten (6) minimized these two shortcomings by basing their estimates on the survival of skin grafted between partially inbred mice and from multiple backcrossed mice to parental strain mice. In both cases the number of surviving skin grafts and the opportunities for crossovers are increased manyfold. Using these techniques Bailey and Mobraaten estimated that strains BALB/cBy and C57BL/6By differ by 28 or 29 H-genes.

To determine if Bailey's estimate, which was significantly larger than previous estimates, was typical of other strain combinations, we have tested additional strain pairs. We used the multiple backcross technique to estimate the number of genes by which DBA/2J and B10.D2 differ (these two strains both possess the $\underline{H-2}^d$ haplotype). These strains were crossed and female offspring backcrossed to B10.D2 males for multiple generations. Skin was grafted to B10.D2 males from females of backcross generations N4 through N8. If the skin grafts were not rejected within 100 days, whenever possible a second B10.D2 male was immunized with 5×10^6 spleen cells from the female donor, skin grafted, and scored for 250 days. The formula $L = 2^{n-1}$ was used to estimate the number of histocompatibility differences between DBA/2J and B10.D2 (L = number of H-genes, n = the generation in which all grafts of a given line were accepted). Based on the responses of unimmunized hosts an H-gene difference of 28 genes was estimated. Based on the responses of immunized hosts an H-gene difference of 63 genes was estimated. In order to estimate the number of H-differences by which C57BL/6J and DBA/2J differ, orthotopic tail skin grafts were exchanged between unimmunized, sex-matched, full siblings of 16 sublines derived from a C57BL/6J-DBA/2J cross. These sublines had been through 13 to 17 sibling matings. The grafts were scored for 100 days. Using the formula $S = (3/4)^{rL} (5/8)^{vL}$ (S = surviving fraction, r and v are constants dependent on the generation number), it was estimated that these strains differed by 34 $\underline{H}$-genes (28 in males, 39 in females) (7). Similarly the number of H-genes by which C57BL/6J and C3H/HeJ differ was estimated to be 32 (37 in males, 27 in females) (7). These results for strain combinations B10.D2 - DBA/2J, C57BL/6J - DBA/2J, and C57BL/6J - C3H/HeJ are of the same order of magnitude as Bailey's estimate for C57BL/6By - BALB/cBy. It is likely that the counts of 28 to 34 for histocompatibility differences between unrelated strains are representative.

Estimates of the total number of H-genes also can be made from the frequency of H-gene mutations. Based on the ratio of H-gene mutation (as determined by rejection of skin grafts exchanged between F_1's) to the mutation rate of other genes, Bailey (8) estimated a total of 720 H-genes. By another method, based on lost type mutations and the number of H-gene differences between the strains being tested, he estimated 430 H-genes. Bailey explained extremely large estimates by the facts that they reflect non-polymorphic (silent) as well as polymorphic H-loci, that they are not dependent on gene segregation and thereby not affected by linkage, and that there may be more than one mutation site at each locus.

Isolation, Mapping, and Allelism of Weak Loci

It was possible to identify and study the H-2 locus in inbred mouse strains which differed by many non-H-2 loci as well as by the H-2 locus because of the strength of the H-2 antigens and because specific antibodies could be produced against H-2 antigens which could then be used for identification. The relative weakness of the non-H-2 antigens and their failure to stimulate good antisera has necessitated the isolation of non-H-2 genes in congenic mouse strains for identification. Through the use of congenic strains 35 minor histocompatibility loci have been reported in addition to loci on the X and Y chromosomes (the latter two not requiring congenic strains). H-1, H-3, H-4, and H-7 through H-13 were described by Snell and co-workers (9-11), H-15 through H-30, H-34 through H-38, and H-X by Bailey (12, 13), H-31, H-32, and H-33 by Flaherty and co-workers (14, 15), H-39 by Artzt (16) and H-Y by Eichwald (17) (Table 1). Sixteen of these loci have been mapped on eight chromosomes (Fig. 1). As many as eight additional H-genes have been demonstrated by the rejection of skin grafts exchanged between non-histocompatibility congenic strains (18). The temporary designations and the congenic strains are indicated in Table 2. H-(JS), H-(Lt), and H-(tn) have not been proven to be distinct from one another; H-(go) and H-(pi) have not been proven to be distinct from one another; and none of these potential loci have been proven to be distinct from the previously described unmapped H-genes.

Although many congenic strains have been produced for the purpose of defining new H-loci, relatively few have been produced to define additional alleles of the known loci. Six alleles of the H-3 locus, four alleles each of the H-1 and H-13 loci, three alleles each of H-8 and H-11, and two alleles of each of the remaining H-loci have been isolated in congenic strain pairs (9-15, 19-26; Graff, unpublished data).

The Cellular Immune Response to Non-H-2 Antigens

The most frequently used technique to study the cellular immune response to non-H-2 antigens has been allograft rejection. The first allografts to be used were tumor, which proved to be a technically easy but insensitive test. Skin grafting techniques were found to be more sensitive because the hosts could be observed for rejection for prolonged periods, whereas tumor recipients would have succumbed. Two major techniques of skin grafting have been used, the grafting of skin onto the flank (27) and the grafting of tail skin orthotopically (28). Median survival times (MST's) tend to

TABLE 1. Congenic Strains Defining the Minor Histocompatibility Loci

H-locus and allele	Strain possessing	H-locus and allele	Strain possessing
$H\text{-}1^a$	B10.C3H-$H\text{-}1^a$, B10.D2-$H\text{-}1^a$	$H\text{-}22^b$	C57BL/6By
$H\text{-}1^b$	B10.129-$H\text{-}1^b$, B6.C-$H\text{-}1^b$	$H\text{-}22^c$	B6.C-$H\text{-}22^c$
$H\text{-}1^c$	C57BL/10Sn, C57BL/6J	$H\text{-}23^b$	C57BL/6By
$H\text{-}1^?$	B10.P(61NX), B10.CE(62NX)	$H\text{-}23^c$	B6.C-$H\text{-}23^c$
$H\text{-}3^a$	C57BL/10Sn, C57BL/6By	$H\text{-}24^b$	C57BL/6By
$H\text{-}3^b$	B10.LP-a, B10-$H\text{-}2^d H\text{-}3^b$	$H\text{-}24^c$	B6.C-$H\text{-}24^c$
$H\text{-}3^c$	B10.C-$H\text{-}3^c$	$H\text{-}25^b$	C57BL/6By
$H\text{-}3^d$	B10.KR-$H\text{-}3^d$	$H\text{-}25^c$	B6.C-$H\text{-}25^c$
$H\text{-}3^e$	B10-pa $H\text{-}3^e a^t$	$H\text{-}26^b$	C57BL/6By
$H\text{-}3^f$	YBR	$H\text{-}26^c$	B6.C-$H\text{-}26^c$
$H\text{-}4^a$	C57BL/10Sn, C57BL/6By	$H\text{-}27^b$	C57BL/6By
$H\text{-}4^b$	B10.129-$H\text{-}4^b$	$H\text{-}27^c$	B6.C-$H\text{-}27^c$
$H\text{-}7^a$	C57BL/10Sn, C57BL/6By	$H\text{-}28^b$	C57BL/6By
$H\text{-}7^b$	B10.C-$H\text{-}7^b$, B6.C-$H\text{-}7^b$	$H\text{-}28^c$	B6.C-$H\text{-}28^c$
$H\text{-}8^a$	C57BL/10Sn, C57BL/6By	$H\text{-}29^b$	C57BL/6By
$H\text{-}8^b$	B10.D2-$H\text{-}8^b$	$H\text{-}29^c$	B6.C-$H\text{-}29^c$
$H\text{-}8^c$	B6.C-$H\text{-}8^c$		
$H\text{-}9^a$	C57BL/10Sn, C57BL/6By	$H\text{-}30^b$	C57BL/6By
$H\text{-}9^b$	B10.C-$H\text{-}9^b$	$H\text{-}30^c$	B6.C-$H\text{-}30^c$
$H\text{-}10^a$	C57BL/10Sn	$H\text{-}31^a$	A/JBoy, C57BL/6JBoyT1a^a
$H\text{-}10^b$	B10.129-$H\text{-}10^b$	$H\text{-}31^b$	A-T1a^b, C57BL/6JBoy
$H\text{-}11^a$	C57BL/10Sn	$H\text{-}31^c$	B6.AK-$H\text{-}2^k$T1a^b
$H\text{-}11^b$	B10.129-$H\text{-}11^b$	$H\text{-}32^a$	A/JBoy
$H\text{-}11^?$	B10.D2-$H\text{-}11^?$	$H\text{-}32^b$	A/JBoy-T1a^b, C57BL/6JBoy
$H\text{-}12^a$	C57BL/10Sn	$H\text{-}33^a$	BALB.TTF-$H\text{-}33^a$/$H\text{-}33^b$
$H\text{-}12^b$	B10.129-$H\text{-}12^b$	$H\text{-}33^b$	BALB/cBoy
$H\text{-}13^a$	C57BL/10Sn	$H\text{-}34^b$	C57BL/6By
$H\text{-}13^b$	B10.CE-$H\text{-}13^b$, B10.LP-$H\text{-}13^b$A^w	$H\text{-}34^c$	B6.C-$H\text{-}34^c$
$H\text{-}13^c$	B10.C3H-$H\text{-}13^c$, B10.KR-$H\text{-}13^c$A^w	$H\text{-}35^b$	C57BL/6By
		$H\text{-}35^c$	B6.C-$H\text{-}35^c$
$H\text{-}15^b$	C57BL/6By	$H\text{-}36^b$	C57BL/6By
$H\text{-}15^c$	B6.C-$H\text{-}15^c$	$H\text{-}36^c$	B6.C-$H\text{-}36^c$
$H\text{-}16^b$	C57BL/6By	$H\text{-}37^b$	C57BL/6By
$H\text{-}16^c$	B6.C-$H\text{-}16^c$	$H\text{-}37^c$	B6.C-$H\text{-}37^c$
$H\text{-}17^b$	C57BL/6By	$H\text{-}38^b$	C57BL/6By
$H\text{-}17^c$	B6.C-$H\text{-}17^c$	$H\text{-}38^c$	B6.C-$H\text{-}38^c$
$H\text{-}18^b$	C57BL/6By	$H\text{-}39$	BTBR/Nev-T
$H\text{-}18^c$	B6.C-$H\text{-}18^c$		BTBR/Nev-$+/t^{w18}$
$H\text{-}19^b$	C57BL/6By		
$H\text{-}19^c$	B6.C-$H\text{-}19^c$	$H\text{-}x^b$	C57BL/6By
		$H\text{-}x^c$	BALB/cBy
$H\text{-}20^b$	C57BL/6By	$H\text{-}x^1$	Lg/Ckc
$H\text{-}20^c$	B6.C-$H\text{-}20^c$		
$H\text{-}21^b$	C57BL/6By	$H\text{-}Y^a$	A/J
$H\text{-}21^c$	B6.C-$H\text{-}21^c$	$H\text{-}Y^b$	C57BL/6

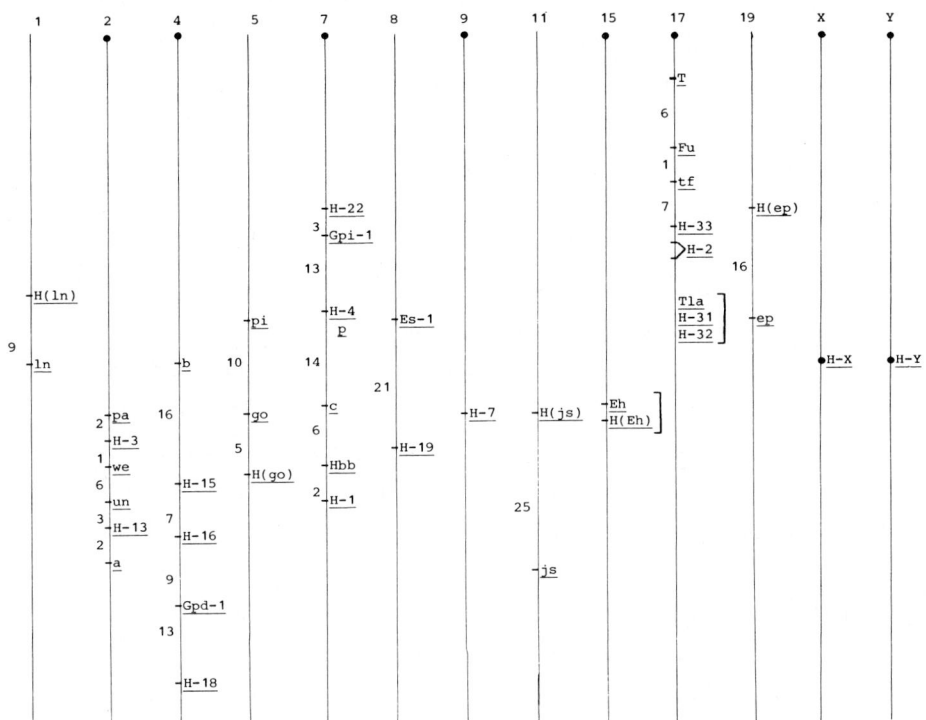

FIGURE 1. Murine linkage map.] means gene order not known. Marker genes on linkage map: a, non agouti (coat color gene); b, brown (coat color gene); c, albino (coat color gene); Eh, hairy ears; ep, pale ears (ears and tail are pale); Es-1, serum esterase; Fu, fused (shortened and kinked tail); go, angora (extra length of guard hairs); Gpd-1, glucose-6-phosphate dehydrogenase; Gpi-1, glucose phosphate isomerase; Hbb, hemoglobin beta chain; js, Jackson shaker; ln, leaden (coat color gene); p, pinkeye (coat color gene); pa, pallid (coat color gene); pi, pirouette (degenerative changes in inner ear); T, Brachyury; tf, tufted (repeated waves of hair loss and regrowth); Tla, thymus leukemia antigen; un, undulated (dorsal kyphosis and kinked tail); we, welharig (wavy coat).

be shorter with the orthotopic tail skin technique, probably because the grafts are smaller.

The magnitude of the immune response by a particular host to a given antigen depends on the degree of "differentness" of the antigen (immunogenicity) and the immune responsiveness of the host. Data are being accumulated documenting variation in host responsiveness to a given antigen. In 1969 Graff and Snell (26) noted a variation in the immune capability of a group of F_1 mice as indicated by their ability to reject non-H-2 disparate skin grafts. Subsequent studies have indicated that genes in the H-2 complex control the speeds of rejection of non-H-2 disparate skin grafts. The immune response to H-Y by female mice has been shown to be under the control of an H-2-linked gene (29, 30). To study the H-2-linked genetic control of immune responsiveness to non-H-2 antigens, Wettstein and Haughton are producing groups of "double congenic" strains. These strains are developed in groups of at least four strains. All of the strains are on a C57BL/10 background, two of the strains share one H-2 haplotype and differ at a given non-H-2 locus, and the other two share a second H-2 haplotype and possess the same non-H-2 difference. With these groups of double congenic mice, Wettstein and Haughton (31, 32) have demonstrated that the magnitude of the allograft rejection reaction against the H-4.2 and H-7.2 antigens is effected by the H-2 haplotype of the host, indicating the existence of immune response genes in H-2 complex. Through the use of crossover strains it appears that the immune response genes for different non-H-2 antigens may be located in different places in the H-2 complex.

To properly study the immunogenicity of histocompatibility antigens, variations in host immune responsiveness must be minimized. This can most effectively be accomplished by using only one host strain in a given experiment. If this is not

TABLE 2. Temporary Designations for Tentative Histocompatibility Loci

H-locus	Congenic possessing allele strains
H-(ep)^a	C57BL/6J
H-(ep)^b	C57BL/6-ep
H-(Eh)^a	C57BL/6J
H-(Eh)^b	C57BL/6-Eh
H-(go)^a	C57BL/6J
H-(go)^b	C57BL/6-go
H-(js)^a	C57BL/6J
H-(js)^b	C57BL/6-js
H-(ln)^a	C57BL/6J
H-(ln)^b	C57BL/6-ln
H-(lt)^a	C57Bl/6J
H-(lt)^b	C57BL/6-lt
H-(pi)^a	C57BL/6J
H-(pi)^b	C57BL/6-pi
H-(tn)^a	C57BL/6J
H-(tn)^b	C57BL/6-tn

Eight strains differing from C57BL/6J at various mutant marker genes affecting morphological and behavioral traits have been shown to be histoincompatible with C57BL/6J at histocompatibility loci linked to the marker genes. The H-genes have been temporarily symbolized by the nearby marker genes. H-(js), H-(lt), and H-(tn) have not been proven to be distinct from one another. Likewise, H-(go) and H-(pi) have not been proven to be distinct from one another (17).

possible, one should use host strains with a common background and preferably the same H-2 haplotype. The subsequently described studies have been carried out with these guidelines in mind. The total immunogenicity (as indicated by skin allograft rejection) of the antigens of H-2 complex plus adjacent histocompatibility genes has been shown to be about equal to the cumulative non-H-2 immunogenicity (Table 3). The individual non-H-2 antigens are weaker, falling on a continuum from the strong $H-4^b$ antigen (which is slightly weaker than H-2D) to the extremely weak H-9 antigens. Great variations exist in the strengths of the antigens controlled by the alleles of many of the loci (Table 3) (20, 21). As data accumulate on the H-2 complex, it is apparent that the "H-2 immunogenicity" is really the cumulative immunogenicity of multiple histocompatibility antigens. In addition to the immunogenicity of H-2K, H-2A, H-2C, H-2D, there are now added H-31, H-32, H-33, and H-39. It would appear that when the appropriate crossover and congenic strains are produced the immunogenicity of the individual antigens of the H-2 complex and adjacent genes may prove to be comparable to that of the individual non-H-2 antigens.

To quantitate more accurately the cellular immune response to histocompatibility antigens of varying strengths, B10.A mice were injected with varying numbers of lymphocytes differing at the H-2 locus, and C57BL/10Sn mice were injected with varying numbers of lymphocytes differing at the H-3, H-4, H-7, and H-13 loci. The B10.A and C57BL/10Sn mice were then grafted with skin from the lymphocyte donors at varying intervals thereafter. The speed of rejection of the skin grafts was taken as an indication of the level of immunity. A first set rejection was taken to indicate that the antigen dose was too small to be recognized (a subthreshold dose); a second set re-

TABLE 3. The MST's (in Days) of Body Skin Grafts Made Across Various H-Barriers

MST		H-barrier		MST
12	B10.A	H-2 Complex	C57BL/10	13
11	B10.D2	H-2 Complex	C57BL/10	11
12	B10.D2 3	H-2 Complex	B10.BR	
11.4	A/Sn	Mult. non-H-2	B10.A	10.8
11	DBA/2	Mult. non-H-2	B10.D2	10
	129/Sn	Mult. non-H-2	C57BL/10	11
	BALB/cJ	Mult. non-H-2	B10.D2	10.2
13	B10.BR	H-2D	B10.A	16
120	B10.129-$\underline{H-4}^b$	H-4	C57BL/10	20
26	B10.C-$\underline{H-7}^b$	H-7	C57BL/10	33
30	B10.D2-$\underline{H-8}^b$	H-8	C57BL/10	37
21	B10.LP$\underline{a}$	H-3	C57BL/10	50
91	B10.129-$\underline{H-10}^b$	H-10	C57BL/10	71
67	B10.LP-A^w$\underline{H-13}^b$	H-13	C57BL/10	73
78	B10.129-$\underline{H-11}^b$	H-11	C57BL/10	164
25	B10.129-$\underline{H-1}^b$	H-1	C57BL/10	>250
259	B10.129-$\underline{H-12}^b$	H-12	C57BL/10	>300
>400	B10.C-$\underline{H-9}^b$	H-9	C57BL/10	>300

jection was taken to indicate that an immunizing dose had been given; prolonged survival was taken to indicate that the immune system had been overwhelmed (I avoid the word "tolerance" because I wish to make no implications concerning mechanism). The responses of mice tested one week after antigen challenge indicated that the ability to immunize and overwhelm relates directly to immunogenicity: the weaker the antigen, the larger the amount of antigen needed to immunize and the smaller the amount of antigen needed to overwhelm. With H-2 antigens and the strongest non-H-2 antigens, small doses of allogeneic lymphocytes sensitized (as indicated by accelerated rejection of subsequent skin grafts) and the largest doses used did not produce non-reactivity. With weaker antigens the threshold immunizing dose increased and the dose producing non-reactivity decreased until, with extremely weak antigens, no single dose sensitized and relatively small numbers of cells overwhelmed (33, 34).

Following the injection of H-2 disparate lymphocytes the immune response to the H-2 antigen was apparent one day later, peaked three days later, and waned 14 days later, persisting in a low grade manner for 98 days. Following the injection of H-4 disparate lymphocytes the immune response to the H-4 antigens appeared two days later, peaked four days later, and waned at 14 days, persisting at a moderate level until 98 days. Following the injection of H-7 disparate lymphocytes the immune response to the H-7 antigen first appeared seven days later and reached a persistent maximum at 28 days, persisting at that high level for 98 days. Following the injection of H-13 disparate lymphocytes the immune response to the H-13 antigen was suppressed for 14 days. Immunity first appeared 35 days after injection and disappeared by day 56 (Fig. 2). This assay has the limitation of being very slow (the length of the assay is the duration of the graft survival time) and relatively inaccurate in estimating latency. Nevertheless, comparisons between the patterns of the responses to antigens of different strengths can be made. Clearly, the weaker the antigen the longer the latent period. As in previous studies of Snell et al. (35) using Winn's assay, strong cellular immunity to H-2 antigens is short lived. If cellular immunity persists at a high level only as long as antigen is present, the short duration of strong cellular immunity against H-2 antigens and the long duration of cellular immunity against H-4 and H-7 antigens may reflect the rate of clearance of the antigen. The

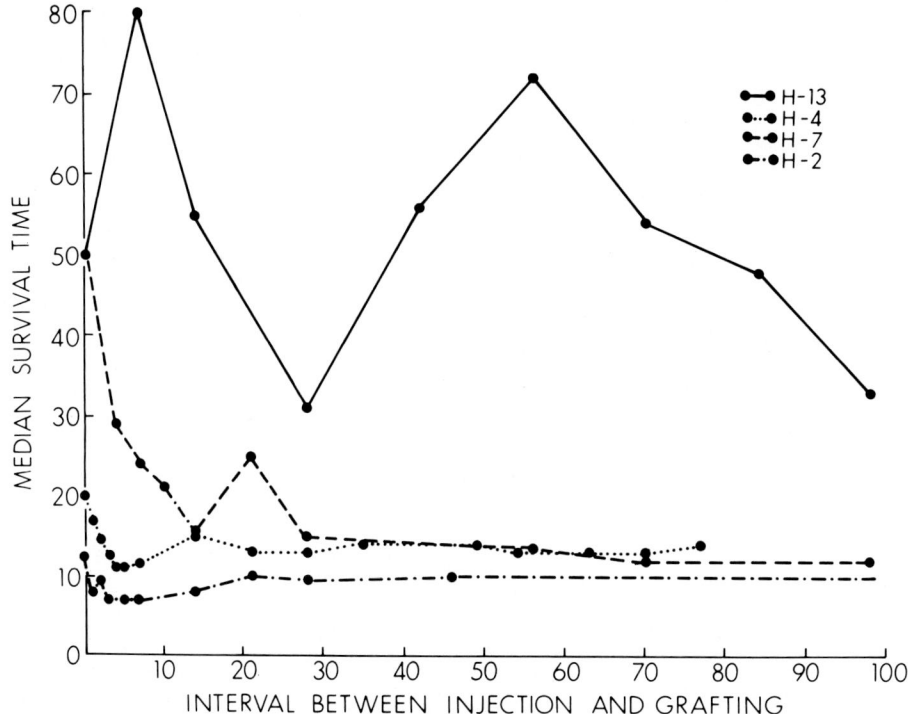

FIGURE 2. The median survival times (MST) of each group are plotted against the time intervals (in days) between injection of 5×10^6 thymocytes and skin grafting.

fluctuating response against the weak H-13 antigen probably reflects the small size of the clone capable of responding.

The immunogenicity of transplantation antigens can be cumulative under some circumstances and not cumulative under others. The immunogenic relationship between multiple transplantation antigens on a single cell has been studied using congenic strains differing at one or more H-loci (36). The immune response against a cell containing two antigens theoretically can be directed at either or both antigens; if the magnitude of the immune response against both antigens is no stronger than the magnitude of the response to the stronger antigen alone, then it can be said that the weaker antigen has made no contribution to total immunogenicity and there is no cumulative effect. If, however, the magnitude of the response to the double difference is stronger than the response to the stronger antigen alone, then it can be stated that there is a cumulative effect. When grafts were exchanged between strains differing by two loci, the greatest cumulative effects were noted when the rejection rates were of similar magnitudes. If the immunogenicities of the individual antigens were different, no cumulative effect was noted. If a third antigen was added to two others, its immunogenicity contributed only if there was a similarity in the rates of rejection of the single and double differences. It would appear that if one continues to add weaker antigens, they would be cumulative only as long as the rate of rejection of the multiple differences was sufficiently similar to the rate of rejection of the last antigen to be added. Although B10.D2 and DBA/2 differ by about 34 minor H-antigens, it is probable that only the two or three of the strongest antigens affect the magnitude of the immune response. These observations held true whether H-2 antigens were involved or not. The mechanism of the cumulative effect is not known. It has been hypothesized (36) that if two antigens of disparate immunogenicity existed on the same cell, the stronger antigen would lead to sensitization and cell destruction before the weak antigen could trigger an immune response in its clone. If the immunogenicities were similar, both clones would be triggered. Nothing is known about the mechanisms of immunologic destruction of cells bearing single antigenic differences versus cells bearing multiple differences. The observed results can be explained on the basis of independent action on each antigen. This does not mean that an effect beyond this independent action does not exist (37).

The H-2 antigens elicit near-maximal responses which obliterate many characteristics of allograft rejection that are apparent with the weaker responses to non-H-2 antigens. Although the size of grafts exchanged across H-2 barriers does not affect graft rejection, the size of grafts exchanged across non-H-2 barriers does. Both the ability to sensitize and the time required for graft destruction are size dependent with larger grafts sensitizing more rapidly and taking longer to reject (38, 39). Rejection of grafts from hemi-allogeneic donors occurs significantly more slowly than rejection of grafts from totally allogeneic donors (40). These observations tend to indicate that both the affector and effector mechanisms are dose dependent. The female immune response is often stronger than the male response to the same antigen (reviewed in 41). This is quite apparent in the immune response to many non-H-2 antigens as indicated by shorter MST's in females than in males. This difference is obliterated following gonadectomy and adrenalectomy. In fact, the strength of male rejection and to a lesser extent the strength of female rejection were greater than the strength of unaltered mice. It has been hypothesized that the strengthened immune response was due to the removal of endogenous immunosuppressive adrenal and gonadal steroids and that the difference between male and female rejection times was due to quantitative and qualitative differences in the endogenous male and female adrenal and gonadal steroids.

In Vitro Assays for Non-H-2 Antigens

Until recently, in vitro assays generally have been unsuccessful in demonstrating non-H-2 antigens. Perhaps as a result of improved technology, positive results have been obtained. Tests using humoral immune techniques to demonstrate non-H-2 transplantation antigens have rarely been successful. Winn, Stevens, and Snell (42) produced hemolytic antibodies directed against H-1 and H-3 antigens. Snell and Graff (unpublished data) produced monospecific hemagglutinating antibodies directed against H-1 and H-7 by exchanging skin grafts between congenic strains differing by these genes and, following rejection, injecting donor strain lymphocytes. The hemagglutinating antibodies produced had extremely high titers but lost activity rapidly following freezing. Subsequent attempts to produce these antisera were unsuccessful. Zink and Heyner (43) produced polyspecific non-H-2 antisera by injecting B10.D2/n spleen and lymph node cells into (BALB/cJxDBA/2J)F_1 mice. Anti H-3 and H-8 activity was demonstrated using hemagglutination, immunofluorescence, and mixed hemabsorption techniques with B10.LP-a and B10.D2(57N) target cells, respectively. It is of particular note that none of these studies have been able to demonstrate lymphocytotoxic activity. The indirect radioimmunoassay has been used successfully to demonstrate TL antigen (44) and tumor-specific antigens (45). In pilot studies in our laboratory target cells were reacted with putative antisera and either ^{125}I-labeled goat-anti-mouse globulin or ^{125}I-labeled staphylococcal protein A as binding agents. Although early studies were encouraging, we have not yet succeeded in applying this technique to non-H-2 antigens.

Although Mangi and Mardiney (46) have reported weak mixed lymphocyte reaction to individual non-H-2 antigens, most investigators have been unsuccessful in doing so. Recent personal communications have indicated success (Wettstein, personal communication) and failure (Dunlop, personal communication) with this technique. The cell-mediated lysis (CML) assay has proven more effective, Gordon et al. (47) using CML against the H-Y antigen and Bevan (48) using CML against H-3, H-13, H-7, and H-8 antigens.

CONCLUSION

In summary, many polymorphic cell membrane structures under the control of multiple genes scattered throughout the genome are specifically recognizable by the immune system. Although these structures are collectively called by the name histocompatibility antigens, and the controlling genes are collectively called by the name histocompatibility genes, they may have no more in common than being membrane structures that are immunologically recognizable. Actually we know nothing about the functions of these structures, which may be different from one another and different from the functioning of the antigens of the H-2 complex. Although these studies were stimulated by a need to know about allograft rejection and histoincompatibility, the knowledge gained may have greater implications for cell biology in general and the function of the cell membrane specifically.

REFERENCES

1. Loeb, L. (1901). J. Med. Res. 1:28.
2. Tyzzer, E.E. (1909). J. Med. Res. 21:519.
3. Little, C.C. (1914). Science 40:904.
4. Snell, G.D. (1948). J. Genet. 49:7.
5. Little, C.C., and Tyzzer, E.E. (1916). J. Med. Res. 33:393.
6. Bailey, D.W., and Mobraaten, L.E. (1969). Transplantation 7:394.
7. Graff, R.J., and Brown, D.H. (1978). Immunogenetics, in press.
8. Bailey, D.W. (1970). Transplant. Proc. 2:32.
9. Snell, G.D., and Stevens, L.C. (1961). Immunology 4:366.
10. Snell, G.D., and Bunker, H.P. (1964). Transplantation 2:743.

11. Snell, G.D., and Bunker, H.P. (1965). Transplantation 3: 235.
12. Bailey, D.W. (1963). Transplantation 1:70.
13. Bailey, D.W. (1975). Immunogenetics 2:249.
14. Flaherty, L., and Wachtel, S.S. (1975). Immunogenetics 2:81.
15. Flaherty, L. (1975). Immunogenetics 2:325.
16. Artzt, K., Hamburger, L., and Flaherty, L. (1977). Immunogenetics 5:477.
17. Eichwald, E.S., and Silmser, C.R. (1955). Transplant. Bull. 2:148.
18. Bailey, D.W., and Bunker, H.P. (1972). Mouse News Letter 46:18.
19. Graff, R.J., Brown, D., and Snell, G.D. (1973). Transplant. Proc. 5:299.
20. Graff, R.J., Hildemann, W.H., and Snell, G.D. (1966). Transplantation 4:425.
21. Graff, R.J., and Snell, G.D. (1968). Transplantation 6:598.
22. Hildemann, W.H. (1970). Transplant. Rev. 3:5.
23. Hildemann, W.H., and Cooper, E.L. (1967). Transplantation 5:707.
24. Snell, G.D., Graff, R.J., and Cherry, M. (1971). Transplantation 11:525.
25. Gasser, D.L. (1976). Immunogenetics 3:271.
26. Graff, R.J., and Snell, G.D. (1969). Transplantation 8:861.
27. Billingham, R.E., and Silvers, W.K. (1960). In "Transplantation of Tissues and Cells," p. 149. Wistar, Philadelphia.
28. Bailey, D.W., and Usama, B. (1960). Transplant. Bull. 7:424.
29. Bailey, D.W., and Hoste, J. (1971). Transplantation 11:404.

30. Gasser, D.L., and Silvers, W.K. (1971). Transplantation 12:412.
31. Wettstein, P.J., and Haughton, G. (1977). Immunogenetics 4:65.
32. Wettstein, P.J., and Haughton, G. (1977). Immunogenetics 5:85.
33. Graff, R.J. (1971). In "Immunogenetics of the H-2 System" (Liblice-Prague, ed.), p. 293. Karger, Basel.
34. Graff, R.J. (1971). Curr. Top. Surg. Res. 3:363.
35. Snell, G.D., Winn, H.J., and Kandutsch, A.D. (1961). J. Immunol. 87:1.
36. Graff, R.J., Silvers, W.K., Billingham, R.E., Hildemann, W.H., and Snell, G.D. (1966). Transplantation 4:605.
37. Bailey, D.W. (1971). Transplantation 11:419.
38. Lapp, W.C., and Bliss, J.Q. (1966). Transplantation 4:754.
39. Lappe, M.A., Graff, R.J., and Snell, G.D. (1969). Transplantation 5:83.
40. Galton, M. (1967). Transplantation 5:154.
41. Graff, R.J., Lappe, M.A., and Snell, G.D. (1969). Transplantation 7:105.
42. Winn, H., Stevens, L.C., and Snell, G.D. (1958). Transplant. Bull. 5:18.
43. Zink, G.L., and Heyner, S. (1977). Immunogenetics 4:257.
44. Esmon, N.L., and Little, J.R. (1976). J. Immunol. 117:911.
45. Ting, C.C., and Herberman, R.B. (1971). Int. J. Cancer 7:499.
46. Mangi, R.J., and Mardiney, M.R., Jr. (1971). Transplant. Proc. 3:873.
47. Gordon, R.D., Simpson, E., and Samuelson, L.E. (1975). J. Exp. Med. 142:1108.
48. Bevan, M. (1976). Immunogenetics 3:177.

SEROLOGICAL DEFINITION OF CELL SURFACE ANTIGENS
OF MOUSE LEUKEMIA

Lloyd J. Old
Elisabeth Stockert

Memorial Sloan-Kettering Cancer Center
New York, New York

The discipline of tumor immunology, like so many other biological fields of contemporary importance, owes its origin to the innovators whom we honor at this workshop. Without their development and characterization of inbred strains, it would be difficult to imagine how the principles that underlie modern immunological approaches to cancer could have been established. The demonstration of tumor-specific antigens in chemical and virus-induced tumors, the development of methods to analyze cellular and humoral immunity to tumors, and the finding that tumor immunity could be bolstered by microbial agents, such as BCG, represent the legacy from work with the inbred mouse that has engendered the vast interest in applying similar approaches to human cancer.

To the cancer immunologist, cell surface antigens are of paramount importance, for it is novel antigens incorporated into the tumor cell surface that can serve as transplantation antigens and provoke immunological rejection. Although such transplantation techniques provided the first evidence for tumor-specific surface antigens, serological techniques have far more potential for analyzing the surface composition of tumor cells. With advances in methods to demonstrate cellular immune reactions in vitro, these can be expected to take their place with serological techniques in the analysis of cell surface antigens of tumors. Present knowledge about these antigens, however, has come almost exclusively from the application of serological techniques, and of these, the cytotoxic test has proved most successful in the identification and analysis of surface antigens, some of which would have been impossible to discover by techniques involving graft rejection.

The cytotoxic test, originally developed by Gorer and O'Gorman to detect antibody to H-2 products, is a simple technique in which cells are lysed by antibody in the presence of a suitable source of complement (1). The test has evolved through a series of modifications and, in its present form, is an analytical method of unrivaled specificity and sensitivity for the detection of cell surface antigens. Because leukemia cells can be easily obtained in free cell suspension and are exquisitely sensitive to cytotoxic antibody, they have been a favorite object of study. The great majority of mouse leukemias are T cells of thymic origin, and this has allowed the serologist to study the leukemia cell side by side with its normal counterpart, the thymocyte. As a consequence of these advantages, more is known about the surface antigens of normal and malignant T cells than about any other cell population of the mouse (2-4).

Four general classes of cell surface antigens have been distinguished on mouse leukemias: 1) conventional H-2 alloantigens, that are present on virtually all cells of the adult mouse; 2) differentiation alloantigens, e.g., Thy-1 and the Lyt series, and several products of the H-2-Ir and Q complex whose presence signify selective gene activation in cells undergoing T cell differentiation; 3) MuLV-related antigens, e.g., GCSA, G_{IX}, $G_{(RADA1)}$, that owe their origin to genetic information of endogenous murine leukemia viruses; and 4) TL (thymus-leukemia) antigens that occur as differentiation alloantigens restricted to normal thymocytes in TL^+ strains of mice and as leukemia-specific antigens in mouse strains not normally expressing TL antigens (TL^- strains).

Before discussing the MuLV and TL systems in more detail, some mention should be made of the immunization procedures that were devised to produce antibody to these cell surface antigens (Table 1). To define non-H-2 specificities, the challenge was to develop methods of immunization or testing that would eliminate the contribution of H-2 antibodies. This has been accomplished in three principal ways. The first can be illustrated by the approach that led to the detection of TL and GCSA antigens and involves immunization with H-2 incompatible leukemia cells and testing the resulting antiserum on syngeneic leukemia cells (5, 6). The rationale

TABLE 1. Immunization Procedures

	H-2 incompatible immunization/ syngeneic target cell	H-2 compatible immunization	Syngeneic immunization
Immunizing cell	A strain spontaneous leukemia ↓	AKR thymocyte ↓	W/Fu MuLV leukemia ↓
Recipient strain	C57BL/6 ↓	C3H ↓	W/Fu or (W/Fu x BN)F_1 ↓
Test cell	C57BL/6 x-ray-induced leukemia	AKR thymocyte	129 thymocyte
Cell surface antigen detected	TL	Thy-1	G_{IX}

behind this immunization was that the allogeneic and syngeneic leukemia cells share a common antigen. In the case of TL, C57BL mice were immunized with an A strain spontaneous leukemia and the antiserum was tested on a C57BL x-ray-induced leukemia, thus eliminating reactions due to H-2 and other alloantibodies. The second method involves donor-recipient combinations that are H-2 compatible and this was first used to produce antibody to θ, or as it is now known, the Thy-1 antigen (7). Syngeneic immunization, although theoretically the method of choice to produce antibody with leukemia specificity, has failed to uncover leukemia-specific antigens in spontaneous, x-ray or chemically induced mouse leukemias. Two antigenic systems, however, were defined with antisera raised by syngeneic immunization; the FMR complex of antigens associated with leukemias induced by Friend, Moloney, and Rauscher leukemia viruses (8, 9) and the MuLV-related G_{IX} antigen, originally identified with sera produced in W/Fu rats immunized with syngeneic MuLV-induced leukemia cells (10). The ease with which antibody can be raised to these antigens is undoubtedly related to the strong immunogenicity of FMR mouse leukemias and MuLV-induced rat leukemias in syngeneic hosts. These leukemias can undergo regression after initial growth, in contrast to the invariable growth of transplants of spontaneous leukemias or leukemias induced by physical or chemical agents in their strain of origin.

In addition to antisera prepared by deliberate immunization, naturally occurring antibodies in normal mouse serum are becoming increasingly valuable as reagents to define the spectrum of cell surface antigens specified by murine leukemia viruses (Table 2). With the exception of antibody to GCSA, which has not as yet been found in the serum of normal mice, antibody to the other MuLV-related specificities is

TABLE 2. Cell Surface Antigens Related to Naturally Occurring MuLV

Designation	Source of antibody	Relation to MuLV structural component	Relation to MuLV class	Reference
GCSA	C57BL anti-AKR leukemia	p15, p30	All classes	6, 11, 12
G_{IX}	(W/Fu × BN)F_1 anti-W/Fu MuLV-induced leukemia (C57BL-G_{IX}^+ × 129)F_1 normal mouse serum	gp70	Ecotropic	10, 12–15
G(RADA1)	Swiss mouse normal serum	gp70	Ecotropic	16
G(ERLD)	(C57BL × 129)F_1 normal mouse serum	gp70	Xenotropic	17
G(AKSL2)	(C3H × AKR)F_1 normal mouse serum		MCF	18

found with characteristic frequency in different mouse strains. Thus far, F_1 hybrids have been the best source of these antibodies, and this most probably has to do with heterozygosity at <u>Ir</u> loci which broadens the range of antigens that can be recognized.

The list of MuLV-related specificities recognized by mouse antibody is rapidly growing and is coming to parallel the remarkable diversity of MuLV recognizable by virological and biochemical methods. The use of the letter G to designate cell surface antigens related to naturally occurring MuLV was intended to honor Ludwig Gross, the discoverer of this class of oncornaviruses. Until a more definitive nomenclature can be devised for these antigens, it has been proposed that new MuLV-related cell surface specificities of the G class be named after the prototype normal or leukemic cell used in defining the antigen, and this has been done in the case of $G_{(RADA1)}$, $G_{(ERLD)}$, and $G_{(AKSL2)}$. Because the GCSA and G_{IX} designations are widely used, it would seem unwise to rename them according to this new convention at the present time.

Each of the five MuLV-related surface antigens can be distinguished on the basis of their strain distribution, tissue distribution, and appearance in leukemias and other tumors of strains that normally do not express these antigens. Their specification by MuLV is shown in two general ways: <u>relation to MuLV structural components</u>, either by absorption studies with purified MuLV proteins or by immunochemical characterization of the cell surface molecule, and <u>antigen induction following MuLV infection of permissive cells</u>. In this way, GCSA has been shown to be a glycosylated polyprotein related to p15 and p30 viral core proteins, and G_{IX}, $G_{(RADA1)}$, and $G_{(ERLD)}$ have been shown to be type-specific determinants of gp70 molecules. In assays of antigen induction by MuLV, GCSA appears to be a general marker for MuLV infection, with GCSA expression being induced by ecotropic, xenotropic, and amphotropic MuLV. In the case of the gp70-related antigens, G_{IX} and $G_{(RADA1)}$ induction is a common property of ecotropic MuLV, particularly those with N-tropism, and $G_{(ERLD)}$ is closely related to xenotropic MuLV in induction assays. The most recently defined MuLV-related specificity, $G_{(AKLS2)}$, is related to the dualtropic MuLV, MCF 247, a virus thought to arise by recombination between ecotropic and xenotropic MuLV (19), and recognition of the $G_{(AKLS2)}$ system provides a way to examine the natural history of this virus in the mouse.

Of the various MuLV-related cell surface specificities, the G_{IX} trait has been the subject of the most comprehensive immunogenetic analysis (Table 3). In normal mice, G_{IX} behaves like an alloantigen, being present in some mouse strains and not in others. It also has the characteristics of a differentiation antigen, being present in some tissues and absent in others, and this feature is most evident in strains such as 129, where thymocytes are the only G_{IX}^+ lymphoid cells.

TABLE 3. G_{IX} Cell Surface Antigen

G_{IX} in normal mice	Thymocyte alloantigen: G_{IX}^+ and G_{IX}^- strains
	Three G_{IX}^+ phenotypes: G_{IX}^3, G_{IX}^2, G_{IX}^1
	Mendelian inheritance: Specified by two unlinked genes Gv-1/Gv-2
Relation of G_{IX} to MuLV	Induced by infection with ecotropic and MCF MuLV in vitro
	G_{IX}-gp70 on normal thymocytes related to xenotropic MuLV
G_{IX} in leukemic mice	G_{IX}^+ leukemias occur in G_{IX}^- and G_{IX}^+ strains

Thymocytes from different G_{IX}^+ strains show characteristic quantitative differences in G_{IX} expression and, as absorption capacity follows a ratio of 3:2:1, the three phenotypes have been termed G_{IX}^3, G_{IX}^2, and G_{IX}^1. Segregation data are consistent with a two-gene specification of the G_{IX} trait, and these two genes have been designated Gv-1 and Gv-2. Despite considerable effort, the chromosomal loci of these genes remain unknown (4, 20). The relation of G_{IX} to MuLV has been shown in a variety of ways: 1) in vivo infection of G_{IX}^- mouse and rat strains with MuLV leads to G_{IX} appearance, 2) in vitro infection of permissive cells with certain ecotropic and MCF MuLV leads to G_{IX} appearance, and 3) the G_{IX} determinant on the surface of thymocytes and MuLV-infected cells resides on a molecule with antigenic and biochemical characteristics of MuLV-gp70. Peptide maps of the G_{IX}-gp70 on normal thymocytes suggest a relationship to xenotropic MuLV (21), even though MuLV-induction assays have shown a close association of the G_{IX} trait with ecotropic MuLV (12). This would suggest that thymocyte G_{IX} represents the product of an ancestral gene coding for a recombinant gp70 molecule that has become fixed in the species, with the G_{IX} determinant originally contributed by ecotropic MuLV and the remainder of the gp70 molecule by xenotropic MuLV. The fact that the MCF 247 MuLV, a virus with both ecotropic and xenotropic properties, codes for G_{IX} is consistent with this idea.

From studies of leukemic mice, we know that all mice have the genetic information specifying G_{IX}, even though they may not express G_{IX} during normal life as a consequence of differentiation signals or MuLV activation. The basis for this statement is the observation that G_{IX}^+ leukemias occur in G_{IX}^- strains of mice as well as in G_{IX}^+ strains, and this may occur in the absence of MuLV replication. This activation of normally silent genetic information as a consequence

TABLE 4. TL (Thymus-Leukemia) System of Cell Surface Antigens

TL in normal mice	Thymocyte alloantigen: TL^+ and TL^- strains
	Two TL phenotypes: TL.1,2,3 and TL.2
	Mendelian inheritance: Specified by Tla locus on chromosome 17
Relation of TL to MuLV	None known
TL in leukemic mice	TL^+ leukemias occur in TL^- and TL^+ strains

of leukemogenesis was first recognized during our analysis of another class of cell surface molecules, the TL system of antigens (5). TL antigens and the G_{IX} antigen have many features in common, with the notable exception that there is presently no evidence linking MuLV to the TL system.

In normal mice, TL is found exclusively on thymocytes, no other normal tissue expressing the antigen (Table 4). Mouse strains can be typed TL^+ or TL^- on the basis of the presence or absence of TL antigen on thymocytes and, like G_{IX}, antigen-negative mice have no alternative antigens specified by an alternative allele. TL is inherited as a Mendelian dominant trait, and linkage studies have placed the TL locus, designated Tla, on chromosome 17 < 2 units from the D end of the H-2 complex. The key feature of the TL system in regard to malignancy is the anomalous occurrence of TL^+ leukemias in strains with the TL^- phenotype (Table 5). This can best be seen by comparing the TL phenotype of normal thymocytes and of leukemia cells in mice with different Tla haplotypes. The three Tla haplotypes determine three TL phenotypes in normal mice. A strain mice with the Tla^a haplotype express three specificities of TL, TL.1, TL.2, and TL.3, on their

TABLE 5. TL Phenotypes of Normal Thymocytes and Leukemia Cells

Tla haplotype	Prototype strain	Phenotype	
		Normal thymocyte	TL$^+$ leukemia
Tla^a	A	TL.1,2,3	TL.1,2,3
Tla^b	C57BL/6	TL$^-$	TL.1,2,4
Tla^c	BALB/c	TL.2	TL.1,2 or TL.1,2,4

normal thymocytes. C57BL/6 mice, the prototype strain with the Tla^b haplotype, express no TL antigens on normal thymocytes. BALB/c mice with the Tla^c haplotype express TL.2 alone on normal thymocytes. In mice with the Tla^a haplotype, TL$^+$ leukemias resemble normal thymocytes and no evidence for appearance of anomalous TL components has been found. Leukemias arising in mice with the Tla^b or Tla^c haplotype may be either TL$^-$ or TL$^+$, and TL$^+$ leukemias invariably express anomalous TL components; TL.1, TL.2 and TL.4 in the case of C57BL leukemias and TL.1 or TL.1 and TL.4 in the case of BALB/c leukemias.

The anomalous appearance of TL antigens has been attributed to the universal presence of structural genes for TL antigens in the mouse. According to this view, regulatory genes determine whether TL is expressed on normal thymocytes. Thus, segregation for the TL trait in normal mice is based on expression vs. nonexpression alleles at the regulatory locus rather than presence vs. absence of TL structural genes. Leukemogenesis leads to an alteration in this regulatory mechanism in TL$^-$ mice, with consequent activation or derepression of TL genetic information and the appearance of TL antigens on the surface of leukemia cells. Antigen systems with the three distinct features of TL -- namely, genetic linkage to the major histocompatibility complex, restriction to normal thymocytes, and anomalous appearance on leukemias -- have not been found in any other species, but it would be surprising if the mouse were unique in this regard.

With the range of surface markers that have been identified on T cells of the mouse, the surface phenotype of normal thymocytes and leukemias of thymic origin is becoming well characterized. A comparison of the surface antigens of three leukemias that have been extensively studied with the corresponding normal thymocyte population is shown in Table 6.

The thymic origin of the AKR spontaneous leukemia AKSL2 is shown by the Thy-1 marker, despite the fact that no Lyt antigens could be detected on this leukemia. As is true of most spontaneous leukemias of AKR mice, the leukemia cells express no anomalous TL components. The five MuLV-related specificities found on the leukemia cells are also present on normal thymocytes. Thus, with regard to TL and MuLV-related antigens, the surface phenotype of this AKR leukemia does not differ from normal thymocytes, at least in a qualitative sense.

The A strain x-ray-induced leukemia RADA1 also lacks Lyt antigens, but its T cell origin is indicated by Thy-1 and by the three TL specificities (TL.1, 2, 3) that appear on RADA1 as differentiation alloantigens. The other surface feature that distinguishes RADA1 from normal A strain thymocytes is the anomalous appearance of the MuLV-related specificity $G_{(RADA1)}$ on the leukemia cell.

The surface phenotype of the C57BL x-ray-induced leukemia ERLD illustrates the phenomenon of anomalous

TABLE 6. Cell Surface Phenotype of Normal Thymocytes and Leukemia Cells

	H-2 D K	Thy-1	Lyt 1 2 3	TL 1 2 3 4	GCSA	G_{IX}	MuLV G(RADA1)	G(ERLD)	G(AKSL2)
AKR thymocytes	+ +	+	+ + +	− − − −	+	+	+	+	+
AKR spontaneous leukemia AKSL2	+ +	+	− − −	− − − −	+	+	+	+	+
A strain thymocytes	+ +	+	+ + +	+ + + −	−	+	−	+	−
A strain x-ray-induced leukemia RADA1	+ +	+	− − −	+ + + −	−	+	[+]	+	−
C57BL/6 thymocytes	+ +	+	+ + +	− − − −	−	−	−	+	−
C57BL/6 x-ray-induced leukemia ERLD	+ +	+	+ + +	[+] [+] − [+]	−	−	−	+	−

☐ represent leukemia-specific antigens in the strain of origin.

TL appearance. In contrast to the TL^- phenotype of
normal C57BL thymocytes, leukemia ERLD has a TL.1, 2,
4 phenotype. With regard to other surface antigens,
ERLD and C57BL thymocytes are identical.

The recognition that cell surface antigens such as TL and
$G_{(RADA1)}$ may be leukemia specific in some strains, yet be
normally expressed as differentiation alloantigens or MuLV-
related antigens in other strains, has been an important
contribution of basic immunogenetics to tumor immunology. As
yet, no transformation-specific surface antigen restricted
to leukemia cells of the mouse has been found. TL.4 comes
closest to fulfilling this characteristic, having never been
found on any normal cell of any mouse strain. Recent work,
however, has shown that anomalous TL components, including
TL.4, are expressed early in the preleukemic phase of x-ray
leukemogenesis, prior to the emergence of fully autonomous
cells, and should therefore be considered markers for preleukemic changes rather than as transformation-specific
traits (22).

Table 7 summarizes current knowledge about the categories
of surface antigens of mouse leukemia cells (4). The term
"derepression antigens" has been used to distinguish antigens
appearing on the surface of tumor cells that are coded for
by normally silent genetic information. The TL and MuLV-
related antigens appearing on leukemias of TL^- and $MuLV^-$
strains would be prime examples of derepression antigens.
Tumor antigens coded for by genes active only in embryonic
or fetal life would also belong to this category, but, despite
considerable interest in the possibility of such tumor antigens, their existence remains to be proven. Another category
of cell surface antigens of particular interest to the tumor
immunologist is the individually distinct or unique tumor-
specific antigen, first demonstrated by transplantation techniques in chemically induced sarcomas of the mouse (23-25)

TABLE 7. Categories of Serologically Demonstrable Surface Antigens on Mouse Leukemias of Thymic Origin

Conventional alloantigens	H-2D, H-2K
Differentiation alloantigens	Lyt-1,2,3,4, Thy-1, TL.1,2,3
MuLV-related antigens	GCSA, G_{IX}, $G_{(RADA1)}$, $G_{(ERLD)}$, $G_{(AKSL2)}$, X.1, FMR
Derepression antigens	TL.1,2,4 in TL^- strains and MuLV-related antigens in strains not normally expressing these antigens
MTV-related antigens	ML
Species antigens	MSLA
Transformation-specific MuLV antigens	None defined
Embryonic and fetal antigens	None defined
Individually distinct (unique) antigens	None defined
Idiotype receptors	None defined

and now known to be present on several other tumor types of mouse and rat. These antigens are characterized by a remarkable polymorphism, no two tumors, even if induced in the same mouse, sharing identical antigens. The origin of these unique antigens has been the subject of considerable speculation, with both genetic and epigenetic theories having been advanced, but their nature remains as obscure now as it was when they were discovered. With the recognition that MuLV exists in a far more polymorphic state than originally envisioned, the possibility that these unique antigens owe their origin to MuLV genes must be reconsidered, especially since recombinational events between different classes of MuLV and between MuLV and host genes, which appears likely as a source of MuLV variation, would be expected to give rise to an enormous repertoire of new antigens. Further understanding awaits serological and biochemical definition of these antigens, and advances in both these directions have recently been made with a methylcholanthrene sarcoma of BALB/c mice (26, 27). As yet, little effort has gone into the serological detection of unique antigens on leukemia cells. Their presence could be easily obscured by cross-reacting systems such as TL or MuLV-related antigens. With the background of information we now have concerning the surface phenotype of leukemia cells, planned immunizations devised to detect new systems of surface antigens, including the individually distinct type, should make the next phase of immunogenetic analysis a most revealing one.

REFERENCES

1. Gorer, P.A., and O'Gorman, P. (1956). Transplant. Bull. 3:142.
2. Old, L.J., and Boyse, E.A. (1973). Harvey Lect. 67:273.
3. Boyse, E.A., and Old, L.J. Harvey Lect., in press.

4. Old, L.J., and Stockert, E. (1977). Ann. Rev. Genet. 11:127.
5. Old, L.J., Boyse, E.A., and Stockert, E. (1963). J. Natl. Cancer Inst. 31:977.
6. Old, L.J., Boyse, E.A., and Stockert, E. (1965). Cancer Res. 25:813.
7. Reif, A.E., and Allen, J.M.V. (1964). J. Exp. Med. 120:413.
8. Old, L.J., Boyse, E.A., and Stockert, E. (1964). Nature 201:777.
9. Klein, E., and Klein, G. (1964). J. Natl. Cancer Inst. 32:547.
10. Stockert, E., Old, L.J., and Boyse, E.A. (1971). J. Exp. Med. 133:1334.
11. Snyder, H.W., Jr., Stockert, E., and Fleissner, E. (1977). J. Virol. 23:302.
12. O'Donnell, P.V., and Stockert, E. (1976). J. Virol. 20:545.
13. Obata, Y., Ikeda, H., Stockert, E., and Boyse, E.A. (1975). J. Exp. Med. 141:188.
14. Obata, Y., Stockert, E., Boyse, E.A., Tung, J-S., and Litman, G.W. (1976). J. Exp. Med. 144:533.
15. Tung, J-S., Vitetta, E.S., Fleissner, E., and Boyse, E.A. (1975). J. Exp. Med. 141:198.
16. Obata, Y., Stockert, E., O'Donnell, P.V., Okubo, S., Snyder, H.W., Jr., and Old, L.J. (1978). J. Exp. Med. 147:1089.
17. Obata, Y., Stockert, E., DeLeo, A.B., O'Donnell, P.V., Snyder, H.W., Jr., and Old, L.J. Manuscript in preparation.
18. Stockert, E., DeLeo, A.B., O'Donnell, P.V., Obata, Y., and Old, L.J. Manuscript in preparation.
19. Hartley, J.W., Wolford, N.K., Old, L.J., and Rowe, W.P. (1977). Proc. Natl. Acad. Sci. 74:789.

20. Stockert, E., Boyse, E.A., Sato, H., and Itakura, K. (1976). Proc. Natl. Acad. Sci. 73:2077.
21. Tung, J-S., O'Donnell, P.V., Fleissner, E., and Boyse, E.A. (1978). J. Exp. Med. 147:1280.
22. Stockert, E., and Old, L.J. (1977). J. Exp. Med. 146:271.
23. Gross, L. (1943). Cancer Res. 3:326.
24. Foley, E.J. (1953). Cancer Res. 13:835.
25. Prehn, R.T., and Main, J.M. (1957). J. Natl. Cancer Inst. 18:769.
26. DeLeo, A.B., Shiku, H., Takahashi, T., John, M., and Old, L.J. (1977). J. Exp. Med. 146:720.
27. Natori, T., Law, L.W., and Appella, E. (1977). Cancer Res. 37:3406.

GENES OF THE Tla REGION: THE NEW
Qa SYSTEM OF ANTIGENS[1]

Lorraine Flaherty

Division of Laboratories and Research
New York State Department of Health
Albany, New York

For the past several years, we have been studying the Tla region of the mouse, located on chromosome 17 immediately to the right of (distal to) the major histocompatibility complex. We first became interested in this region after our initial discovery that it may code for several cell surface antigens in addition to TL (1, 2). Since then we have made a determined effort to define and more precisely locate these newly discovered cell surface loci and determine their characteristics. So far, six cell surface loci have been mapped to this region; two which determine skin and tumor graft rejection -- H-31 and H-32 (2) -- and four which determine differentiation antigens on lymphoid cells -- Tla (3), Qa-1 (4), Qa-2 (5), and Qa-3 (6). In addition, two genes which determine isozyme variations -- Pgk-2 (7) and Ce-2 (8, 9) -- have also been located here.

---

[1] This work was supported in part by NIH Grant AI 12603 awarded by the National Institute of Allergy and Infectious Diseases and NIH Grant CA 23027 awarded by the National Cancer Institute, DHEW.

In this report, I will present the general characteristics of the serologically defined cell surface loci with particular emphasis on the newly discovered Qa series of antigens, pointing out the similarities and differences between them. I will also present data on the strain and substrain distributions of some of these antigens.

Characteristics of the Tla Region Cell Surface Loci

Table 1 and Figure 1 give a general description and mapping of these loci. Fuller descriptions are given below.

Tla. The Tla locus was first discovered in 1963 by Old and co-workers (3). It determines a series of thymus-leukemia specific antigens called TL. They found that there were three alleles determining four TL antigens (4). The fourth allele and fifth antigenic specificity were found by Flaherty et al. (12) in an analysis of the antibodies contained within an anti-H-2 serum.

MAP OF THE *Tla* REGION

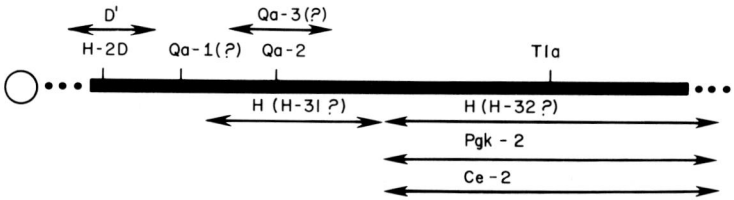

FIGURE 1. Map of the Tla region. The question marks after Qa-1 and Qa-3 indicate that the location of these genes is in doubt. The question marks after H-31 and H-32 indicate that it is not known whether the H (histocompatibility) loci mapped to these sites are identical to H-31 and H-32.

TABLE 1. Tla Region and Its Differentiation Antigens[a]

Locus	Alleles	Antigens	Tissue distribution			
			Thymus	LNC	Spleen	Bone marrow
Tla	a	TL.1,2,3,5	+	–	–	–
	b	– – –				
	c	TL.2				
	d	TL.1,2,3				
Qa-1	a	Qa-1	+	+	+	–?
	b	– – –				
Qa-2	a	Qa-2	+	+	+	+
	b	– – –				
Qa-3	a	Qa-3	–	+	+	–
	b	– – –				

[a]References are: 4, 6, 11, 12; + = presence of antigen; – = absence of antigen.

These TL antigens have several unusual properties (for review see ref. 11). First, they undergo a process known as antigenic modulation in which the presence of TL antibody or its Fab fragment induces the phenotypic loss of TL from the cell surface. Second, they appear on leukemias of mice which do not ordinarily express these antigens on their thymuses. Moreover, this leukemic TL expression is genetically determined since only certain TL phenotypes can appear on a leukemia from a given Tla genotype. And third, TL antigens have a reciprocal quantitative relationship with the antigens determined by a closely linked locus, H-2D, such that a TL^+ cell expresses less H-2D antigen than a TL^- cell.

Because of these characteristics and others, Boyse and Old (13) have suggested that there are actually two components to the Tla locus, one "structural" and one "expressional." The structural component directly specifies the TL protein while the expressional component determines the presence or absence of it. No crossing-over has ever been observed within the Tla locus separating these components. (For a further discussion of these two components and the mapping of them to the Tla locus, see ref. 13.)

Biochemically, the molecules expressing the TL antigens have a molecular weight of 45,000 with a β_2-microglobulin subcomponent making them similar to the H-2D and H-2K molecules (14).

Qa-1. The Qa-1 locus determines the presence or absence of the Qa-1 antigen (4). It appears on thymocytes as well as on a precise subpopulation of peripheral T lymphocytes. This tissue distribution distinguishes it from the known set of TL antigens which are found exclusively in the thymus.

The Qa-1 locus (or loci) may be complex. Stanton and Boyse (4) originally observed that different anti-Qa-1 sera gave slightly different strain distributions, consequently they have suggested that at least one of their anti-Qa-1 sera

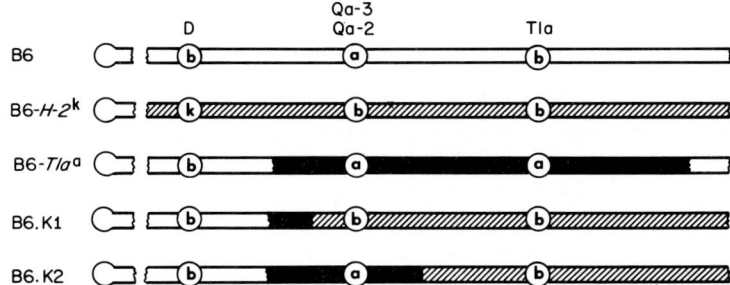

FIGURE 2. Chromosomes of Tla region congenic strains.

may have more than one antibody against peripheral lymph node cells.

Using the B6.K1 and B6.K2 recombinant strains, they have preliminarily mapped the Qa-1 locus to the chromosome stretch between H-2D and Qa-2. (B6.K1, B6.K2, and B6-Tla^a are all Qa-1$^+$ while B6 and B6-H-2^k are Qa-1$^-$; see Figure 2.) However, in a further serological analysis of the Qa-1 antigen and its strain distribution, Stanton has found that a change in the source of rabbit complement or antiserum batch alters the activity of the anti-Qa-1 serum against B6.K1 and B6.K2 (15).[2] These results suggest that there is more than one activity in anti-Qa-1 sera and thus makes the mapping of the Qa-1 locus difficult. We have therefore placed a question mark after Qa-1 in our chromosomal map of the Tla region and feel that

---

[2] Where these strains were originally strongly Qa-1$^+$ with one source of complement and antiserum, they later appeared to be either weakly positive or negative. In our laboratory we have also confirmed these findings. Using a different anti-Qa-1 serum, (A.SW x A.TL) anti-A.TH, we have found that certain sources of rabbit complement will show B6.K1 and B6.K2 as weakly positive where others will show these strains to be negative (9).

the main Qa-1 activity may be due to a locus more closely linked to Tla.

The Qa-1 antigen has not been biochemically analyzed.

Qa-2. The Qa-2 locus located between H-2D and Tla determines the presence or absence of the differentiation antigen Qa-2. We originally described this locus as a complex determining at least two antigenic specificities (5). By further serological analysis, we have now subdivided this locus into two loci, Qa-2 and Qa-3 (6). Qa-3 will be described in the next section.

As with the Qa-1 antigen, Qa-2 is on a restricted population of lymphocytes as seen by its low expression on thymocytes ($\sim$ 20% are Qa-2$^+$) and its high expression on lymph node cells (>65% are Qa-2$^+$). It is also present in the spleen and bone marrow. By cytotoxicity testing, it is predominantly expressed on Thy-1$^+$ lymphocytes although at least some Thy-1$^-$ cells are Qa-2$^+$ (6). In addition, it is present on some T cell leukemias; both TL$^+$ and TL$^-$ leukemias can be Qa-2$^+$.

Qa-2 is also associated with a CML (cell-mediated lympholysis) locus. Forman and Flaherty (16) have found that Qa-2^b lymphocytes will kill Con A blasts from a Qa-2^a mouse in a secondary in vitro CML test. This killing correlates completely with our strain distribution for the serologically detectable Qa-2 antigen.

Qa-3. The Qa-3 locus was separated from Qa-2 on the basis of both the strain and tissue distributions of the Qa-2 and Qa-3 antigens which they determine. Qa-3 is more limited in its tissue distribution than Qa-2 and is only present on lymph node and spleen cells. It is absent from at least two inbred strains of mice which type Qa-2$^+$ (Qa-2^a) (6). These results indicate that Qa-3 is different from Qa-2 and is determined by a different locus, Qa-3. So far we have not detected any recombinants between Qa-2 and Qa-3, and therefore the precise position of Qa-3 is not known (see Figure 1).

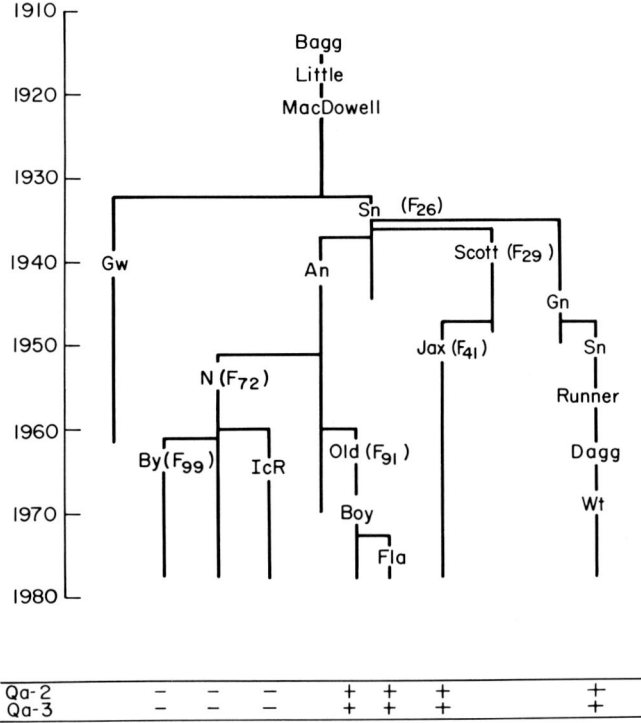

FIGURE 3. Derivation of and Qa typing for several BALB/c sublines. Abbreviations used are: Sn = G.D. Snell; Gw = J.W. Gowen; An = H.B. Andervont; Gn = E.L. and M.C. Green; N = National Institutes of Health; JAX = Jackson Laboratory; By = D.W. Bailey; IcR = Institute for Cancer Research, Philadelphia; Boy = E.A. Boyse; Wt = W.K. Whitten; Fla = L. Flaherty.

Using immunoprecipitation techniques, Michaelson et al. (17) have shown that an antiserum which contains both Qa-2 and Qa-3 activity precipitates a molecule of approximately 45,000 daltons with a β_2-microglobulin subcomponent. Subsequent studies have preliminarily shown that the main molecule precipitated by this serum is the Qa-3 molecule (Michaelson et al., personal communication). The molecular weight of the Qa-2 molecule is still not known.

BALB/c Sublines and Their Qa Phenotypes

In the course of testing different strains of mice for their Qa phenotype, we have found that certain BALB/c sublines differ at both the Qa-2 and Qa-3 loci (6).

The BALB/c strain was derived from a colony of albino mice maintained by Bagg. In the early 1920s, MacDowell obtained this colony from Little and proceeded to inbreed it. In 1932, it was given to Snell in the F_{26} generation, who subsequently gave it to Andervont, Green, and Scott. It was then widely distributed to many laboratories. Figure 3 gives a partial pedigree of the BALB/c strain with our current knowledge of its distribution (18, 19, 20).

In our Qa-2 and Qa-3 typings, we have found that the BALB/c sublines have two Qa phenotypes -- all lines derived from Green or Scott are Qa-2$^+$Qa-3$^+$, while most of the lines derived from Andervont are Qa-2$^-$Qa-3$^-$. One BALB/cAn-derived line, BALB/cBoy (and therefore also all strains derived from it such as BALB/cFla), is Qa-2$^+$Qa-3$^+$.

The simplest explanation for this genetic divergence is that either a mutation or deletion occurred in the BALB/cAn line. It is unlikely that it is due to foreign genes still segregating from the original Bagg albino stock since the BALB/c strain was at least in its F_{29} generation when given to Andervont. We have also typed these sublines for the neighboring loci, H-2D, D', and Tla, and found them to be identical (Table 2). If the Qa-2 and Qa-3 phenotypes were maintained from the original heterogeneity in the Bagg albino stock, it would be very unlikely that neighboring genes would

TABLE 2. Comparison of BALB/cBy to BALB/cFla

	Antiserum	Titer of antiserum on[a]	
Name	Strain combination	BALB/cBy	BALB/cFla

Serology

Name	Strain combination	BALB/cBy	BALB/cFla
D-4	(B10.AKM x 129) anti-B10.A	1/160	1/160
D'	BALB/c-H-2^{db} anti-BALB/c	1/160	1/160
TL	(A-Tla^b x B6) anti-ASL1	1/400	1/400
D-28	(B10.BR x A.CA) anti-A.SW[b]	1/100	1/300
Qa-2	B6.K1 anti-ERLD	0	1/200
Qa-3	B6.K1 anti-E abs. with EL4[c]	0	1/100

Skin grafting

Recipient	Donor	Rejection times (days)[d]
BALB/cBy	BALB/cFla	24,24,24,26,26,26,28,--,--,--
BALB/cFla	BALB/cBy	27,27,32,32,cr,--,--,--,--,--

[a] Target cell = lymph node lymphocytes for all sera except anti-TL. Target cell for anti-TL serum = thymocytes.

[b] This antiserum was found to contain anti-Qa-2 (and/or Qa-3) activity -- hence the difference between BALB/cBy and BALB/cFla (6).

[c] For method of absorption, see ref. 6.

[d] Observation time = 100 days; cr = graft underwent crisis; - = not rejected.

not also be affected. Therefore, our hypothesis of a mutation or a deletion seems to be the most tenable one at the present time. The skin grafting results presented in Table 2 between BALB/cBy and BALB/cFla also indicate that there are other genetic differences between these two lines. This is not surprising since these strains are at least 60 generations apart.

If our hypothesis that a mutation or deletion occurred in the BALB/cAn subline is correct, it would indicate that this genetic event (or two consecutive events) modified both Qa-2 and Qa-3.

Relationships Between the Tla Region Loci

In our typings of a number of inbred and recombinant strains of mice, we have noticed several correlations between the alleles within the Tla region and between the alleles of the Tla region and the H-2 region (Table 3).

First, there is a strong correlation between the Qa-1 genotype and the Tla genotype. Most, if not all, tested strains which are Qa-1^a are either Tla^a or Tla^d, while all

TABLE 3. Characteristic H-2 and Tla Region Genotypes for Some Common Inbred Strains[a]

H-2	Qa-2	Qa-3	Qa-1	Tla	Strains
a	a	a	a	a	A
b or bc	a	a	b	b	C57BL/6, C57BL/10
	a	a	b	c	129, C57L
d	b	b	b	c	BALB/cBy
	a	a	b	c	BALB/cJ
	a	a	a	a	NZB
f	b	b	a	d	A.CA, B10.M
k	b	b	b	b	AKR, C3H/An, RF
	b	b	a	a	C58, C57BR
m	a	a	a	a	B10.AKM
p	b	b	a	a	BDP/J
q	a	b	b	b	DBA/1
	a	b	a	a	SWR
	a	a	?	a	BUB/BnJ
r	b	b	b	b	RIII
s	?	?	a	a	SJL
	a	a	b	b	A.SW

[a] References are: 4, 6, 9, 11, 12, 21.

tested strains which are $Qa-1^b$ are either $\underline{Tla}^b$ or $\underline{Tla}^c$. This may simply be because Qa-1 and Tla are tightly linked. Since many of the inbred strains have a common ancestry (21) it would be likely that the alleles of these two loci would then be preserved together. At present the position of Qa-1 in relationship to Tla is somewhat in doubt, and therefore this hypothesis must be considered a viable one. Alternatively, of course, there may be some functional relationship between these two genes. Currently, there are not enough data to suggest which of these hypotheses is correct.

With few exceptions, there is a strong correlation between the strain distribution of the H-2D.28 specificity and the Qa-2 antigen. All tested inbred strains which have been described as H-2D.28⁻ are also Qa-2⁻ ($\underline{H-2}^{f,k,p,r}$). We have investigated this correlation and found that one reason for it is probably due to the contamination of at least one anti-H-2.28 typing serum with anti-Qa-2 activity (6). By absorption and direct cytotoxicity testing, we have found that the anti-H-2.28 serum (B10.BR x A.CA) anti-A.SW contains a substantial amount of anti-Qa-2 (and/or Qa-3) activity. Thus a strain would only appear negative with this typing antiserum if it were also negative for Qa-2 and appear positive if it were positive for either H-2.28 or Qa-2.

Structurally the Qa-3 and TL molecules are very similar to the H-2K, H-2D, and D' molecules; they are all approximately 45,000 daltons in size and associated with β_2-microglobulin. Immunoprecipitation techniques have so far been unsuccessful in determining the molecular weights of the Qa-1 and Qa-2 molecules. These similar molecular weights might indicate that these genes have a common evolutionary ancestry.

The most interesting relationship to us is that between the expression of Qa-2 and Qa-3. In typing a large number of inbred and recombinant strains plus a few wild haplotypes, we have only observed three Qa phenotypes -- $Qa\text{-}2^+Qa\text{-}3^+$, $Qa\text{-}2^-Qa\text{-}3^-$ and $Qa\text{-}2^+Qa\text{-}3^-$ (and not $Qa\text{-}2^-Qa\text{-}3^+$). This is also true for the tissue and tumor distributions of these antigens. There seems to be an obligatory requirement for Qa-2 in order for Qa-3 to be expressed. This is similar to the relationship between two of the TL antigens, TL.2 and TL.3 (22). In all known cases, there is no strain or leukemia which expresses TL.3 without expressing TL.2.

I would like to comment on one last point. Even though our information is still limited on the Qa loci, we have observed that in contrast to the neighboring H-2 complex, the genes of the Tla region, Qa-1, Qa-2, Qa-3, and Tla, do not appear to be highly polymorphic and each appears to have an allele which is serologically undetectable. In the cases of Qa-2, Qa-3, and Tla we have made several efforts to induce antibodies against these alternative gene products especially those potentially governed by null alleles but have been unsuccessful. This is also true of the CML locus associated with Qa-2. Forman and Flaherty (16) have found that this reaction is unidirectional; $Qa\text{-}2^b Qa\text{-}3^b$ lymphocytes will kill $Qa\text{-}2^a Qa\text{-}3^a$ cells but the reciprocal is not true. The level of cell-mediated cytotoxicity against $Qa\text{-}2^a Qa\text{-}3^a$ cells also does not vary significantly from strain to strain, indicating that, if polymorphism exists, all the differing gene products behave similarly. We realize that congenic immunizations are sometimes difficult (23) and that Ir genes are sometimes important in detecting mouse alloantigens (24); however, it is still interesting to speculate that the Tla region might have several true null alleles which do not code for allogeneic cell surface antigens.

Research on the Tla region and the differentiation antigens which it determines is relatively new. Three of these antigens were only discovered within the last two years and have not been fully characterized. Even with our limited knowledge of this genetic region, it is becoming apparent that this region is complex and might have some unique properties. We hope that it will provide us with information not only on the genetic control of the lymphoid cell surface but also on its differentiation.

ACKNOWLEDGMENTS

The author wishes to thank Ted Hansen for the serum, BALB/c-$\underline{H\text{-}2}^{db}$ anti-BALB/c and Don Bailey for his help in constructing the BALB/c pedigree.

REFERENCES

1. Boyse, E.A., Flaherty, L., Stockert, E., and Old, L.J. (1972). Transplantation 13:431.
2. Flaherty, L., and Wachtel, S.S. (1975). Immunogenetics 2:81.
3. Old, L.J., Boyse, E.A., and Stockert, E. (1963). J. Natl. Cancer Inst. 31:977.
4. Stanton, T.H., and Boyse, E.A. (1976). Immunogenetics 3:525.
5. Flaherty, L. (1976). Immunogenetics 3:533.
6. Flaherty, L., Zimmerman, D., and Hansen, T.H. (1978). Immunogenetics, in press.
7. Eicher, E.M., Cherry, M., and Flaherty, L. (1978). Molecular and General Genetics, in press.
8. Hoffman, H.A., and Grieshaber, C.K. (1976). Biochem. Genet. 14:59.
9. Flaherty, L. Unpublished observations.
10. Boyse, E.A., Old, L.J., and Stockert, E. (1965). In "Immunopathology IVth International Symposium" (P. Grabar and P.A. Miescher, eds.), p. 23. Schwabe and Co., Basel.

11. Old, L.J., and Stockert, E. (1977). Ann. Rev. Genet. 17:127.
12. Flaherty, L., Sullivan, K., and Zimmerman, D. (1977). J. Immunol. 119:571.
13. Boyse, E.A., and Old, L.J. (1971). Transplantation 11:561.
14. Vitteta, E.S., Uhr, J.W., and Boyse, E.A. (1972). Cell. Immunol. 4:187.
15. Stanton, T. Personal communication.
16. Forman, J., and Flaherty, L. (1978). Immunogenetics, in press.
17. Michaelson, J., Flaherty, L., Vitetta, E., and Poulik, M.D. (1977). J. Exp. Med. 145:1065.
18. Inbred Strains of Mice No. 10. Mouse News Letter 57 (Companion Issue) (1977).
19. Old, L.J. Personal communication.
20. Bailey, D. Personal communication.
21. Klein, J. (1975). In "Biology of the Histocompatibility-2 Complex." Springer-Verlag, New York.
22. Boyse, E.A., Stockert, E., and Old, L.J. (1969). In "International Convocation on Immunology" (N.R. Rose and F. Milgrom, eds.), p. 353. Karger, Basel.
23. Shen, F.W., Boyse, E.A., and Cantor, H. (1976). Immunogenetics 2:591.
24. Fuji, H., Zaleski, M., and Milgrom, F. (1971). J. Immunol. 106:56.

RECOMBINANT INBRED STRAINS: USE IN GENE MAPPING[1]

Benjamin A. Taylor

The Jackson Laboratory

Bar Harbor, Maine

Genetic polymorphism is very extensive in cross-fertilizing species (1, 2), a fact that is reflected in multiple genetic differences between inbred strains. Mammals have sufficient DNA to code for thousands of genes, and significant fractions of samples of loci exhibit polymorphic variation. Thus, unrelated strains of mice differ by hundreds, possibly thousands, of genetic loci. A variety of biochemical and immunological techniques have proved useful for detecting genetic polymorphism (3, 4). Emphasis in mouse genetics has shifted away from the study of mutants to the study of genetic polymorphisms. There are several reasons why it is desirable to locate these polymorphic genes in the linkage map of the mouse. Traditional methods of linkage testing are slow and uncertain of yielding positive results. A systematic approach is clearly desirable. Recombinant inbred strains represent such an approach.

---

[1]These studies were supported in part by contract NO1 CP33255 within the Virus Cancer Program of the National Cancer Institute, and by NIH research grant GM 18684 from the National Institute of General Medical Sciences. The Jackson Laboratory is fully accredited by the American Association for Accreditation of Laboratory Animal Care (AAALAC).

RI Strains and Linkage Detection

Recombinant inbred (RI) strains are derived by systematic inbreeding beginning with the F_2 generation of the cross of two preexisting inbred (progenitor) strains (5). Multiple independent strains are derived without selection. Once inbred, such a set of RI strains can be thought of as a stable segregant population. Unlinked genes are randomized in the F_2 generation and are therefore equally likely to be fixed in parental or recombinant phases. Linked genes will tend to become fixed in the same (parental) combinations as they entered the cross. Thus inbreeding preserves part of the linkage disequilibrium generated when two inbred strains are crossed. Once inbred, the RI strains are typed with respect to the numerous genetic differences that distinguish the progenitor strains. Each locus has a particular pattern of inheritance called the strain distribution pattern (SDP). Comparisons are made between different SDPs; a significant excess of parental genotypes, with respect to two SDPs, signals the possibility of genetic linkage. The enormous advantage of this approach is that the data are cumulative. Each RI strain needs to be typed only once for a particular locus. The discoverer of a new variant needs only to type the RI strains for that particular locus. If there is no apparent linkage with any of the available markers, the choice is made to either continue linkage testing by traditional crosses, or to wait and hope that as more markers are added to the system that linkage will be revealed. The RI results may exclude linkage to certain chromosomes or parts of chromosomes, thus narrowing further linkage testing.

The chief limitation of the method is that it is only useful for mapping genes that differ in the progenitor strains. If several different sets of RI strains are available, then most polymorphisms will segregate in one or more sets. However, this is no help in mapping new mutations. Another limitation is the poor reproduction encountered in some RI strains.

Estimation of Recombination Frequencies

To quantitate the use of RI strains for linkage analysis, one must relate the probability of fixing a recombinant genotype in an RI strain (R) as a function of the recombination frequency (r) in a single meiosis. This relationship was derived by Haldane and Waddington (6). For the case of brother-sister inbreeding, $R = 4r/1+6r$. (It is of interest that if RI strains are derived by parent-offspring inbreeding $R = 3r/1+4r$; for selfing, $R = 2r/1+2r$.) Note that as r approaches zero, R approaches $4r$. This can be interpreted to mean that in the development of an RI strain prior to fixation of one allele, it will have been transmitted through a heterozygote four times on the average, each occasion representing an opportunity for recombination with a neighboring gene. Another deduction we can make is that in RI strains we can expect 0.04 crossovers per centimorgan. Thus a chromosome 100 cM in length would have, on the average, four exchanges. Thus in effect, RI strains "expand" the map four-fold. This is an advantage if one is looking for rare recombinants, but a disadvantage when the objective is linkage detection.

By the principle of inverse estimation we can obtain an estimate of r, $\hat{r} = \hat{R}/4-6\hat{R}$ where $\hat{R}$ is the ratio of recombinant strains relative to the total number of RI strains (n). The estimated variance of $\hat{r}$ is given by $r(1+2r)(1+6r)^2/4n$ (7). We can use the reciprocal of the ratio of the variance of $\hat{r}$ from RI data to the variance of $\hat{r}$ from an equal number of backcross mice, $4(1-r)/(1+2r)(1+6r)^2$, to compare the relative efficiency of the two methods per independent marker. From

Figure 1 we see that RI strains are more efficient up to $r = 0.125$, but rapidly become less efficient for greater recombination values. Such a comparison does not reflect the inherent advantage of RI strains, that the data are cumulative permitting the incorporation of more markers, nor the great saving in effort afforded because it is necessary to test only for the new locus. Of course it is relatively easy to increase the number of backcross mice, but the costs may rise if the locus of interest is difficult to type of if many markers are to be followed.

If no recombinants between two loci are detected among n RI strains, the upper confidence limit of r is estimated by solving the following equation for r:

$$\alpha = (1 - 4\underline{r}/1+6\underline{r})^{\underline{n}}$$

where $1 - \alpha$ is the confidence level, e.g., for 95% confidence limit $\alpha = 0.05$. The value of r that satisfies this equation is the upper confidence limit of the estimate.

Power of RI Strains for Linkage Detection

I have attempted to quantify the power of RI strains for linkage detection through the use of Haldane's concept of the radius swept (8). The statistical power of a method is defined as the probability of rejecting the null hypothesis when it is false. (Note that this is defined to equal $1 - \beta$, where β is the probability of failing to reject the null hypothesis when it is false. The following calculations are based on adopting a significance level (α) of 0.01. Having chosen a significance level, we can calculate from the binomial distribution the maximum number of permissible recombinants, such that the null hypothesis of no linkage would be rejected for a given number of RI strains (Table 1).

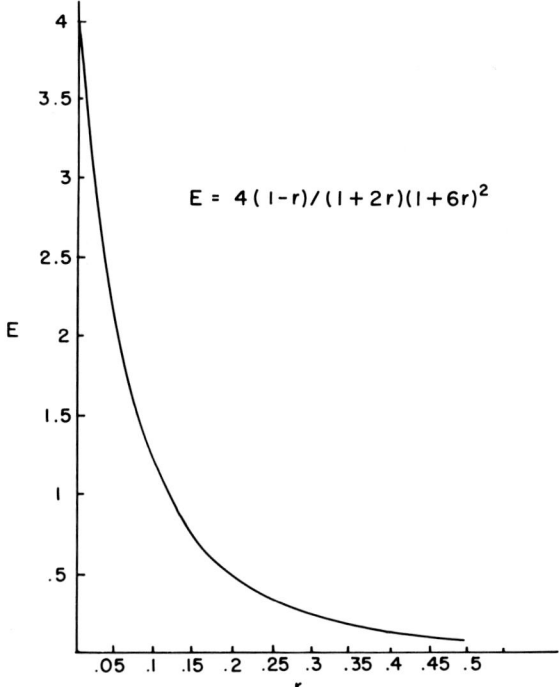

FIGURE 1. Efficiency of RI strains relative to a backcross for estimation of recombination fraction.

Thus with seven RI strains we would pursue linkages only if no recombinant were observed. With 11 RI strains we would use one recombinant as the cut-off point, etc. The power function, $1 - \beta$, is equal to

$$\sum_{k=0}^{a}(4\underline{r}/1+6\underline{r})^{k}(1-4\underline{r}/1+6\underline{r})^{n-k}$$

where $\underline{n}$ is the number of RI strains, and $\underline{a}$ is the permissible number of recombinants for a given $\underline{n}$. This function varies between one (when $\underline{r} = 0$) and α (when $\underline{r} = 0.5$). Figure 2 plots this function for four different values of $\underline{n}$. The radius swept, $\underline{s}$, is defined as the integral of the power function over all possible distances. For ease of computa-

TABLE 1. Distance Swept by a Single Marker Gene as a Function of the Number of RI Strains ($\alpha \leq 0.01$), Compared with a Backcross of the Same Size

Number of RI strains or backcross progeny	Permissible number of recombinants	Distance swept (d_n) in centimorgans	
		RI strains	Backcross
7	0	9.3	23.9
11	1	12.8	32.3
14	2	16.2	38.9
17	3	18.7	43.4
19	4	22.5	48.9
22	5	23.8	51.1
25	6	24.7	52.8
28	7	25.5	54.4

tion I have assumed the complete interference model $r = d$, $0 \leq d \leq 0.5$ where d is the map distance in morgans. This approximation is good for small values of n but will underestimate the contribution of loose linkage to s, when n is large. The radius swept can be thought of as the average length of chromosome on either side of a marker locus for which linkage would be detected. Thus each independent marker tests $2s$ of the total genome provided it is not close to the end of a chromosome. The value of s for different numbers of RI strains was determined by numerical integration over the interval of (0, 0.5). The total distance swept ($d_n = 2s_n$) is given in Table 1. The results emphasize the fact that the RI method will generally only detect close linkages.

We can use the distance swept (d_n) by a single locus in n RI strains to estimate the probability (P) of detecting

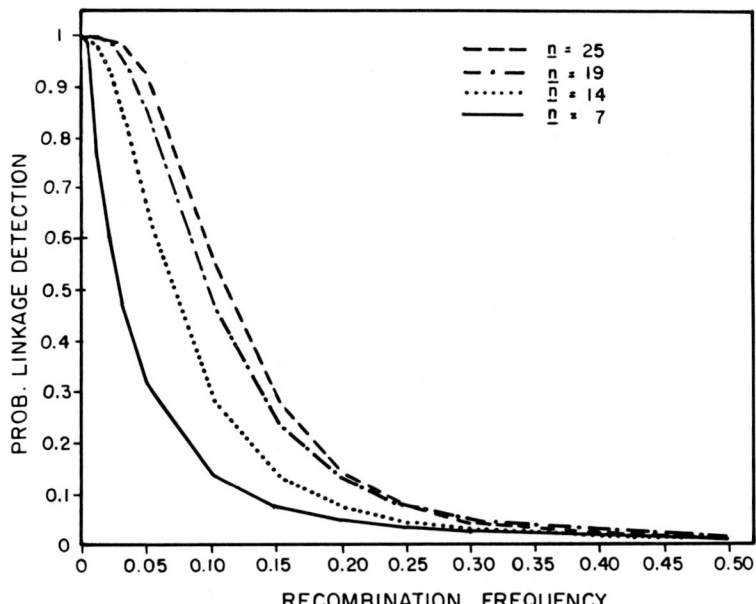

FIGURE 2. The probability of detecting linkage using $\underline{n}$ RI strains as a function of the recombination frequency. The significance level is chosen to be less than 0.01.

linkage between a new locus and one or more of $\underline{m}$ markers randomly distributed over a genome with total length $\underline{D}$; $\underline{P} = 1 - (1 - \frac{d_n}{D})^{\underline{m}}$. This neglects the fact that markers near the ends of chromosomes sweep distances less than $\underline{d_n}$. The total length of the mouse genome has been estimated to be 1600 cM (9, 10). Table 2 shows the value of $\underline{P}$ for different combinations of $\underline{n}$ and $\underline{m}$. It should be noted that $\underline{d_n}/\underline{D}$ is the $\underline{a}$ $\underline{priori}$ probability of detecting linkage between any two loci, assuming a random distribution of loci throughout the map. When $\underline{n}$ is seven, this ratio is only 0.0058, while the probability of getting a false lead is $(1/2)^7$ or 0.0078. This means that if only seven RI strains are available, and if one only pursues perfect matches, that 57% of these would be expected to be false. Even with larger sets of RI strains

TABLE 2. The Probability of Detecting Linkage with m Markers and n RI Strains

m \ n	7	14	19	25
10	.057	.097	.132	.144
20	.110	.184	.247	.267
40	.208	.330	.432	.463
60	.295	.457	.572	.607
80	.373	.557	.678	.712
100	.442	.638	.757	.789
120	.503	.705	.817	.845
140	.558	.759	.862	.887
160	.606	.804	.896	.917
200	.688	.869	.941	.955

a fair proportion of significant matches are expected to be false positives. We can see from Table 2 that with 25 RI strains, 200 markers are required to be 95% confident of detecting linkage of a new marker.

Gene Clustering

If the polymorphic loci that distinguish the progenitor strains are clustered, then the likelihood of detecting linkage will be increased for small values of m, but clustering may retard the approach to saturation as m becomes large. In practice, some degree of clustering is to be expected. For example, all immunoglobulin structural genes are expected to fall into three clusters coding for heavy, kappa, and lambda chains. Localization of chiasmata would also enhance clustering (11). Insofar as the progenitor strains are related, certain chromosomal segments may be identical by descent, just as pairs of RI strains will share blocks of genes with each other. Major evolutionary changes in the karyotype of the species, such as the fixation of reciprocal translocations, may result in the loss of polymorphisms in the chromosomes involved. Certain chromosomal regions may be deficient in genes and comprise redundant DNA of unknown function.

Gene Mapping

In addition to detecting linkage and estimating recombination frequency, it is desirable to determine the linear order of genes on the chromosomes. The ease with which multiple markers can be scored in RI strains lends itself to gene ordering. In conventional mapping, once linkage has been detected, one or more additional crosses are often needed to define the correct gene order with respect to the new linkage and neighboring loci. In the case of polymorphic loci which often require specialized techniques for typing, it may not be practical for a single investigator to carry out the tests needed to determine gene order. RI strains offer the possibility of ordering such loci. Gene ordering depends on the fact that double crossovers are relatively rare compared to single crossovers. Interference is defined as a deficiency of double crossovers relative to the product of the frequency of single crossovers in adjacent intervals. Since exchange points along a chromosome fixed in an RI strain may have occurred in different individuals during the development of the strain, interference is likely to be weak. I have carried out preliminary computer simulations of RI strain formation to investigate the distribution of crossover points along a chromosome. One model was of a chromosome 50 cM in length, using the mapping function $\underline{r} = \underline{d}$, $0 \leq \underline{d} \leq 0.5$. In other words, the probability of a crossover in any meiosis was 0.5, and the location of the crossover point was random. No multiple crossovers were allowed.

Therefore, interference was complete. The number of exchanges/10 cM segment of the chromosome was tabulated in several hundred runs. The mean number of exchanges agreed well with the expectation (0.4). The number of segments with 0, 1, 2, 3, ... exchanges agreed remarkably well with a Poisson distribution with mean of 0.4, implying that the complete interference model in meiosis generates RI chromosomes that resemble chromosomes that would be generated in a situation with no interference. This model, if valid, greatly simplifies prediction of the distribution of crossover points. Thus we would predict that exchange points would be randomly distributed along the chromosome. This means that, although the likelihood of two exchange points occurring between two closely linked genes is necessarily small, occasionally it will happen. Therefore, we should expect to occasionally obtain three point data that are ambiguous regarding gene order or, more rarely, the data may be misleading in regard to gene order. A corollary would be that one is not <u>assured</u> that because an RI strain has become fixed for two closely linked markers from the same progenitor that the entire segment between these markers was inherited intact.

Although seven is near the minimum number of RI strains sufficient for detecting linkage, it is worth emphasizing that even a single RI strain can be very valuable. A single recombinant is sufficient to show that two loci are distinct or to indicate gene order. Thus RI strain data may indicate gene order of loci that were shown to be linked by other means. Therefore, existing miscellaneous RI strains should be maintained and characterized genetically.

Existing RI Strains

Bailey (5) was the first to develop RI strains of mice.

From the cross of BALB/cBy and C57BL/6By he derived seven RI strains. Although seven is near the minimal number of RI strains sufficient to detect linkage, a perfect match or near match is a good enough hint to encourage further testing. Using these CXB RI strains, Bailey and his collaborators have found approximately 20 linkages (12-19).

I began making several sets of RI strains shortly after arriving at the Jackson Laboratory in 1969. The status of these strains is given in Table 3. Most of these strains have now attained a high degree of inbreeding such that heterozygosity is encountered only rarely. The BXD RI strains are outstanding in regard to the number of strains (twenty-four), the number of loci at which they have been typed (sixty-five), and the number of chromosomes with at least one marker (fourteen). All four of these sets have been useful for gene mapping and a total of 20 linkages have been detected. Some of these are listed in Table 4. There are presently 32 loci for which there are SDPs but no evidence for any linkage. Thus the success rate appears to be about 40%. The power of RI strains for gene ordering is illustrated by the example of seven markers distributed over approximately 40 cM of chromosome 9 in the BXD RI strains (20, 21, and unpublished observations). A minimum of 34 exchange points are identified by these seven loci. Each exchange point potentially defines the locations of yet-to-be discovered loci in this region. I am also in the process of developing some additional miscellaneous RI strains. These are, in general, for special purposes, few in number, incompletely inbred, and not extensively tested as yet. A number of other workers outside the Jackson Laboratory are presently developing RI strains. Although it seems likely that a point of diminishing returns will be reached regarding additional sets of RI strains, it is too soon to know if it has yet been reached.

TABLE 3. Status of Four Sets of RI Strains

Progenitors	RI strain designation	Number of strains	Generations of BXS mating	Number of loci typed	Number of chromosomes[a]
AKR/J x C57L/J	AKXL	21[b]	12-35	56	12
SWR/J x C57L/J	SWXL	7	13-34	54	11
C57BL/6J x DBA/2J	BXD	24	25-39	65	14
C57BL/6J x C3H/HeJ	BXH	13	28-34	53	14

[a] The number of chromosomes known to bear at least one of the loci typed in the various RI strains.

[b] Two sublines of seven strains were established after 10 generations of BXS mating.

TABLE 4. Linkages Detected or Further Quantified Using RI Strains

Chromosome	Locus name (gene symbol)	RI strains used				References
		AKXL	BXD	BXH	SWXL	
1	Adrenal lipid depletion (ald)	x				22
1	Mixed lymphocyte stimulator (Mls)	x	x			23
1	Renin regulator (Rnr)	x	x		x	24
4	Response to bacterial lipopolysaccharide (Lps)			x		25, 26
4	Friend virus-1 (Fv-1)		x	x		27
4	B-lymphocyte alloantigen-2 (Lyb-2)		x			28
6	Kappa light chain IEF variant (PC8)	x				29
7	Coumarin hydroxylase (Coh)		x		x	30
7	Tamase (Tam-1)		x			31
7	Embryonic hemoglobin y-chain	x	x	x		32
7	Erythrocyte HPRT alteration (Hma)		x			33
9	Leucine arylaminopeptidase (Lap-1)		x			20
9	Esterase-13 (Es-13)				x	34
11	Galactokinase (Glk)	x		x		35
14	Formamidase-5 (For-5)		x	x		36
--	Prealbumin-1, immunoglobulin heavy chain (Pre-1, Ig-1)	x	x			37

The number of loci typed in the RI strains has been increasing by about ten per year for the past several years. The technique of isoelectric focusing, either alone or combined with conventional electrophoresis, promises to reveal numerous protein polymorphisms not seen with electrophoresis alone. The application of these techniques to RI strains is expected to result in a quantum jump in the number of loci typed. Recent developments in DNA chemistry and the ability to clone may make it possible to recognize strain differences at the DNA level. Presumably, when more workers become aware of RI strains and their potential usefulness, they will avail themselves of this tool. Therefore I am optimistic that in the next few years we will see the RI method come into its own in mouse genetics.

REFERENCES

1. Lewontin, R.C., and Hubby, J.L. (1966). Genetics 54:595.
2. Harris, H. (1966). Proc. Roy. Soc. Lond. 164:298.
3. Roderick, T.H., Ruddle, F.H., Chapman, V.M., and Shows, T.B. (1971). Biochem. Genet. 5:457.
4. Snell, G.D., and Cherry, M. (1972). In "RNA Viruses and Host Genome in Oncogenesis" (P. Emmelot and P. Bentvelzen, eds.). North-Holland, Amsterdam.
5. Bailey, D.W. (1971). Transplantation 11:325.
6. Haldane, J.B.S., and Waddington, C.H. (1931). Genetics 16:357.
7. Green, E.L. Unpublished.
8. Haldane, J.B.S. (1956). J. Genet. 54:327.
9. Carter, T.C. (1955). J. Genet. 53:21.
10. Searle, A.F., Berry, R.J., and Beechey, C.V. (1970). Mutat. Res. 9:137.
11. Lyon, M.F. (1977). Genet. Res. 28:291.

12. Bailey, D.W., and Hoste, J. (1971). Transplantation 11:404.
13. Merryman, C.F., Maurer, P.H., and Bailey, D.W. (1972). J. Immunol. 108:937.
14. Blomberg, B., Geckeler, W., and Weigert, M. (1972). Science 177:178.
15. Oliverio, A., Eleftheriou, B.E., and Bailey, D.W. (1973). Physiol. Behav. 10:893.
16. Eleftheriou, B.E., and Kristal, M. (1974). J. Reprod. Fertil. 38:41.
17. Wikstrand, C.J., Haughton, G., and Bailey, D.W. (1974). Cell. Immunol. 10:238.
18. DeMaeyer, E., DeMaeyer-Guignard, J., and Bailey, D.W. (1975). Immunogenet. 1:438.
19. Bailey, D.W. Personal communication.
20. Womack, J., Lynes, M.A., and Taylor, B.A. (1974). Biochem. Genet. 13:511.
21. Meisler, M.H. (1976). Biochem. Genet. 14:921.
22. Taylor, B.A., and Meier, H. (1976). Genetical Res. 26:307.
23. Festenstein, H., Bishop, C., and Taylor, B.A. (1977). Immunogenet. 5:357.
24. Wilson, C., Wilson, J., and Taylor, B.A. (1978). Mouse News Letter 58:49.
25. Watson, J., Riblet, R., and Taylor, B.A. (1977). J. Immunol. 118:2088.
26. Watson, J., Kelly, K., Largen, M., and Taylor, B.A. (1978). J. Immunol. 120:422.
27. Taylor, B.A., Bedigian, H.G., and Meier, H. (1977). J. Virol. 23:106.
28. Taylor, B.A., and Shen, F.W. (1977). Immunogenet. 4:597.
29. Claflin, J.L., Taylor, B.A., Cherry, M., and Cubberley, M. (1978). Immunogenet. (accepted).

30. Taylor, B.A., and Wood, A.W. (1976). Mouse News Letter 54:41.
31. Skow, L., and Taylor, B.A. (1977). Mouse News Letter 57:19.
32. Stern, R.H., Russell, E.S., and Taylor, B.A. (1976). Biochem. Genet. 14:373.
33. Day, C., and Nesbitt, M. (1977). Mouse News Letter 57:10.
34. Womack, J.E., Taylor, B.A., and Barton, J.E. (1978). Biochem. Genet. (submitted).
35. Mishkin, J.D., Taylor, B.A., and Mellman, W.J. (1976). Biochem. Genet. 14:635.
36. Cumming, R.B., Walton, M.V., Fuscoe, J.C., Taylor, B.A., Womack, J.E., and Gaertner, F.H. (1978). Biochem. Genet. (submitted).
37. Taylor, B.A., Bailey, D.W., Cherry, M., Riblet, R., and Weigert, M. (1975). Nature 256:644.

Inbred Strains: Subline Differences

Herbert C. Morse III, Chairman

DIFFERENCES AMONG SUBLINES OF INBRED MOUSE STRAINS

Herbert C. Morse III

National Institute of Allergy and Infectious Diseases
National Institutes of Health
Bethesda, Maryland

In many fields other than biology, conclusions reached about comparative results coming from different laboratories are not open to question on the basis of the reagents employed -- one organic chemist's supply of toluene is equivalent to another's. It has been an implicit belief among many involved in mouse genetics that a similar constancy is inherent in the use of highly inbred mice for experimental purposes -- one investigator's C3H is equivalent to another's. The elegant theoretical analysis of Bailey (1, 2, and this volume) as well as multiple reports of genetic differences among sublines of inbred strains of mice (3-7) clearly demonstrate that this belief is unfounded.

As outlined by Bailey, there are three possible causes for subline differences: 1) contamination from outcrossing, 2) incomplete inbreeding, and 3) mutation. Probable examples of each of these causes are easily found. The demonstration that C57BL/Ks differs from all other C57BL sublines at its entire $H-2$ complex as well as at least three other histocompatibility loci (8) probably reflects unknown outcrossing.

Existing differences among the major sublines of the C3H family -- those of Strong, Bittner, and Andervont -- are un-

doubtedly due to incomplete inbreeding before distribution.
This understanding is based upon Andervont's observation that,
on receiving C3H mice from Strong at F_{15}, they were still
segregating for coat color. Andervont inbred C3H's, selecting for agouti, for another 17 generations before giving
Heston mice in 1942 (H.B. Andervont, personal communication).

Finally, the studies of Flaherty and Rosenstreich (this
volume) on Qa antigens and LPS-responsiveness of BALB/c and
C3H sublines, respectively, strongly suggest that mutations
are responsible for the observed differences in these mice.

In studying the strain distribution of XenCSA, a cell-
surface antigen related to the major glycoprotein of xenotropic murine leukemia viruses (Chused and Morse, this volume),
we tested several sublines of different inbred strains for
expression of this antigen. As shown in Table 1, sublines of
C3H, C57BL, and A differ markedly in the amount of XenCSA
detected on the surface of thymocytes and spleen cells. High
expression of this antigen on lymphoid cells of C57BL/10SnGrf
as compared to other C57BL strains probably reflects a mutational event occurring in the relatively recent history of
this subline. The basis for other differences among sublines
is unknown.

The existence of such genetic differences between sublines of inbred strains can be viewed in two ways. First,
they may render results obtained in different laboratories
using different sublines of one strain impossible to compare.
This could result in a considerable waste of research efforts.
On the other hand, detection of differences between closely
related sublines provides the equivalent of congenic lines
for analyses of gene function. In either case, it is obviously imperative that investigators employing inbred strains
provide sufficient information in their papers so that the
ancestral relationship of their subline to others can be

TABLE 1. Differences in Expression of XenCSA on Lymphocytes of Sublines of Inbred Mice

Family	Subline	XenCSA[1] (Mean fluorescence)	
		Thymus	Spleen
C3H	C3H/AnfCum	_88_	_114_
	C3H/HeN	_83_	_201_
	C3H/HeJ	_81_	_157_
	C3H/H	_81_	_157_
	C3H/FgLw	56	_136_
	C3H/BiMai	48	_128_
	C3H/DiSnGrf	46	_97_
	C3H/St	27	37
C57	C57BL/10SnGrf	_96_	_200_
	C57BL/6J	50	_80_
	C57L/J	42	47
	C57BR/cdJ	39	42
	C57BL/10ScSnJ	26	25
	C57BL/St	18	33
	C57BL/10J	15	27
A	A/J	_67_	_76_
	A/SnGrf	_67_	_69_
	A/St	36	38

[1] XenCSA levels were determined by methods described by Chused and Morse (this volume). Mean fluorescence values >60 indicate high expression for XenCSA and are underlined. Numbers indicate geometric means for two or more assays of thymocytes or spleen cells from each subline. Each family of inbred mice demonstrates major differences in expression of this single cell surface antigen. The genetic basis for these differences is uncertain.

readily determined. The latter point brings into focus the question of nomenclature. Problems related to appropriate designations of strains, sublines, and genes are dealt with more extensively in the paper by Dr. Lyon which follows.

The remainder of this section is devoted to brief reports of differences detected among sublines of strains C3H, AKR, CBA, C57BL, and BALB/c.

REFERENCES

1. Bailey, D.W. (1959). J. Hered. 50:26.
2. Bailey, D.W. (1977). Ciba Found. Symp. 52:291.
3. Rechicigl, R., Jr., and Heston, W.E. (1963). J. Natl. Cancer Inst. 30:855.
4. Acton, R.T., Blankenhorn, E.P., Douglas, T.C., Owen, R.X., Hilgers, J., Hoffman, H.A., and Boyse, E.A. (1973). Nature New Biol. 245:8.
5. Maurer, P.H., Merryman, C.F., and Jones, J. (1974). Immunogenetics 1:398.
6. Taniguchi, M., Tada, T., and Tokuhisa, T. (1976). J. Exp. Med. 144:20.
7. Olsson, M., Lindahl, G., and Ruoslaht, E. (1977). J. Exp. Med. 145:819.
8. Graff, R.J. (1970). Transplant. Proc. 2:15.

STANDARDIZED GENETIC NOMENCLATURE FOR MICE:
PAST, PRESENT, AND FUTURE

Mary F. Lyon

MRC Radiobiology Unit
Harwell, Didcot, Oxon, U.K.

Nomenclature is one of those systems, analogous to digestion, circulation or respiration, which should function smoothly without our awareness of it. Nevertheless, like those physiological functions it is essential, and since unlike them it is man-made by corporate decisions, it is very important that nomenclature should sometimes be discussed. Perhaps the most convenient way to consider the subject is a historical or developmental sequence.

The history of committees dealing with the nomenclature of mouse genetics and inbred strains has been reviewed by Snell (1) and by Staats (2) and is presented in tabular form in Table 1. Although inbred strains have been of such paramount importance to mouse genetics and were begun so early in the development of the subject, the first nomenclature committee dealt not with inbred strains but with mouse gene nomenclature. It was formed as early as 1939, when there were only 31 known gene loci and seven linkage groups. This early formation of a nomenclature committee was an important factor in its success since, if nomenclature is allowed to develop in an unorganized way for any length of time, it becomes very difficult to gain the acceptance of scientists

TABLE 1. Some Events in the History of Committees Dealing with Mouse Genetics Nomenclature

Year	Event
1939	Formation of first Committee on Mouse Genetics Nomenclature, Drs. Dunn, Gruneberg, and Snell
1949	First issue of Mouse News Letter
1952	Committee on Standardized Nomenclature for Inbred Strains of Mice
	First listing in Cancer Research
1958	Committee on Standardized Genetic Nomenclature for Mice
1959	First issue of Inbred Strains of Mice
1970	Dr. Margaret Green succeeds Dr. George Snell as Chairman of Committee
1971	Affiliation of Committee with International Committee on Laboratory Animals
	First issue of Mouse Membrane Alloantigen News
1975	Dr. Mary Lyon succeeds Dr. Margaret Green as Chairman

for the changes necessary to produce order. Present day geneticists thus owe a considerable debt to the foresight of those who took part in organizing the first committee. One of the prime movers was G.D. Snell.

Another important factor in the success of any nomenclature committee is the wide dissemination of its findings and in this respect also the first committee made a very wise move in founding the Mouse News Letter, which first appeared in 1949, and is still a major organ today for disseminating information on mouse genetical nomenclature.

Inbred strains as such first entered the nomenclature scene in 1952 when a Committee on Standardized Nomenclature for Inbred Strains of Mice was formed. This group succeeded in gaining wide acceptance not only on the nomenclature for inbred strains, but also on the genetic criteria for the establishment of new inbred strains and for congenic strains, which are now of such great importance and had then recently been introduced, with G.D. Snell again a leading figure. Like the Committee on Gene Nomenclature, the Committee on Inbred Strain Nomenclature attached much importance to the dissemination of its findings and began the listing of inbred strains and their nomenclature in <u>Cancer Research</u> which has been carried out regularly by Joan Staats since then (3).

In 1958 the two committees for nomenclature of inbred strains and for gene nomenclature were merged to form the Committee on Standardized Genetic Nomenclature for Mice which is the body responsible for coordinating and regulating all problems concerned with the nomenclature of strains, genes, and chromosome anomalies of the mouse today. The merging of the strain and gene nomenclature committees into a single body was of course highly appropriate, as the contribution of inbred strains, congenic strains, and recombinant strains to knowledge of gene loci has been so great. There were in 1977 approximately 545 gene loci with known variants (4), and of these Staats (3) listed over 120, or 22%, as being polymorphic in inbred strains.

In 1971 the Committee on Standardized Genetic Nomenclature for Mice (hereafter called the Nomenclature Committee) sought and gained affiliation with the International Committee on Laboratory Animals, which is a dependent body of UNESCO and WHO. Hence, the Nomenclature Committee has official international status. Its members are all scientists working actively in the field. They are drawn from a number of different countries and represent as many areas of expertise in mouse genetics as possible.

The work is carried out largely by correspondence, with occasional meetings timed to coincide with international genetics meetings. When particular problems arise ad hoc subcommittees may be formed to prepare recommendations which are then circulated to scientists active in the area of concern for their comments. Finally the members of the Committee vote on the recommendations (Table 2).

The present rules for nomenclature fall under three main headings:

1. Inbred strains (including congenic and recombinant inbred strains).
2. Gene nomenclature.
3. Designation of chromosome anomalies.

However, it is clear that knowledge of these three areas is closely interrelated, and one must in fact consider all parts of the nomenclature together (5).

TABLE 2. Some Work of Mouse Genetics Nomenclature Committees

Year	Publication
1940	Rules for gene nomenclature (24)
1952	Standardized nomenclature for inbred strains (7)
1963	Revised rules for gene nomenclature (8)
1972	Standard karyotype of the mouse (9)
1973	Guidelines for biochemical variants (10)
1974	Rules for designation of chromosome anomalies (11)
1976	Nomenclature for inbred strains preserved by freezing (12)

From time to time it is necessary to make changes or additions to the rules. In drawing up its suggestions the Committee bears in mind various principles:
1. The nomenclature must be simple to use.
2. There should be no ambiguity.
3. The symbols should convey the maximum information.
4. The nomenclature must have general acceptance.
5. The system must be adaptable to future advances in knowledge.

In dealing with simplicity in use, a consideration to be borne in mind is that the symbols should be suitable for use in computerized information retrieval systems, with the minimum of modification. Some computers do not handle lower case letters. However, the use of upper and lower case initial letters of gene symbols to indicate dominance or recessivity is standard practice for all organisms (8-12), and more modern computers do handle lower case letters, and hence mouse nomenclature follows the standard practice. On the other hand, subscripts and superscripts are still difficult to handle by computer and hence these are kept to a minimum.

An example of the need to avoid ambiguity lies in the use of substrain symbols for inbred strains. Some of the well known strains have now been inbred for approaching 200 generations. Bailey (13) has shown that different branches of these strains may come to differ genetically quite considerably as a result of mutational drift, and with the increasing knowledge of polymorphisms in inbred strains more and more differences among substrains are in fact being found. Thus, it is important that different workers aiming for comparable results should use not only the same strain but the same substrain, and the symbolism should be so devised that it is clear which substrain was used. It is likely that changes will be made in the rules for inbred strain nomenclature in the near future to make the designation of substrains clearer.

A further likely change, aimed at providing maximum information, is in the symbolism for congenic lines. Bailey (14) has pioneered the system of including in the name of the strain symbols for both the differential allele(s) and the donor strain which provided the allele, e.g., B10.129-$\underline{H\text{-}12}^b$, a strain with the genetic background of C57BL/10Sn (= B10) but which differs from that strain in a differential allele ($\underline{H\text{-}12}^b$) derived from strain 129/J. This system clearly provides more information than one in which only the donor strain or the differential allele is listed, since if the donor strain is known it may be possible to predict which alleles at neighboring loci have been carried along in the differential chromosomal segment. Consequently, Bailey's nomenclature is likely to be recommended in future.

Conversely, to promote simplicity in use, the inclusion of many details of manipulative processes such as fostering or egg transfer is likely to be discouraged. Obviously, information on such processes is valuable as an indication of the likely presence or absence of vertically transmitted viruses, etc. However, when the original rules were formulated, no one envisaged that strains would be subjected to more than one, or perhaps two, such processes. Now, with the widespread use of fostering in SPF techniques some strains have been fostered many times, and to include such details in the name makes the symbol unwieldy. The advent of freeze preservation as another manipulative process makes the prospect even worse. Therefore, the Committee is likely to recommend that in future only the most recent manipulative process should be mentioned in the name.

The Committee is of course reluctant to make changes to avoid any risks of confusion and ambiguity, and because frequent changes might not be generally acceptable to scientists.

However, the points already mentioned provide examples of the ways in which scientific advances force the need for change.

It may be profitable to try to foresee what changes may be necessary in nomenclature and the work of the Nomenclature Committee in the future. First, one may expect to see a continuing very rapid growth in knowledge of the field. The number of known gene loci in the mouse is showing exponential growth at present, with the numbers almost doubling in 10 years (Fig. 1). The number of mapped loci and the numbers

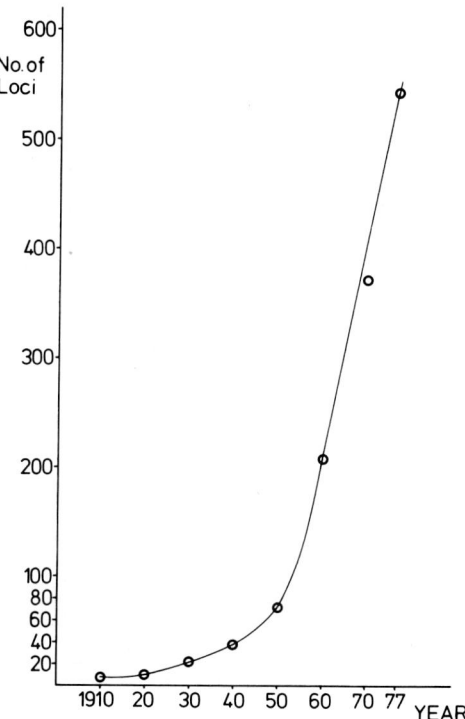

FIGURE 1. Numbers of mouse loci with known variants either published (early years) or listed in Mouse News Letter (later years).

of loci known to be polymorphic in inbred strains are likewise increasing very rapidly (3, 15).

This means that computer storage of the information on these matters is now essential. It also means much increased work for those who undertake the collection of the information for the Committee. Joan Staats has been secretary of the Committee and has collected and collated all the information on inbred strains since the first listing in 1952. Tony Searle has been editor of Mouse News Letter, and thereby a collector of information on gene loci since 1969. A great debt is owed to both of them for this work, and perhaps the Committee should consider what might be the difficulties of replacing them if and when they should wish to retire, in view of the ever-increasing work load.

Not only may one expect to see many more known gene loci, but the knowledge of these loci is likely to become far more detailed. There have been reports already of variants which are believed to involve operators, or cis-acting regulators, of structural loci (16), and many more of these are likely to be found. Gene complexes, too, will probably need to be incorporated in the nomenclature. At present the H-2 complex is the only well established gene complex (17), but there are other possible candidates, such as the closely linked genes Ph, Rw and W on chromosome 5 (18).

Knowledge of genetic mapping is also likely to become far more detailed, the location of gene loci relative to chromosomal G-bands will become known, possibly with the aid of polymorphisms of G-bands in inbred strains. This more detailed mapping may mean that in congenic strains, not merely a differential allele, but the full length of the transferred chromosomal segment may be known. Thus, at some time in the future it may be necessary to change the symbolism for congenics once again, to incorporate this information. Certainly the symbols for chromosome anomalies are likely to be changed quite soon to give details of exact positions of chromosome breaks. Yet another possibility is that variants may be found in repetitive DNA, but this point is highly speculative at present (19).

The points mentioned so far have concerned mouse genetics sensu stricto, but there are at least two ways in which the Committee is likely to have to broaden its field of interest. On the one hand, there is the question of uniformity of nomenclature among species. There is widespread interest in the evolution and phylogeny of gene loci and, particularly in the case of structural loci for enzymes, homologous loci have been studied not only in mouse, man, and various species of mammals, but also in birds, fish, and insects (20-23). There is obviously a need for liaison among the nomenclature committees for various organisms and possibly the International Genetics Federation could play a role here.

Since the nomenclature of mouse genetics is so well organized, relative to that in some other species, perhaps wider dissemination of the work of the mouse Nomenclature Committee would be valuable. Certainly there is a great need for this wider dissemination of information to those who work with mice but are not geneticists. Genetics is now such a central subject in biology, and the mouse is such a widely used organism, that genetic variants are being used, and new variants are being discovered, in such disparate fields as biochemistry, cancer research, endocrinology, immunology, virology, and so on. Unfortunately many workers in these fields are not fully conversant with genetic terminology, and particularly with mouse genetics nomenclature. Thus, the Nomenclature Committee has an important role in the future, not only in keeping abreast with probable rapid advances in knowledge, but also in disseminating its information as widely as possible in the biological field.

REFERENCES

1. Snell, G.D. (1974). Mouse News Letter 50:7.
2. Staats, J. (1966). In "Biology of the Laboratory Mouse," 2nd Ed. (E.L. Green, ed.), p. 45. McGraw-Hill, New York.
3. Staats, J. (1976). Cancer Res. 36:4333.
4. Searle, A.G. (1977). Mouse News Letter 56:4.
5. Lyon, M.F. (1977). Immunogenetics 5:393.
6. Committee on Genetic Symbols and Nomenclature (1957). International Union of Biological Sciences Series B No. 30, p. 1.
7. Committee on Standardized Nomenclature for Inbred Strains of Mice (1952). Cancer Res. 12:602.
8. Committee on Standardized Genetic Nomenclature for Mice (1963). J. Hered. 54:159.
9. Committee on Standardized Genetic Nomenclature for Mice (1972). J. Hered. 63:69.
10. Committee on Standardized Genetic Nomenclature for Mice. (1973). Biochem. Genet. 9:369.
11. Committee on Standardized Genetic Nomenclature for Mice. (1974). Mouse News Letter 51:2.
12. Committee on Standardized Genetic Nomenclature for Mice. (1976). Mouse News Letter 54:2.
13. Bailey, D.W. (1978). In "Origins of Inbred Mice" (H.C. Morse III, ed.). Academic Press, New York.
14. Bailey, D.W. (1977). Inbred Strains of Mice 10:26.
15. Roderick, T.H., and Davisson, M.T. (1977). Ann. Rep. Jackson Lab. 48:49.
16. Paigen, K., Swank, R.T., Tomino, S., and Ganschow, R.E. (1975). J. Cell Physiol. 85:379.

17. Klein, J., Bach, F.H., Festenstein, F., McDevitt, H.O., Shreffler, D.C., Snell, G.D., and Stimpfling, J.H. (1974). Immunogenetics 1:184.
18. Searle, A.G., and Truslove, G.M. (1970). Genet. Res. 15:227.
19. Lyon, M.F., and Mason, I. (1977). Genet. Res. 29:255.
20. Harris, H. (1975). "The Principles of Human Biochemical Genetics," 2nd Ed. North Holland, Amsterdam.
21. Lush, I.E. (1966). "The Biochemical Genetics of Vertebrates Except Man." North Holland, Amsterdam.
22. Masters, C.J., and Holmes, R.S. (1972). Biol. Rev. 47:309.
23. Courtright, J.B. (1976). Adv. Genet. 18:249.
24. Dunn, L.C., Grüneberg, H., and Snell, G.D. (1940). J. Hered. 31:505.

THE BIOLOGICAL FUNCTION OF THE LPS GENE

David L. Rosenstreich

Laboratory of Microbiology and Immunology
National Institute of Dental Research
National Institutes of Health
Bethesda, Maryland

C3H/HeJ mice are resistant or poorly responsive to all the known biological effects of endotoxin (reviewed in 1). This defect is due to a mutation in a single gene that occurred some time between 1960 and 1965 (2). The LPS gene has been mapped to the fourth chromosome near the gene that codes for brown coat color (b) (3), and exhibits co-dominant inheritance, so that F_1 hybrid progeny of a cross between C3H/HeJ mice with an LPS-responsive strain are intermediate responders to LPS (2).

Although LPS exhibits activity in a variety of cellular and humoral systems, it is thought that the primary expression of the LPS gene is in cells since the B-lymphocytes, T-lymphocytes, macrophages (1), and fibroblasts (4) of C3H/HeJ mice all respond abnormally to LPS in vitro, while at least one humoral function, LPS-induced complement activation, is normal in these mice (5). Since a single gene mutation has resulted in the complete loss of LPS-sensitivity in these mice, and since this mutation seems to be expressed only in cells, it is very likely that all the biological effects of

LPS are in fact mediated via the action of endotoxin on some cell or cells.

We were interested in understanding the biological role of the LPS gene, and one approach to this question was to determine the ability of C3H/HeJ mice to deal with a variety of challenges such as viruses, bacteria, or malignant cells.

Mice are susceptible to lethal challenge with Herpes simplex virus-1 (HSV-1). Surprisingly, we found that approximately 100 times more HSV-1 was required to kill a C3H/HeJ than was required to kill a C3H/HeN mouse (6). This resistance was related to the inability of HSV-1 to grow within C3H/HeJ peritoneal cells since there was no difference in susceptibility of these strains when HSV-1 was inoculated intracerebrally.

In contrast, we have found that C3H/HeJ mice are remarkably susceptible to lethal injection with Salmonella typhimurium. C3H/HeJ mice will die after inoculation with approximately 1,000 times fewer S. typhimurium organisms than are required to kill C3H/HeN mice.

Macrophages will kill tumor cells in vitro if they are obtained from mice infected with organisms such as the BCG strain of Mycobacterium tuberculosis. However, macrophages from BCG-infected C3H/HeJ mice are not tumoricidal, in contrast to those from C3H/HeN mice (7). Furthermore, we have found that the ability of macrophages to be rendered tumoricidal is controlled by the LPS gene (8).

Thus, in addition to their well characterized defect in LPS-responsiveness, C3H/HeJ mice appear to respond differently to a number of environmental challenges that are unrelated to the exogenous administration of endotoxin (Table 1). These findings raise an interesting question in relation to the true biological function of the LPS gene. One possible explanation is that endotoxin is a ubiquitous compound whose

TABLE 1. Biological Activities in Which C3H/HeJ Mice
Differ from C3H/HeN Mice

I. Related to the exogenous administration of LPS
  A. <u>Immune mechanisms</u>
     Mitogenicity
     Immunogenicity
     Adjuvanticity
     Interferon production
     Colony stimulating factor production
     Nonspecific resistance to infection
  B. <u>Mechanism unknown</u>
     Extravascular accumulation of leukocytes
     Lethality
     Abortion
     Tumor necrosis
     Protection against x-irradiation

II. <u>Unrelated to the exogenous administration of LPS</u>
  A. Resistance to HSV-1
     Resistance to <u>S. typhimurium</u>
     Macrophage tumoricidal capacity

presence is required by living organisms to maintain the immune system in an "activated" state. If this were the case, then the lymphoreticular cells of C3H/HeJ mice would normally be less "activated" than those of other LPS-responsive mice. This could certainly account for their inability to kill tumor cells, replicate HSV, or adequately kill gram-negative organisms. On the other hand, the LPS molecule itself may be unrelated to this phenomenon. The true function of the LPS gene may be to regulate the activated state of lymphoreticular cells possibly via some serum component that is similar to LPS in structure.

REFERENCES

1. Rosenstreich, D.L., Glode, L.M., Wahl, L.M., Sandberg, A.L., and Mergenhagen, S.E. (1977). In "Microbiology - 1977" (D. Schlessinger, ed.), p. 314. ASM, Washington, D.C.
2. Glode, L.M., and Rosenstreich, D.L. (1976). J. Immunol. 117:2061.
3. Watson, J., Kelly, K., Largen, M., and Taylor, B.A. (1974). J. Immunol. 120:422.
4. Ryan, J.L., and McAdam, K.P.W.J. (1977). Nature 269:153.
5. Moeller, G.L., Terry, L., and Snyderman, R. (1978). J. Immunol. 120:116.
6. Kirchner, H., Hirt, H.M., Rosenstreich, D.L., and Mergenhagen, S.E. (1978). Proc. Soc. Exp. Med. Biol. 157:29.
7. Ruco, L.P., and Meltzer, M.S. (1978). J. Immunol. 120:329.
8. Ruco, L.P., Meltzer, M.S., and Rosenstreich, D.L. (1978). J. Immunol., in press.

EFFECT OF MACROPHAGE MIGRATION INHIBITORY FACTOR
ON PERITONEAL EXUDATE CELLS OF C3H/HeN AND C3H/HeJ MICE

Aldo Tagliabue
Luigi Ruco
James L. McCoy
Monte S. Meltzer
Ronald B. Herberman

National Cancer Institute
National Institutes of Health
Bethesda, Maryland

It has been shown that a mutation occurred in C3H/HeJ mice rendering the B lymphocytes of those animals defective in their response to LPS. Further studies suggested that in C3H/HeJ mice, macrophages also had defective biological responses directly related to LPS; moreover, C3H/HeJ macrophages showed failure to be activated by lymphokines to kill tumor cells in vitro. In an attempt to determine whether the defect(s) of C3H/HeJ macrophages could also affect their response to other lymphokines, we compared the responsiveness of mineral oil-induced peritoneal cells (PEC) from C3H/HeN and C3H/HeJ mice to migration inhibition factor (MIF). Supernatants of antigen-stimulated immune spleen cells were tested against PEC, employing the indirect agarose microdroplet technique. No migration inhibition with PEC from C3H/HeJ mice was detected, whereas PEC from C3H/HeN mice were significantly inhibited by even a 1/32 dilution of MIF supernatant. However, responsiveness to MIF of C3H/HeJ PEC could be induced. In fact, in vivo intraperitoneal injections of Mycobacterium bovis, strain BCG, seven days before the in vitro assay could render the C3H/HeJ PEC responsive to MIF. The usual lack of responsiveness to MIF by C3H/HeJ macrophages seemed to be related to some form of suppression, since mixture of PEC from C3H/HeN mice with 10% PEC from C3H/HeJ mice resulted in undetectable migration inhibition at any of the MIF dilutions employed. The degree of suppression of the responsiveness of MIF appeared to be related to the percentage of C3H/HeJ PEC added, with 5% C3H/HeJ PEC partially suppressing the C3H/HeN PEC response, and 2.5% C3H/HeJ PEC having only borderline effects. We found that T cells did not appear to be involved in the suppression, since pretreatments with anti-theta serum plus complement of PEC of both the C3H substrains affected neither the defective responsiveness to MIF of C3H/HeJ PEC nor the suppression caused by these PEC. The percentage of PEC phagocytizing latex particles after injection with mineral oil was 60-70% for the C3H/HeN mice and only 30-40% for the C3H/HeJ mice, whereas no differences have been found when the phagocytosis of ^{51}Cr-antibody coated erythrocytes was analyzed. These differences in PEC subpopulations did not seem to be reflected by the pattern of cell migration in the absence of MIF, since this was similar for C3H/HeN and C3H/HeJ PEC. Further studies are in progress in an attempt to elucidate the mechanism(s) of action responsible for the defective responsiveness to MIF of PEC from C3H/HeJ mice and for their suppressive activity on C3H/HeN PEC.

DIFFERENT HEMATOPOIETIC RESPONSES TO ENDOTOXIN IN DIFFERENT SUBLINES OF C3H MICE

Sallie S. Boggs

D. R. Boggs

R. A. Joyce

University of Pittsburgh School of Medicine
Pittsburgh, Pennsylvania

One or more mutations in the C3H mouse background have produced a C3H/HeJ strain with absent or markedly reduced responses to endotoxin (ET) with respect to mortality, inflammatory response, production of colony-stimulating factor (CSF), and increase in in vitro granulocyte colony-forming cells. Antibody and mitogenic responses of B lymphocytes have been reported to be absent if ET extracted by the Westphal procedure (ETW) was used while present, although greatly reduced, if ET extracted by the TCA-Boivin method (ETB) was used. We have compared mortality and various hematopoietic responses to ETB and ETW in C3H/HeJ and C3HeB/FeJ and C3H/HeN (CR). The lethality of HeJ mice was 0 after 1,000 or 500 μg ETB, while that of eB/FeJ mice was 100% for both doses and HeN was 60% after 500 μg and 100% after 1,000 μg. Hematopoietic responses were studied in irradiated and nonirradiated mice. In C3HeB/FeJ (eB) and C3H/HeN (HeN) mice, 25 μg of ETW or ETB given one day before midlethal radiation between 550-800 rads produced hematopoietic responses like those previously described for normal mice of other non-C3H strains.

Specifically, when ET-treated, irradiated eB or HeN mice were killed at various times after radiation, there was a 3- to 20-fold increase in endogenous spleen colony count (E-CFU), and concomitant increases in spleen weight, and in spleen iron uptake. At earlier times there was a modest increase in splenic erythroid colonies and a dramatic early wave of granulocytic spleen colonies as compared to irradiated controls (12-40 granulocytic colonies per spleen section compared to 2-6 colonies per spleen section in irradiated controls). The recovery and overshoot of marrow granulocytes occurred four days earlier and onset of regeneration of transplantable stem cells (CFU-S) was one to two days earlier in the "normal" mice given ETB or ETW than in non-injected irradiation controls. None of these effects were seen in C3H/HeJ mice given ETW, but when they were given ETB, some of the responses were seen. These included an initial but not persistent early increase in CFU-S, increases in E-CFU, spleen weight and spleen iron uptake, and erythroid colonies in the "normal" range. There were, however, no increases in granulocytic spleen colonies and no early increase in marrow granulopoiesis in C3H/HeJ mice given ETB or ETW before irradiation.

In nonirradiated "normal" C3HeB/FeJ mice 5 µg ETB or ETW increased blood CSF so that approximately 60 granulocytic colonies were stimulated per 10^5 normal marrow cells plated and 25 µg ETB or ETW resulted in CSF which stimulated 155-185 colonies/10^5 cells plated. Blood CSF after ETB or ETW in C3H/HeJ mice stimulated no colony growth at the 5 µg dose and only 10-13 colonies/10^5 of the same marrow cells plated. ETB and ETW caused "flushing" of neutrophils from the bone marrow of C3H/HeJ mice in a qualitatively normal fashion. These differences in the various hematopoietic responses will be further investigated to help determine control factors regulating hematopoiesis.

RELATIONSHIP OF ENDOTOXIN TOXICITY AND RESPONSE OF BONE MARROW DERIVED HEMATOPOIETIC CELLS

Sallie S. Boggs
Robert W. Baker
Gretchen N. Schwartz
Kenneth D. Patrene

University of Pittsburgh School of Medicine
Pittsburgh, Pennsylvania

Several substrains of C3H show markedly different responses to endotoxin. C3H/HeJ mice have markedly reduced endotoxin-induced mortality, macrophage responses, B-lymphocyte responses, and hematopoietic responses in irradiated mice. C3HeB/FeJ mice have much larger "normal" responses to endotoxin. By exchanging bone marrow (BM) in these two substrains, we hoped to determine whether the hematopoietic cellular response or the non-transplantable environmental cell response was the cause of the various differences observed. It has already been shown by others that the low or absent B-lymphocyte responses of the C3H/HeJ are due to alterations in the cells. Similarly, the lack of production of the granulocyte colony-stimulating factor (CSF) by macrophages and the inhibitor prostaglandin E after endotoxin is a defect of the cell. Our first concern was whether mortality after large doses of endotoxin was related to responses by the cells of the hematopoietic system. In a preliminary study, C3H/HeJ or

C3HeB/FeJ mice were exposed to a lethal dose (950 rads) of γ radiation and salvaged by injection of 6-9 x 10^6 BM cells. After the donor BM had replaced that of the host (> 45 days), animals were given an intraperitoneal injection of 500 μg Salmonella typhosa endotoxin. This dose invariably kills all the C3HeB/FeJ (eB) mice and spares all the C3H/HeJ (He) mice. The results showed: eB + eB BM, 17/17 dead or 100%; eB + He BM, 23/26 dead or 88%; He + He BM, 0/6 or 0% dead; and He + eB BM, 6/9 or 67% dead. These preliminary results suggest that mortality after endotoxin is partly due to hematopoietic cell responses and partly due to environmental factors. This is in agreement with the data of others suggesting some of the responses are under single gene control (CSF production, B-lymphocyte responses) and some are under multigene control (mortality and splenic granulocyte stem cell responses). In this light, we have begun to consider variations in lethal response for various strains. Data to date are as follows: after 500 μg Salmonella typhosa LPS-B; eB, 20/20 dead; HeJ, 0/20 dead; HeN, 3/5 dead; HeJ-(+/+ of the W^x line), 6/7 dead and HeJ-(W^x/+), 6/8 dead.

GENERATION OF AN H-2 RESTRICTED CML RESPONSE
BETWEEN AKR SUBSTRAINS

Marion M. Zatz

National Cancer Institute
National Institutes of Health
Bethesda, Maryland

AKR/Cu mice have previously been reported to reject AKR/J primary lymphoma cells (1). The AKR/J and AKR/Cu strains are known only to differ at a segment of chromosome 9, resulting in Thy 1.1 and Thy 1.2 phenotypes, respectively. Although this is the only known antigenic difference between these substrains, the Thy antigen was believed not to be responsible for tumor rejection (1). The present studies (2) document reciprocal resistance to transplantation of spontaneous lymphoma cells between these high leukemia AKR/J and AKR/Cu substrains, and reveal that spleen cells from such primed mice generate an H-2 restricted cell-mediated lympholytic (CML) response *in vitro* when cultured with cells of the appropriate AKR substrain.

This CML response has the following characteristics: 1) priming *in vivo* and culture with immunogen *in vitro* are required; 2) sensitization *in vivo* and restimulation *in vitro* can be elicited with AKR cells of thymus, spleen, and lymphoma origin; 3) lymphoma and T- and B-cell-enriched, mitogen-stimulated target cell populations are lysed by the effectors; 4) the effector cells are Thy positive; 5) the parameters of the CML response are similar to those reported for the H-Y (3) and minor H antigen (4) systems.

Using a variety of mouse strains as sources of target cell populations, the data of Table 1 show that 1) the Thy antigen is irrelevant to the CML response, thus AKR/Cu (Thy 1.2) AKR/J (Thy 1.1) effectors lyse CBA and B10.BR, but not C3H cells; 2) B10.BR cells either bear multiple antigens, or a cross-reacting antigen, since they are recognized and lysed by sensitized cells of both AKR substrains; 3) AKR/J and CBA/J cells share the same or similar critical antigen(s) as do AKR/Cu and C3H cells; 4) a non-H-2 antigen(s) is being recognized since lysis of $H-2^k$ CBA, C3H, and B10.BR cells occurs, but at least partial homology at the H-2 locus is required, as shown by the failure to lyse B10 cells; and 5) target cell homology at either the D or K end of the H-2 locus is sufficient for lysis by AKR/J effector cells, resulting in lysis of C3H.OH, B10.A, and B10.AKM target cells, whereas homology at the D-end of the H-2 locus is required for lysis by AKR/Cu effector cells, thus no lysis on B10.AKM target cells. Similar results and interpretations are obtained when these strains are used as a source of in vitro immunogens or blocking cells.

This differential H-2 restriction of the CML response could reflect either differential modification of the K- and D-end H-2 molecules by the discrepant AKR/J and AKR/Cu antigen(s), or differential immune responsiveness of these AKR substrains (? non-H-2 IR genes) to modified H-2K and D-end products on cells of the reciprocal substrain.

To summarize, the data indicate that one or more minor histocompatibility antigens distinct from the Thy antigen, prime AKR/J and AKR/Cu mice for a secondary CML response, during the course of rejection of the reciprocal strain's tumor cells. Both normal T and B cells as well as primary lymphoma cells bear the antigen(s). Although previous studies

TABLE 1. Strain Distribution of Donors of Target Cells Lysed by AKR Effector Cells

Target[a]	[Thy/H-2]	% Lysis by effectors	
		AKR/J α AKR/Cu	AKR/Cu α AKR/J
AKR/J	(1.1/kkkkkk)	1.2	48.4
AKR/Cu	(1.2/kkkkkk)	49.2	2.8
CBA/J	(1.2/kkkkkk)	6.0	54.4
C3H/HeJ	(1.2/kkkkkk)	30.3	2.4
C3H.OH	(1.2/dddddk)	33.0	4.6
B10.BR	(1.2/kkkkkk)	40.2	58.4
B10.A	(1.2/kkkddd)	27.0	1.9
B10.AKM	(1.2/kkkkkq)	33.8	6.8
B10	(1.2/bbbbbb)	-1.5	4.4
A/J	(1.2/kkkddd)	15.5	9.6
A/Thy 1.1	(1.1/kkkddd)	14.0	-2.9

[a] PHA stimulated spleen cells were used as targets at an effector/target ratio of 50/1.

1) suggest that the histocompatibility antigen(s) is linked to the Thy antigen and 2) find no evidence for genetic disparity at nine other linkage groups including chromosome 17 (H-2), the possibility remains that the antigen(s) is coded by a locus outside of the section of chromosome 9, linkage group II, which is known to differ between these two substrains. Backcross studies will be required to prove or disprove linkage to the Thy antigen and chromosome 9.

REFERENCES

1. Acton, R.T., Blankenhorn, T.C., Douglas, T.C., Owen, R.D., Hilgers, J., Hoffman, H.A., and Boyse, E.A. (1973). Nature New Biol. 245:8.

2. Zatz, M.M. (1978). Cell. Immunol. 36:251.
3. Gordon, R.D., Simpson, E., and Samuelson, L.E. (1975). J. Exp. Med. 142:1108.
4. Bevan, M.J. (1975). J. Exp. Med. 142:1349.

DERIVATION AND CHARACTERISTICS OF THE CBA/N SUBLINE

Donald E. Mosier

Laboratory of Immunology
National Institute of Allergy and Infectious Diseases
National Institutes of Health
Bethesda, Maryland

The CBA/N subline expresses an X-linked immunological defect which appears to be confined to the bone marrow-derived (B) lymphocyte lineage (1-8). Although the precise nature of the defect is poorly understood, the functional consequences suggest that a maturation arrest of B-lymphocyte development has taken place (9, 10).

The CBA/N subline (also known as CBA/HN mice in earlier publications) was derived from the CBA/Harwell line. In 1966, mice with the spontaneous mutation for foam cell reticulosis (fm^+) were imported to the NIH. The fm mutation arose in offspring of two to three breeding pairs of CBA/H mice in 1963 and is a homozygous lethal autosomal recessive gene (11). CBA/N mice were derived from fm^+/fm^- heterozygotes by brother-sister mating of fm^-/fm^- offspring whose phenotype was confirmed by progeny testing. Only a small number of such mice were bred, and in 1968 the subline almost was lost save for one pregnant female. All subsequent CBA/N mice are derived from that one litter of six females and one male. By 1972, when a defect in IgM serum levels was first noted (1), the subline had undergone 20 generations of brother-sister mating.

The phenotypic abnormalities of the CBA/N subline are summarized in Table 1. Most, if not all, of their functional behavior can be explained by a deletion of a mature B-lymphocyte subset. These mice provide an interesting model

TABLE 1. B-Lymphocyte Phenotype and Functional Defects of the CBA/N Subline

	CBA/N	CBA/CAnN
Surface phenotype		
surface immunoglobulin	IgM > IgD	IgD > IgM
Lyb3	---	50% +
Lyb5	---	50% +
C3 receptor	low	high
Mls determinants	low	high
Antibody responses		
Thymic-independent 1		
TNP-Brucella abortus	+	+
TNP-lipopolysaccharide	+	+
TNP-polyacrylamide bead	+	+
Thymic-independent 2		
TNP-Ficoll	−	+
TNP-dextran	−	+
Type III pneumococcal polysaccharide	−	+
Thymic-dependent		
SRBC	++	++
DNP-KLH	+	++
DNP-OVA	+	++
B cell colony formation	−	+

for a restricted immunologic defect and for studying the regulation of B lymphocyte differentiation. They have allowed the definition of a set of alloantigens which are expressed only on a subset of normal B cells, namely, those which appear to be absent in the CBA/N strain (12, 13). The mice are available to interested investigators through the NIH.

REFERENCES

1. Amsbaugh, D.F., Hansen, C.T., Prescott, B., Stashak, P.W., Barthold, D.R., and Baker, P.J. (1972). J. Exp. Med. 136:931.
2. Scher, I., Frantz, M.M., and Steinberg, A.D. (1973). J. Immunol. 110:1396.
3. Scher, I., Steinberg, A.D., Berning, A.K., and Paul, W.E. (1975). J. Exp. Med. 142:637.
4. Scher, I., Ahmed, A., Steinberg, A.K., and Paul, W.E. (1975). J. Exp. Med. 141:788.
5. Scher, I., Sharrow, S.O., Wistar, R., Jr., Asofsky, R., and Paul, W.E. (1976). J. Exp. Med. 144:494.
6. Cohen, P.L., Scher, I., and Mosier, D.E. (1976). J. Immunol. 116:301.
7. Mosier, D.E., Scher, I., and Paul, W.E. (1976). J. Immunol. 117:1363.
8. Finkelman, F.D., Smith, A.H., Scher, I., and Paul, W.E. (1975). J. Exp. Med. 142:1316.
9. Kincade, P.W. (1977). J. Exp. Med. 145:249.
10. Mosier, D.E., Zitron, I.M., Mond, J.J., Ahmed, A., Scher, I., and Paul, W.E. (1977). Immunol. Rev. 37:89.
11. Lyon, M.F., Hulse, E.V., and Rowe, C.E. (1965). J. Med. Genet. 2:99.

12. Ahmed, A., Scher, I., Sharrow, S.O., Smith, A.H., Paul, W.E., Sachs, D.H., and Sell, K.W. (1977). J. Exp. Med. 145:101.

13. Huber, B., Gershon, R.K., and Cantor, H. (1977). J. Exp. Med. 145:10.

CBA/Ki VS. CBA/St/Ki:
FIFTEEN YEARS OF OBSERVATIONS

Annabel G. Liebelt
Kirschbaum Memorial Mouse Colony
Northeastern Ohio Universities
College of Medicine
Rootstown, Ohio

Two inbred strains of CBA mice have been bred and studied for 15 years in our laboratory. Both were given to us by Dr. L.C. Strong -- the first, CBA/Ki, to Dr. Kirschbaum in 1941, and the second, CBA/St/Ki, to me in 1963. They have been inbred, brother-sister, for approximately 80 and 53 generations, respectively, since 1952 and 1963. During this time, the geographic location of the Ki colony has changed as my husband and I have moved: 1952-1954 at the University of Illinois College of Medicine, Chicago; 1954-1971 at Baylor College of Medicine in Houston; 1971-1974 at the Medical College of Georgia in Augusta; 1974 to the present at Northeastern Ohio Universities College of Medicine (temporarily housed at Akron General Medical Center) 1978- now located permanently in Rootstown, Ohio.

The external appearance of mice of the two strains is similar, but the strains differ in many ways.

1. Histocompatibility. Within each strain normal skin autografts and isografts were uniformly successful, while allografts between the strains were always sloughed between two and four weeks (up to 200 mice). Tumor transplants of

several different types likewise were successful within each strain, but only rarely crossed strain barriers (1). That is, of seven CBA/St/Ki tumors, none grew in CBA/Ki. Of 13 CBA/Ki tumors, many of which had been transplanted numerous generations, only two tumor lines had progressive growths in CBA/St/Ki [4/20 mice of 2663 stomach carcinoma line from T-75 on, and 21/21 in 2468 uterine tumor line from T-285 on (approximately 300 mice used)] (1). In addition, the CBA/Ki tumor 2663 is in the T-360 transfer now, still transplants successfully in the strain of origin, and maintains its lipid-mobilizing characteristics which result in the anorexia, cachexia, and death of the animals with only a relatively small subcutaneous transplant (2).

2. Spontaneous tumors. Occurrence of three types of tumors has been examined. Tumors of the reticular system, primarily systemic leukemia, occur in low incidence in both strains and both sexes (male = M, female = F): CBA/Ki - F=3.1% (27/871) at 578 days (194-780), M=1.8% (14/766) at 554 days (210-743); CBA/St/Ki - F=5.7% (15/261) at 627 days (456-992), M=4.0% (9/224) at 606 days. Hepatomas occurred more frequently in males than females of both strains: CBA/Ki - F=0.6% (5/871) at 607 days (481-796), M=8.4% (64/766) at 593 days (259-1016), CBA/St/Ki - F=0, M=8.0% (18/224) at 656 days. On the other hand, mammary cancer is rare in CBA/Ki, but occurs in CBA/St/Ki on the average of 50% (3). The incidence and age of death in females were: Texas 39% (43/111) at 479 days (189-957); Georgia 48% (43/90) at 457 days (242-700); and 44% (35/80) at Ohio - 367 days (221-584) (still in progress). The incidence in CBA/St/Ki virgin females is: Georgia - 27% (18/66) at 598 days (460-729); Ohio - in progress - so far approximately 20%.

Milk from lactating females of these strains was examined by Dr. L. Dmochowski, who found morphological evidence of mammary tumor virus of the B type in CBA/St/Ki and atypical particles in the CBA/Ki strain. No leukemia particles were seen (4).

3. Goldthioglucose-induced obesity varies between the two strains. A series of doses ranging from 0.2 to 1.5 mg/g body weight were tested by one injection i.p. to mice of both sexes at 2-3 months of age (5). Only CBA/Ki mice respond consistently, that is, a dose of 0.35 mg/g body weight results in no toxicity, no death, and 100% obesity. Associated with the obesity are a transitory increase in food consumption, hypothalamic lesions, and alterations in endocrine function. Mice of the CBA/St/Ki strain do not respond in the same manner, since many mice die or become ill following the injection or do not become obese at a given dose level.

4. Retinal degeneration (rodless retina). In mice of the CBA/Ki strain the retina at birth is indistinguishable from the retina of other strains including CBA/St/Ki. By 15 days the outer nuclear layer is reduced to a few nuclei in thickness; that is, there is loss of the rod and outer molecular layers and a reduction of the external nuclear layer; by 35 days the retina lacks photoreceptors (6). Mice of CBA/St/Ki strain have an intact photoreceptor layer.

5. Feeding patterns. In CBA/Ki mice the feeding pattern is atypical with periodic dominance of food consumption in the period of 8 a.m. to 4 p.m. and is due to a circadian rhythm with a period of greater than 24 hours. On the other hand, mice of the CBA/St/Ki strain have a more typical feeding pattern with the constant dominance of food consumption between 4 p.m. and 8 a.m., that is, a nocturnal pattern (7, 8).

TABLE 1. Differences Between Sublines of CBA Mice[a]

History	CBA/Ki	CBA/St/Ki
1. Source	Strong	Strong
1. Year of entry	Before 1952 (records from 1941)	1963
3. Life in Illinois	1952-54	
" Texas	1954-71	1963-71
" Georgia	1971-74	1971-74
" Ohio	1974-	1974-
4. Generations in colony	F? + 80	F? + 53
Characteristics		
1. Retinal degeneration	+	−
2. GTG-induced obesity	Consistent	Inconsistent
3. Feeding patterns	Atypical	Typical
4. Histocompatibility		
Normal tissues	+ ⟵−⟶	+
Tumorous	+ ⟵−⟶	+
4. Spontaneous tumors	%	%
Liver ♀	0.6	−
♂	8.4	8.0
Reticular tissue ♀	3.1	5.7
♂	1.8	4.0
Mammary gland ♀	<1%	50%
6. Milk viral particles − MTV	Atypical	B
Leukemia	0	0

[a] Strains listed in:
  1) Inbred Strains of Mice, No. 10 (M.N.L.), Staats, 1977;
  2) Standardized Nomenclature...Sixth Listing, Staats, 1976.

SUMMARY

Since Strong first developed the CBA strain(s), numerous mouse colonies throughout the world have been breeding CBA mice, many of which have reported differences in mammary cancer incidence. We have reported our observations over a number of years on two CBA strains which differ in multiple characteristics, but which have remained stable.

REFERENCES

1. Liebelt, A.G., and Liebelt, R.A. (1967). In "Methods in Cancer Research," Vol. I, p. 143. Academic Press, New York.
2. Liebelt, R.A., Gehring, G., Delmonte, L., Schuster, G., and Liebelt, A.G. (1974). Ann. N.Y. Acad. Sci. 230:547.
3. Liebelt, A.G., and Liebelt, R.A. (1967). In "Carcinogenesis: A Broad Critique" (University of Texas, M.D. Anderson Hospital), p. 315. Williams & Wilkins, Baltimore.
4. Dmochowski, L., Langford, P.L., Williams, W.C., Liebelt, A.G., and Liebelt, R.A. (1968). J. Natl. Cancer Inst. 40:1339.
5. Liebelt, R.A., Sekiba, K., Liebelt, A.G., and Perry, J.H. (1960). Proc. Soc. Exp. Biol. Med. 104:689.
6. Caley, D.W., Johnson, C., and Liebelt, R.A. (1972). Am. J. Anat. 133:179.
7. Ishiki, D.M. (1968). Thesis for M.S., Baylor University College of Medicine, Houston, Texas.
8. Liebelt, R.A., Ishiki, D., and Liebelt, A.G. (1973). Fed. Proc. 32:383.

DIFFERENCES BETWEEN C57BL STRAINS AT AN ERYTHROCYTE ANTIGEN LOCUS[1]

Marianna Cherry
Donald W. Bailey
George D. Snell

The Jackson Laboratory
Bar Harbor, Maine

Three different strain combinations have produced alloantisera that agglutinate red cells from mouse strains C57BL/6J and C57BL/6By, but do not react with cells from strains C57BL/10Sn, C57BL/10J, or C57BL/10Gn-Rn/+ (Rn/+ or +/+). Erythrocytes from 35 H-2 and non-H-2 congenic lines on the C57BL/10Sn background are unreactive, whereas red cells from all other inbred strains tested to date, including related strains C57L/J, C57BR/cdJ, C57BL/KsJ, and C58/J are reactive. This pattern of serological reactivity is different from that of any other known erythrocyte antigen system of the mouse and suggests residual heterozygosity or a mutation in an early progenitor of the C57BL/10 substrain family.

We have recently developed a congenic line in which a histocompatibility (H) allele from strain C57BL/10Sn was introduced onto a C57BL/6J background by 16 successive generations of backcross followed by sib mating. Cells from mice

---

[1] Support for this work was provided by grants and contracts CA-01329, AI-12257, and AI-13130.

selected at the first intercross generation for homozygosity at the introduced H̲ allele were serotyped with one of the alloantisera characterized above. They were unreactive or phenotypically like cells from strain C57BL/10Sn. Thus it appears that the erythrocyte antigen described here is determined by a locus closely linked to, or perhaps identical with, a histocompatibility locus that distinguishes between strains C57BL/6J and C57BL/10Sn.

SUBLINE DIFFERENCES IN BEHAVIORAL RESPONSES TO PHARMACOLOGICAL AGENTS

Beatriz Moisset

Psychology Department
Temple University
Philadelphia, Pennsylvania

Mutations that modify the response to a drug or that cause some minor change in behavior are likely to remain unnoticed indefinitely. Thus, separate colonies that originated from an inbred strain may differ in many genes that could be detected only by using pharmacological or behavioral techniques.

The stimulant drug d-amphetamine causes a large increase in locomotor activity in C57BL/6J and C57BL/10J mice, but has only a moderate effect on C57BL/6By (1). Analysis of F_1 and backcross matings suggests a one-gene model. A dominant mutation seems to have taken place in the C57BL/6By subline. A genetic difference in response to d-amphetamine may reflect a difference in norepinephrine receptor sensitivity and could be useful in studying the neurochemical regulation of locomotor activity.

There are large differences in the response to the narcotic effects of ethanol between inbred strains of mice. BALB/cJ mice have been repeatedly reported as sensitive to ethanol (long sleep time) (2-4). A mutation(s) appears to have occurred in BALB/cJ mice, changing from highly sensitive to resistant to ethanol. Sleep time after 4 g/kg of ethanol is

35.91 ± 5.59 minutes as compared to previous reports that vary from 66 to 110 minutes. This possible mutation may have occurred quite recently and it is possible that BALB/cJ mice are still segregating for this gene. BALB/cBy mice, on the other hand, are very sensitive to ethanol. Their sleep time is 90.68 ± 7.52 minutes. This value resembles that of previous authors using BALB/cJ. This factor affecting the narcotic effects of ethanol has little or no effect on ethanol preference in a free choice between 10% ethanol and water. Both sublines of the BALB/c strain show a low preference, thus indicating a dissociation between these two physiological phenomena. A mutation has been reported that affects ethanol preference (5). C57BL/6A shows a lower preference than C57BL/6J. To our knowledge, no research has been done on whether these two sublines differ in ethanol sleep time.

Subline differences like the ones reported above may be useful in investigating one gene's effects in a fairly similar background. Pharmacogeneticists and behavioral geneticists may benefit from a search of mutations in sublines of inbred strains.

REFERENCES

1. Moisset, B. (1977). Psychopharmacology 53:263.
2. Damjanovich, R.P., and Mac Innes, J.W. (1973). Life Sci. 13:55.
3. Belknap, J.K., Mac Innes, J.W., and MacClearn, G.E. (1972). Physiol. Behav. 9:453.
4. Randall, C.L., and Lester, D. (1974). J. Pharm. Exp. Therap. 188:27.
5. Poley, W. (1972). Behav. Genet. 2:245.

FURTHER INFORMATION ON SUBLINE DIFFERENCES[1]

Thomas H. Roderick

The Jackson Laboratory

Bar Harbor, Maine

Several investigators have reported genetic differences between sublines of highly inbred strains of mice. Bailey at this workshop has given considerable attention to the theoretical aspects of subline differentiation as these aspects may apply to the pedigrees of various strains and their sublines (1). Our particular study has for the most part concerned biochemical variants which are known to be polymorphic in feral populations and among strains which are relatively unrelated.

The data shown below have been obtained over about a 10-year period on strains which have been available at the Jackson Laboratory. In all cases, at least two individuals were examined from each strain to establish the allelotype of each locus. The loci chosen for study were dipeptidase-1 (Dip-1), three esterases (Es-1, Es-2, and Es-3), autosomal glucose-6-phosphate dehydrogenase (Gpd-1), glucose phosphate isomerase-1

---

[1]This work was supported in part by grant BMS75-03397 from the National Science Foundation and by contract ES4-2159 with the National Institute of Environmental Health Sciences. The Jackson Laboratory is fully accredited by the American Association for Accreditation of Laboratory Animal Care.

(Gpi-1), hemoglobin beta chain (Hbb), isocitrate dehydrogenase-1 (Id-1), lactate dehydrogenase regulator (Ldr), supernatant malic enzyme (Mod-1), two phosphoglucomutases (Pgm-1 and Pgm-2) and retinal degeneration (rd). The last locus, which affects the retina of the eye, was chosen because the mutant allele is so frequently found among laboratory strains of mice, although it is probably not polymorphic in feral populations. In all analyses involving gel electrophoresis, controls with alternative alleles were present on each gel for purposes of comparison and ascertainment. References concerning electrophoretic procedures can be found in earlier studies (2). The strains listed for study have been described in various publications of Staats (3, 4).

RESULTS

The data are displayed in three tables. Table 1 comprises strains and sublines of the C57 groups whose relationships are shown in Figure 1. The relationships of the major groups are shown in the charts of Bailey (1). The several B-lines shown directly under C57BL/6J were derived from C57BL/6J and inbred by brother-sister mating. The RX-GE was a closed colony of mice derived from C57BL/10Gn and exposed to high doses of x-irradiation (5). Table 2 shows similar data for the DBA, BALB/c, A, and related strains. Their relationships can be found in the pedigree chart of Staats (6). For our purposes, it is sufficient to point out that the A lines are related to BALB/c and that the C3H and CBA lines were derived from a cross between early progenitors of the DBA and BALB/c lines. Table 3 shows similar data for four unrelated groups. Their lack of relationship is based more on our lack of understanding of their ancient ancestry rather than on our knowledge of their origins in feral populations. The S-lines were all brother-sister bred and were recent offshoots of SJL/J.

TABLE 1. Alleles at Polymorphic Loci in the C57 and Related Inbred Strains

Chromosome	1	1	4	4	5	5	6	7	7	8	8	9	11
Gene Strain	Id-1	Dip-1	Pgm-2	Gpd-1	Pgm-1	rd	Ldr	Gpi-1	Hbb	Es-1	Es-2	Mod-1	Es-3
C57BL/KsJ	a	a	a	a	a	+	a	b	s	a	b	b	a
C57BL/6By	a	a	a	a	a	+	a	b	s	a	b	b	a
C57BL/6J	a	a	a	a	a	+	a	b	s	a	b	b	a
BH1	a	a	a	a	a		a	b	s	a	b	b	a
BH2	a	a	a	a	a		a	b	s	a	b	b	a
BC1	a	a	a	a	a		a	b	s	a	b	b	a
BC2	a	a	a	a	a		a	b	s	a	b	b	a
BL1	a	a	a	a	a		a	b	s	a	b	b	a
BL2	a	a	a	a	a		a	b	s	a	b	b	a
C57BL/10ScSn	a	a	a	a	a	+	a	b	s	a	b	b	a
C57BL/10J	a	a	a	a	a	+	a	b	s	a	b	b	a
C57BL/10Gn	a	a	a	a	a	+	a	b	s	a	b	b	a
RX-GE	a	a	a	a	a	+	a	b	s	a	b	b	a
C57BR/cdJ	b	a	a	a	a	+	a	a	s	a	b	b	a
C57L/J	b	a	a	a	a	+	a	a	s	a	b	b	a
C58/J	a	a	a	a	a	+	a	a	s	b	b	b	c

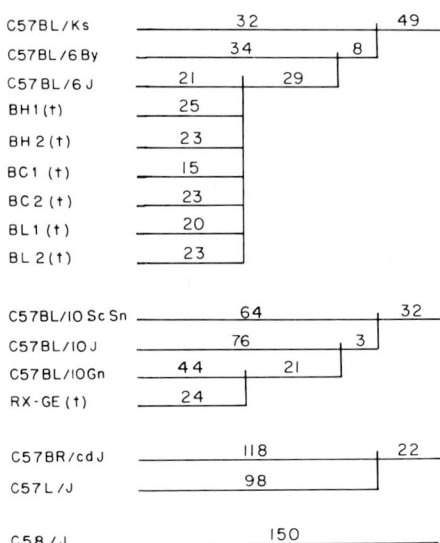

FIGURE 1. The relationships of strains of the C57 group. Numbers refer to generations of brother-sister matings between branching points. In some cases estimates were made based on known year of separation and known number of generations propagated per year for individual lines. More distant ancestral connections between these groups can be found in Staats (6) and Bailey (1).

† refers to lines that are extinct or probably extinct.

In all strains studied there was no evidence of within-strain genetic heterogeneity for those loci studied. This finding has been confirmed repeatedly by us with other enzyme polymorphisms. The lack of genetic heterogeneity within strains is not surprising probably to those attending this workshop, but it has been cause for surprise to investigators of other organisms, particularly those who study Drosophila. The constancy of genotype between distantly related sublines also supports the concept of genetic homozygosity of the strains studied.

TABLE 2. Alleles at Polymorphic Loci in the DBA, BALB/c, A, and Related Inbred Strains

Chromosome	1	1	4	4	5	5	6	7	7	8	8	9	11
Gene Strain	Id-1	Dip-1	Pgm-2	Gpd-1	Pgm-1	rd	Ldr	Gpi-1	Hbb	Es-1	Es-2	Mod-1	Es-3
DBA/1fB6Hu	b	b	a	a	b	+	a	a	d	b	b	a	c
DBA/1J	b	b	a	a	b	+	a	a	d	b	b	a	c
DBA/2J	b	b	a	b	b	+	a	a	d	b	b	a	c
DBA/2DeJ	b	b	a	b	b	+	a	a	d	b	b	a	c
DBA/2WyDi	b	b	a	b	b	+	a	a	d	b	b	a	c
CBA/J	b	b	a	b	a	rd	a	b	d	b	b	b	c
CBA/CaJ	b	b	a	b	b	+	b	b	d	b	b	b	c
CBA/CaGn-se^{Ca}	b	b	a	b	b	+	b	b	d	b	b	b	c
C3H/HeJ	a	b	a	b	b	rd	a	b	d	b	b	a	c
C3HeB/FeJ	a	b	a	b	b	rd	a	b	d	b	b	a	c
C3HeB/Hu-Sl^{J}	a	b	a	b	b	rd	a	b	d	b	b	a	c
BALB/cJ	a	a	a	b	a	+	a	a	d	b	b	a	a
BALB/cBy	a	a	a	b	a	+	a	a	d	b	b	a	a
BALB/cScHu	a	a	a	b	a	+	a	a	d	b	b	a	a
A/J	a	b	a	b	a	+	a	a	d	b	b	a	c
A/HeJ	a	b	a	b	a	+	a	a	d	b	b	a	c
A/WySn	a	b	a	b	a	+	a	a	d	b	b	a	c

TABLE 3. Alleles at Polymorphic Loci in Sublines of Four Unrelated Groups

Chromosome	1	1	4	4	5	5	6	7	7	8	8	9	11
Gene Strain	Id-1	Dip-1	Pgm-2	Gpd-1	Pgm-1	rd	Ldr	Gpi-1	Hbb	Es-1	Es-2	Mod-1	Es-3
CE/J	a	b	a	a	a	+	a	a	s	b	b	a	c
CE/WyDi	a	b	a	a	a	+	a	a	s	b	b	a	c
MA/J	a	b	a	b	b	+	a	a	s	b	b	a	c
MA/MyJ	a	b	a	b	b	+	a	a	s	b	b	a	c
129/Sv	a	b	a	a	a	+	a	a	d	b	b	a	c
129/J	a	b	a	a	a	+	a	a	d	b	b	a	c
SJL/J	b	b	a	b	b	rd	a	a	s	b	b	a	c
SH2	b	b	a	b	b	rd	a	a	s	b	b	a	c
SC1	b	b	a	b	b	rd	a	a	s	b	b	a	c
SC2	b	b	a	b	b	rd	a	a	s	b	b	a	c
SL1	b	b	a	b	b	rd	a	a	s	b	b	a	c
SL2	b	b	a	b	b	rd	a	a	s	b	b	a	c

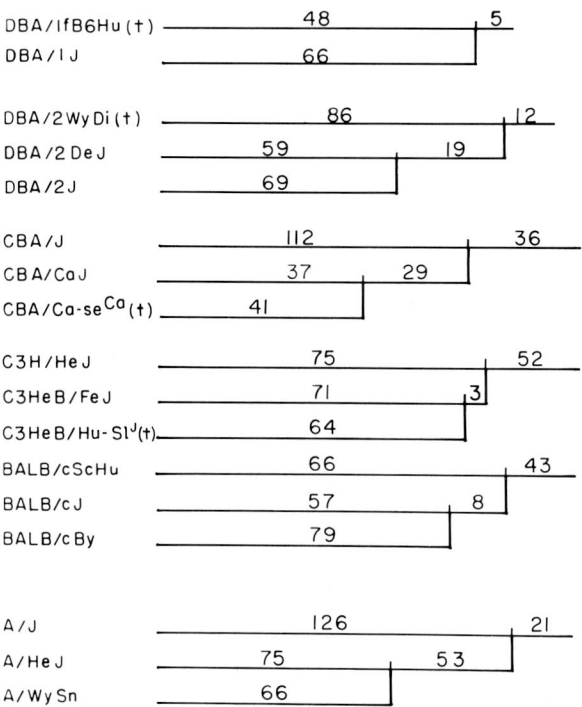

FIGURE 2. The relationships of strains of the DBA, BALB/c, A, and related strains. See caption for Figure 1.

DISCUSSION

The usefulness of each of the 13 loci studied depends on their respective degree of polymorphism in feral populations and among commonly used laboratory strains. For instance, a locus which displays a variant only rarely in feral populations or only in one or two laboratory strains would be less useful for this analysis than a locus where two alleles were in approximately equal frequencies among feral mice and among inbred strains. This reservation would pertain to analyses of both residual heterozygosity and mutation using this approach. Thus, loci Pgm-2, Ldr, Es-1, and Es-2 are less useful in differentiating strains than the other nine loci (2).

The data in Tables 1 and 3 support the concept of consistency of sublines even as distantly separated as 200 generations. The genetic similarities between the major groups

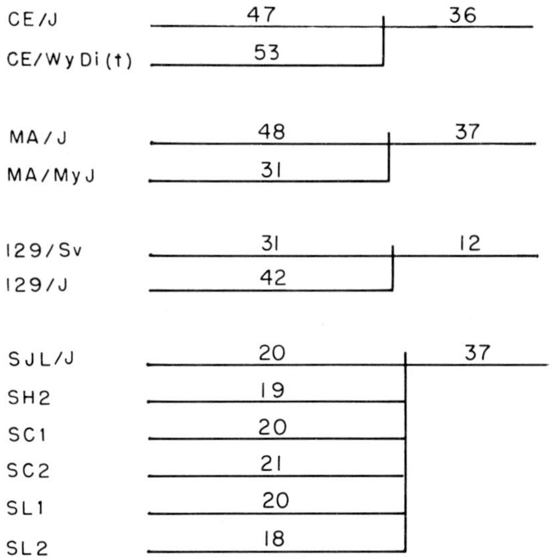

```
CE/J           ____47____|___36___
CE/WyDi(t)     ____53____|

MA/J           ____48____|___37___
MA/MyJ         ____31____|

129/Sv         ____31____|___12___
129/J          ____42____|

SJL/J          ____20____|___37___
SH2            ____19____|
SC1            ____20____|
SC2            ____21____|
SL1            ____20____|
SL2            ____18____|
```

FIGURE 3. Relationships of sublines within four unrelated groups of strains. See caption for Figure 1.

support the reality of their relationships as given by their purported pedigrees and further support the concept of genetic homogeneity and stability. THese data confirm the close relationship of C57BL/Ks with other C57BL lines. This is in contrast to the data on histocompatibility differences shown by Graff (7) and leaves unresolved the question of origin or genetic contamination of this strain.

Table 2 provokes discussion. CBA/J differs from its related CBA/Ca lines at two loci, Pgm-1 and rd. And yet, CBA/J is identical with the C3H lines at these two loci as well as all others studied. Dr. Robert S. Bader who studied dentition patterns in these strains in the 1960s said that without doubt CBA/J and CBA/CaJ had similar dentition patterns suggesting their close relationship, but he did not mention his comparison of these patterns with C3H (Bader, personal communication). Green and Kaufer (8) found histocompatibility and other differences between CBA and CBA/Ca lines and stated that it would be advisable to regard the CBA strain as composed of at least two major sublines which should not be regarded as the same strain. Our data support their findings and recommendation. It is difficult to believe that residual heterozygosity would account for the differences at these two loci and at histocompatibility loci and for major differences in radiation sensitivity (cited in ref. 8). We interpret these data as supporting the hypothesis that the CBA line was at some time during its propagation in the United States since 1933 (the time of separation of the lines) inadvertently crossed with C3H with which it has often been housed, and with which it is outwardly identical.

Table 2 also reveals that DBA/1 and DBA/2 lines differ at the Gpd-1 locus. This is not surprising since these lines were not derived from a common inbred pair. We would interpret this difference as due to residual heterozygosity.

Sublines are for most experimental purposes a nuisance because repeatability of experimental results depends in part on repeatability of the genotype of the experimental animal. So there are continuing attempts to minimize the number of sublines and the distances of relationships of experimental animals of a single inbred strain. With this paper, I wish to point out that there are good reasons for propagating at least some distantly related sublines. First, the analysis of subline differences to determine what heterogeneity persists or is evolving between lines is important to our concept of appropriate experimental animals to use. We need more information. If differences between sublines do appear, one can sometimes, especially with more than two branching sublines, determine from the pattern of alleles whether the variation is due to mutation or to residual heterozygosity. Thus, it is important to compare our observations of subline differentiation with Bailey's (1) theoretical expectations.

Second, one can use distantly related sublines for estimations of an <u>upper</u> limit of mutation rates under the assumption that most differences will be due to mutation and not residual heterozygosity. The denominator for estimating the rate for any locus can be based on a function of the number of individuals in each pedigree recognizing that new mutations, if neutral, will be fixed approximately one fourth of the time in a pedigree of sib-matings.

And third, if mutations do appear between closely related sublines, one can use the differing stocks in much the same way that congenic stocks have been used. Their utility is especially valuable in understanding the effects of a single locus without the problem of additional background genetic heterogeneity.

REFERENCES

1. Bailey, D.W. (1978). In "Origins of Inbred Mice" (H.C. Morse III, ed.). Academic Press, New York.
2. Roderick, T.H., Ruddle, F.H., Chapman, V.M., and Shows, T.B. (1971). Biochem. Genet. 5:457.
3. Staats, J. (1976). Cancer Res. 36:4333.
4. Staats, J. (1977). "Inbred Strains of Mice No. 10." (This publication is issued as a companion to <u>Mouse News Letter</u> and has appeared revised every other year since 1961. The earlier issues are indispensable for constructing pedigrees of inbred strains of mice).
5. Green, E.L., and Les, E.P. (1964). Genetics 50:497.
6. Staats, J. (1966). In "Biology of the Laboratory Mouse" (E.L. Green, ed.), p. 1. McGraw-Hill, New York.
7. Graff, R.J. (1970). Transplant. Proc. 2:15.
8. Green, M.C., and Kaufer, K.A. (1965). Transplantation 3:767.

Wild Mice: Classification, Protein Polymorphisms

Michael Potter, Chairman

COMMENTS ON THE RELATIONSHIP OF INBRED STRAINS
TO THE GENUS MUS

Michael Potter

National Cancer Institute
National Institutes of Health
Bethesda, Maryland

It is difficult with existing data to trace the origins of the vast majority of inbred strains of Mus musculus to feral or commensal wild populations. There are some clear examples where wild mice were introduced into existing colonies. Abbie Lathrop and Leo Loeb, in 1918 (1), for example obtained wild mice from Vermont and Michigan and bred them to other mice in the famous collection maintained at Granby, Massachusetts. Lathrop and Loeb, in 1915, imported mice from Europe and purchased others from dealers and made many matings between different stocks (2). They supplied mice to the Bussey Institute and were thus a source of the ancestors of some of the inbred strains of today. For example, ♂ C52 and ♀ C57 and C58 in Little's laboratory at Cold Spring Harbor were descended from mice raised at the Granby mouse farm. The Lathrop colony then may well be the American mouse melting pot, a possibility that adds further obscurity to tracing the origins of inbred strains. The records from other sources documenting the domestication of wild mice are also lacking. Keeler (3), in his classic monograph on the history of the laboratory mouse, cites extensive evidence of repeated domestications of mice in both Europe and Asia over a period of several centuries. C.C. Little was apparently also involved in importing mice from distant geographic localities into laboratory colonies (4, 5). Mouse fanciers no doubt added to the mixing of mice of different origins and, to complicate matters even more, there were exchanges of fancy mice between Europe and Asia in the last century. All this indicates the ancestors of many of the present day strains of mice may have originated from different geographic areas and no doubt from different subspecies.

Subspecies mixing can then bring together genes that have evolved away from each other over an extensive period of evolutionary time. While such a process may not affect viability under the conditions of domestication, it could lead to unnatural combinations of characters. Recent evidence indicates that C-type RNA (oncornaviruses or retroviruses) are integrated into the mouse genome at different chromosomal sites in different inbred strains (6). If this process is recent, i.e., occurs post-speciation or is continually occurring, then mixing of genomes could also bring together novel viral genomes.

The almost hopeless mixing of ancestors of most of the existing inbred strains today makes it difficult to trace the origins of specific genes or gene complexes. It will be of value to those interested in evolutionary problems to establish new inbred strains from taxonomically identified subspecies of Mus musculus. A number of potentially interesting stocks have been introduced into the laboratory from the wild and these should be very useful sources of new inbred strains. M.E. Wallace of the University of Cambridge has established stocks from Peru, San Francisco, and Skokholm Island (7). A number of strains of Mus musculus molossinus have been established in Japan (8). And in the United States,

Mus musculus molossinus (imported by the author from Kyushu) has been inbred at the Jackson Laboratory (T. Roderick, personal communication) and partially inbred at the National Cancer Institute. In addition, Mus musculus castaneus from Bangkok and Mus musculus musculus from Denmark (see Chapman, this volume) have been established and some inbreeding is in progress.

Other species in the genus Mus that are close relatives of Mus musculus add another dimension to the understanding of the species. The taxonomic of the genus Mus is still incomplete, and based mainly on morphology. The greatest concentrations of Mus are in Southeast Asia, Malasia, New Guinea, and Central and East Africa (15). The ancestors of these two species concentrations probably originated in Southeast Asia, and migrated into new regions, e.g., Africa and island populations in the East Indies occurred millions of years ago. Now, parallel diversifications make the study of the relation of the species more complex.

Insights into relationships between evolving species in the genus Mus and subspecies of Mus musculus are being provided by advances in karyology, DNA hybridization, and the comparative primary and antigenic structure of homologous proteins. The species in the genus Mus may offer unusual opportunities. The discovery of Robertsonian fusion in the pgymy mice of Africa (the superspecies Mus Leggada) by Matthey (9, 10) and in Mus poschiavinus from the Swiss Alps by Gropp and colleagues (11) has revealed that species are developing in these areas.

Greater insight into the interrelationships of the various species of Mus will no doubt evolve from the newer morphologic methods of chromosome identification. Knowledge of the relationships between species in Africa and Asia has proceeded rapidly from the work of Matthey and Petter (12), Petter and

Genst (13), Jotterand (14), and Missone (15). New interest is developing in the speciation of the genus Mus in Europe (see Marshall and Sage in this volume).

Recent progress in studying Southeast Asian mice has been stimulated largely through the efforts of Joe T. Marshall who trapped, identified, and sent breeding stocks of different Mus species from Thailand to various workers in the U.S. The work of Joe Marshall has greatly enriched our biological source of information on the relatives of Mus musculus by providing closely related but genetically isolated species such as Mus caroli and Mus cervicolor. Mus caroli and Mus cervicolor do not breed with Mus musculus nor with each other. All three species have 40 telocentric chromosomes (16). A study of the relationship of enzymes, immunoglobulins, hemoglobins, and histocompatibility antigens in these three species should provide great insight into the origins of M. musculus. Mus cervicolor and Mus caroli have provided new types of retroviruses (see Callahan, this volume). A classification of the genus Mus in Southeast Asia, based on extensive field trips and taxonomic studies, has been published by Joe Marshall (16).

A list of species in the genus Mus is given in Table 1 based on descriptions of Marshall (16), Matthey (10), Matthey and Petter (12), Petter and Genst (13), and Jotterand (14). Since the author is not a systematist, this list is presented to provide an indication of the size of the genus, and not as a proposed classification.

Mus musculus molossinus (Kyushu)

Our interest in the Mus subspecies problem emerged in studies on immunoglobulin allotype genetics. Following the discovery of Kelus and Moor-Jankowski (17) that there were strong antigenic differences associated with heavy chain gene products (allotypes) in different strains of mice, Herzenberg

TABLE 1. Subspecies of Mus musculus (4)

Wild form	Commensal	Geographic distribution
M. m. wagneri (Eversmann)	M. m. brevirostris	Europe
	M. m. domesticus	
	M. m. orientalis	No. Africa
	M. m. bactrianus	Russia, Afghanistan
	M. m. praetextus	Turkey, Iran
	M. M. tytleri	India
	M. m. homourus	India
	M. m. urbanus	India
	M. m. castaneus	India
M. m. spicelegus	M. m. musculus	Europe
M. m. manchu	M. m. molossinus	Japan
M. m. spretus		Europe

(18) and Lieberman and Dray (19) described a number of Ig allotypic markers in mice. Using myeloma proteins, it was possible to assign these antigens to specific C_H genes. This study soon revealed that the IgC_H genes (there are now eight different C_H genes in the mouse) were closely linked together on one chromosome (18, 20, 21). Allelic forms of six of the genes are now known (34). Genetic studies indicated that the C_H genes were closely linked together and that crossing over was not observed in samples of about 2,000 progeny (20, 21). Because of this close linkage and lack of crossing over, we decided to look at wild mouse allotypes for evidence of recombinations. We found a possible case in a wild mouse population

trapped in Kitty Hawk, North Carolina, where two markers always found separately in inbred strains were in the same haplotype and probably the same structural gene (IgC_HG2a) (22) in these wild mice. In this study of wild mice we obtained specimens from many different sources throughout the U.S. and even some mice from Europe (22, 23). Among the strong allotype markers that have subsequently been shown to be alleles of IgG2a we did not find the strong allotype "2" (G^2) in U.S. or European samples. (The "2" marker is found in strains C57BL, HR, LP, SJL/J, SM and STR/1.) The possibility that the "2" marker might be associated with Asian mice was suggested by available circumstantial evidence.

First, it is a well documented fact that the Japanese waltzing mouse (a fancy mouse) was imported into Europe from Asia probably during the last century (3). Japanese waltzing mice were used in a number of laboratories in the U.S. in the early part of this century. Lathrop and Loeb (1) had Japanese waltzing mice in their colony and mated these mice to various other stocks and strains. Japanese waltzing mice were smaller than the usual laboratory mice and were suspected of being a different subspecies. They were first thought to be Mus wagneri (the wild mouse from Northern Asia, China, Tibet, etc.). Schwarz (24), however, in 1942 argued that these mice were not of Chinese but were in fact of Japanese origin and hence derived from the wild M. m. molossinus (the common Mus musculus species in Japan). Through the kindness of Dr. Hiroshi Kobayashi of the University of Hokkaido, the author contacted Professor Fusanori Hamajima from the Department of Parasitology, Kyushu University. Professor Hamajima had made extensive studies of the natural habitats of the native Japanese rice-field mouse, Mus musculus molossinus (25). These mice live in straw stacks in barns or other human constructions in the winter and migrate into the rice paddies in the summer where they live in the banks of the paddy fields. They have a restricted breeding season in the spring and autumn (25). Professor Hamajima very graciously trapped 11 mice and sent them to me in Bethesda where my wife and I bred them in the basement of our home before they were introduced into the laboratory. The original shipment which arrived in 1967 contained 10 very small light black agouti mice, several of which had snowy white bellies, and one large mouse. The identity of the large mouse was never established and all of the matings were made with the small mice. These mice have bred vigorously. They had the singular honor of being imported to Bar Harbor, Maine, where they were introduced into the Jackson Laboratory by Dr. Thomas Roderick. Roderick et al. (26) studied the ratio of the weight of the brain to the spinal cord in inbred strains of mice and M. m. molossinus and found that M. m. molossinus has the highest value.

The immunoglobulin allotypic analysis of the M. m. molossinus mice proved to be most interesting (23). First, we found the missing "2" marker in these mice. Second, all of the other allotypic markers associated with the IgC_H complex locus of the C57BL were also identified. Third and most unusual was the presence of a marker associated with IgG2a of the BALB/c mouse, the $G^{1,6,7,8}$. Immunoelectrophoretic studies indicated that the Ig carrying the $G^{1,6,7,8}$ and G^2 were on different molecules, indicating the M. m. molossinus (Kyushu) locus carried an extra gene.

	G2a	G2b	G1	A
C57BL	G^2	$H^{9,16}$	F^s	A^{15}
M. m. molossinus	$G^{1,6,7,8}, G^2$	$H^{9,16}$	F^s	A^{15}
BALB/c	$G^{1,6,7,8}$	$H^{9,11}$	F^f	$A^{12,13,14}$

The novel IgC_H locus could then have evolved by unequal crossing over. This complex haplotype has not been found in any other inbred strain.

We have studied another marker in M. m. molossinus that has proved to be equally extraordinary and this is the electrophoretic pattern of the major urinary protein complex (MUP). As has been known for many years (27), mice excrete a large amount of highly acidic proteins in the urine (0.5 to 1.5 mg/ml). These proteins have a molecular weight of 17,900 and exhibit considerable electrophoretic heterogeneity (28). There is a sex dimorphism in the pattern so that males and females of the same strain can often be distinguished (28). The sex pattern can be reversed by gonadectomy and treatment with appropriate hormones. The mouse urinary proteins (MUP) are produced in the liver (29). Their exact function has not yet been determined although some recent evidence by Ruoslahti and colleagues (30) suggests MUP may be a nuclear non-histone protein. Two major phenotypic forms of one of the components controlled by the MupA locus, the 1 and 2 components, are apparently allelic variations (28). All of the inbred strains so far studied are either of the "1" type (the most common) or the "2" type. When M. musculus molossinus urine was electrophoresed, a completely new type was seen that had much slower mobility than any component seen in the inbred strains (31) (Fig. 1). Mus musculus castaneus trapped by Joe Marshall was also studied and these mice have the same new slow component and still another new type. Dev et al. (32) have compared the chromosomes of M. m. molossinus with C57BL by Quinacrine (Q) and Centromeric, heterochromatin (C) banding patterns. They found no differences in Q banding but very different C banding patterns. The latter were much greater than found among inbred strains (33) providing further evidence of genetic distance between M. m. molossinus and laboratory mice. Genetic differences between subspecies may be more extensive than simply an accumulation of allelic forms. Subspecies could differ from each other,

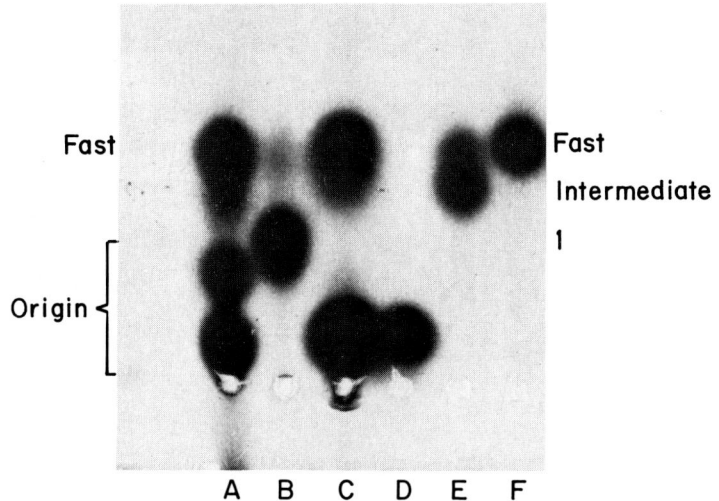

FIGURE 1. Agar gel electrophoresis of the nondialyzable material from urine of (A, B) <u>Mus musculus molossinus</u>. (C, D) <u>Mus musculus castaneus</u>, (E, F) <u>Mus caroli</u>. (G) BALB/c. Samples B, D, and F were from female mice, the others from males. Anode is at top.

by having structurally different complex loci. The variation of IgC_H (23), MUP (31), brain:spinal cord weight ratios (26) and C-banding (32, 33) suggests this possibility. Retrovirus integration (6) is yet another basis for a difference, however, retrovirus loci have not yet been mapped in subspecies such as <u>M. m. molossinus</u>.

CONCLUSIONS

The remarkable advances that are being made in the descriptive genetics of the laboratory mouse make it very desirable to understand this species as it naturally evolved. The inbred strains of mice that are in common use today are probably derived from several subspecies backgrounds and thus cannot be used to trace gene origins. The construction of

TABLE 2. Some Recent Data on the Geographical Origin, Karyotypes of Species, and Subspecies in the Genus Mus

Geographic location	Genus Mus subgenera	Karyotypic data[b]	References
Southeast Asia	Pyromys (spiny mice)	M. shortridgei[c] (46-49) M. saxicola (22) (24-26) M. platythrix (26) (30) M. phillipsi M. fernandoni	16
	Coelomys (shrew mice)	M. mayori M. pahari (48)[c] M. famulus M. crociduroides M. vulcani	16
	Mus (house, ricefield mice)	M. caroli (40)[c] M. cervicolor (40)[c] M. cookii (40)	16 16
India		M. booduga/dunni (40)[c]	12, 16
Europe, Asia, Africa		M. musculus (40)[c]	
Europe		M. spretus (hispanicus) ?	d
Europe		M. poschiavinus (26)	11
Africa	M. (Leggada)		10
	Primitive Type	M. minutoides f.c.1 (36/36)[b] M. minutoides f.c.5 (34/36) M. minutoides indutus (36/36) M. setulosus (36/36) M. tenellus (36/36)	10, 14
	Translocation Type[a]	M. minutoides f.c.1 M. minutoides f.c.6 (18/36) M. minutoides d.c.7 (22/36) M. minutoides minutoides (18,19/36) M. minutoides musculoides (31,32, 33,34/36) M. minutoides bellus (30/32) M. triton f.c.1 (32/34, 32/33) M. triton f.c.2 (20,21/32)	10, 14
		M. oubanguii (28) (F.N. = 30 to 34)	13, 14
		M. goundae (16,17,18,19) F.N. = 30	13, 14

[a] X and Y chromosomes are translocated a pair of autosomes.

[b] Fractional numbers in parentheses = diploid number/F.N. (fundamental number of the female). The lower diploid numbers reflect the process of Robertsonian fusion.

[c] Mice currently available in the U.S.

[d] See Sage this volume.

new inbred strains, or the use of stocks from taxonomically identified subspecies from different geographic locations might provide a useful source of information on the origin of alleles and complex genetic loci. This information should allow us to better understand the flexibility of a single genome throughout millions of years of evolution. The systematics of the living species of the genus Mus are still incomplete. Many exciting new advances have been made in the last 25 years, promising that we shall better understand these fascinating animals. This is a time when species of Mus and subspecies of Mus musculus are being adapted to the laboratory from many parts of the world.

Some unusual findings with the IgC_H complex locus and the MUP complex locus in Mus musculus molossinus (Kyushu) were summarized.

REFERENCES

1. Lathrop, A.E.C., and Loeb, L. (1918). J. Exp. Med. 28: 475.
2. Lathrop, A.E.C., and Loeb, L. (1919) J. Exp. Med. 22: 713.
3. Keeler, C.E. (1931). "The Laboratory Mouse: Its Origin, Heredity and Culture." Harvard University Press, Cambridge.
4. Schwarz, E., and Schwarz, H.K. (1943). J. Mammology 24:59.
5. Green, C.W. (1930). Am. Nat. 64:540.
6. Rowe, W.P., Hartley, J.W., and Bremner, T. (1972). Science 178:860.
7. Wallace, M.E. (1971). Environ. Pollut. 1:175, Elsevier, England.

8. Nishimura, M., Kondo, K., Nakamura, H., and Watanabe, T. (1973). In "Exp. Animals" Vol. 22. Suppl. Proceedings of the ICLA Asian Pacific Meeting of Laboratory Animals, Tokyo and Inuyama.
9. Matthey, R. (1966). Revue Suisse Zool. 73:585.
10. Matthey, R. (1970). Revue Suisse Zool. 77:625.
11. Gropp, A., Winking, H., Zech, L., and Müller, H. (1972). Chromosoma 39:265.
12. Matthey, R., and Petter, F. (1968). Revue Suisse Zool. 75:461.
13. Petter, F., and Genst, H. (1970). Mammalia 33:118.
14. Jotterand, M. (1972). Revue Suisse Zool. 79:287.
15. Missone, X. (1969). Ann. Musee Royal Afr. Central 172:1.
16. Marshall, J.T., Jr. (1977). Bull. Am. Mus. Natl. Hist. 158, Art. 3:177.
17. Kelus, A., and Moor-Jankowski, J.K. (1961). Nature 191:1405.
18. Herzenberg, L.A. (1964). Cold Spring Harbor Symp. Quant. Biol. 29:455.
19. Lieberman, R., and Dray, S. (1964). J. Immunol. 93:584.
20. Potter, M., and Lieberman, R. (1967). Cold Spring Harbor Symp. Quant. Biol. 32:187.
21. Potter, M., and Lieberman, R. (1967). Adv. Immunol. 7:91.
22. Lieberman, R., and Potter, M. (1966). Science 154:535.
23. Lieberman, R., and Potter, M. (1969). J. Exp. Med. 130:519.
24. Schwarz, E. (1942). Science 95:46.
25. Hamajima, F. (1962). Science Bull. of the Faculty of Agriculture, Kyushu Univ. 20:61 (in Japanese).
26. Roderick, T.H., Wimer, R.E., Wimer, C.C., and Schwartzkroin, P.A. (1973). Brain Res. 64:345.
27. Finlayson, J.S., and Baumann, C.A. (1958). Am. J. Physiol. 192:69.

28. Finlayson, J.S., Potter, M., and Runner, C.C. (1963). J. Natl. Cancer Inst. 31:91.
29. Finlayson, J.S., Asofsky, R., Potter, M., and Runner, C.C. (1965). Science 149:981.
30. Ruoslahti, E., Engvall, E., Jalanko, H., and Comings, D.E. (1977). J. Exp. Med. 146:1054.
31. Finlayson, J.S., Potter, M., Shinnick, C.S., and Smithies, O. (1974). Biochem. Genetics 11:1974.
32. Dev, V.G., Miller, D.A., Tantravahi, R., Schreck, R.R., Roderick, T.H., Erlanger, B.F., and Miller, O.J. (1975). Chromosoma 53:335.
33. Miller, D.A., Tantravahi, R., Dev, V.G., and Miller, O.J. (1976). Genetics 84:67.
34. Lieberman, R. (1978). Springer Seminars in Immunopathol. 1:7.

BRIEF REVIEW OF EUROPEAN HOUSE MICE

Joe T. Marshall

National Fish and Wildlife Laboratory
National Museum of Natural History
Washington, D.C.

The house mouse, Mus musculus, is widely distributed throughout Eurasia and northern Africa where it displays spectacular geographic variation in color and proportions as an adaptation to regional climates and soils. Its geographic races contain alleles beyond the wildest expectations of those geneticists familiar only with the albino. Now that the fascinating wild forms are being colonized in laboratories, the reasonable demand to know their scientific names is embarrassing the museum taxonomist, who must admit that the nomenclature of the species is utterly chaotic. I have dealt with the relatively simple situation in Asia (1), where the only trouble was the confounding of Mus caroli with M. musculus in much of the literature [stemming from Allen (2)]. Variation in Asian M. musculus follows geographic clines and the one subspecies carried by commerce, Mus musculus castaneus, has colonized cities in precisely that area lacking native outdoor populations of house mice. Even in Nepal [contra Schwarz and Schwarz (3)] the dark-bellied castaneus of Kathmandu is not in extended contact with white-bellied M. m. homourus. The latter lives at a much higher altitude, the only outdoor mouse around Kathmandu being the unrelated Mus cervicolor.

In Europe and around the Mediterranean it appears that house mice have been carried by commerce, and have secondarily invaded areas already occupied by "country mice." Perhaps the newcomers are able to gain a foothold only because they start out in buildings. The result of interbreeding between city and country mice is a bewildering polymorphism and local variation in coloration and tail length that has spawned many scientific names and endless taxonomic confusion. Furthermore, the literature on European mice indicates that it is possible to find two recognizably different kinds of house mice, one in the houses and the other in the adjacent fields.

I have been asked by this symposium to evaluate the literature and nomenclature of European house mice. Although I have examined many specimens, including types of these European taxa, I have no field experience with wild (outdoor) kinds of house mice. Furthermore, of the hundreds of museum specimens of European house mice I have examined, no more than a handful are furnished with the vital information on their labels as to whether they were caught inside or outside of the house! It seems we must start all over again by collecting properly authenticated specimens.

The best place to start, since the theories of Schwarz and Schwarz (3) have been discredited [by Jones and Johnson (4), Setzer (5), Ranck (6), and Cockrum (7)], is with the excellent study by Zimmermann (8). Zimmermann shows that the Elbe River separates two entirely different subspecies of the house mouse. To the east and north a small, short-tailed, white-bellied Mus musculus musculus is capable of completing its annual cycle entirely in the fields. West from the Elbe is the large, long-tailed, dark-bellied Mus musculus domesticus confined to houses. The need now is for someone to extend the work of Klaus Zimmermann outside the boundaries of Germany. Here I shall report on a brief look at some of the specimens that point to the direction such study might take.

First, I believe I have discovered a skull trait that distinguishes pure Mus musculus domesticus. It is the straight-edged, narrow zygomatic plate (Fig. 43 in ref. 1) which I contrasted with that of the Asiatic assemblage (Fig. 46). I was incorrect in assigning the straight-edged plate to the entire musculus (European) group; I now see that most pure populations have a plate like that of castaneus. Tendencies toward straightness of the zygomatic plate might be useful, therefore, along with long tail, dark belly, and large size in assessing the amount of admixture of domesticus into a wild population. Such a problem can be much better solved genetically, of course, but morphological characters are necessary for use with museum voucher specimens. [I am unable to detect the slightest hint of reduction in molar surface and muzzle length, claimed as specialization of indoor mice by Schwarz and Schwarz (3)].

Tables 1 and 2 present a series of samples of house mice. Summarizing, the greatest morphological contrast between supposed outdoor and indoor mice is in the north. There it should be easiest to detect mixing: the indoor mouse of Germany is much darker on the back, belly, and feet, and is considerably larger and longer-tailed than its contrastingly pale, small, short-tailed, white-footed outdoor relative in Uppsala, Sweden (the type locality whence Linne named the taxon). Southward through France and into Spain, the "indoor" form gets much paler, and contrasts only slightly with the outdoor taxon, spretus. Nevertheless, spretus maintains a striking individuality separate from

TABLE 1. Some Average Measurements of "Pure" Populations of Western House Mice

	Underparts	n	Skull	Rostrum	Toothrow	Head and body	Tail	Hind foot	Ear	Weight
Spain	dark	5	21.44 ± 0.36	5.02 ± 0.08	3.26 ± 0.11	84.8 ± 1.3	82.8 ± 5.2	19.0 ± 1.2	15.0 ± 0.8	
Germany	dark	14	21.80 ± 0.60	5.13 ± 0.24	3.31 ± 0.11	90.2 ± 7.4	89.5 ± 5.9	18.6 ± 1.0	13.7 ± 0.5	
Sweden	pale	11	20.62 ± 0.53	4.65 ± 0.23	3.26 ± 0.12	82.5 ± 3.7	66.7 ± 5.2	18.1 ± 1.2		
Spain	pale	19	20.41 ± 0.69	4.81 ± 0.28	3.24 ± 0.13	77.0 ± 6.4	62.4 ± 7.4	16.5 ± 0.9	13.7 ± 0.8	
Libya	white	8	21.51 ± 0.53	5.20 ± 0.23	3.40 ± 0.13	76.0 ± 5.7	80.4 ± 7.8	18.9 ± 0.8	?18.0 ± 2.9	
Tunisia	pale	20	21.29 ± 0.76	5.02 ± 0.37		80.5 ± 4.8	64.5 ±10.9*	16.1 ± 1.0	13.4 ± 1.0	13.7 ± 2.9
Iran	white	10	21.72 ± 0.86	5.40 ± 0.37	3.35 ± 0.08	82.6 ± 6.5	85.7 ± 6.0	18.8 ± 0.8	14.0 ± 0.9	

*Notice the great variation in tail length of this mixed population.

TABLE 2. Prevailing (Majority) Coloration of Samples of House Mice

Place	Habitat	Dorsal	Ventral	Feet	Zygomatic plate	Racial allocation
Spain	indoor?	sandy brown	brown/gray	various	various	mixed
France	indoor?	brown	brown/slate	brown	various	domesticus
Germany: west of Elbe	indoor?	fuscous	brown/slate	dark brown	straight-edged	pure domesticus
Sweden: Uppsala	outdoor?	brown	buff/gray	white	convexly curved	pure musculus by definition
Kaliningrad	outdoor?	brown	gray	white	convex	mostly musculus
France	"habitations" for one	brown	buff/gray	white	various	mixed (long tails)
Spain	outdoor?	sandy brown or sandy ochre	gray	white	convex	pure spretus
Italy, Sicily, Greece		(various, but includes many sandy brown)			various	mixed
Libya	vegetation near oases	sandy brown	white/gray	white	convex	praetextus
Tunisia	Atlas foothills	(various, but mostly pale)			various	mixed
Iran: Baluchistan	buildings	sandy yellow	white	white	various	bactrianus

presumed contiguous indoor mice, unlike its neighbors in Italy, Sicily, and Greece. Dr. Sage is going to give you a genetic and ecologic explanation of this fact. So a pure Hispanic population is a cornerstone of any genetic study of the European house mice.

Nevertheless, we must build a genetic base with the two most contrasted populations, those with the least complications, namely the pure, outdoor musculus of Sweden and indoor domesticus of Germany west of the Elbe. Also, the Hungarian "hillock mouse" or "mound-building mouse," hortulanus, should receive attention because of its reputed unique social organization and behavior (W. Prychodko in litt.).

No matter what genetic triumphs will supersede the pedestrian efforts of museum taxonomists, I must insist that without a few identifiable voucher specimens (skin with perfect skull) your work is meaningless zoologically. A glance at my Asian mouse paper (1) will show you how names change and change until somebody examines the type-specimen. To cite a mere scientific name in a controversial genus like Mus is almost meaningless. You have to say, "This is a virologic study of Mus caroli from Okinawa, the species with long tail, short nasals, and brown, pro-odont incisors represented by my specimen number 593 in the Smithsonian Institution." With Mus musculus you would have to state also whether your colony or specimens came from a field, house, or natural vegetation.

REFERENCES

1. Marshall, J.T., Jr. (1977). Bull. Amer. Mus. Nat. Hist. 158:175.
2. Allen, G.M. (1927). Amer. Mus. Novitates No. 270, pp. 1-12.

3. Schwarz, E., and Schwarz, H.K. (1943). J. Mammalogy 24:59.
4. Jones, J.K., Jr., and Johnson, D.H. (1965). Univ. Kansas Publ. Mus. Nat. Hist. 16:357.
5. Setzer, H.W. (1957). J. Egyptian Public Health Assoc. 32:41.
6. Ranck, G.L. (1968). Smiths. Inst., U.S. Nat. Mus. Bull. 275:264.
7. Cockrum, E.L. "Mammals of Tunisia" (ms).
8. Zimmermann, K. (1949). Zool. Jahrb. f. Systematic 78:301.

GENETIC HETEROGENEITY OF SPANISH HOUSE MICE
(MUS MUSCULUS COMPLEX)[1]

Richard D. Sage

Museum of Vertebrate Zoology
Berkeley, California

In 1943, Schwarz and Schwarz (1) published a summary of their extensive studies on the systematics of the house mouse, Mus musculus L. In their view, the house mouse was one species comprised of several recognizable subspecific units. These subspecies exhibited distinctive morphological features and ecological behavior. The biogeographical history of these units was seen as a series of wild forms giving rise to populations of commensal types associated with the first agricultural civilizations of man that developed throughout the Palearctic region. This interpretation was accepted for many years, in spite of the problem in Europe where often two phenotypically distinctive types (i.e., subspecies) of mice can be found living in the same area (2, 3). The sympatry of subspecies runs counter to the modern paradigm of the biological-species concept (4). Under this construct, subspecies are considered as geographically separated populations of individuals that are

---

[1] The National Science Foundation provided funds for the laboratory work through grant DEB 72-02545.

reproductively compatible among themselves. Thus discrete phenotypic classes in one region would not be expected because of the dilution effects of interbreeding. An alternate, but unlikely, explanation for the presence of two sympatric phenotypes of one species might be the existence of an ecological and morphological polymorphism, such as described for cichlid fishes (5).

In the interest of learning more about the genetic structure of house mouse populations where two subspecies were said to coexist, I collected samples from several localities in Spain. Here M. m. brevirostris Waterhouse (1837) and M. m. spretus Lataste (1883) (following the nomenclature of 1) were said to occur in sympatry throughout most of the Iberian peninsula. In this paper I describe the results of an electrophoretic study of the proteins of these animals. Mice collected in England and animals from an inbred lab strain served as outside reference samples. A more detailed report on the morphology and karyology of the wild animals is in preparation.

Nomenclature of Spanish Mice

Electrophoretic analysis clearly demonstrated that two species of Mus are present in Spain. They live together at many localities and show no evidence of exchanging genes through interbreeding. This directly invalidates the classification scheme of Schwarz and Schwarz (1), thereby requiring a revised nomenclature in referring to these animals. For the moment the use of the name M. musculus for the commensal form is appropriate. However, the use of the name M. spretus for the aboriginal mice [terminology of Bruell (6)] in Spain does not seem to be correct. Originally these Spanish animals were described as the subspecies M. spicilegus hispanicus by Miller (7). The nominate form, M. s.

spicilegus, occurs in eastern Europe and to some unknown
distance into southern Russia [see Serafinski (8) for another opinion about these taxa). There is presently a hiatus
of nearly one thousand kilometers between the ranges of
these aboriginal mice. Zoogeographical considerations suggest that the Iberian form is unlikely to be closely related
to this more continental, interior population. Schwarz
and Schwarz (1), however, considered these animals to be
related to the form M. m. spretus of northern Algeria. Zoogeographically this inferred relationship is more probable
as many other Spanish species show similar relationships to
the North African fauna. However, the Spanish mammalogist
Cabrera, who knew the aboriginal mice in Spain to be a different species from the sympatric commensal mice, presented
evidence (9), including four features of pelage color, body
size, and dentition, that demonstrated the distinctiveness
of the Algerian spretus animals from Spanish and Moroccan
mice that he had trapped and considered closely related.
For these reasons it seems that the Spanish aboriginal mouse
should be considered a species distinct from other Mus populations to the east and south. Mus hispanicus (7) appears
to be the oldest name referring specifically to these animals, and I will employ this name in the remainder of the
paper.

MATERIALS AND METHODS

During September 1977, 88 mice were trapped at six localities (Fig. 1): 1) England, Berkshire, Abingdon, Culham
College (8 M. musculus); 2) Spain, Catalunya, Puigcerdá,
110 km NNW Barcelona (2 M. musculus); 3) Spain, Catalunya,
Balenya, 46 km NNE Barcelona (11 M. musculus, 40 M. hispanicus); 4) Spain, Catalunya, San Fausto, 10 km NE Barcelona (3 M. musculus); 5) Spain, Catalunya, Barcelona (6 M.

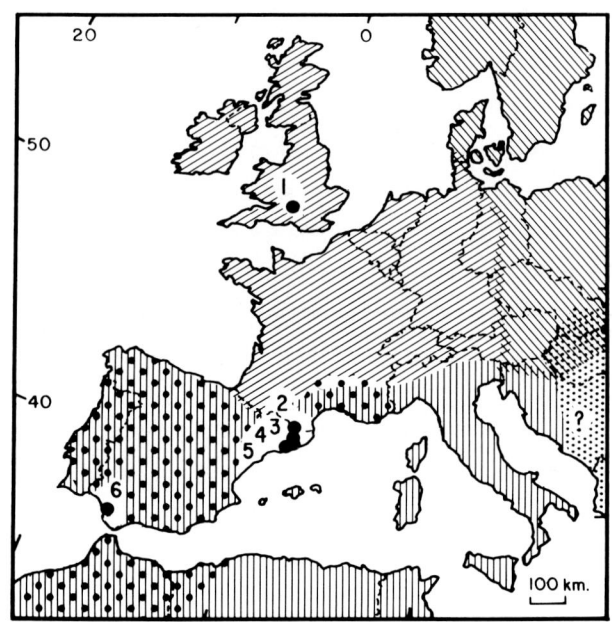

FIGURE 1. The approximate geographic limits of house mice (Mus) in western Europe and northwestern Africa (1, 35): M. hispanicus (large dots); M. m. brevirostris (vertical lines); M. m. domesticus (right-slash lines); M. m. musculus (left-slash lines); and M. m. spicilegus (small dots).

musculus); and 6) Spain, Andalucia, Puerto Real (0-16 km E), 10+ km E Cadiz (1 M. musculus, 17 M. hispanicus). Both snap traps and Sherman live traps were placed in various types of human habitations, in agricultural fields, and in natural environments. Captured animals were weighed and measured using standard procedures (10). For all individuals the heart, liver, kidneys, and a piece of leg muscle were saved. The testes and epididymides were taken from all adult males. Blood was collected in heparinized tubes from some specimens which were caught alive. Red cells were allowed to settle before removing plasma. The cells were then washed once or twice with an isotonic saline solution. Tissues and blood samples were stored during the field work in liquid nitrogen (-196°C). In the laboratory, the tissue samples and resultant extracts were maintained frozen at -76°C. Extracts were prepared by mincing the tissue with scissors, adding approximately an equal amount of deionized water to the material, and centrifuging at 27,000 x g for 40 minutes. The heart-kidneys (Ht+K) and testes-epididymides (T+E) were made into pooled extracts, but the other materials were kept as monospecific samples. Horizontal starch gel electrophoresis was carried out following the general procedures described in (11, 12). The staining recipes were similar to those described in these preceding two references. Electrostarch (O. Hiller, Madison, Wisconsin) was used throughout the study. Starch from Lot 302 was used in assaying the lactate dehydrogenase and diaphorase loci, and Lot 307 was used in the rest of the work. The buffer systems and electrophoresis conditions used are identified in Table 1.

Tissue extracts were prepared from C57BL mice obtained from the University of California Cancer Research Laboratory. The allelomorphic classes shown at all loci from these samples were used as standards of comparison for the rest of the material. Relative mobility of non-standard allelomorphs was measured by dividing the absolute mobility distance of the variant allele from the origin by the distance moved by the standard on the same gel. This quotient was used to designate all of the mobility class allelomorphs that were found during the study. An extract of liver and kidney mitochondria was prepared from C57BL mice (13) and used to identify enzymes bound to these subcellular structures.

TABLE 1. Buffer Systems and Electrophoresis Conditions Used in the Study of Spanish Mus

	Buffer type	Running conditions	Literature source
1.	Tris-maleic-EDTA, pH 7.4	100 v./4 hr.	11
2.	Tris-maleate-NaOH, pH 7.0	100 v./6 hr.	29
3.	Tris-EDTA-borate, pH 8.0	200 v./4 hr.	11
4.	Tris-EDTA-borate, pH 9.0	400 v./4 hr.	30
5.	Continuous Tris-citrate, pH 8.0	130 v./4 hr.	11
6.	Continuous Tris-citrate, pH 7.0	180 v./3-4 hr.	30
7.	Phosphate-citrate, pH 7.0	100 v./4 hr.	11
8.	Phosphate, pH 6.7	130 v./4 hr.	11
9.	Tris-hydrochloric acid, pH 8.5	250 v./1.5 hr.	11
10.	Lithium hydroxide, pH 8.1	300 v./3 hr.	11

Samples of testes, plasma, and hemolysates were unavailable for all specimens, thus loci surveyed from these tissue or blood components are represented by a smaller number of genes than stated in the sample size column of Table 3. The numbers of animals surveyed for loci from these extracts are as follows: M. musculus, locality 1 (T+E=4, P=4, H=3); locality 2 (T+E=1, P=2, H=2); locality 3 (T+E=3, P=6, H=3); locality 4 (T+E=2, P=1); locality 5 (T+E=2, P=2); locality 6 (T+E=1); and for M. hispanicus, locality 5 (T+E=13, P=5, H=5); locality 6 (T+E=4, P=3, H=8). To include the loci scored from plasma (ES_3, ES_5, PT_A) in the computations of genetic distance between populations, the gene frequencies at Pto. Real for M. musculus were assumed to be the same as at Barcelona.

Skins and skeletons of the wild animals collected for this study have been deposited in the collection of the Museum of Vertebrate Zoology. Unground tissues and extracts, representing all of the alleles that were found, will be deposited as voucher specimens in the frozen tissue collection at the same institution.

RESULTS

Twenty-three identified and three unidentified enzyme systems and seven non-enzymatic proteins, controlling a presumptive total of 56 structural gene loci, were surveyed (see Table 2). The enzyme nomenclature used in this table follows as closely as possible that of Harris and Hopkinson (12). Gene frequencies for variable loci are presented in Table 3.

Description of Selected Loci and Alleles

<u>Lactate Dehydrogenases</u>. The sperm type enzyme (LDH_C) runs cathodal to the other two lactate dehydrogenases. Variants observed at LDH_C in <u>M. hispanicus</u> are the first reported for the genus. The heterozygote at this locus shows the typical five-banded pattern characteristic of a tetrameric molecule. The tetramers at this locus do not appear to interact and form heterotetramers with the other lactate dehydrogenase isozymes.

<u>Glyceraldehyde-Phosphate Dehydrogenases</u>. The presence of mercaptoethanol and NAD is necessary in the gel to produce isozymes with good banding qualities. No variants were seen in the locus running anodally ($GAPDH_1$) in this study or in other work in progress on mice from Pacific islands. In scoring the second, cathodal locus a problem similar to that observed with the ADH isozymes is encountered, i.e.,

TABLE 2. Listing of Enzymes and Proteins Surveyed in Spanish and English Mus

Enzyme or protein systems	Locus abbrev.	Buffer system/ tissue type*	Remarks
1. Alcohol dehydrogenase (E.C.1.1.1.1)	ADH	9/L	Migrates cathodally; $\underline{Adh}^{54}$ is partially blocked by achromatic activity of SOD_B
2. Glycerol-3-phosphate dehydrogenase (E.C.1.1.1.8)	GPD	3/Ht+K	
3. Sorbitol dehydrogenase (E.C.1.1.1.14)	SORDH	6/Ht+K	
4. Lactate dehydrogenase (E.C.1.1.1.27)	LDH_A	3/Ht+K	
	LDH_B	3/Ht+K	
	LDH_C	4/T+E	Equals sperm type, LDH-X
5. Malate dehydrogenase (E.C.1.1.1.37)	MDH_S	5/Ht+K	
	MDH_M	5/Ht+K	Migrates cathodally
6. Malic enzyme (E.C.1.1.1.40)	ME_M	5/Ht+K	
7. Isocitrate dehydrogenase (E.C.1.1.1.42)	ICD_S	6/Ht+K	
	ICD_M	6/Ht+K	Migrates cathodally
8. Phosphogluconate dehydrogenase (E.C.1.1.1.44)	PGD	1/L	10 mg NADP added to gel before aspirating
9. Glyceraldehyde-phosphate dehydrogenase (E.C.1.2.1.12)	$GAPDH_1$	7/L	NAD and 2-mercaptoethanol added to gel before aspirating (31)
	$GAPDH_2$	7/L	Migrates cathodally
10. NADH Diaphorase (E.C.1.6.2.2)	DIA_1	4/T+E	
	DIA_3	4/H	
	DIA_4	4/T+E	
	DIA_5	4/H	
11. Superoxide dismutase (E.C.1.15.1.1)	SOD_A	3/Ht+K	Identified by Tegelström (32)
	SOD_B	3/Ht+K	Identified by Tegelström (32) Migrates cathodally
13. Glutamate-oxaloacetate transaminase (E.C.2.6.1.1)	GOT_S	5/Ht+K	
	GOT_M	5/Ht+K	Migrates cathodally

Table 2 (continued)

Enzyme or protein systems	Locus abbrev.	Buffer system/ tissue type*	Remarks
14. Hexokinase (E.C.2.7.1.1)	HK	3/Ht+K	
15. Phosphoglucomutase	PGM$_1$	6/Ht+K	Nomenclature of Shows et al. (15)
(E.C.2.7.5.1)	PGM$_2$	6/Ht+K	" " " " "
16. Esterases (E.C.3.1.1.1)	ES$_3$	9/H	Nomenclature of Wheeler and Selander (17)
	ES$_2$	10/P	" " " "
	ES$_1$	10/P	" " " "
	ES$_B$	10/P	" " " "
	ES$_5$	10/P	" " " "
	ES$_A$	9/H	" " " "
17. Acid phosphatase	ACP$_E$	6/L	Erythrocytic type enzyme
(E.C.3.1.3.2)	ACP$_1$	6/L	Nomenclature of Lalley and Shows (33)
18. Peptidases	PEP$_C$	10/L	Equals DIP-1 of Ruddle and Nichols (34)
	PEP$_A$	10/L	
	PEP$_B$	10/L	Equals TRIP-1 of Ruddle and Nichols (34)
	PEP$_S$	10/L	
	LAP	6/L	Using Leucyl-β-naphthyl-amide as substrate
19. Adenosine deaminase (E.C.3.5.4.4)	ADA	6/Ht+K	
20. Aconitase (E.4.2.1.3)	ACON$_S$	6/M	
	ACON$_M$	6/M	Migrates cathodally
21. Enolase (E.C.4.2.1.11)	ENO$_1$	6/Ht+K	
	ENO$_2$	6/Ht+K	
22. Mannose phosphate isomerase (E.C.5.3.1.8)	MPI	5/Ht+K	
23. Glucosephosphate isomerase (E.C.5.3.1.9)	GPI	8/L	Migrates cathodally
24. Unidentified enzymes	Enzyme-1	5/Ht+K	See Results: Nucleoside phosphorylase
	Enzyme-2	6/Ht+K	See Results: Phosphoglucomutases
	Enzyme-3	5/Ht+K	See Results: Mannose phosphate isomerase

Table 2 (continued)

Enzyme or protein systems	Locus abbrev.	Buffer system/ tissue type*	Remarks
25. Nonenzymatic proteins	ALB	10/M	Equals albumin
	MYO	10/M	Probably is myoglobin
	PT-1	10/M	Slow migrating muscle protein
	PT-2	10/M	" " " "
	PT_B	10/P	Nomenclature of Wheeler and Selander (17)
	PT_A	10/P	" " " "
	HBB	2/H	Equals beta chain of hemoglobin

*Numbers identify buffer systems listed in Table 1. Tissue abbreviations correspond to: plasma (P); hemolysate (H); heart-kidney (Ht+K); liver (L); muscle (M); and testes-epididymides (T+E) extracts.

TABLE 3. Gene Frequencies of Loci Where There Are Differences Between Samples

Species	Locality	Sample size	ADH		LDH$_A$		LDH$_C$		ME$_M$	
			1.00	0.54	1.00	0.88	0.82	2.04	1.00	2.34	1.68	1.00

Species	Locality	Sample size	1.00	0.54	1.00	0.88	0.82	2.04	1.00	2.34	1.68	1.00
1. M. musculus	C57BL	--	+		+				+		+	+
2. "	Abingdon	8	1.0		1.0				1.0	0.06	0.94	
3. "	Puigcerdá	2	1.0		1.0				1.0		0.75	0.25
4. "	Balenya	11	1.0		1.0				1.0		0.68	0.32
5. "	San Fausto	3	1.0		1.0				1.0		0.83	0.17
6. "	Barcelona	6	1.0		1.0				1.0		0.58	0.42
7. "	Pto. Real	1	+		+				+		+	+
8. M. hispanicus	Balenya	40		1.0	0.28	0.05	0.68	0.12	0.88	0.14	0.86	
9. "	Pto. Real	17		1.0	0.91		0.09	0.12	0.88	0.41	0.59	

TABLE 3. Gene Frequencies of Loci Where There Are Differences Between Samples (cont.)

Species	Locality	Sample size	ICD_S		$GAPDH_1$		$GAPDH_2$		DIA_5	
			1.30	1.00	1.00	0.54	1.00	0.60	1.00	1.73
1. *M. musculus*	C57BL	--		+	+		+		+	
2. "	Abingdon	8		1.0	1.0		1.0		0.50	0.50
3. "	Puigcerdá	2	1.0		1.0		1.0		0.75	0.25
4. "	Balenya	11	0.27	0.73	1.0		1.0		0.17	0.83
5. "	San Fausto	3	0.17	0.83	1.0		1.0		no sample	
6. "	Barcelona	6		1.0	1.0		1.0		"	"
7. "	Pto. Real	1	+		+		+		"	"
8. *M. hispanicus*	Balenya	40		1.0		1.0		1.0	0.30	0.70
9. "	Pto. Real	17		1.0		1.0		1.0	0.31	0.69

TABLE 3. Gene Frequencies of Loci Where There Are Differences Between Samples (cont.)

Species	Locality	Sample size	NP				GOT$_M$			PGM$_1$			PGM$_2$	
			1.18	1.00	0.85	0.72	1.00	0.66		1.00	0.83		1.00	0.68
1. M. musculus	C57BL	--		+			+			+			+	
2. "	Abingdon	8		1.0			0.62	0.38		1.0			1.0	
3. "	Puigcerdá	2		1.0			0.75	0.25		1.0			1.0	
4. "	Balenya	11		0.95	0.05		0.45	0.55		1.0			0.77	0.23
5. "	San Fausto	3		1.0			0.50	0.50		1.0			0.50	0.50
6. "	Barcelona	6		1.0			1.00			1.0			0.58	0.42
7. "	Pto. Real	1	+	+			+	+		+	+		+	
8. M. hispanicus	Balenya	40			0.01	0.99	1.0			1.0	1.0		1.0	
9. "	Pto. Real	17			0.09	0.91	1.0			0.18	0.82		1.0	

TABLE 3. Gene Frequencies of Loci Where There Are Differences Between Samples (cont.)

Species	Locality	Sample size	ES_3		ES_2		ES_1			ES_B	
			1.09	1.00	1.00	null	1.00	0.97	0.96	1.00	0.95
1. *M. musculus*	C57BL	--		+	+		+			+	
2. "	Abingdon	8		1.0	0.50	0.50		1.0		1.0	
3. "	Puigcerdá	2	1.0		0.50	0.50		1.0		1.0	
4. "	Balenya	11	1.0		0.50	0.50		1.0		1.0	
5. "	San Fausto	3	no sample		1.0			1.0		1.0	
6. "	Barcelona	6	no sample		0.67	0.33		1.0		1.0	
7. "	Pto. Real	1	no sample		+		no sample			+	
8. *M. hispanicus*	Balenya	40	1.0		0.10	0.90			1.0		1.0
9. "	Pto. Real	17	1.0			1.0			1.0		1.0

TABLE 3. Gene Frequencies of Loci Where There Are Differences Between Samples (cont.)

| | | Sample | ES_5 | | PEP_C | | | PEP_B | | PEP_S | |
	Species	Locality	size	1.00	null	1.07	1.02	1.00	1.08	1.00	1.29	1.00
1.	*M. musculus*	C57BL	--	+				+		+		+
2.	"	Abingdon	8	0.38	0.62		1.0		0.06	0.94		+
3.	"	Puigcerdá	2	0.50	0.50	1.0				1.0	0.25	0.75
4.	"	Balenya	11	0.25	0.75	0.50	0.50			1.0		1.0
5.	"	San Fausto	3	1.0		0.67	0.33			1.0		1.0
6.	"	Barcelona	6	0.25	0.75	0.83	0.17			1.0		1.0
7.	"	Pto. Real	1	no sample			+			+		+
8.	*M. hispanicus*	Balenya	40		1.0	0.06	0.94			1.0		1.0
9.	"	Pto. Real	17		1.0		1.0			1.0	0.12	0.88

TABLE 3. Gene Frequencies of Loci Where There Are Differences Between Samples (cont.)

Species	Locality	Sample size	$ACON_S$			$ACON_M$		MPI		GPI	
			1.83	1.45	1.00	1.00	0.35	1.00	0.88	1.00	0.57
1. _M. musculus_	C57BL	--			+	+		+		+	
2. "	Abingdon	8		0.25	0.75	1.0		1.0		0.75	0.25
3. "	Puigcerdá	2		0.25	0.75	1.0		1.0			1.0
4. "	Balenya	11	0.09	0.14	0.77	1.0		1.0		0.18	0.82
5. "	San Fausto	3		0.83	0.17	1.0		1.0		0.33	0.67
6. "	Barcelona	6		0.25	0.75	1.0		1.0		0.25	0.75
7. "	Pto. Real	1			+	+		+		+	
8. _M. hispanicus_	Balenya	40	0.03	0.66	0.31		1.0	1.0			1.0
9. "	Pto. Real	17	0.06	0.62	0.32		1.0	0.97	0.03		1.0

TABLE 3. Gene Frequencies of Loci Where There Are Differences Between Samples (cont.)

Species	Locality	Sample size	Enzyme$_2$		ALB		PT$_1$		PT$_2$		PT$_A$		
			1.00	0.92	1.00	0.99	1.25	1.00	0.30	1.00	1.08	1.06	1.00
1. *M. musculus*	C57BL	--	+		+			+		+		+	
2. "	Abingdon	8		1.0	1.0			1.0		1.0	0.50		0.50
3. "	Puigcerdá	2		1.0	1.0			1.0		1.0	0.25		0.75
4. "	Balenya	11	0.32	0.68	1.0			1.0		1.0	0.33		0.67
5. "	San Fausto	3	0.83	0.17	1.0			1.0		1.0			1.0
6. "	Barcelona	6	0.42	0.58	1.0			1.0		1.0	0.50		0.50
7. "	Pto. Real	1		+	+			+		+		no sample	
8. *M. hispanicus*	Balenya	40	1.0			1.0	0.01	0.99	1.0			1.0	
9. "	Pto. Real	17	0.97	0.03		1.0		1.0	1.0			1.0	

there is interference in the staining of the more anodal allele (Gadph$_2$·^{60}) due to the reactions of SOD$_B$, and this slower allele is partially obscured. Variants at this locus, showing a typical dimeric structure, have been seen in Pacific island mice.

NADH Diaphorases. Studies of diaphorases in mice have not been reported so it is necessary to describe here the results of a survey of tissue and substrate specificities that were done in conjunction with this study. The diaphorase patterns seen in humans have been described (14). The best resolution and greatest mobility of the slower isozymes was seen with the Tris-EDTA-borate buffer system (Buffer 4 of Table 1). Extracts of the four tissue preparations, the two blood components, and the mitochondria were run on the same gel. The resultant banding patterns showed a heterogeneity of isozymes among the various tissue types. A minimum of five loci are thought to be controlling these isozymes based on their occurrences in different tissues, their differential response to the substrates NADH and NADPH, and the variation observed in the shapes of the bands. The most anodal system, designated as DIA$_1$, was seen only in the testes-epididymides extract. This isozyme migrated well in front of the hemoglobin (i.e., approximately 7 cm from the origin), and appeared as a thin band. The second fastest isozyme, designated as DIA$_2$, showed the strongest staining activity in heart-kidney extract, but was also present at low levels in hemolysate and testes preparations. This isozyme appeared just cathodally to the SOD$_A$ achromatic region. It also stained strongly with the NADPH as a substrate, which the other isozymes did not do. It formed a broad diffuse band in the M. musculus and a more narrow, slower band in the M. hispanicus. It was not studied any further because of this poor banding quality. The next-fastest isozyme, designated as DIA$_3$, ran just cathodal to DIA_2. It occurred in the heart-kidney, liver, hemolysate, and mitochondrial extracts. Because of interference from the strongly staining DIA_2 in some tissues, it was possible to score this isozyme only in hemolysate samples. This isozyme was the only one appearing in the mitochondrial preparation. The next locus, designated as DIA_4, migrated about 10 mm cathodal to DIA_3. It occurred in all samples but the plasma and mitochondria. This locus was surveyed in the testes preparations where there was considerable variation in staining intensity between individuals. Whether this variation represents a physiological or genetic source of variation is unknown. Only one mobility class was observed in the two species. The slowest isozyme migrated about 10 mm anodally from the origin. It was designated as DIA_5 and was observed only in the hemolysate sample. There was variation in banding patterns among the individuals, and this was attributed to genetic causes. Presumptive heterozygotes showed a two-banded phenotype suggesting a monomeric subunit structure.

The above results are in partial agreement with the findings on human diaphorases (14) where it is reported that both "sperm" and "red cell" isozymes are encoded by separate loci. These may be homologous to the DIA_1 and DIA_5 loci described here. They (14) considered that their "tissue" isozyme was produced by the same "red cell" diaphorase locus. The fact that in this study there were no concordant changes in mobilities of the DIA_3 or DIA_4 isozymes in animals which were variable at the DIA_5 locus suggests that these former two isozymes are not coded for by the same erythrocytic locus. The higher activity of the DIA_3 isozyme in the mitochondrial preparation is suggestive of an independent locus that codes for an enzyme with a specific subcellular location. The broader substrate specificity of the DIA_2 isozyme (not studied here) provides support for considering this to be the product of still another locus.

Nucleoside Phosphorylase. The nucleoside phosphorylase enzyme migrated about 55 mm anodally. The heterozygotes showed a four-banded pattern, characteristic of a trimeric structure. In the regular extracts the allozymes are difficult to see because of a great amount of diffuse staining in the general region of the active molecules. The bands can be seen separately by diluting the extracts to about one-fifth of the normal concentration. A short distance anodally from the origin a second staining region was visible. Staining activity was found to be dependent on the presence of NADP, either in the gel itself or mixed into the staining mixture. Without all of the ingredients called for in the staining recipe plus the NADP the bands did not appear. This system was considered a separate unidentified enzyme (Enzyme-1). It was invariant within and between species.

Phosphoglucomutases. The pattern of isozymes observed when staining for phosphoglucomutases in kidney extracts was complex. A series of bands were seen, not all of which are phosphoglucomutases. Some of the bands represented isozymes controlled by other known enzyme loci, and one system was seen whose enzymological identity was not determined (Enzyme-2). Near the origin two systems are visible and these were subsequently found to be equivalent to the aconitase loci. More anodally the strongly staining PGM_2 enzyme is present, with a series of one or two weaker subbands extending still more anodally. In the most anodal region of staining activity two principal bands are observed in the C57BL sample. These correspond to the isozymes referred to as $1A^3$ and $1A^4$ in Figure 2(9) of Shows, Ruddle and Roderick (15).

In this study and in work done on Pacific island mice, mobility changes of these two bands were observed to occur independently. When the coupling enzyme G-6PDH is left out of the staining mixture the isozyme $1A^3$ disappears, along with the PGM_2 bands, leaving only the $1A^4$ band on the upper part of the gel. Accordingly, it is deduced that the $1A^3$ component is the PGM_1 isozyme. The other band is called Enzyme-2 in this work. By using these differential staining characteristics, it was shown that the slower allele, $\underline{Pgm_2}^{.83}$, migrates to a less anodal position than either of the two allelomorphs of Enzyme-2. In such cases the relative position of the isozymes of the two loci are reversed. Mobility variation at the unidentified locus (Enzyme-2) is considered to be due to genetic causes. Presumptive heterozygotes showed a two-banded phenotype.

Esterases. Esterases were scored only in the plasma and hemolysate samples. Because of the comparatively small number of samples of these types of extracts the survey of esterase variation must be considered preliminary. The assumption of genetic homology of the "null" allele pattern at the ES_2 and ES_5 loci in the two species is only presumptive, since the absence of staining might be due to a number of different reasons.

Acid Phosphatases. The two acid phosphatases that were scored show very different substrate affinities. The ACP_E enzyme is seen as a fluorescent band when 4-methylumbelliferyl phosphate is used as a substrate. The ACP_1 locus does not fluoresce, but is seen by staining with alpha-naphthyl phosphate as the substrate. Gels used for assaying these loci contained 15% glycerol by volume which acts to enhance the relative staining activity of acid phosphatases (G. Sensebaugh, personal communication).

Peptidases. Isozymes of four peptidase loci were seen in liver samples. Based on the studies of substrate affinities, Lewis and Truslove (16) identified three of the isozymes with homologous systems in humans. The slowest migrating isozyme seen in the liver extracts is not present in hemolysates, and was not studied by these authors. It seems likely that this fourth isozyme is equivalent to the enzyme called PEPS (12). In humans and mice this isozyme is present in liver but not in red cells, shows the least electrophoretic mobility, and reacts with both leucyl-alanine and leucyl-glycyl-glycine as substrates. Because of these similarities, they are assumed to be homologous. Mobility differences were observed at this locus in a few individuals and this was attributed to genetic variation. One animal showed a single, faster migrating band. Other animals showed a two-banded phenotype, with the faster band being somewhat weaker in staining intensity than the slower band. This asymmetry of staining might indicate a simple post-translational alteration of the enzyme molecule in the homozygous condition, except for the presence in one individual of the faster migrating, single-banded phenotype. This phenotype is not expected from such a non-genetical change in molecular structure.

Aconitases. Variation of these enzymes does not appear to have been studied in wild mice. The samples were scored using muscle extracts, but the enzymes also stain well in heart-kidney and liver samples. One problem encountered during the staining reaction is the initial coloring, presumably caused by ICD_M, near the origin of the cathodal slice of the gel. If the gel is left to stain, the $ACON_M$ bands begin to show through and can be scored without any further difficulty. Variation due to a genetic polymorphism was present at the $ACON_S$ locus. Heterozygotes showed a two-banded phenotype.

Enolases. These enzymes also appear not to have been studied in samples of wild mice. The presence of two loci controlling the observed banding patterns is inferred from variation in staining intensities of two isozymes in different tissue types. In extracts of muscle the faster band stained about three times as intensely as the slower band, but in liver extracts the staining intensity of the bands was reversed. In the heart-kidney samples the staining intensity was about equal for both bands. There was no staining observed in plasma, hemolysate, or mitochondrial extracts.

Mannose Phosphate Isomerase. Only one heterozygous animal was found in this study, although in the Pacific island M. musculus the level of polymorphism at this locus is much higher. In addition to the MPI isozyme found about 35 mm anodally, a second anodal isozyme was present about 3 mm from the origin. This band stained as intensely as the MPI isozyme, and is believed to represent the product of some unidentified enzyme locus (Enzyme-3). It was invariant and had the same mobility in both species.

Non-enzymatic Proteins. Seven proteins were assayed on gels stained with amido-black. In muscle extracts four proteins were studied: albumin, myoglobin, and two slower migrating proteins. Albumin is the most anodal, strongly staining protein in these extracts. The next principal band is identified as myoglobin, based on a positive staining response to a benzedine solution, which indicates the presence of a molecule containing a heme-group. This protein forms a comparatively broad band, and it migrates a couple of millimeters slower than hemoglobin on the lithium buffer system. The next fastest muscle protein band is a very thin one that migrates just cathodal to the PT_A band found in plasma. This muscle protein is designated PT_1. One putative variant animal showed a two-banded phenotype. The last protein scored from muscle extracts appears as a thin band migrating just cathodal to the previously described protein. It is designated PT_2. There were fixed mobility differences between the two species at this locus. The behavior of this protein was variable between gels. On some gels there was a sharp band present, but on another gel it would appear as a diffuse staining region. In such cases runs were repeated until clearly resolved bands were produced.

In plasma samples two proteins [PT_A and PT_B of (17)] were studied. Variation in electrophoretic mobility of the PT_A bands is thought to be due to genetic causes. Some samples showed strongly staining bands with a series of one or two weaker subbands, and these were considered homozygotes. Another phenotype seen was that of a two-banded pattern with each band showing equal amounts of staining activity, and these were scored as being heterozygotes.

In hemolysate the hemoglobin was studied on the Tris-maleate buffer system. The banding pattern seen here showed conclusively that none of the animals contained the allele Hbb^d. This allele is present at a low frequency in both English (18) and Danish (19) populations of M. musculus. This absence is presumably due to sampling error because of the small number of hemolysate samples available.

Population Variability

Of 56 loci studied, 27 (48%) were invariant, showing the same allelomorph in both species. Nineteen additional loci (34%) shared allelomorphs that were polymorphic in at least one sample, and ten loci (18%) showed total allelic dissimilarities between the species. The mean heterozygosity per locus (H) for each sample is given in Table 4. These results show that the two species differ considerably in the average individual heterozygosity. The M. musculus animals are heterozygous at about twice (6-10%) the number of loci as M. hispanicus individuals (3.5-4.5%). Among the M. musculus populations the English mice have a slightly lower level of variability than animals from Spain.

These measures of variability for M. musculus fall within the ranges reported for English and Danish populations (18, 19) of the subspecies M. m. domesticus. Twenty-one of the 22 loci studied (18) in large samples of English mice were included in the present survey. At none of these loci were alleles found in the Abingdon sample that do not appear equivalent to variants described in that work.

Genetic Relationships

The genetic relationships among the populations was determined using the distance measure (D) of Nei (20), which is defined as the average number of codon differences per locus. Because hemolysate samples were not collected from some localities, no gene frequency estimates were available for four loci (i.e., DIA_3, DIA_5, ES_3, HBB) for all populations. The estimates of gene frequency used in the calculations came from data for the remaining 52 loci. The results of the computations are presented in Table 5. The D-value matrix was clustered with a complete linkage technique (21) and is depicted in Figure 2. These results show three interesting features: the very marked genetic difference between M. musculus and M. hispanicus; the great similarity among M. musculus populations, even when separated by great distances and topographic barriers; and the distinctiveness of the inbred laboratory mouse from its putative ancestor.

The occurrence of numerous fixed allelic differences between wild and commensal populations of mice at Balenya and Pto. Real shows that these animals do not interbreed

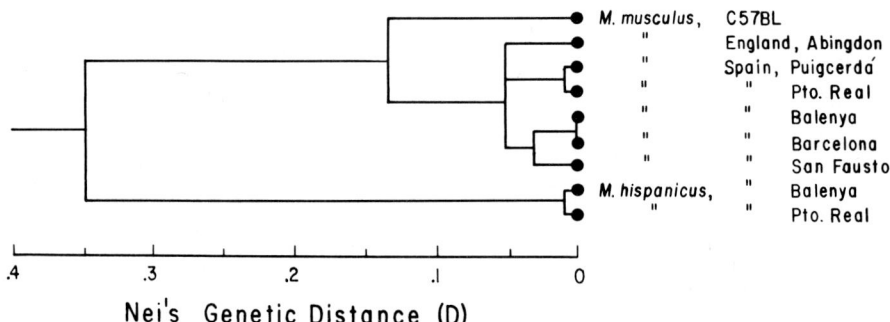

FIGURE 2. Genetic distance (D) for samples of an inbred lab strain and native mice (Mus) from England and Spain, based on 52 structural gene loci.

and represent distinct species. The genetic distance values (D=0.3-0.4) between M. musculus and M. hispanicus reflects the great amount of divergence between these lineages. Intraspecific, interpopulation values of genetic distance can be considered as controls for the expected random sampling variance due to the numbers of animals per sample, the number of loci studied, or to founding events (see below). In this study these values range from 0.01 to 0.04 and are about one order of magnitude smaller than the interspecific genetic distances. The interspecific distances observed between these mice are similar to those reported for other rodents [(20); M. Smith and J.L. Patton, personal communication].

The interpopulation distance values among the M. musculus samples are small and comparable to the values observed in a wide array of animal taxa (22). Similar levels of interpopulational genetic distance within, but not between, two

TABLE 4. Mean Heterozygosities (H) of English and Spanish Mus

Species	Locality	Sample size*	% Polymorphic loci	H (± S.E.)
M. musculus	England, Abingdon	8	18	0.062 (0.0226)
"	Spain, Puigcerdá	2	--	0.075 (0.0269)
"	" , Balenya	11	25	0.097 (0.0260)
"	" , San Fausto	3	--	0.071 (0.0241)
"	" , Barcelona	6	20	0.085 (0.0266)
"	" , Pto. Real	1	--	0.098 (0.0421)
M. hispanicus	" , Balenya	40	20	0.034 (0.0145)
"	" , Pto. Real	17	20	0.044 (0.0164)

*Sample size is smaller than this number for loci surveyed in plasma, hemolysate, and testis-epididymides extracts.

subspecific lineages of Danish house mice have been reported (19). Genetic distances of this order of D (approximately 0.05) were observed between samples of M. m. domesticus from different farms within a five-kilometer radius in England (18) and it was suggested that such differences were to be expected from stochastic processes altering gene frequencies during the founding of local populations. Similar small distance values were observed between the English M. musculus and populations from Spain, in spite of the great geographical distance separating this population from the others and the water barrier separating the English and continental land masses. This similarity in proteins between nominally different subspecies of house mice (M. m. domesticus and M. m. brevirostris) in western Europe contrasts greatly with the observed situation in northern and eastern Europe, where M. m. domesticus and M. m. musculus are found (see Figure 1). In Denmark, where genetic studies comparable to those reported here were

TABLE 5. Genetic Distance (D-Values) (Upper Right Triangular Matrix) and Standard Error of D (Lower Left Triangular Matrix) Between Samples of a Lab Mouse and Native House Mice from Spain and England

Sample number and locality	Sample number								
	1	2	3	4	5	6	7	8	9
1 M. musculus, C57BL	---	0.11	0.10	0.11	0.08	0.13	0.13	0.40	0.38
2 M. musculus, Abingdon	0.043	---	0.02	0.03	0.05	0.04	0.02	0.32	0.30
3 M. musculus, Puigcerdá	0.036	0.009	---	0.00	0.03	0.01	0.01	0.31	0.29
4 M. musculus, Balenya	0.041	0.018	0.005	---	0.03	0.03	0.04	0.34	0.32
5 M. musculus, San Fausto	0.035	0.021	0.016	0.014	---	0.05	0.05	0.34	0.33
6 M. musculus, Barcelona	0.050	0.033	0.013	0.025	0.022	---	0.01	0.36	0.34
7 M. musculus, Pto. Real	0.054	0.027	0.017	0.029	0.028	0.024	---	0.31	0.30
8 M. hispanicus, Balenya	0.097	0.084	0.082	0.087	0.089	0.092	0.084	---	0.01
9 M. hispanicus, Pto. Real	0.095	0.080	0.078	0.084	0.087	0.089	0.084	0.008	---

done (19), populations of the two subspecies had genetic distances computed to be D=0.19. The contrast in magnitude of genetic divergence indicates the considerable heterogeneity contained within the currently recognized subspecific taxa of M. musculus in Europe, and shows how misleading this form of nomenclature can be in denoting genetic relationships among populations.

Selander and Yang (23), working on house mice in the United States, were unable to detect major gene frequency differences between northern and southern populations which have been classified as belonging to M. m. domesticus and M. m. brevirostris, respectively (1). Their results are in agreement with the findings of the present study of native populations of mice from which the United States animals may well have originated, in that no major allelic differences appear to distinguish these taxa.

The last point of interest emerging from this work is the relatively great genetic distance of the inbred mouse from what might be considered its native progenitor. The C57BL mice show consistently greater distance from the wild populations of England and Spain (D=0.10-0.13) than do any of the latter among themselves. While the homozygosity of an inbred mouse would be expected to introduce a distance component in any comparison with a genetically variable population, because of the impossibility of expressing any of the original variability that the lineage may have contained, a good part of the distance value of the C57BL mice is due to the presence of two "private" alleles ($Es_1^{1.00}$ and $Pep_C^{1.00}$) that were not observed in any of the native European mice.

The C57BL lineage of inbred strains is genetically quite distinct from the bulk of inbred mice (24). It is likely that they owe part of their distinctiveness to a contribution from the oriental subspecies of house mouse (M. m. molossinus), via the Japanese fancy mice available in the pet trade in the early part of this century (25). Determination of the genes segregating in samples of free-living mice in Japan and Korea would provide an answer to this question. It is apparent, however, that most of the genes in this inbred strain could have been derived from native populations of western European mice.

DISCUSSION

This study was initiated because according to the modern paradigm of the biological species concept the presence of coexisting subspecies is contrary to predictions. The results indeed show that the Spanish mice are divided into two genetically isolated gene pools and must be considered as separate species taxa. Other recent studies of native house mice have revealed distinctive species concealed under the classificatory scheme of Schwarz and Schwarz (1). Marshall (26) showed that the species Mus caroli in eastern Asia was not a synonym of M. m. homourus. Gropp et al. (27) showed that M. poschiavinus, which had been subsumed into M. m. domesticus, was a valid species and was characterized by seven Robertsonian fusions between acrocentric chromosomes. In addition to the above demonstrated cases of reproductively isolated species populations within the range of the house mouse, Schwarz and Schwarz (1) report the coexistence of other subspecies of mice in central and eastern Asia. There is every reason to expect that these also represent valid species taxa. All of this indicates that the present scheme (1) for the classification of the house mouse is invalid and that a revised systematic treatment is needed.

With the decline and fall of old taxonomic arrangements, those interested in the native house mice are left with a blank slate on which to reconstruct the evolutionary history of these animals. However a new paradigm develops, it will differ radically from the earlier one (1) in which much of the differentiation of the group was attributed to the coevolution of mice with man and his agriculture. The time scale for the divergence among some lineages (e.g., M. musculus and M. hispanicus) will have to be framed in Plio-Pleistocene times, based on the size of the genetic distance values reported in this paper. Even the assumption of specific morphological changes as being associated with the development of the commensal habit with man may be incorrect. Tchernov (28) reported that the mandible shape of M. musculus fossils in deposits in Israel remained identical to recent materials (commensal forms), throughout the Pleistocene. Thus the skull morphology typical of commensal mice antedates agricultural practices in the region and suggests that the phenotype was determined by selective forces independently of the commensal behavior pattern.

The new systematics of the house mice will differ from the earlier attempt in that both the external phenotype as well as the genotype will be examined. With the development of more powerful analytical methods for studying changes in the genome, an increasing interest in native house mice as a source of additional genetic material for laboratory studies, and an awareness that multidisciplinary approaches will yield a more satisfactory interpretation of such a historical process, the house mouse will continue to make valuable contributions to our understanding of organic evolution.

SUMMARY

On the Iberian peninsula sympatric populations of two subspecies of house mice (M. musculus) were reported by Schwarz and Schwarz (1). The genetic structure of these mouse populations from localities in Spain was studied using electrophoretic techniques. Wild mice from England and an inbred laboratory strain (CB7BL) were also studied as outside reference points. Enzymes and other proteins, presumptively controlled by 56 structural gene loci, were identified in animals of both subspecific phenotypic classes. The results show that these two kinds of Spanish mice differ at many loci and there is no sign of intermediate genotypes. The two populations must be considered as full species: M. m. brevirostris and M. hispanicus. Average heterozygosity in M. m. brevirostris was twice as great as in M. hispanicus. There were no major allelic differences between English and Spanish populations of M. musculus, although they are currently recognized as different subspecies. The inbred mouse was genetically distinct from wild populations of its putative ancestors. The difference is thought to result from genes introduced into the lineage from oriental populations of M. musculus. The adequacy of the current paradigm of house mouse classification is discussed.

ADDENDUM

After this manuscript was completed, a paper describing biochemical differences between "long-tailed" (commensal) and "short-tailed" (aboriginal) mice in southern France was discovered (36). These forms are the same as the M. musculus and M. hispanicus populations described in the present work. Professor Thaler (personal communication) suggests that Cabrera (9) was incorrect in considering the Algerian M. spretus a different species from the wild animals he found in Morocco and Spain (here considered as M. hispanicus). If future field work in the region of the type locality in Algeria shows that there is only one type of wild mouse in this whole area then, because of nomenclatorial considerations, the proper name for the populations in Spain will be M. spretus.

ACKNOWLEDGMENTS

I am grateful to P. Alberch and A.G. Searle for their help during the field work. Ms. M. Frelow provided excellent technical assistance in the laboratory. For comments and suggestions during the preparation of the manuscript, I thank J. Hanken, W.Z. Lidicker, and J.L. Patton.

REFERENCES

1. Schwarz, E., and Schwarz, H.K. (1943). J. Mamm. 24:59.
2. van den Brink, F.H. (1968). "A Field Guide to the Mammals of Britain and Europe." Houghton Mifflin Co., Boston.
3. Saint-Girons, M.C. (1973). "Les Mammiféres de France et du Benelux." Doin, Paris.
4. Mayr, E. (1963). "Animal Species and Evolution." Harvard University Press, Cambridge.
5. Sage, R.D., and Selander, R.K. (1975). Proc. Natl. Acad. Sci. U.S.A. 72:4669.
6. Bruell, J.H. (1970). In "Contributions to Behavior-Genetic Analysis" (G. Lindzey and D.D. Thiessen, eds.), p. 261. Appleton-Century-Crofts, New York.
7. Miller, G.S. (1909). Ann. Mag. Nat. Hist., Ser. 8, iii, 415.
8. Serafinski, W. (1965). Ekol. Polska, Ser. A, 13:305.
9. Cabrera, A. (1923). Bol. Real Soc. España Hist. Nat. 23:429.
10. Ingles, L.G. (1947). "Mammals of the Pacific States." Stanford University Press, Stanford.

11. Selander, R.K., Smith, M.H., Yang, S.Y., Johnson, W.E., and Gentry, J.R. (1971). Studies in Genetics VI, Univ. Texas Publ. 7103.
12. Harris, H., and Hopkinson, D.A. (1976). "Handbook of Enzyme Electrophoresis in Human Genetics." North-Holland Publ. Co., Amsterdam.
13. Johnson, D., and Lardy, H. (1967). In "Methods in Enzymology. X. Oxidation and Phosphorylation" (R.W. Estabrook and M.E. Pullman, eds.), p. 94. Academic Press, New York.
14. Fisher, R.A., Edwards, Y.H., Putt, W., Potter, J., and Hopkinson, D.A. (1977). Ann. Hum. Genet. 41:139.
15. Shows, T.B., Ruddle, F.H., and Roderick, T.H. (1969). Biochem. Genet. 3:25.
16. Lewis, W.H.P., and Truslove, G.M. (1969). Biochem. Genet. 3:493.
17. Wheeler, L.L., and Selander, R.K. (1972). Studies in Genetics VII, Univ. Texas Publ. 7213.
18. Berry, R.J., and Peters, J. (1977). Proc. Roy. Soc. Lond. B197:485.
19. Selander, R.K., Hunt, W.G., and Yang, S.Y. (1969). Evolution 23:379.
20. Nei, M. (1975). "Molecular and Population Genetics and Evolution." North Holland Publ. Co., Amsterdam.
21. Sneath, P.H.A., and Sokal, R.R. (1973). "Numerical Taxonomy." W.H. Freeman and Co., San Francisco.
22. Avise, J. (1976). In "Molecular Evolution" (F.J. Ayala, ed.). Sinauer Associates, Inc., Sunderland.
23. Selander, R.K., and Yang, S.Y. (1970). In "Contributions to Behavior-Genetic Analysis" (G. Lindzey and D.D. Thiessen, eds.), p. 293. Appleton-Century-Crofts, New York.
24. Taylor, B.A. (1972). J. Hered. 63:83.
25. Keeler, C.E. (1931). "The Laboratory Mouse." Harvard University Press, Cambridge.

26. Marshall, J.T. (1977). Bull. Amer. Mus. Nat. Hist. 158:173.
27. Gropp, A., Winking, H., Zech, L., and Muller, H. (1972). Chromosoma (Berl.) 39:265.
28. Tchernov, E. (1968). "Succession of Rodent Faunas During the Upper Pleistocene of Israel." Verlag Paul Parey, Hamburg.
29. Petras, M.L., and Martin, J.E. (1969). Biochem. Genet. 3:303.
30. Ayala, F.J., Powell, J.R., Tracey, M.L., Mourão, C.A., and Pérez-Salas, S. (1972). Genetics 70:113.
31. Wright, D.A., Siciliano, M.J., and Baptist, J.N. (1972). Experientia 28:888.
32. Tegelström, H. (1975). Hereditas 81:185.
33. Lalley, P.A., and Shows, T.B. (1977). Genetics 87:305.
34. Ruddle, F.H., and Nichols, E.A. (1971). In Vitro 7:120.
35. Zimmermann, K. (1949). Zool. Jahrb. F. Systematik 78:301.
36. Britton, J., Pasteur, N., and Thaler, M.L. (1976). C. R. Acad. Sc. Paris, ser. D 283:515.

BIOCHEMICAL POLYMORPHISMS IN WILD MICE[1]

Verne M. Chapman
Department of Molecular Biology
Roswell Park Memorial Institute
Buffalo, New York

The development of electrophoretic techniques coupled with histochemical staining on gels for specific enzyme reactions has provided a powerful tool for evaluating qualitative differences in primary gene products. These techniques have been widely used in mouse genetics to describe new loci and to define the linkage of the structural genes for numerous proteins and enzymes (1). This work has resulted in a substantial growth and refinement of the genetic map of the mouse over the past several years.

The ability to qualitatively identify the products of structural genes has also provided a means of examining wild or feral mice for genetic variation. In this regard, several workers interested in the genetic structure of natural populations and the issues of population genetics have surveyed mouse populations in various parts of the world [reviewed by Chapman and Selander (2)]. Several generalizations about genetic variation in mice can be drawn from these surveys.

---

[1] This work was supported by NIH Grants GM-19521 and GM-24125 from the U.S. Public Health Service.

First, most mouse populations sampled are highly polymorphic. Typically, about 30 to 40% of the loci examined are variable. Second, while a few loci show a striking tendency to be polymorphic in many of the samples, there is a pronounced difference between populations in the loci that are polymorphic and in the alleles that are present. Third, there are many allelic forms found in wild mice that have not been observed among laboratory strains and stocks of mice. Furthermore, in many cases, genetic variation for a particular gene has been found only in wild mice or in mice recently derived from the wild. That is, the standard inbred strains are monomorphic. The contribution of wild or wild-derived mice to genetic variation at specific loci can be determined from various listings of biochemical genetic variation in the mouse (3, 4).

The purpose of this paper is to focus on the wild mouse as a valuable genetic resource that can be used in conjunction with the laboratory strains for many aspects of mouse genetics. In addition to the characters that can be qualitatively determined such as electrophoretic mobility differences, it is also possible to use wild mouse populations to look for quantitative variation in enzyme activity levels. In this regard, wild mice have been studied for quantitative variation in α-amylase (5), β-galactosidase (Nielsen and Chapman, unpublished), and androgen inducibility of kidney β-glucuronidase (Nielsen and Chapman, in preparation). Work on these systems is of particular interest because it extends the use of wild mouse populations from a source of structural variants to asking questions concerning variation in gene regulation in natural populations. It also involves using wild populations to look for recombination events between closely linked elements which might be difficult to realize in laboratory experiments. Finally, because wild mice can serve a variety of purposes in biochemical genetics, I would also like to briefly consider different breeding programs for maintaining and using wild-derived mice in the laboratory.

Biochemical Genetic Variation in Wild Mice

Extensive surveys of wild mouse populations for biochemical genetic variation have been conducted in Great Britain, the Faroe, Shetland and Orkney archipelagoes (6), in Denmark (7), and in North America (8-10). Estimates of genetic variation vary from 5 of 17 loci examined (29%) (10) to 17 of 41, 41% (7). These findings and the results of limited samplings of the Asian subspecies M. m. castaneus (11) from Thailand and M. m. molossinus from Japan clearly demonstrate that substantial amounts of genetic variation can be found within and between wild populations of M. musculus.

A more striking result is that the potential array of genetic variation in the mouse gene pool is so large. This variation is partially observed among inbred strains and stocks of mice (1, 3, 4, 12). To date, about 90-100 biochemical loci have been characterized in the mouse. A partial list of those loci which are characterized by variation from wild mice is shown in Table 1, part 1. A second group of loci are listed where uncommon alleles have been observed in wild mice.

Many of the biochemical loci in the mouse are characterized by quantitative variation and primarily in surveys of inbred mice. Thus, of the total loci characterized, wild-derived variation accounts for over a third of the total reported. However, if only those loci are considered where wild mice have been examined, the contribution of wild-derived alleles to the total genetic variation is more than 60%. These results clearly indicate that the continued

TABLE 1. Polymorphism from Wild Trapped Mice

Loci polymorphic in wild mice		Loci with unusual alleles in wild mice
Adh-1	Got-1	Amy-1
Ags	Ipo-1	Amy-2
Alb-1	Mor-1	Es-2
Erp-1	Np-1	Es-5
Es-6	Pgd	Gdc-1
Es-7	Pgk-1	Gpt-1
Es-8	Prt-2	Gur
Es-9	Prt-3	Gus
Es-11		Id-1
		Mod-1

Number of established biochemical markers in the mouse: 88.

Proportion of markers established by polymorphism from wild mice: 17/88, 20%.

Total variation derived from wild mice: 27/88, 30%.

Chapman, V.M., et al.: Inbred and Genetically Defined Strains of Laboratory Animals, Table III; 27. FASEB.

Chapman, V.M. (unpublished).

Nielsen, J.T. (personal communication).

growth of the biochemical genetic map can be significantly aided by a systematic sampling of wild mouse populations.

The heterogeneous nature of wild mice is both an advantage and a limitation. In general, we have been confined to genetic characters that can be examined qualitatively such as the electrophoretic mobility of proteins. In the past, much of the electrophoretic variation has been detected by means of starch gel techniques. Most of the phenotypes have been codominant and presumably represent variation in the structural gene. Recent work on two enzymes, acid phosphatase (13, 14) and α-mannosidase (15, 16), shows that electrophoretic variation can also be present as a recessive-dominant trait. These mutants presumably represent variation in post-translational modification of proteins.

These observations indicate that certain precautions are warranted in surveys of wild mice for new electrophoretic variation. To distinguish between structural loci and processing mutants it is essential to know the kinds of phenotypes present and their distribution in a sample. Furthermore, conventional genetic crosses with standard inbred strains are essential to establish the mode of inheritance and linkage.

In the past, we have been primarily concerned with structural gene variation. However, the development of more sophisticated electrophoretic techniques involving systematic alterations of acrylamide concentrations holds considerable promise for identifying a wide array of processing mutants that might not have been detected in conventional starch gel electrophoretic systems (17, 18). A reexamination of both laboratory and wild-derived mice using this technology should substantially increase the number of biochemical markers known in the mouse genome.

Sampling Wild Mice for Variation in β-Glucuronidase Regulation and Structure

The molecular biology and genetics of β-glucuronidase have been extensively studied in the laboratory mouse (19). An important feature of mouse β-glucuronidase is that it is androgen inducible in the proximal tubule cells of adult mice. Female mice show a 10- to 30-fold increase in kidney β-glucuronidase within seven days following androgen treatment. This increased activity is an increase in the amount of β-glucuronidase which is the result of an increased rate of synthesis.

Three different induction phenotypes are observed among inbred mice. Two of these phenotypes characterized in the strains C57BL/6J and A/J have been studied in detail (20). C57BL/6J is a low inducibility strain (Gur^b) and A/J is a high inducibility strain (Gur^a). These strains differ in induced β-glucuronidase activity by almost threefold at seven days (Table 2). The F_1 values are intermediate. The difference in β-glucuronidase induction segregates as a single gene in backcrosses and the F_2 generation, and it is closely linked to the structural locus, Gus, on chromosome 5. The electrophoretic phenotype of the induced F_1 kidney is consistent with a cis acting regulatory site closely associated with the structural gene.

In the initial genetic crosses 3/115 presumptive recombinants between the structural gene and the regulatory site were observed. However, the mice were killed as part of the test procedure and it was not possible to verify them as true recombinants by breeding tests. In a subsequent test of 1,042 backcross progeny no recombinants were observed (21). These data indicate that the Gur and Gus sites are very closely linked and are probably separated by less than 0.7 recombination units.

Among inbred mice strains those that are Gus^a, fast electrophoretic form, are Gur^a, high inducibility phenotypes. Conversely, nearly all of the strains that are Gus^b, slow electrophoretic phenotype, are associated with the low inducibility phenotype, Gur^b.

In 1975 J. Tønnes Nielsen, the Genetics Institute, University of Aarhus, Aarhus, Denmark, and I examined wild trapped mice from Denmark to determine whether recombinant type chromosomes could be recovered from wild populations. Specifically, we asked the following questions:

TABLE 2. β-Glucuronidase Induction and Structural Gene Phenotypes Among Inbred Strains

Structural allele	Induction allele	Source of allele	Electrophoresis[2]	Act at 7 days[3] (units)
Gus^b	Gur^b	C57BL/6[1] (V.I.S.)	1.0	55
Gus^a	Gur^a	A/J (V.I.S.)	1.1	130
Gus^h	Gur^h	C3H/HeJ	1.0	20

Source: Swank, R.T., Paigen, K., and Ganschow, R. (1973). J. Mol. Biol. 81:225. Nielsen, J.T., and Chapman, V.M., Unpublished.

[1] V.I.S.: various inbred strains.
[2] Relative electrophoretic mobility to Gus^b.
[3] Units: mole hr^{-1} g wet $weight^{-1}$ at 37°C, using substrate 4 methylumbelliferyl β-D-glucuronide.

1) Is β-glucuronidase polymorphic in wild mice?
2) Do recombinant types exist for inducibility and structure, specifically, Gus^a-Gur^b or Gus^b-Gur^a?
3) What array of Gus-Gur combination exists in wild mice? That is, are there new phenotypes not found in laboratory mice?

Our strategy was to androgen induce wild trapped females directly and test for β-glucuronidase activity levels in kidneys at seven days and the electrophoretic phenotypes. Wild males were mated to a C3H strain (Gus^b Gur^b) maintained in Aarhus. C3H × wild F_1 female progeny were tested to determine androgen inducibility levels at seven days and their electrophoretic mobility. Backcrosses, F_2 and F_3 generations, were examined for cosegregation of structural gene variation and inducibility levels.

Two inducibility-structural gene combinations were found that are not present among laboratory mice (Table 3). These include a fast electrophoresis low inducibility type (Gur^b Gus^a) and a new very high induction phenotype associated with fast electrophoresis (Gur^a Gus^a).

Backcross type matings of the C3H x wild F_1 ($Gur^{b/b}$ $Gus^{a/b}$) with the strain C57BL/6J ($Gur^{b/b}$ $Gus^{b/b}$) segregated for electrophoretic mobility 16/32 : 16/32 which is consistent with the expected 1:1 ratio for a single gene. The electrophoretic patterns of the heterozygote are consistent with codominant expression of subunits for the tetrameric β-glucuronidase.

Seven day induction levels of glucuronidase activity did not segregate (Table 3). When induction levels of heterozygotes ($Gus^{a/b}$) and homozygotes ($Gus^{b/b}$) are compared, there is no significant difference. By contrast, induction levels of $Gur^{a/b}$ $Gus^{a/b}$ and $Gur^{b/b}$ $Gus^{b/b}$ females from a congenic backcross are clearly different.

Electrophoretic phenotypes of the heterozygous $Gur^{a/b}$ $Gus^{a/b}$ from the congenic backcross and the $Gur^{b/b}$ $Gus^{a/b}$ from the wild-derived mice also differ. The $Gur^{a/b}$ phenotype is predominantly GUS-A subunits shifting the distribution of activity to 4A and 3A1B multimeric forms. However, in the wild-derived phenotype the electrophoretic pattern is consistent with an equal synthesis of the Gus^a and Gus^b type subunits. That is, a binomial expression of $(.5\ Gus^a + .5\ Gus^b)^4$.

Additional genetic and structural studies are under way with both of the new wild-derived chromosomes to determine whether the two Gur^a electrophoretic forms are the same and whether they differ from the Gus^a characteristic of the strain A/J. These data are essential for determining whether we are dealing with recombinants of Gur and Gus or three separate structure-induction combinations.

TABLE 3. Specific Activity Levels of Kidney β-Glucuronidase at 7 Days of Induction (μmoles hr^{-1} g^{-1} at 37°)

	Electrophoretic phenotypes	
	GUS-B	GUS-AB
1. (C3H x Wild)F$_1$[b] x C57BL/6J	58.3 ± 3.7 (8)	54.7 ± 5.4 (8)
2. (C3H x Wild)F$_1$ x C57BL/6J	56.3 ± 4.2 (8)	50.8 ± 3.3 (8)
Total of 1 and 2[a]	57.3 ± 2.7 (16)	52.7 ± 3.1[NS]1
		(16)
3. Congenic[c] F$_1$ x C57BL/6J	60.5 ± 2.4 (3)	104.5 ± 6.5[***]1 (12)

[a] 1 and 2 are tests of comparable populations which were done at different times.

[b] C3H - Aarhus substrain Gus$^{b/b}$.

[c] Congenic C57BL/6 Gus$^{a/b}$ N8.

NS. GUS-B vs. GUS-AB not significantly different ($P > 0.05$).

***GUS-B vs. GUS-AB significantly different $t = 3.46$; $P < 0.001$.

In this regard, we have also observed two additional Gur-Gus combinations from wild-derived mice which are also shown in Table 4. The Gus^c allele from M. m. castaneus, trapped in Thailand, produced a β-glucuronidase which differs from any produced by laboratory mice. The second allele, Gus^n, was found in wild-derived inbred lines developed by Dr. James Connor, University of Nebraska. The relative electrophoretic mobility of the Gus^n β-glucuronidase is

TABLE 4. β-Glucuronidase Induction and Structural Gene Phenotypes Among Wild Trapped Mice

Structural allele	Source of allele	Electrophoresis[1]	Induction act at 7 days
Gus^c	M. m. castaneus (Thailand)	1.1	40
Gus^n	M. m. domesticus (Pennsylvania)	0.9	200
Gus^a	M. m. musculus (Denmark)	1.1	50
Gus^{a+}	M. m. domesticus (MOR/Cv; Denmark)	1.1	200

[1] Relative electrophoretic mobility to Gus^b.

slower than any inbred phenotype. It is associated with a very high inducibility at seven days. More complete molecular and genetic characterizations of these two allelic forms are under way.

Our experience with sampling wild mice for β-glucuronidase parallels the findings for many loci. Namely, many allelic forms may be present in wild mice that are not present among inbred strains. Furthermore, we have demonstrated that it is possible to survey wild mice for complex phenotypes such as kidney levels of β-glucuronidase following induction. Even if the presumptive recombinants of Gur-Gus reported here ultimately prove to be different chromosome sets of Gur-Gus with different structural sequences of β-glucuronidase, the notion of looking for recombination between closely linked genes in wild populations may be useful, particularly when the expected recombination frequency is less than 10^{-4}. This will be especially important as additional cis acting regulatory elements are observed for other systems besides β-glucuronidase induction.

Breeding Programs for Wild-Derived Mice

In most instances, wild populations of mice are fairly heterogeneous with relatively large amounts of polymorphism in most of the samples taken. Developing inbred strains from wild-derived sources has some value in producing a new array of gene combinations that may not exist among inbred mice. For example, M. m. castaneus differs from C57BL/6J for more than 27 biochemical markers on 13 different chromosomes (Table 5) (Chapman, unpublished). As such, it provides a valuable tool for linkage analysis and as a mutagenesis testing stock. Similar differences may also be present for M. m. molossinus (from Japan) and a strain developed at Roswell Park Memorial Institute called MOR/Cv. However, the development of wild-derived strains is costly in terms of time and money, and there is probably a limit to the number of such strains that fulfill this general need.

As an alternative, I would propose that wild mouse populations with interesting differences from laboratory stocks be kept as random breeding colonies. This will maintain a greater proportion of genetic variation originally present in the wild-derived sample. Reproduction in these stocks will be generally better than in inbreds as well. Where interesting alleles are present for different loci, it is relatively easy to transfer them to a standard inbred strain background. This has the additional advantage of providing a new allele on a known genetic background. This may be of special value in working with quantitative phenotypes.

From a practical standpoint the congenic stocks are easier to handle, i.e., they behave like inbred mice. Also, congenic lines have a special value for linkage tests because

TABLE 5. Loci Segregating in Crosses of C57BL/6 and
M. m. castaneus

Locus	Chromosome	Locus	Chromosome
Id-1	1	Es-9	8
Dip-1	1	Mpi-1	9
a	2	Mod-1	9
Car-1	3	Bgs	9
Gpd-1	4	Lap-1	9
Gus	5	Es-3	11
Pgm-1	5	Amy-1	12
Hbb	7	Amy-2	12
Mod-2	7	Np-1	14
Gpi-1	7	Gdc-1	15
Es-8	7	Gpt-1	15
Es-1	8	Got-1	19
Es-2	8	Hao-2	?

they carry a chromosome segment from the donor strain which may contain different alleles for other linked loci.

SUMMARY

Feral mouse populations are highly polymorphic and contain genetic variation not present among laboratory stocks of mice. Genetic analysis of wild-derived polymorphisms is possible by crossing wild mice with homozygous inbred strains. Backcross and F_2 generations can be readily studied for segregation and linkage analyses.

Complex phenotypes such as quantitative levels of enzyme activity as well as qualitative characteristics such as electrophoretic mobility can be evaluated in crosses with characterized inbred strains. This has been demonstrated for the enzyme loci α-amylase (5) and in work on β-glucuronidase.

Wild mouse populations may also serve as an experimental resource. In the case of closely linked elements, sufficient test progeny may not be economically feasible to evaluate recombination distances less than 0.01 map units apart. Cis acting regulatory elements would be reasonably expected to be within that distance from structural loci. Thus, the identification of populations polymorphic for both structure and regulation may provide a possibility of recovering what would otherwise be rare recombination events in the laboratory.

REFERENCES

1. Hutton, J.J. (1978). In "Origins of Inbred Mice." (H.C. Morse III, ed.). Academic Press, New York.
2. Chapman, V.M., and Selander, R.K. (1978). "FASEB Biological Handbook," in press.
3. Chapman, V.M., Paigen, K., Siracusa, L., and Womack, J. (1978). "FASEB Biological Handbook," in press.
4. Staats, J. (1976). Cancer Res. 36:4333.
5. Nielsen, J.T., and Sick, K. (1978). Hereditas 79:279.
6. Barry, R.J., and Peters, J. (1977). Proc. Roy. Soc. London. B. 197:485.
7. Selander, R.K., Hunt, W.G., and Yang, S.Y. (1969). Evolution 23:379.
8. Selander, R.K., and Yang, S.Y. (1969). Genetics 63:653.
9. Petras, M.L., Remer, J.P., Biddle, F.G., Martin, J.E., and Lauton, R.S. (1969). Canada J. Genet. Cytol. 11:497.
10. Ruddle, F.H., Roderick, T.H., Shows, T.B., Weigle, P.G., Chipman, R.K., and Anderson, P.K. (1969). J. Hered. 60:321.
11. Chapman, V.M. (1973). Mouse News Letter 48:44.

12. Roderick, T.H., Ruddle, F.H., Chapman, V.M., and Shows, T.B. (1971). Biochem. Genet. 5:457.
13. Womack, J.E., and Belsky, S. (1975). Mouse News Letter 52:37.
14. Lalley, P.A., and Shows, T.B. (1977). Genetics 87:305.
15. Dizik, M., and Elliott, R.W. (1977). Biochem. Genet. 15:31.
16. Dizik, M., and Elliott, R.W. (1978). Biochem. Genet., in press.
17. Johnson, G. (1977). Biochem. Genet. 15:665.
18. Johnson, G. (1978). Proc. Natl. Acad. Sci. USA 75:395.
19. Paigen, K. (1978). In "Origins of Inbred Mice." (H.C. Morse III, ed.). Academic Press, New York.
20. Swank, R.T., Paigen, K., and Ganschow, R.E. (1973). J. Mol. Biol. 81:225.
21. Swank, R.T., Paigen, K., Davey, R., Chapman, V., Labarca, C., Watson, G., Ganschow, R., Brandt, E.J., and Novak, E. (1978). Progress in Hormone Res., in press.
22. Quarantillo, B. (1977). M.S. Thesis, Roswell Park Memorial Institute Division, SUNY at Buffalo.

THE EXTENT OF ALLELIC DIVERSITY UNDERLYING
ELECTROPHORETIC PROTEIN VARIATION
IN THE HOUSE MOUSE[1]

Francois Bonhomme[2]
Robert K. Selander

Department of Biology
University of Rochester
Rochester, New York

Although technological developments may soon make it feasible to use amino acid and nucleotide sequences in measuring genic diversity in individuals and populations, thus far the most practical and productive method has involved the indirect assessment of variation in protein structure by the use of electrophoresis -- the allozyme technique introduced to population genetics by J.L. Hubby, R.C. Lewontin, and H. Harris in 1966. Early demonstrations of large amounts of polymorphism and heterozygosity at structural gene loci in fruit flies, house mice, and humans rapidly led to several major developments in population genetics, including a revision of concepts of genetic load

---

[1] Research supported by NIH Grant 2R01 GM-22126.

[2] Present address: Laboratoire d'Electrophorése du C.E.R.E.M., Université des Sciences et Techniques du Languedoc, Montpellier, France.

and a revival of interest in genetic drift and the possibility of nonselective modification of gene pools (concepts, data, and literature summarized in refs. 1-5). As more species were studied, it became apparent that estimates of genic heterozygosity for Drosophila and other invertebrates are, on the average, twice as large as those for mammals and other vertebrates (6). Ecological correlates of this difference were suggested by Selander and Kaufman (7) and Gillespie (8), whereas Nei (4) and Soulé (9) attributed it to an average difference in species number. Another possibility (which has been examined in the work reported here) is that electrophoretically derived measures of heterozygosity provide a misleading picture, mammals and some other vertebrates only appearing to be less variable than invertebrates because in homeotherms there are greater selective constraints on amino acid substitutions affecting the net electrostatic charge of proteins.

The house mouse, Mus musculus, is a "typical" vertebrate in respect to the amount of electrophoretically demonstrable genic variation carried by its natural populations. Estimates of mean individual heterozygosity over 41 loci for different populations vary from 6% to 11%, with an average of 8.0% (10, 11).

Increasingly in recent years it has been realized that, because electromorphs may be a special class of variants (12, 13), our estimates of genic variation may be heavily biased and, hence, unreliable. Even more importantly, recent considerations of various integer-step or "ladder" models postulating that alleles at structural gene loci are clustered in a generally small number of net-charge classes (14) have made it clear that there can be large amounts of allelic diversity underlying individual electromorphs, and that the extent of this diversity may depend on factors other than the proportion of amino acid substitutions that affects the net charge of proteins (15). Even the magnitude of the "hidden" fraction of allelic variation remains unknown (16), yet knowledge of the actual numbers and frequencies of alleles in populations is crucial to further advances in the field of molecular population genetics (17). Attempts to test the neutral theory and its recent derivative, the mutation-equilibrium theory (18, 19) of molecular evolution, have failed largely because the available data pertain to electromorphs rather than actual alleles. For example, it has been argued that the neutral theory cannot account for geographic uniformity of molecular polymorphisms, combined with relatively small numbers of "alleles" (= electromorphs) in species of Drosophila and some other organisms. If migration sufficient to account for the uniformity is invoked (20, 21), effective population size is inflated and large numbers of alleles are expected (22). But what if electromorphs represent highly heterogeneous charge classes of alleles convergently evolving in different populations? Again, Milkman's (23) report of a small effective "allele" (= electromorph) number at each of five loci in Escherichia coli would seem to be incompatible with a theory of neutrality, but the results are not contrary to predictions of the mutation-equilibrium model, which, even for large populations, postulates the generation of a relatively small number of allelically heterogeneous electromorphs.

To detect structural variants of proteins of similar electrophoretic mobility, additional means of analysis, assessing physicochemical properties other than net charge, must be employed. By varying gel concentration and pH and applying a heat-stability test, Singh et al. (17) revealed a total of 37 allelic classes of xanthine dehydrogenase in 146 genomes of Drosophila pseudoobscura, where only six had previously been recognized in the species. (This may be an exceptionally polymorphic locus, as these authors note.) The major effect of varying gel concentration was to enhance the resolution of bands, so that small differences in mobility were increased.

The success of Singh et al. (17), Coyne (24), and others in revealing "hidden" variability in several enzymes in Drosophila by modifying electrophoretic conditions immediately raises the question: Are there equivalent amounts of this type of "hidden" variation as yet undetected in the house mouse? We think not. House mouse proteins already have been electrophoresed in many laboratories on both acrylamide and starch gels under a wide variety of conditions, including gel concentration and pH. For example, we have employed nine buffer types in screening enzymes and other proteins in our laboratory.

Although the recent work on Drosophila has demonstrated the existence of "hidden" variability at several loci, we as yet have no knowledge for any organism of the full extent of allelic heterogeneity of electromorphs. What is needed is a method of determining the actual number of alleles per locus over a randomly selected set of loci. In our recent work on electrophoretic mobility and thermostability of proteins in the house mouse (25), we have employed an experimental approach that potentially can yield reliable estimates of the total allelic diversity at structural gene loci. The results of this work are summarized here.

MATERIALS AND METHODS

Strains of Mice. Our analysis was based on adults of 39 laboratory strains: A/He, Au/Ss, AKR, BALB/c, BDP, BUB/Bn, CBA, C3H/He, C57BL/10, C57BL/Ha, C57BR/cd, C58-waved, CE,

DBA/2, DBA/LiHa, DE, DN, HRS, ICR/Ha, I/ST, JB, LP, MA/My, NZB/BlN, P, PL, RF, RIII/2, SEC/Re, SJL, SM/J, ST/b, SWR, YBR, Yebt, 129, CALIF, MOR, ALBU. All but three of the strains were fully inbred and, hence, monogenic; MOR and ALBU were only partially inbred; and CALIF is a heterogeneous group of hybrids between a wild stock of M. musculus from California and M. m. castaneus from Asia. Additionally, eight individuals each of M. m. castaneus and M. m. molossinus were used in analyzing esterases, PGI, and non-enzymatic proteins.

Tissue Extracts. Samples of plasma, hemolysate, and aqueous extracts of kidney and liver were prepared according to methods described by Selander et al. (26).

Treatment with Thiol Reagents. Twenty inbred strains were screened for variation in the effects of p-chloromercuribenzene sulfonate, cystamine, maleate, oxidized glutathione, and N-ethylmaleimide on the electrophoretic mobilities of 33 enzymes and other proteins. Reagent solutions of 10^{-2} M concentration in a phosphate buffer, pH 6.0, were mixed in equal volume with plasma, hemolysate, or tissue extract and incubated at 37°C for 30 min before electrophoresis.

Heat Denaturation. Prior to electrophoresis, aliquots (30-65 µℓ) of plasma, hemolysate, kidney extract, and liver extract, in 5 ml corked glass culture-tubes were heated for 20 min in a waterbath at temperatures ranging in 3°C steps from 44°C to 65°C.

Electrophoresis and Protein Staining. The procedures (including starch-gel types and stain recipes) were similar to those of Selander et al. (26). Details concerning the six gel-buffer systems used and other aspects of technique are given by Bonhomme and Selander (25).

The 18 proteins (14 enzymes and 4 non-enzymatic proteins), together with the corresponding tissue extracts used in this survey (Tables 1 and 2), were glutamic oxaloacetic transaminase-2 (GOT-2), liver extract; phosphoglucose isomerase (PGI), hemolysate; malate dehydrogenase-1 (MDH-1), kidney; lactate dehydrogenase-A and -B (LDH-1 and LDH-2), kidney; 6-phosphogluconate dehydrogenase (6-PGDH), hemolysate and liver; glucose-6-phosphate dehydrogenase (G-6PDH), hemolysate; phosphoglucomutase-1 and -2 (PGM-1 and PGM-2), hemolysate and liver; esterase-A (EST-A), hemolysate; esterase-1 (EST-1), plasma; esterase-2 (EST-2), plasma; esterase-3 (EST-3), hemolysate; esterase-D (EST-D), plasma; albumin (ALB), plasma; protein-A (PROTEIN-A), plasma; protein-2 (PROTEIN-2), hemolysate; and hemoglobin (HBB), hemolysate.

Work on nine other proteins examined in preliminary studies was discontinued because the results could not be precisely replicated: two peptidases (in hemolysate); malate dehydrogenase-2 (MDH-2), kidney; α-glycerophosphate dehydrogenase (αGPDH), liver; sorbitol dehydrogenase (SDH), liver; alcohol dehydrogenase (ADH), liver; malic enzyme (ME), kidney; isocitrate dehydrogenase-1 (IDH-1), kidney; and glutamic oxaloacetic transaminase-1 (GOT-1), liver.

Procedure. Two surveys were made, each involving extracts from one mouse of each strain. Some 9,000 aliquots were electrophoresed on 450 gels in the course of the study.

RESULTS

Treatment with Thiol Reagents

A method for detecting certain amino acid substitutions leading to the acquisition or loss of cysteine residues was developed by Hopkinson and Harris (27) for use in screening human populations for "silent" ("hidden") variants of peptidase and other enzymes in hemolysates. Allozymes dif-

TABLE 1. Heat-denaturation Profiles for 13 Proteins[a]

Protein	Number of electromorphs	Temperature °C (20 min)					
		50°	53°	56°	59°	62°	65°
Enzymes							
GOT-2	2	+++	+++	+++	+++	++-	---ε
PGI	2	+++	++-	---	---	---	---
MDH-1	1	+++	+++	+--	---	---	---
LDH-1[b]	1	+++	+++	+++	+--ε	---	---ε
LDH-2[b]	1	+++	+++	+++	+++	++-	---ε
6-PGDH	1	+++	+++	++-	---	---	---
G-6PDH	1	+++	+--	---	---	---	---
PGM-1	2	+++	+++	+++	---	---	---
PGM-2	2	+++	+++	+++	++-	+--	---ε
Other proteins							
ALB	2	+++	+++	+++	++-	++-	+--[c]
Protein-A	1	+++	+++	+++	++-	---ε	---
Protein-2	1	+++	+++	---	---	---	---ε
HBB	3	+++	+++	+++	++-	+--	---

[a] Key: +++: full activity (enzymes) or concentration (other proteins); ++-: slight reduction in activity; +--: severe reduction in activity; ---ε: trace of activity; ---: no activity.
[b] Homotetramers.
[c] Trace remaining at 68°C.

fering in number of free reactive sulfhydryl (-SH) groups can be distinguished electrophoretically following treatment with thiol reagents. With each reagent, the reaction product exhibits a characteristic electrophoretic mobility that

TABLE 2. Heat-denaturation Profiles for Five Esterases[a]

Enzyme and allele[b]	Temperature °C (20 min)					
	44°	47°	50°	53°	56°	59°
EST-A						
100R	+++	+++	+++	+--	---ε	---
100S	+++	---ε	---	---	---	---
EST-1						
110S	+++	+++	+++	+++	+--	---
110R	+++	+++	+++	+++	+++	---ε
100	+++	+++	+++	+++	+--	---
95	+++	+++	+++	+++	+++	---ε
EST-2						
100R	+++	+++	+--	---	---	---
100S	+++	---ε	---	---	---	---
90	+++	+++	+--	---	---	---
85	+++	+++	+++	---ε	---	---
75	+++	+++	+++	---ε	---	---
EST-3						
80	+++	+++	---ε	---	---	---
75	+++	---	---	---	---	---
70	+++	+++	---ε	---	---	---
EST-D						
100S	+++	+++	+++	++-	---ε	---
100R	+++	+++	+++	+++	++-	---ε
110	+++	+++	+++	+++	++-	---ε
95	+++	+++	+++	+++	++-	---ε

[a] Key: See Table 1.
[b] Numbers indicate relative mobilities of electromorphs. R = heat-resistant allele; S = heat-sensitive allele.

is either the same as the untreated enzyme or faster or slower, depending on whether the added moiety is neutral, acidic, or basic.

The electrophoretic mobility of 16 of the 33 proteins tested in the house mouse (including PGI, ALB, LDH-1, ME, IDH-1, GOT-1, and MPI) was modified in the "direction" and to the extent predicted by knowledge of the charge characteristics of the various reagents, and that of 17 of the proteins was unaffected. But for none of the proteins was there evidence of interstrain heterogeneity. Some enzymes (e.g., LDH, ME, and IDH) were inhibited by PCMB and/or by one or more other reagents, but, again, there was no interstrain heterogeneity in the reaction.

Heterogeneity in Thermostability of Electromorphs

No interstrain heterogeneity was detected in the thermostability of the electromorphs of any of the four non-enzymatic proteins studied (Table 1), and the electromorphs themselves did not differ in thermostability.

The 14 enzymes analyzed are represented in the 39 strains of mice by a total of 27 electromorphs. As shown in Table 1, for nine of the enzymes heat treatment revealed no heterogeneity within or between electromorphs. Thus, for example, the thermostability profiles of strains homozygous for either fast- or slow-migrating electromorphs (allozymes) of GOT-2 were indistinguishable, and, similarly, the two electromorphs could not be distinguished.

But in the case of five esterases (Table 2), the heat-denaturation technique revealed differences between electromorphs, and at four of these loci (EST-A, EST-1, EST-2, and EST-D), interstrain heterogeneity within electromorphs was detected. For example, at the electrophoretically monomorphic EST-A locus, most strains have an allele encoding a heat-resistant form of the enzyme (100R), which retains full activity at 50°C, but only a trace at 56°C. But the strains AKR and SJL have a heat-sensitive form of the protein that denatures between 44° and 50°C. Progeny tests demonstrated a simple co-dominant Mendelian segregation of the 100R and 100S alleles at the EST-A locus; and we have determined that both alleles (100R and 100S) are segregating in M. m. castaneus. The total number of new alleles detected in our survey was four, one each at four of the esterase loci.

Major changes in configuration of albumin and the other non-enzymatic proteins probably must occur on heating before we are able to detect decreases in concentration of the proteins at their normal positions on gels. But even minor changes in the tertiary (or quaternary) structure of an enzyme might lessen or eliminate its activity. Hence, the power of the heat-denaturation technique to detect heterogeneity among polypeptides undoubtedly is much greater for enzymes than for other proteins. In the following analysis and discussion, we will consider only the data derived from our studies of the 14 enzymes.

Estimating Total Genic Diversity

The total number of allelic classes detected in our survey of 14 enzyme loci was 31, which may be considered as 27 electromorphs + 4 thermomorphs not detected by electrophoresis, or as 20 thermomorphs + 11 electromorphs not detected by heat-denaturation (Table 3).

We will call allopeptides the distinctive polypeptides encoded by a structural gene locus; and the number of these is equivalent to the number of alleles segregating at a locus. Allomorphs are defined as sets of one or more allopeptides having the same properties in regard to the techniques of

TABLE 3. Distribution of Electrophoretic and Thermic Variation at 14 Enzyme Loci

Enzyme	Number of		
	Electromorphs	Thermomorphs	Allomorphs
GOT-2	2	1	2
PGI	2	1	2
MDH-1	1	1	1
LDH-1	1	1	1
LDH-2	1	1	1
6-PGDH	1	1	1
G-6PDH	1	1	1
PGM-1	2	1	2
PGM-2	2	1	2
EST-A	1	2	2
EST-1	3	2	4
EST-2	4	3	5
EST-3	3	2	3
EST-D	3	2	4
Total	27	20	31

detection employed. Since an invariable locus is nevertheless represented by one polypeptide, a relevant measure of total variability is the number of allopeptides minus one; and the comparable measure of the variation revealed by a technique or several techniques is the number of allomorphs detected minus one.

For purposes of computation, let X = the (unknown) actual (total) number of allopeptides (or alleles) at L loci assayed, which in the present case is 14; A = the number of allomorphs demonstrated by joint application of electrophoresis and heat-denaturation = 31; E_v = the number of electromorph variants, which is $E - L = 27 - 14 = 13$; T_v = the number of thermomorph variants, which is $T - L = 20 - 14 = 6$; J_v = the number of allomorph variants detected by joint application of the two techniques, which is $A - L = 31 - 14 = 17$; and U = the (unknown) actual number of undetected allopeptide variants.

If the two techniques employed are orthogonal, such that the probability of an allopeptide variant being recognized electrophoretically is independent of the probability of its being detected by heat-denaturation, it can be shown that

$$U = (J_v - E_v)(J_v - T_v)/(J_v - (J_v - E_v) - (J_v - T_v))$$
$$U = (17 - 13)(17 - 6)/(17 - (17 - 13) - (17 - 6)) = 44/2 = 22$$

Then, the estimated actual number of allopeptide variants, $V = J_v + U = 17 + 22 = 39$. And the estimated actual number of allopeptides or alleles at the 14 loci, $X = V + L = 39 + 14 = 53$ (or $X = A + U = 31 + 22 = 53$). Hence, the estimated mean number of alleles per locus is $X/L = 53/14 = 3.8$, which compares with 1.9 electromorphs per locus; 1.4 thermomorphs per locus; and 2.2 allomorphs per locus.

The power of electrophoresis to detect variants is $E_v/V = 13/39 = 0.33$, and that of heat-denaturation is $T_v/V = 6/39 = 0.15$. Jointly applied, the power of the two techniques is $J_v/V = 17/39 = 0.44$. Of the estimated 53 alleles at the 14 enzyme loci assayed, approximately 50% ($E/X = 27/53$) were detected by electrophoresis, 38% by heat-denaturation, and 58% by joint application of the techniques. The undetected proportion was 42%.

DISCUSSION

A significant finding from our research on the house mouse is that the heat-denaturation technique we have used apparently is rather efficient at differentiating proteins with amino acid differences at the level that characterizes electromorphs. (Whether or not this level is at or near the single substitution is another problem.) In half the pairwise comparisons of electromorphs, a thermostability difference was detected. The generally low level of demonstrable heterogeneity among the 27 electromorphs at 14 enzyme loci over 39 strains of the house mouse is important, since it immediately suggests that there cannot be large amounts of hidden variability at most of the loci examined. Our findings, when compared with those for XDH, ODH, and EST-5 in Drosophila (17, 24, 28) suggest that the differences in effective number of alleles and heterozygosity between Mus and Drosophila are greater than those indicated by electrophoretically derived estimates.

The value of our approach and results depends on the validity of several underlying assumptions. First, we have assumed that the 39 laboratory strains and a small number of individuals of the subspecies M. m. castaneus and M. m. molossinus represent the species Mus musculus as a whole insofar as the common alleles at the 14 enzyme loci studied are concerned. One could argue that this is a safe assumption because most of the high- and moderate-frequency, widespread electromorphs occurring in wild populations also are represented in the group of domestic strains used in our study (29-32). But we cannot determine precisely how many genomes from wild populations are represented in the 39 strains, although the number probably is between 10 and 20. Laboratory strains were derived from wild populations in Europe and the

United States and are a mixture of the subspecies M. m. domesticus, M. m. brevirostris, and M. m. musculus, but their origins cannot now be sorted out. To increase the representation of the species, we included individuals of M. m. castaneus and M. m. molossinus and the CALIF strain.

Second, we have assumed that charge-changing amino acid substitutions are a representative subset of substitutions with regard to their effect on thermostability. At present, we have no evidence bearing on this problem. If in fact thermostability differences are likely to be associated with charge-changing substitutions, we have erred in underestimating the mean number of alleles per locus. But even if the techniques are not strictly independent, our experimental approach can yield valuable insight into the problem of the amount of "hidden" variability in populations by providing information on the relative amounts of heterogeneity within and between electromorphs. Moreover, studies of a variety of organisms will provide a test of the predictions of Nei and Chakraborty (15) regarding the amounts of allelic diversity underlying electromorphs that would be expected according to the neutral and the mutation-equilibrium theories of molecular polymorphism.

On the theory that electromorphs are allelically heterogeneous charge classes that can convergently evolve in populations, several models of step-wise integer production of electromorphs recently have been investigated (14, 33-35). In these models, mean selection coefficients may be equal over electromorphs or may increase with increasing distance of an electromorph from that of the "type allele." But in any event, it is proposed that polymorphism can be maintained in populations by a balance between mutation to slightly deleterious alleles (whose products are manifested as electromorphs at integer states on gels) and selection against the mutants. Provisions of the models are that the product of the effective population size and the mutation rate is large and that the selection coefficient is larger than the mutation rate. Under these conditions, there will be in large populations a build-up of silent heterogeneity underlying electromorphic variation.

Nei and Chakraborty (15, 36, 37) have pointed out that the prevalent notion that electrophoresis detects some fixed proportion (e.g., 1/4) of variant polypeptides is erroneous. Rather, the proportion of alleles detectable by electrophoresis depends on population size. Using the infinite allele model of Kimura and Crow (38) to obtain the expected number of alleles at the codon level (neglecting selection), Ewens' (39) formula for the expected number of alleles in a sample of s genes at equilibrium, and Kimura and Ohta's (41) formula for the expected number of electromorphs per locus, they have calculated the proportion of alleles undetectable by electrophoresis for various values of $L = N_e v$ (where N_e is the effective population size and v the mutation rate at the codon level) and sample sizes (s). If L is smaller than 0.01, the ratio of numbers of silent alleles to electromorphs is very small (e.g., 1.2 alleles per 1.1 electromorphs in samples of 200), but if L is large (e.g., 4.0), the expected number of silent alleles per electromorph in samples of 200 is 8 (42 alleles at the codon level and 5 electromorphs). In general, they argue that L may actually be less than 1, since, as they note, the average heterozygosity for electromorphs is 0.3 or less in all bisexual organisms thus far studied.

The model is simplistic, but it has the advantage of making specific predictions that can be tested, even if we cannot determine the absolute values of N_e for different organisms. The proportion of hidden genetic variability relative to total variability or to that detectable electrophoretically should be large in species with large numbers; whereas in species with small numbers, most of the variability should be detectable electrophoretically. Where there is little or no hidden variability, adjacent electromorphs will differ at little more than one amino acid, on the average.

Chakraborty and Nei (36) have estimated the probability distributions of the number of codon differences between two identical electromorphs and between two electromorphs with a one-step difference in charge ($\underline{M} = 4\underline{N_e}v$, and $\underline{c}$ is assumed to be 1/4). If $\underline{M} = 4\underline{N_e}v$ is 0.1, 87% of electromorphs will differ at one codon only; and there will be essentially no heterogeneity within electromorphs (there will be almost no hidden variation). As $\underline{M}$ increases (e.g., as $\underline{N_e}$ increases) to 10, the ratio of codon difference within electromorphs to those between electromorphs approaches 1, and the average number (and the variance) of codon differences increases to about 5. At $\underline{M} = 10$, only 16% of one-step electromorphs differ in a single codon, and only 22% of identical electromorphs have identical codon structure.

According to Nei and Chakraborty's (36) model, in a sample of 40 genes, a ratio of 3.7 alleles at the codon level to 1.7 electromorphs per locus is expected when $\underline{L} = 0.2$. If, for example, the mutation rate is 10^{-8}, $\underline{N_e}$ would be 5×10^6. Since the mean numbers of alleles and electromorphs per locus estimated for the house mouse (3.8 alleles/1.9 electromorphs) are similar to those in the example, the question arises: Can the effective number for this species be as small as 5×10^6? In other words, is it a reasonable possibility that the house mouse, notwithstanding its present worldwide distribution, is, from an evolutionary standpoint, a relatively small species? We think so for the following reasons: 1) The present worldwide distribution and very large population size of Mus musculus were achieved in historical times; and prior to the Neolithic period (∿4,000 B.P.), the subspecies which now occupy Europe, the Western Hemisphere, Australia, and other areas of the world were confined to the Middle East (see review in ref. 40). Since the effective size of a population is approximately the harmonic mean of numbers of breeding individuals over generations, the relatively recent expansion in size will have had little effect on the effective species number. 2) Because subdivision of Mus musculus is extensive at several levels from the local breeding unit to the subspecies, $\underline{N}_e$ probably is not to be equated with the full species number (see discussion in ref. 9). Soulé (9) suggests that $\underline{N}_e$ for Mus musculus is ∿10^7 and for other mammals from 10^4 to 10^6. Compared with some widely distributed species of Drosophila and other small invertebrates, the effective species number of the house mouse, and of vertebrates in general, undoubtedly is several orders of magnitude smaller.

In sum, to a first approximation, our results seem compatible with predictions of the Nei-Chakraborty theory. For the house mouse, we estimate that, on the average, there are only 3.8 alleles per structural gene locus, and that half the existing allelic variation is detectable electrophoretically. Further testing of the theory, involving comparable studies of organisms having very large species numbers, such as Drosophila pseudoobscura and Escherichia coli, presently are in progress in our laboratory.

SUMMARY

In a survey of variation in both electrophoretic mobility and thermostability of 14 enzymes in 39 laboratory strains

of the house mouse (Mus musculus), 27 electromorphs and 20
thermomorphs were identified. Heat-denaturation detected 4
variants within electromorphs, and electrophoresis detected
11 variants within thermomorphs. The total number of distinctive polypeptides (allomorphs) distinguished by joint
application of the techniques was 31. From these data, and
on the assumption that heat-denaturation and electrophoresis
are independent, orthogonal methods of detecting variation,
it is estimated that the actual total number of alleles at
the 14 enzyme loci is 53, or an average of 3.8 per locus (2.0
per electromorph). Electrophoresis apparently detects one-third of variant polypeptides, thus revealing about 50% of
the alleles at structural gene loci in the house mouse. To
a first approximation, the results of this analysis are consistent with predictions of the Nei-Chakraborty theory regarding the degree of allelic heterogeneity of electromorphs as
a function of species number.

No heterogeneity within or between electromorphs of four
non-enzymatic proteins was detected by the heat-denaturation
technique. And for 33 enzymes and other proteins, treatment
with various thiol reagents failed to reveal heterogeneity
among 16 strains of mice tested.

ACKNOWLEDGMENTS

We are indebted to V.M. Chapman for supplying the CALIF,
MOR, and ALBU strains, a sample of his M. musculus (California) x M. m. castaneus hybrids, and samples of M. m. castaneus
and M. m. molossinus. The JB strain was provided through the
courtesy of E. Caspari. J.L. King suggested the statistical
method of estimating allelic diversity, which is equivalent to,
but simpler than, that employed by Bonhomme and Selander (25).

REFERENCES

1. Kimura, M., and Ohta, T. (1971). "Theoretical Aspects of Population Genetics." Princeton University Press, Princeton, N.J.
2. Le Cam, L.M., Neyman, J., and Scott, E.L. (eds.) (1972). Proc. Sixth Berkeley Symp. Math. Stat. Prob., vol. V. University of California Press, Berkeley.
3. Lewontin, R.C. (1974). "The Genetic Basis of Evolutionary Change." Columbia University Press, New York.
4. Nei, M. (1975). "Molecular Population Genetics and Evolution." North-Holland, Amsterdam.
5. Ayala, F.J. (ed.) (1976). "Molecular Evolution." Sinauer, Sunderland, Mass.
6. Selander, R.K. (1976). In "Molecular Evolution" (F.J. Ayala, ed.), p. 21. Sinauer, Sunderland, Mass.
7. Selander, R.K., and Kaufman, D.W. (1973). Proc. Natl. Acad. Sci. 70:1875.
8. Gillespie, J. (1974). Amer. Natur. 108:831.
9. Soulé, M. (1976). In "Molecular Evolution" (F.J. Ayala, ed.), p. 60. Sinauer, Sunderland, Mass.
10. Selander, R.K., Hunt, W.G., and Yang, S.Y. (1969). Evolution 23:379.
11. Selander, R.K., and Yang, S.Y. (1969). Genetics 63:653.
12. Markert, C.L. (1968). Proc. N.Y. Acad. Sci. 151:14.
13. Kimura, M. (1971). Theoret. Pop. Biol. 2:174.
14. Ohta, T., and Kimura, M. (1973). Genet. Res. 22:201.
15. Nei, M., and Chakraborty, R. (1976). J. Mol. Evol. 8:381.
16. Johnson, G.B. (1976). Genetics 83:149.
17. Singh, R.S., Lewontin, R.C., and Felton, A.A. (1976). Genetics 84:609.
18. Ohta, T. (1974). Nature 252:351.

19. Ohta, T. (1976). Theoret. Pop. Biol. 10:254.
20. Kimura, M., and Maruyama, T. (1971). Genet. Res. 18:125.
21. Maruyama, T., and Kimura, M. (1974). Nature 249:30.
22. Ayala, F.J. (1972). In "Proc. Sixth Berkeley Symp. Math. Stat. Prob." Vol. V. (L.M. LeCam, J. Neyman, E.L. Scott, eds.), p. 211. University of California Press, Berkeley.
23. Milkman, R. (1975). "Isozymes" Vol. IV, p. 273. Academic Press, New York.
24. Coyne, J. (1976). Genetics 84:593.
25. Bonhomme, F., and Selander, R.K. (1978). Biochem. Genet., in press.
26. Selander, R.K., Smith, M.H., Yang, S.Y., Johnson, W.E., and Gentry, J.B. (1971). Stud. Genet. 6 (Univ. Texas Publ. 7103):49.
27. Hopkinson, D.A., and Harris, H. (1969). Ann. Hum. Genet. 33:81.
28. McDowell, R.E., and Prakash, S. (1976). Proc. Natl. Acad. Sci. 73:4150.
29. Roderick, T.H., Ruddle, F.H., Chapman, V.M., and Shows, T.B. (1971). Biochem. Genet. 5:457.
30. Taylor, B.A. (1972). J. Hered. 63:83.
31. Wheeler, L.L., and Selander, R.K. (1972). Stud. Genet. 7 (Univ. Texas Publ. 7213):269.
32. Nichols, E.A., and Ruddle, F.H. (1973). J. Histochem. Cytochem. 21:1066.
33. King, J.L. (1974). Genetics 76:607.
34. Ohta, T., and Kimura, M. (1975). Amer. Natur. 109:137.
35. King, J.L., and Ohta, T. (1975). Genetics 79:681.
36. Chakraborty, R., and Nei, M. (1976). Genetics 84:385.
37. Chakraborty, R. (1977). J. Mol. Evol. 9:313.
38. Kimura, M., and Crow, J.F. (1964). Genetics 48:725.

39. Ewens, W.J. (1972). Theoret. Pop. Biol. 3:87.
40. Hunt, W.G., and Selander, R.K. (1973). Heredity 31:11.
41. Kimura, M., and Ohta, T. (1975). Proc. Natl. Acad. Sci. 72:2761.

CYTOGENETICS[1]

Orlando J. Miller
Dorothy A. Miller

Departments of Human Genetics and Development
and Obstetrics and Gynecology
College of Physicians and Surgeons
Columbia University
New York, New York

Mouse chromosomes can be examined by three types of staining methods. The classical approach, in which the entire chromosome is more or less uniformly stained by orcein, Giemsa or Feulgen, can be used to determine the number of chromosomes (2n=40 in <u>Mus musculus</u>), the position of the centromere (at one end of each <u>musculus</u> chromosome), and the presence and position of secondary constrictions. This approach has been largely replaced by the general banding methods (Q, G and R), which can be used for the same purposes but also produce a consistent pattern of light and dark bands along the length of each chromosome. The banding pattern is different for each mouse chromosome and thus permits its accurate identification (1, 2). In addition, special banding methods can be used to observe particular

---

[1] This work was supported in part by grants from the U.S. Public Health Service (CA 12504) and the National Foundation-March of Dimes.

regions of chromosomes, such as the centromeric heterochromatin or nucleolus organizer regions. The special methods are most effective when combined with a general banding method by which the chromosomes are identified. Each of these three types of staining has been used to demonstrate differences between species or strains of Mus (Table 1).

The general banding techniques are, and will remain for the foreseeable future, the most powerful tools in cytogenetic analysis. The first of these methods to be discovered was quinacrine fluorescence (Q-) banding (3), but the same banding patterns are revealed by Giemsa staining after treatment with trypsin (G-banding). Based on the general banding patterns, a standard karyotype of the mouse has been adopted (2) and a nomenclature for individual bands proposed (4). Figure 1 shows a karyotype of Mus musculus stained with quinacrine. The intensity of fluorescence varies along the length of each chromosome, producing a banding pattern that is consistent and highly characteristic for each chromosome, and which remains unaltered if a segment of chromosome is transferred to a different location. There are no Q-brilliant regions in musculus. Some of the chromosomes in the cell shown in Figure 1 have a secondary constriction, which probably represents the nucleolus organizer region. More specific stains for this region are discussed below.

The ability to identify each mouse chromosome has been utilized in a number of ways. The Q-banding technique was used to identify the chromosomes in a series of translocations involving known linkage groups, making possible the assignment of each linkage group to a specific chromosome, as well as the location of the centromeric end of linkage groups (1, 5, 6). Q- and G-banding were used to determine where the albino locus is located on chromosome 7 by locating a band that was deleted in the c^{25H} deletion, which is about 6 cM long (7). Methods are being developed to study the

TABLE 1. Methods That Reveal Chromosomal Differences Between Species, Subspecies or Strains of Mus

Type	Banding method
Translocation or inversion	Unbanded, Q, G, R
Secondary constriction	Unbanded, Q, G, R
Variant satellite DNA	Q, C, Hoechst 33258, anti-5-methylcytidine
18S + 28S rDNA sites	Silver, in situ hybridization

chromosomes of prometaphase cells, which have a greater number of bands (8); these would make possible the recognition of deletions even smaller than the c^{25H}. G-banding techniques have been used to identify the presence of specific chromosome changes in malignancies, for example, the association of trisomy 15 with leukemia in the AKR mouse (9).

Figure 2 is a karyotype of Mus cervicolor (10). The banding patterns of the chromosomes of musculus and cervicolor are the same except for one striking difference: each cervicolor chromosome has a brilliant fluorescent centromeric region separated from the remainder of the chromosome by a dull region. This difference reflects an evolutionary change in the satellite DNA (11, 12), which is concentrated in this region of mouse chromosomes. Musculus has a single satellite DNA that is dully stained by quinacrine; cervicolor has two satellite DNAs that have different molecular properties and stain differently, one being Q-brilliant and the other Q-dull. As shown in Figure 2, these satellite DNAs are located in adjacent regions near the centromere.

If we assume that there are 30,000 genes in the mouse, the average chromosome would have about 1,500 genes; how-

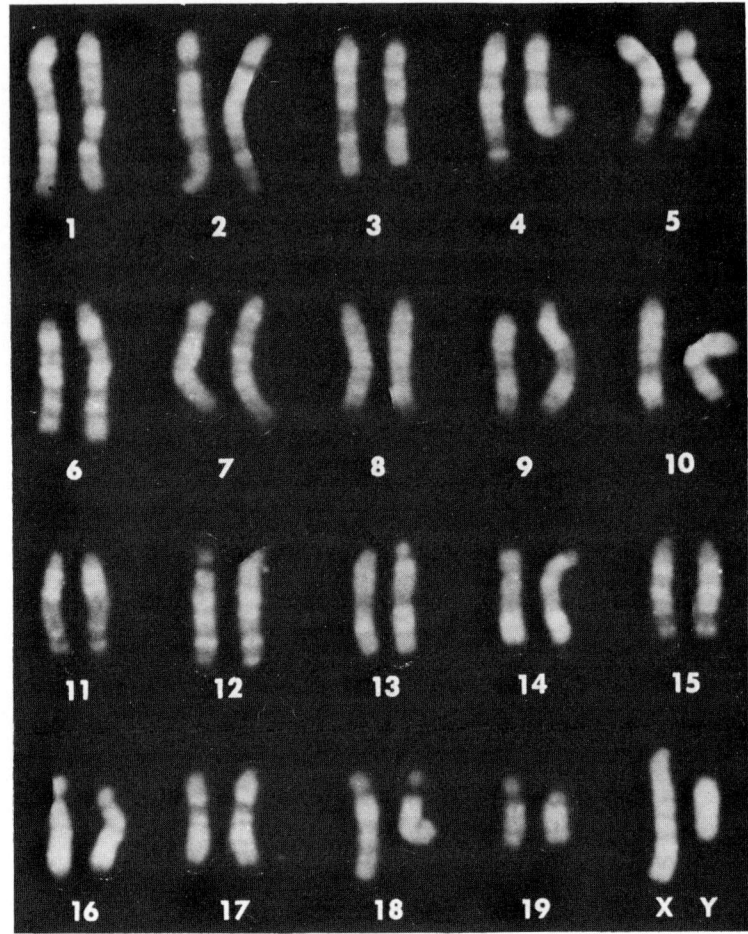

FIGURE 1. Chromosomes of Mus musculus stained with quinacrine. Secondary constrictions are present near the centromere of numbers 12, 16, 18, and 19.

ever, it has enough DNA to code for 50 to 100 times that many genes. Much of this DNA may be the transcribed spacers coding for the heterogeneous nuclear RNA that is degraded in RNA processing. More may be nontranscribed spacer regions, which can be quite heterogeneous (13) and may be clustered. There is some evidence that the Q-bright or G-dark bands

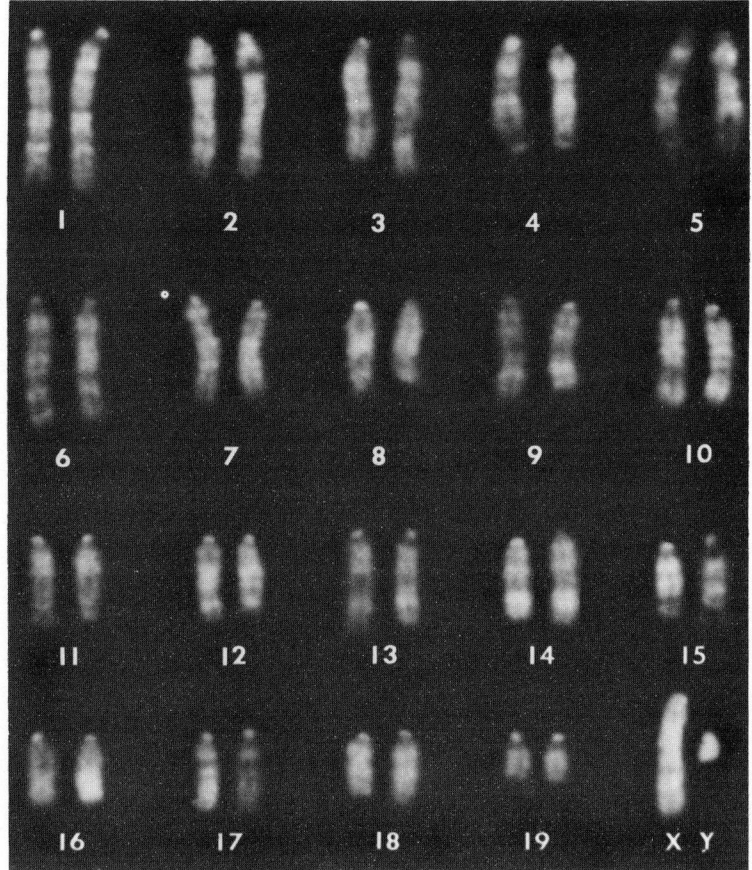

FIGURE 2. Chromosomes of <u>Mus</u> <u>cervicolor</u> stained with quinacrine. The banding patterns are the same as those of <u>M</u>. <u>musculus</u> except in the centromeric region. Reprinted from ref. 10 by permission of Academic Press.

consist primarily of such DNA, with the genes concentrated in the bands that are dully stained by these methods (the reverse or R-bands). As methods of DNA sequencing and analysis are improved, variations in nontranscribed DNA may provide major sources of information about differences between mouse sublines.

All species of Mus examined thus far have the same general chromosome banding patterns (10, 14-16) (Fig. 3). This is not unexpected for the closely related subspecies Mus musculus musculus and M. musculus molossinus, but it is surprising that the chromosomes of M. cervicolor, M. caroli, M. cookii, M. booduga fulvidentris, and M. dunni show no sign of inversions, reciprocal translocations, or centric fusion translocations. Inversions have been specifically looked for in wild mice that are fairly closely related to musculus, but none have been found (T. H. Roderick, personal communication). Although the unbanded Giemsa-stained karyotype of M. dunni (Fig. 4) clearly differs from that of M. musculus in having short arms on virtually every autosome, as well as a metacentric X and a very large Y, G- and C-banding have shown that the differences are entirely due to the presence of short arms composed of constitutive heterochromatin, with no detectable change in the banding pattern of any chromosome's long arm (16). This is an important observation because differences of this magnitude observed in the pre-banding era were usually interpreted as the result of inversions or translocations.

The translocations that have been observed in wild mice, as well as those that have occurred spontaneously in laboratory stocks, have all been of the centric fusion type (6, 17-19). The tobacco mouse, Mus poschiavinus, has 2n=26 chromosomes, including seven pairs of biarmed chromosomes (17). These have been identified by their Q-banding pattern (Fig. 5) and in this case, too, the patterns are identical to those of musculus chromosomes, showing that the biarmed chromosomes were produced by centric fusion. Multiple centric fusion chromosomes have been found in homozygous form in mice from several mountain valleys in Italy and Switzerland (17-19). Comparison of the various populations indi-

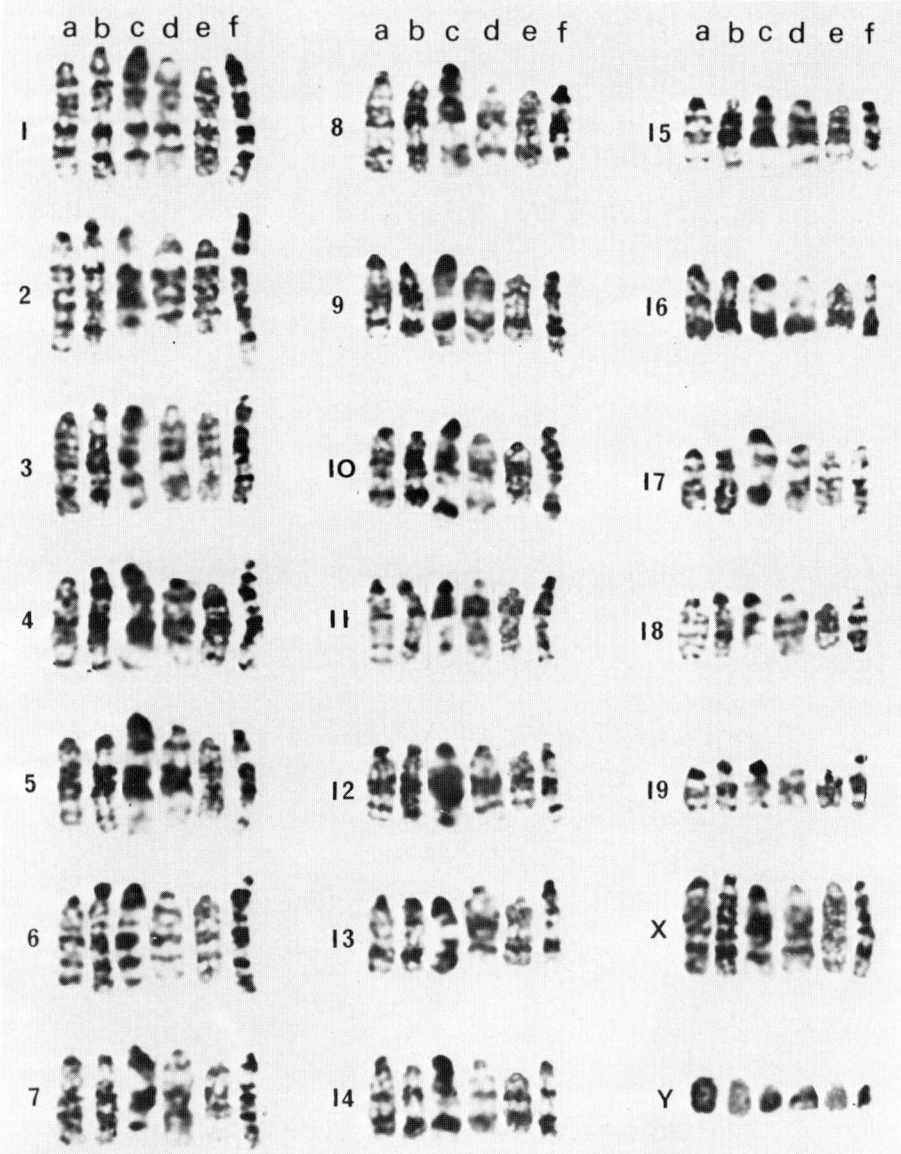

FIGURE 3. Comparison of the G-band pattern of the chromosomes of a) M. musculus musculus, b) M. musculus molossinus, c) M. caroli, d) M. cervicolor, e) M. cookii, and f) M. booduga fulvidentris. The bands are the same in each species. Reprinted from ref. 15 by permission of S. Karger AG, Basel.

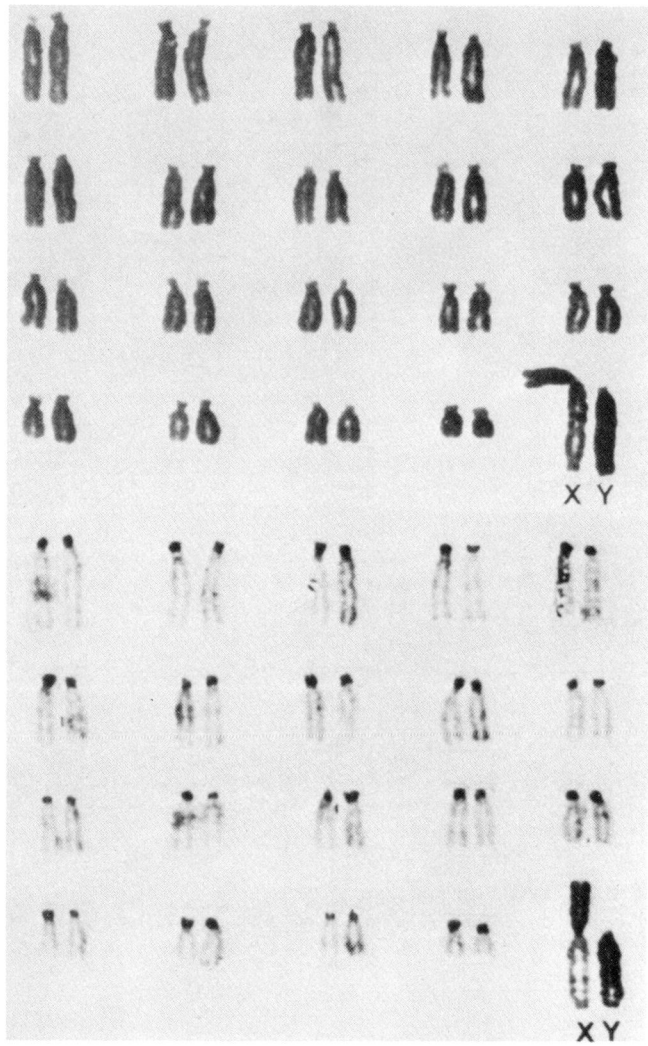

FIGURE 4. Chromosomes of M. dunni stained with conventional Giemsa (top) and to show C-banding (bottom). The short arms of each chromosome (including the X), and most of the Y, are heterochromatic. Reprinted from ref. 16 by permission of S. Karger AG, Basel.

cates that, in general, different chromosomes are represented in the fusion chromosomes of mice from different locations.

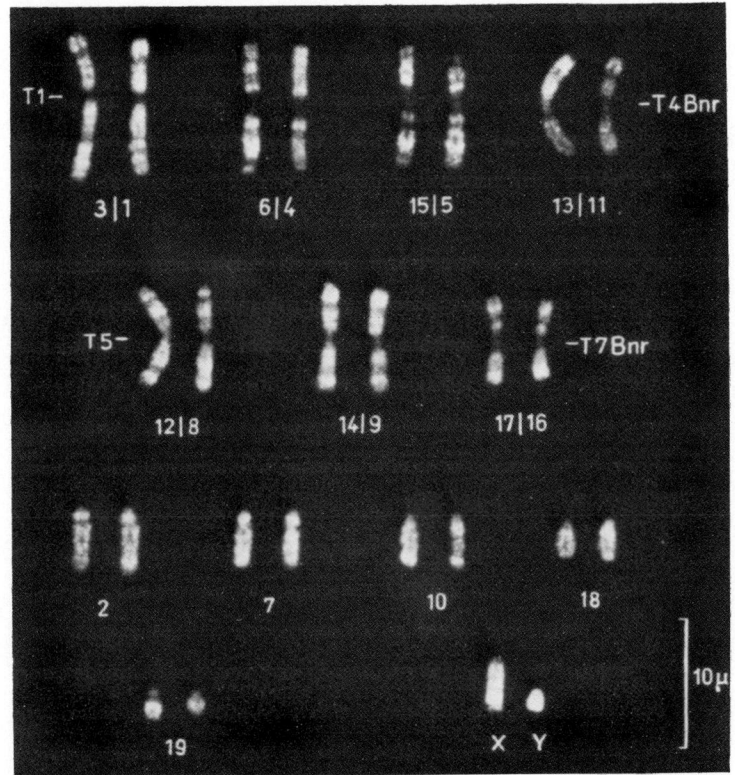

FIGURE 5. Karyotype of M. poschiavinus (2n=26). The Q-band patterns are the same as M. musculus, but seven pairs of autosomes have undergone centric fusion. Reprinted from ref. 17 by permission of Springer-Verlag, New York.

Figure 6 shows pairing of meiotic chromosomes in an F1 between a mouse of the T Rov 1-6 stock (caught near Roveredo in the Val Mesolcina) and poschiavinus (caught in the Valle di Poschiavo about 75 miles away). Only one of the fusion chromosomes is identical in the two stocks of mice, and virtually every chromosome is included in one or another of the 13 biarmed chromosomes.

C-banding involves a rather nonspecific type of staining reaction: treating chromosome preparations with alkali and then staining with Giemsa. The C-band method stains only the

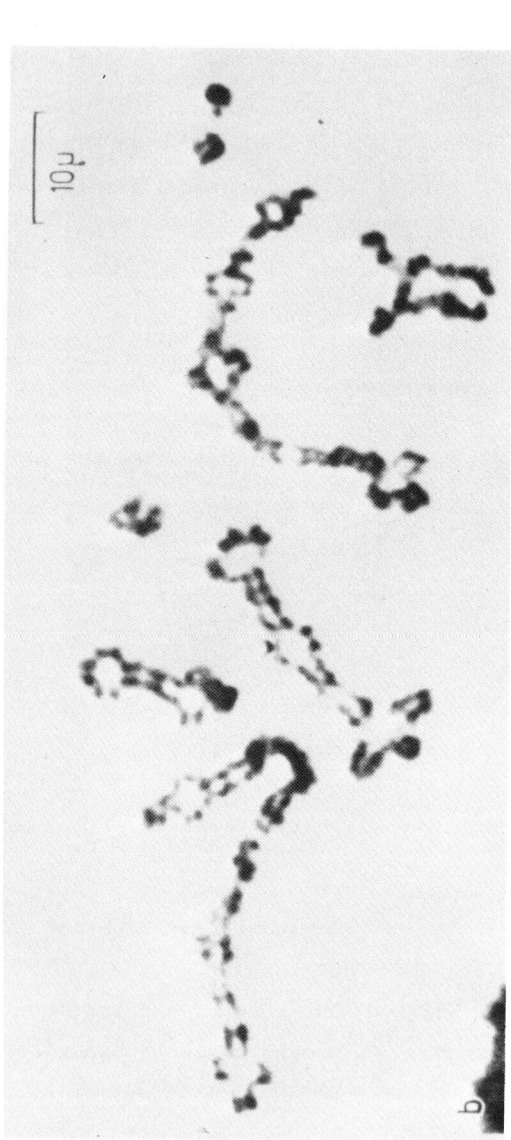

FIGURE 6. Meiotic cell of a (T Rov 1-6 x M. poschiavinus)F1 stained with Giemsa. All but one of the centric fusion chromosomes of the parents involve different chromosomes. Reprinted from ref. 17 by permission of Springer-Verlag, New York.

centromeric heterochromatin, and a general banding technique must be used on the same cells in order to identify the chromosomes. When this is done, differences are seen between wild mice and laboratory strains (20) and between even closely related sublines. C57L/J and C57BL/6J, for example, differ markedly in the amount of C-band material on five different pairs of chromosomes, and 129 and AKR differ in the amount on four pairs of chromosomes (21). In the inbred lines examined thus far, eight chromosomes have been shown to have such variants (22) and if we add those found in molossinus virtually every chromosome has a C-band variant. In an F1 between one molossinus strain and C57BL/6J, differences were observed between eight pairs of homologs (14) (Fig. 7). Since these variants are at the centromeric end of the chromosome, they provide useful markers in monitoring stocks for changes in this region. C-banding also facilitates meiotic analysis because the position of each centromere can be located; this is particularly useful in analyzing reciprocal translocations (23). Figure 8 shows that gross differences in the amount of C-band material on homologous chromosomes in a musculus x molossinus F1 did not interfere with meiotic pairing 914).

The C-band regions of musculus can also be stained using Hoechst 33258 (24). Hsu et al. (15) have reported that Hoechst 33258 produces bright fluorescence in the C-band regions of musculus and molossinus chromosomes but not in those of other species of Mus. Changes in satellite DNA could be responsible for this, since Hoechst 33258 shows enhanced fluorescence when bound to some AT-rich DNA sequences, such as those known to be present in musculus satellite DNA.

Greater specificity in detecting classes of repetitive DNA can be obtained by using antibodies to specific DNA bases, e.g., anti-5-methylcytidine. These antibodies react

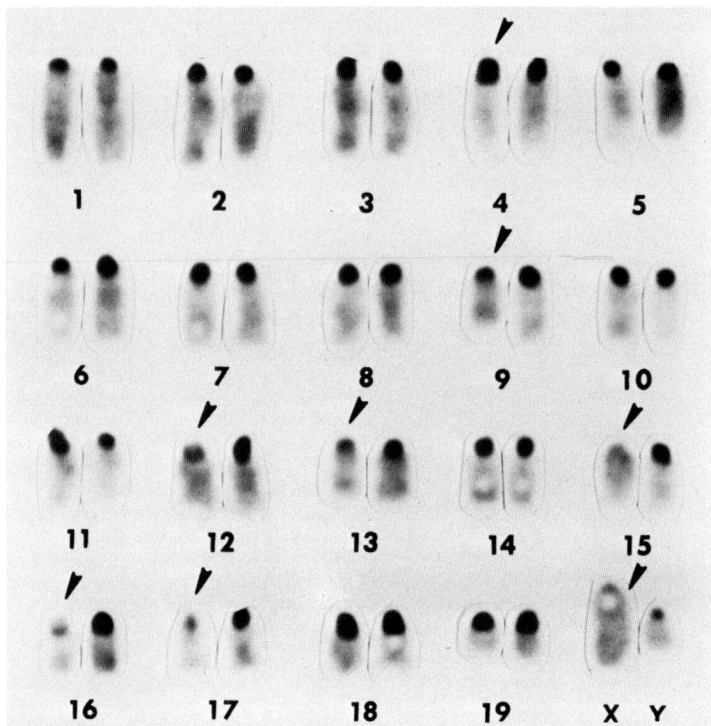

FIGURE 7. Karyotype of (musculus x molossinus)F1 stained to show C-bands. Arrows indicate homologous chromosomes with marked differences in the amount of centromeric heterochromatin. Reprinted from ref. 14 by permission of Springer-Verlag, New York.

only with single-stranded DNA so that, by using selective methods of denaturation of DNA in conjunction with specific antibodies, one can visualize specific classes of repetitious DNA sequences in mouse chromosomes (14, 25, 26). In musculus, satellite DNA contains much more 5-methylcytosine than does main band DNA (27). When chromosomes are UV-irradiated, the major effect is the production of thymine dimers, which creates single-stranded regions primarily in AT-rich regions such as the satellite DNA. One observes intense binding of anti-5-methylcytidine to the satellite DNA because it is both

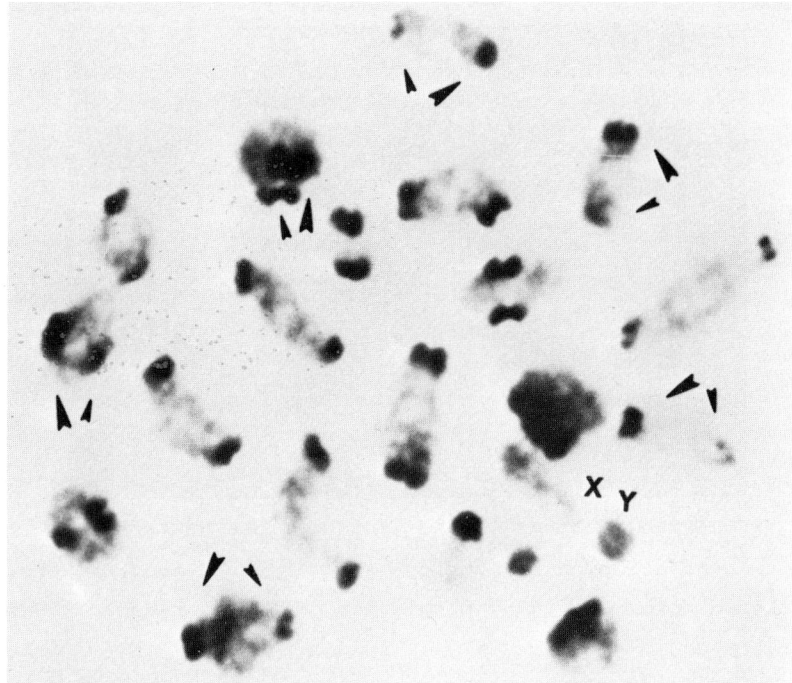

FIGURE 8. Meiotic cell from a (musculus x molossinus)F1 stained to show C-bands. Differences in amount of C-band material have not interfered with pairing. Reprinted from ref. 14 by permission of Springer-Verlag, New York.

methylated and single stranded under these conditions (Fig. 9). However, little binding of the antibody occurs in main band DNA because in this case the 5-methylcytosine is located in GC-rich sequences which remain double stranded (26). We have not yet applied this method to mouse species other than musculus; it might detect changes in nucleotide sequences in satellite DNA and permit distinctions other banding techniques do not allow.

Another method that has been used to distinguish between strains of musculus is silver staining of nucleolus organizer regions (NORs) (28) which identifies sites of active 18S

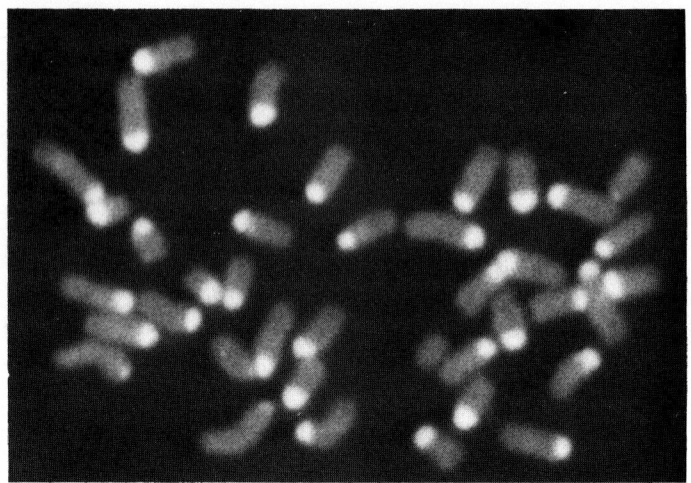

FIGURE 9. Metaphase of Mus musculus stained to show the concentration of 5-methylcytosine in the C-band regions. Reprinted from ref. 26 by permission of Springer-Verlag, New York.

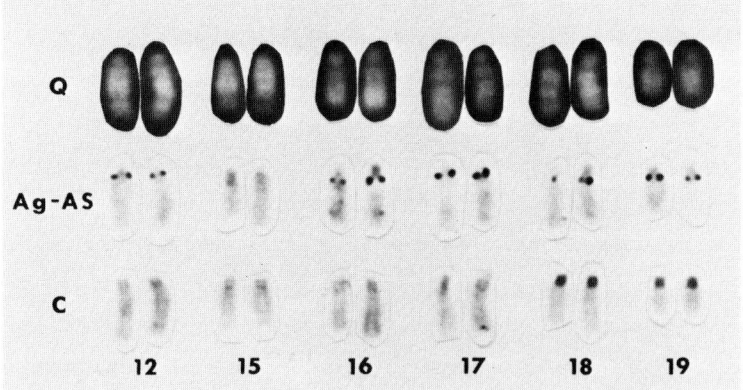

FIGURE 10. Partial karyotype of molossinus stained to show Q-bands (top row), Ag-stained NORs (middle row) and C-bands (bottom row). Ag-stain was present on five pairs of chromosomes irrespective of whether there was C-band material. Reprinted from ref. 30 by permission.

and 28S ribosomal RNA (rRNA) genes. For example, Ag-stained NORs are present on numbers 12, 15, and 18 in both BALB/cJ and C57/BL6J, but, of the two strains, only BALB/cJ has NOR staining on chromosome 16 and only C57BL/6J has an NOR on number 19 (29, 30). Figure 10 shows that NORs are present on most of the same chromosomes in Mus musculus molossinus. In this animal there are NORs on chromosomes 12, 16, 17, 18, and 19 and it is clear that the presence of an NOR is unrelated to the presence or absence of C-band material (30). In other strains of mice in situ hybridization with radioactively labeled rRNA has been used to show that rDNA is present on five of the same pairs of chromosomes (all but number 17) (31, 32). The finding that number 17 can be included in the list is of interest because the rDNA cluster must be located relatively close to the H2 major histocompatibility locus.

Mouse-human somatic cell hybrids are important tools in current genetic studies. Mouse-human hybrids have been used to demonstrate that silver stain detects active, but not inactive, rRNA gene clusters. Hybrids that have lost human chromosomes have only mouse NORs silver stained and produce only mouse rRNA, even though the human NOR chromosomes are still present (33, 34). Conversely, hybrids that have selectively lost mouse chromosomes have only human NORs silver stained and only human rRNA is synthesized, even though the mouse NOR chromosomes are still present (29, 34). Mouse-human hybrids have also provided a powerful tool for mapping the human genome. In just eight years the number of genes assigned to human autosomes has grown from one to nearly 150 (35), most of them assigned using somatic cell genetic analysis of interspecific hybrids. The same methods that have proved so useful for mapping the human genome are now being applied to the mouse, and the human is on the way

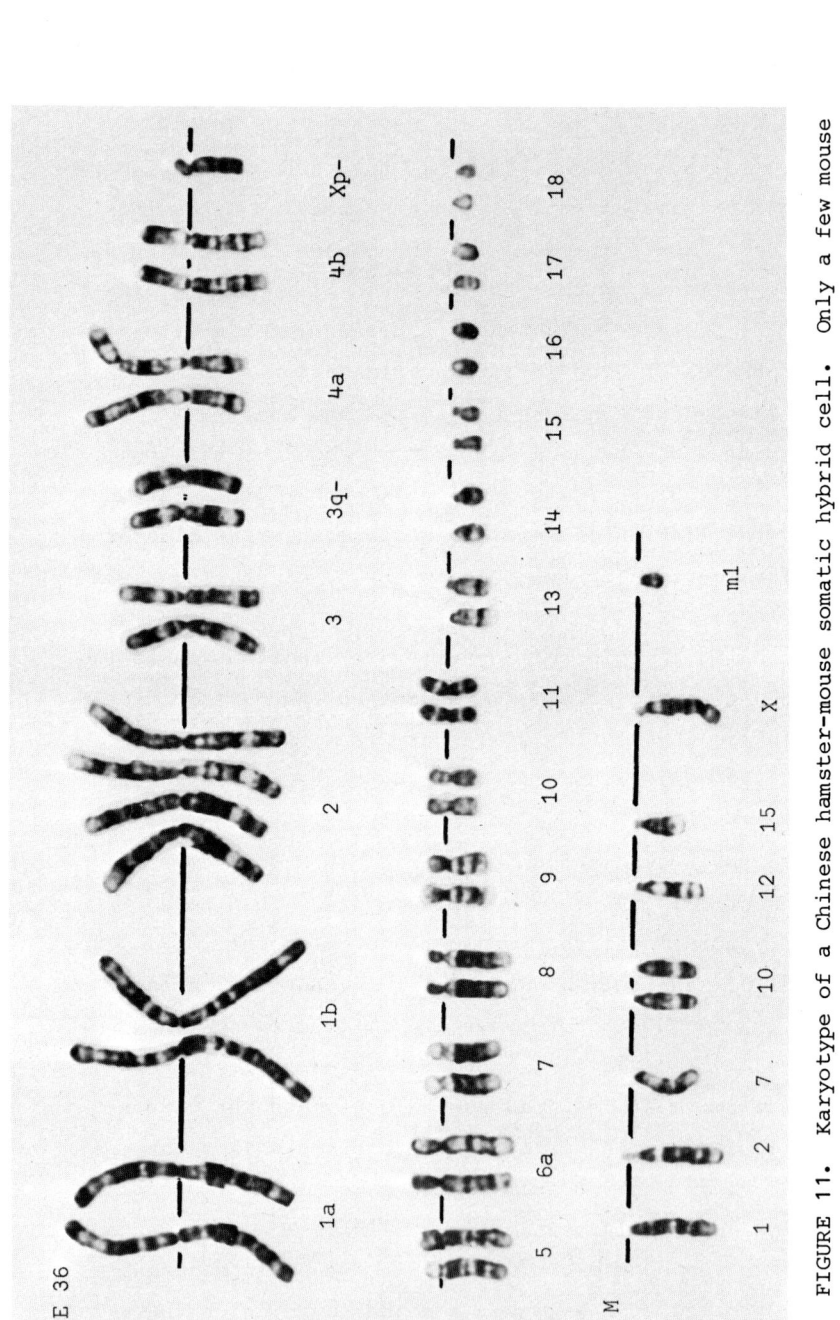

FIGURE 11. Karyotype of a Chinese hamster-mouse somatic hybrid cell. Only a few mouse chromosomes are present (bottom row); such hybrids have been used to assign genes to mouse chromosomes. Reprinted from ref. 36 by permission of S. Karger AG, Basel.

to becoming an excellent model system for those interested in mouse genetics. Chinese hamster-mouse hybrids usually lose mouse chromosomes (Fig. 11), and by correlating the presence of a mouse isoenzyme, antigen or other genetic marker with the presence of a particular mouse chromosome one can assign the relevant gene to its chromosome (36). More than a dozen biochemical markers have been assigned to mouse chromosomes in this way, and hundreds more can be. A method has also been worked out by which one can determine the linear order of linked genes using somatic cell hybrids (37). Prior to hybridization, the cells of the parent species to be mapped are heavily irradiated to produce chromosome breaks. The closer together two linked genes are, the higher the dose of radiation required to produce a break between them with loss of the fragment that lacks a centromere. Thus, by correlating the dose of radiation with the frequency of separation of linked loci, one can determine both linear order and relative distance between genes along each chromosome. Finally, using the ability to distinguish homologous chromosomes, mouse-mouse somatic cell hybrids have been used to map both structural and regulatory genes to mouse chromosomes (38, 39).

A special banding method, Giemsa staining at pH 11 (G11), has proved particularly useful in the analysis of interspecific hybrid cells because mouse chromosomes stain strongly, except at the centromere, whereas primate chromosomes generally stain faintly (40). This has made it possible to identify translocations between entire chromosomes of two species, as well as of much shorter segments of chromosome. Furthermore, feeding isolated chromosomes to cells in culture leads to incorporation of very short segments of genetic material, sometimes only long enough to contain two closely linked genes, e.g., those coding for thymidine kinase and galactokinase in both human (41) and mouse (42, 43) genomes. The ability to produce viable cells in culture whose chromosomes can contain DNA of two species may be relevant to geneticists who work with whole animals. The work of Mintz and her collaborators (44) indicates that it is possible to make chimeric mice using, as one parent, in vitro cultured teratocarcinoma cells. When the teratocarcinoma cells were mutagenized and a mutant ($HPRT^-$) subline selected in vitro, mice containing some $HPRT^-$ cells were produced (41). If such $HPRT^-$ teratocarcinoma cells were converted to wild type ($HPRT^+$) with DNA of a different species, it might be possible to produce mouse chimeras that contained blocks of genes from a species other than Mus: chimeras in the real sense! One can even imagine, for example, putting the human HLA major histocompatibility complex on a mouse background, or carrying out similar manipulations affecting the mouse genome in a profound way by artificial introgression of alien genetic material.

REFERENCES

1. Miller, O.J., Miller, D.A., Kouri, R.E., Allderdice, P.W., Dev, V.G., Grewal, M.S., and Hutton, J.J. (1971). Proc. Natl. Acad. Sci. USA 68:1530.
2. Committee on Standardized Genetic Nomenclature for Mice. (1972). J. Hered. 63:69.
3. Caspersson, T., Farber, S., Foley, G.E., Kudynowski, J., Modest, E.J., Simonsson, E., Wagh, U., and Zech, L. (1968). Exp. Cell Res. 49:219.
4. Nesbitt, M.N., and Francke, U. (1973). Chromosoma 41:145.
5. Miller, D.A., Kouri, R.E., Dev, V.G., Grewal, M.S., Hutton, J.J., and Miller, O.J. (1971). Proc. Natl. Acad. Sci. USA 68:2699.

6. Miller, O.J., and Miller, D.A. (1975). Ann. Rev. Genet. 9:285.
7. Miller, D.A., Dev, V.G., Tantravahi, R., Miller, O.J., Schiffman, M.B., Yates, R.A., and Gluecksohn-Waelsch, S. (1974). Genetics 78:905.
8. Yunis, J.J. (1976). Science 191:1268.
9. Dofuku, R., Biedler, J.L., Spengler, B.A., and Old, L.J. (1975). Proc. Natl. Acad. Sci. USA 72:1515.
10. Dev, V.G., Miller, D.A., Miller, O.J., Marshall, J.T., Jr., and Hsu, T.C. (1973). Exp. Cell Res. 79:475.
11. Sutton, W.D., and McCallum, M. (1972). J. Mol. Biol. 71:633.
12. Rice, N.R., and Straus, N.A. (1973). Proc. Natl. Acad. Sci. USA 70:3546.
13. Arnheim, N., and Southern, E.M. (1977) Cell 11:363.
14. Dev, V.G., Miller, D.A., Tantravahi, R., Schreck, R.R., Roderick, T.H., Erlanger, B.F., and Miller, O.J. (1975). Chromosoma 53:335.
15. Hsu, T.C., Markvong, A., and Marshall, J.T. (1978). Cytogenet. Cell Genet. 20:304.
16. Markvong, A., Marshall, J.T., Pathak, S., and Hsu, T.C. (1975). Cytogenet. Cell Genet. 14:116.
17. Gropp, A., Winking, H., Zech, L., and Muller, H. (1972). Chromosoma 39:265.
18. Capanna, E., Gropp, A., Winking, H., Noack, G., and Civitelli, M.-V. (1976). Chromosoma 58:341.
19. Lehmann, E.V., and Radbruch, A. (1977). Experientia 33:1025.
20. Forejt, J. (1973). Chromosoma 43:187.
21. Dev, V.G., Miller, D.A., and Miller, O.J. (1973). Genetics 75:663.
22. Miller, D.A., Tantravahi, R., Dev, V.G., and Miller, O.J. (1976). Genetics 84:67.

23. Allderdice, P.W., Dev, V.G., Miller, D.A., Jagiello, G.M., and Miller, O.J. (1973). Chromosoma 41:103.
24. Natarajan, A.T., and Gropp, A. (1972). Exp. Cell Res. 74:245.
25. Miller, O.J., Schnedl, W., Allen, J., and Erlanger, B.F. (1974). Nature 251:636.
26. Schreck, R.R., Dev, V.G., Erlanger, B.F., and Miller, O.J. (1977). Chromosoma 62:337.
27. Salomon, R., and Kaye, A.M. (1970). Biochim. Biophys. Acta 204:340.
28. Goodpasture, C., and Bloom, S.E. (1975). Chromosoma 53:37.
29. Miller, O.J., Miller, D.A., Dev, V.G., Tantravahi, R., and Croce, C.M. (1976). Proc. Natl. Acad. Sci. USA 73:4531.
30. Dev, V.G., Tantravahi, R., Miller, D.A., and Miller, O.J. (1977). Genetics 86:389.
31. Henderson, A.S., Eicher, E.M., Yu, M.T., and Atwood, K.C. (1974). Chromosoma 48:155.
32. Elsevier, S.M., and Ruddle, F.H. (1975). Chromosoma 52:219.
33. Miller, D.A., Dev, V.G., Tantravahi, R., and Miller, O.J. (1976). Exp. Cell Res. 101:235.
34. Croce, C.M., Talavera, A., Basilico, C., and Miller, O.J. (1977). Proc. Natl. Acad. Sci. USA 74:695.
35. McKusick, V.A., and Ruddle, F.H. (1977). Science 196:390.
36. Francke, U., Lalley, P.A., Moss, W., Ivy, J., and Minna, J.D. (1977). Cytogenet. Cell Genet. 19:57.
37. Goss, S., and Harris, H. (1977). J. Cell Sci. 25:17.
38. Hashmi, S., and Miller, O.J. (1976). Cytogenet. Cell Genet. 17:35.
39. Benoff, S., and Skoultchi, A.I. (1977). Cell 12:263.

40. Friend, K.K., Dorman, B.P., Kucherlapati, R.S., and Ruddle, F.H. (1976). Exp. Cell Res. 99:31.
41. Willecke, K., Lange, R., Kruger, A., and Reber, T. (1976). Proc. Natl. Acad. Sci. USA 73:1274.
42. McBreen, P., Orkwiszewski, K.G., Chern, C.J., Mellman, W.J., and Croce, C.M. (1977). Cytogenet. Cell Genet. 19:7.
43. Kozak, C.A., and Ruddle, F.H. (1977). Somatic Cell Genet. 3:121.
44. Dewey, M.J., Martin, D.W., Jr., Martin, G.R., and Mintz, B. (1977). Proc. Natl. Acad. Sci. USA 74:5564.

Wild Mice: Viruses, T Locus, Histocompatibility Antigens

Verne M. Chapman, Chairman

POPULATION GENETICS OF T/t COMPLEX MUTATIONS[1]

Dorothea Bennett

Laboratory for Developmental Genetics
Sloan-Kettering Institute for Cancer Research
New York, New York

Mutations at the T/t complex in the mouse have been under study now for just over 50 years, and have proved valuable tools for analyzing such diverse situations as embryogenesis, spermatogenesis, and population dynamics in wild mice.

The first mutation at this region of chromosome 17 was reported in 1927 by Dobrovolskaia-Zavadskaia (1). She described a dominant gene, $\underline{T}$, that resulted in heterozygotes that were short-tailed, and homozygotes that died during prenatal development. Subsequent breeding of T-bearing mice to apparently normal animals that were (significantly for the rest of this story) in at least one case derived from wild-caught mice revealed the presence of recessive ($\underline{t}$) mutations. These genetic variants interacted with the dominant $\underline{T}$ to produce a new phenotype, complete taillessness (2). These first recessive mutations detected were also lethal when homozygous and behaved as genetic alleles to $\underline{T}$; thus

---

[1]The author's data in this paper were obtained with the valuable assistance of Janice Cookingham and supported by ERDA contract EE-77-S-02-4159.

tailless lines of the same ancestry bred as balanced lethal systems:

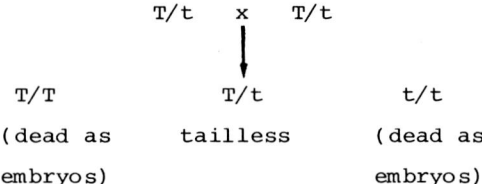

        T/T   T/t   t/t
     (dead as   tailless   (dead as
     embryos)   embryos)

L.C. Dunn and his colleagues at Columbia took up the study of the T/t system in the 1930s and soon showed that these mutations differed from conventional genes in a number of properties that are outlined below.

A "Pseudoallelic" System

The genetic analysis of the first tailless stocks studied showed that lethal t-factors of independent origin could differ from one another sufficiently to show a degree of genetic complementation. The test for this was simple and straightfoward; matings of tailless mice from different lines did not behave as balanced lethal crosses, but rather produced two classes of offspring as diagrammed below for the original cross that defined this phenomenon (3).

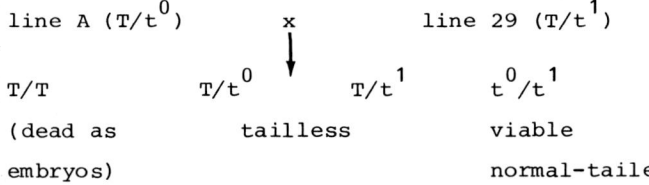

Complementation is generally not perfect, and t^x/t^y compounds are often both morphologically abnormal and subviable. At the present writing, we have studied more than 50 recessive lethal t-mutations of independent origin. By complementation analysis we found them to fall into six complementation groups as follows:

$$\begin{array}{l}
\underline{t^0} \\
\underline{t^9} \\
\underline{t^{12}} \\
\underline{t^{w1}} \\
\underline{t^{w5}}
\end{array}
\quad
\begin{array}{l}
(\underline{t^1}, \underline{t^6}, \underline{t^{30}}, \underline{t^{w4}}) \\
(\underline{t^4}, \underline{t^{w18}}, \underline{t^{w30}}, \underline{t^{w52}}) \\
(\underline{t^{w32}}) \\
(\underline{t^{w3}}, \underline{t^{w12}}, \underline{t^{w20}}, \underline{t^{w21}}, \underline{t^{w71}}, \underline{t^{w72}}) \\
(\underline{t^{w6}}, \underline{t^{w10}}, \underline{w^{w11}}, \underline{t^{w13}}, \underline{t^{w14}}, \underline{t^{w15}}, \underline{t^{w16}}, \underline{t^{w17}}, \underline{t^{w38}}, \\
\underline{t^{w39}}, \underline{t^{w41}}, \underline{t^{w46}}, \underline{t^{w47}}, \underline{t^{w74}}, \underline{t^{w75}}, \underline{t^{w80}}, \underline{t^{w81}}, \underline{t^{w93}}, \\
\underline{t^{w94}}, \underline{t^{w97}})
\end{array}$$

$\underline{t^{w73}}$

Each complementation group has furthermore been shown to produce sharply different effects and to act at different times during embryonic development (ref. 4 for review). These observations suggest that although $\underline{t}$-mutations have some of the attributes of genetic alleles, they cannot be considered true alleles, but rather fall into the category (defined only by our ignorance of their real relationships or structure) called "pseudoalleles." Complicating the issue further are a group of semilethal $\underline{t}$-factors ($\underline{t^{w2}}$, $\underline{t^{w8}}$, $\underline{t^{w36}}$, $\underline{t^{w49}}$, $\underline{t^{w101}}$, $\underline{t^{w102}}$) which have many of the attributes (see below) of lethal $\underline{t}$-mutations, but cannot be defined by complementation analysis, because the homozygous $\underline{t^{SL}}/\underline{t^{SL}}$ genotype is in any case potentially viable, although viability ranges from 50% to as low as 2%, depending on the specific semilethal factor.

Transmission Ratio and Sterility Effects in Males

One of the most remarkable effects of $\underline{t}$-mutations is an abrogation of Mendel's rules in males. This departure from conventional genetics became obvious early in the study of the T/t complex, when animals of genotype $\underline{T}/\underline{t^0}$ were outcrossed to normal ($\underline{+}/\underline{+}$) mates. Tailless females produced the expected proportions of 50% short-tailed ($\underline{T}/\underline{+}$) and 50% normal tailed ($\underline{t^0}/\underline{+}$) offspring, but tailless males transmitted $\underline{t^0}$ to a surprisingly high degree; about 80% of their offspring were normal tailed. Tests of $\underline{t}^0/+$ males mated to $\underline{T}/+$ females showed that the effect, called transmission ratio distortion, was directly attributable to the $\underline{t}^0$ factor, not to any interaction with T (3). All evidence since that time has continued to demonstrate that transmission ratio distortion is a general feature of lethal and semilethal $\underline{t}$-factors (the single exception being the $\underline{t}^9$ complementation group). A number of cases have been described of specific $\underline{t}$-mutations with transmission ratios of the lethal gene that are over 95%, so it is clear that this situation represents an extreme departure from conventional genetic systems. The mechanism of this distortion does not apparently depend on meiotic disturbances nor on degeneration of one type of sperm, nor on such factors as embryo selection. The best guess at the moment is that $\underline{t}$-bearing sperm have some superiority in fertilizing ability that is conferred by haploid gene function after meiosis (ref. 4 for review). Other effects of $\underline{t}$-factors in males have also been described; both homozygotes for semilethal $\underline{t}$'s and males carrying two different lethal complementing ($\underline{t}^x/\underline{t}^y$) mutations are completely sterile. In both these cases, pronounced morphological abnormalities occur in developing spermatids (5, 6), so the basis for sterility appears to be quite different from that of distorted transmission ratio. Nevertheless, the two phenomena are presumably related in some way.

Suppression of Recombination

Both lethal and semilethal $\underline{t}$-factors in heterozygous condition suppress recombination over a long stretch of chromosome. The recombination suppression effect may vary slightly from $\underline{t}$-haplotype to $\underline{t}$-haplotype, but in general effectively includes the distance from the locus of T to, but not much beyond, the H-2 region (7). Although this obviously leads to suggestions that t-mutations represent inversions or deletions of some magnitude, numerous high resolution cytological studies have failed to provide any evidence for chromosome aberrations (ref. 4 for review).

Generation of "New" t-Factors from Pre-existing Ones

The initial studies of the genetic behavior of t-mutations suggested that they were "hypermutable," since they frequently (about 1/500 gametes) produced what Dunn and his colleagues referred to as "exceptions." Exceptions were detected in balanced lethal crosses by the appearance, usually singly, of normal-tailed offspring, which proved on genetic analysis to carry two different complementing t-factors (8). Once chromosome markers became available and were incorporated into t-bearing stocks, it became apparent that the great majority of such exceptions were accompanied by recombination. Most studies have made use of the marker tf (tufted), thus:

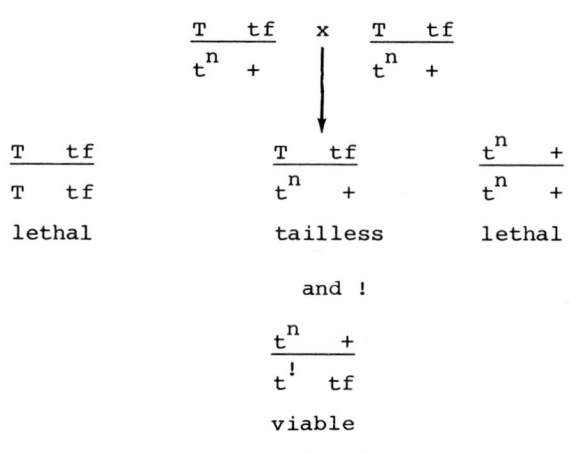

The majority of exceptional chromosomes generated by lethal or semilethal t-factors have proved to be viable when homozygous, and in fact to retain as their only t-like property the ability to interact with T to produce taillessness.

No revertants to complete wild-type have ever been detected, however. Over 100 exceptions have been analyzed, and of these 96 were viable and only nine lethal. All of the viable t-factors generated in stocks marked with tf were recombinants. These data lead quite inescapably to the conclusion that lethal or semilethal t-haplotypes are at the very least bilocal and consist of a minimum of two separable elements, one that is allelic with T and responsible for interaction with it to produce taillessness, and one near tf that is involved with lethality, recombination suppression, and effects on sperm function. Detailed analysis of recombination-derived t-factors suggests even more complicated models are possible (9), but these are sufficiently speculative not to warrant discussion in the present context.

Relatively few lethal t-mutations have been derived as exceptions from pre-existing factors. No doubt many that do arise are lost before birth because of poor complementation with the parent allele, and others that occur in male $t^n/t^!$ compounds are lost to analysis because of the inevitable sterility associated with that genotype. For these reasons it is virtually impossible to get reliable estimates of the spectrum or frequency of newly generated lethal t's. Nevertheless, those that have been analyzed show an interesting unidirectional pattern of occurrence. As seen in the diagram below, t^{w5} has never been generated from a preexisting t, but has given rise to virtually all others. Likewise, t^9, which has never itself given rise to a lethal, can be produced by most lethals.

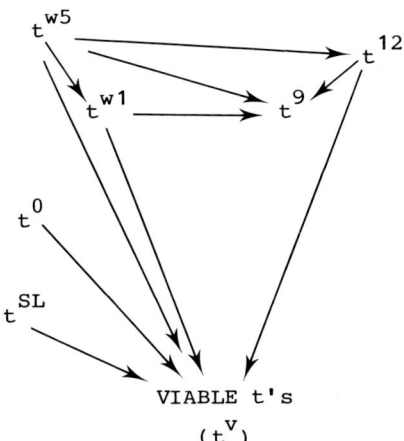

The event by which one lethal $\underline{t}$-factor converts to a different one is not well defined. Only one lethal ($\underline{t^{w32}}$) has been generated in stocks marked with $\underline{tf}$ where recombination could be assayed. In this one case, the origin of $\underline{t^{w32}}$ was clearly not accompanied by crossing-over between $\underline{T}$ and $\underline{tf}$ (10). This suggests that viables and lethals may originate by different mechanisms (11), with lethals perhaps arising by mutation rather than by recombination. It is worth emphasizing here that no instance of mutation from a wild-type gene to a $\underline{t}$-factor of any sort has ever been observed; all $\underline{t}$-mutations studied have either been derived from wild populations or obtained as exceptions from pre-existing $\underline{t}$-factors.

$\underline{t}$-FACTORS IN WILD POPULATIONS

The first exploration for $\underline{t}$-mutations in wild populations of mice was apparently initiated because L.C. Dunn saw a possible correlation among three factors: the original detection of $\underline{t}$-mutations by Dobrovalskaia-Zavadskaia in the descendents of wild mice, the high transmission ratio of $\underline{t}$-factors through males, and their apparent hypermutability in the laboratory. He obtained representatives of a closed colony of mice originated by Schneider (12) from animals caught wild in suburbs of Philadelphia and New York. In the original study 10 males were tested for the possible presence of a t-mutation by mating to T/+ females. Of the eight which actually bred, five tested positive for the presence of a t-factor by producing tailless offspring. The transmission ratio of the t-mutation(s) found in this population was, as predicted, high (86%) (13).

This sample was only the first of many in which t-factors were found in wild populations of mice. Their distribution is ubiquitous, and t-factors have been found all over the world: North America, Europe, Australia, Asia, and South America.

To date, as summarized by Klein (14) with additions of unpublished data from Dunn and Bennett, 52 separate populations have been sampled for their content of t-factors; 27 were found to contain t-factors and 31 were negative. The proportion of negative populations is undoubtedly artificially high, because of the small sample size available for most of them; generally under 10 and more often 2-4 animals were tested. Actually only two of the negative populations can be considered as subjected to a valid test. A sample of 23 animals from a farm in Calgary, Canada, contained no t-factor, although other nearby farms had mice that did (15). Likewise, 40 animals from Great Gull Island in Long Island Sound were analyzed over a period of several years without detecting a t-factor and that population can be securely considered to be free of t-mutations (16). The fact that this was an island population may be a significant factor (see below). Table 1 gives information on sample sizes, numbers of animals tested, and gene frequency. All of the t-haplotypes found in these wild populations were lethal or

TABLE 1. Distribution of t-Factors in Wild Populations

Sample size	Number of negative populations sampled	Number of t-bearing populations sampled
1-3	10	9
4-6	9	4
7-10	4	5
11-20	6	5
20+	2[a]	4
Total animals	167	249
t-heterozygote frequency	0/167	87/249 = 0.35
t-gene frequency	0	0.175
Combined t-heterozygote frequency omitting[a]	87/353 = 0.25	
Combined t-gene frequency omitting[a]	0.125	

[a] Populations considered securely negative on the basis of adequate tests.

semilethal, but the distribution of complementation groups was very unequal; some populations contained more than one complementation group. Table 2 shows that almost twice as many examples of the t^{w5} group were found than all others combined, while semilethal haplotypes and representatives of the t^{w1} complementation group comprised most of the other t-mutations that were defined. One member of the t^0 group and the only example known of t^{w73} were also found. Neither t^{12} nor t^9 were represented in these samples. Again, every

TABLE 2. Complementation Groups of t-Factors in Wild Populations

Complementation group	Number of independent captures
t^{w5}	21
t^{SL}	6
t^{w1}	5
t^{0}	1
t^{w73}	1

t-haplotype found proved to have a high male transmission ratio, always over 80% and not infrequently in the 95-98% range.

Tested populations that did contain a t-factor had an overall heterozygote frequency of 0.35, which is of course extraordinarily high for a lethal or semilethal gene. Even if the assumption is made that most so-called "negative" populations were insufficiently tested, and these data (with the exceptions of the two quite securely negative populations, see a in Table 1) are pooled with those from positive populations, an estimated heterozygote frequency of 0.25 is obtained for mouse populations in general. Thus, the conclusion must be that lethal or semilethal t-haplotypes are a natural polymorphism in wild mice, with gene frequencies of 10% or more.

This situation is not completely without precedent since, for example, the gene for sickle cell anemia, which is virtually a homozygous lethal, has comparable gene frequencies in some human populations, where it appears to be maintained by the advantage conferred on heterozygotes in respect to protection from malarial parasitism (17). Although no strong evidence has been found for any conventional relative advantage in fitness in t-heterozygotes (18), an obvious heterozygous advantage, although of a rather unique sort, occurs in t-heterozygotes due to male transmission ratio distortion.

There have been several attempts to analyze in theoretical terms the apparent ability of transmission ratio distortion to maintain a polymorphic frequency of lethal or semilethal genes in wild populations of mice, and most have been instructive in various ways. The first formally valid approach was proposed by Bruck (19) who generated a deterministic model based on the Hardy-Weinberg parameters of infinite population size and random mating. His equation took into account the necessary factors of embryonic lethality of homozygotes and segregation distortion in males only, but led to a prediction that lethal t-gene frequencies at equilibrium would be far higher (between 30 and almost 50%) than those actually found in nature (about 10%). A somewhat more sophisticated version that took semilethality into account was later produced by Dunn and Levene (20) and suffered from the same inability to predict the gene frequencies in natural populations. The failure of these deterministic solutions to correspond to the real world was actually very informative, because it suggested that mouse populations probably were subject to stochastic rules that were imposed because of small effective population sizes where random genetic drift would be an important force. Actually, there were already data being collected, that are now much more extensive (ref. 15 for review), to show that mouse breeding populations are divided into extremely small subunits called demes. The average deme may consist of about 8 or 10 mice, usually not more than 12, and is relatively isolated from other groups since demes are usually governed by a dominant male who excludes intruders. The isolation of demes is not fixed or permanent, however, since the absence or death of the dominant male is likely to be quickly followed by the immigration of another.

The suggestion that effective breeding populations are indeed small was bolstered very much by stochastic models proposed by Lewontin and Dunn (21) for the behavior of *t*-factors in wild mice. These interesting studies involved computer simulations of small breeding populations of mice, which were set up with designated numbers of parents of desired genotypes and transmission ratio; a "Monte Carlo" method of gamete selection from the resulting gene pool was used to generate the next generation, and so on. One of the most telling points that emerged from this study was that, under these conditions population sizes of 25 males and 25 females, or even 10 males and 10 females, behaved in a manner that approached Hardy-Weinberg predictions of a *t*-gene frequency of about 40%. Thus, for the mouse, 20 appears to represent an infinite population size! When the simulated populations were reduced to a size of about eight, it became apparent that any given population was virtually certain, given enough time, to lose its *t*-factor and reach fixation for the wild type. The length of time a population maintained a *t*-haplotype was heavily dependent on the value set for transmission ratio, with only very high ratio factors persisting for appreciable lengths of time. On the other hand, it also appeared that a single male with a high ratio factor could be highly effective in introducing that factor into an otherwise negative deme. Taken together these observations suggested that mouse populations as a whole were divided into small demes, each of which, if it had a *t*-factor in it, was nevertheless progressing ultimately to fixation of the wild type alternate, but each of which, if free of *t*-factors at the moment, was subject to a certain probability

of "re-infection" by the immigration of a heterozygous male.
In this context it is significant that one of the two securely
t-free populations so far discovered was the Great Gull Island population, which exists on an essentially uninhabited
island, where inward migration would be very restricted,
although not impossible since boats do visit the island.
Anderson et al. (22) and Bennett and associates (23) used
this population as a kind of natural laboratory to study the
effects of introducing males heterozygous for a wild-derived
lethal t-haplotype. In 1956 and 1957, $+/t^{w11}$ males were
released in one small area near the center of the island,
and samples taken yearly between 1959 and 1962 and again
in 1966 showed not only that t^{w11} had persisted in the population, but that it had spread virtually throughout the
island (Fig. 1). However, in 1969, 1970, and 1972, additional samples studied were free of t-mutations. The total
number of animals sampled was large enough (39) to suggest
strongly that complete fixation of the wild type had been
reached. Thus, it is likely that the natural population on
Gull Island fulfilled the predictions of the stochastic
models of Lewontin and Dunn for 1) easy infection by $t/+$
heterozygous males and 2) eventual loss from an isolated
population. A definitive solution of this point has unfortunately not been possible, because in 1974 representatives
of the American Museum of Natural History in New York, who
supervise the island and study populations of terns that
nest there, permitted the escape of pet-store mice into the
natural population and thus rendered it useless for our purposes.

Stochastic models discussed above led to gene frequency
predictions that were relatively low and far more compatible
with data from actual wild populations than were the predictions of deterministic models. In general terms, the actual

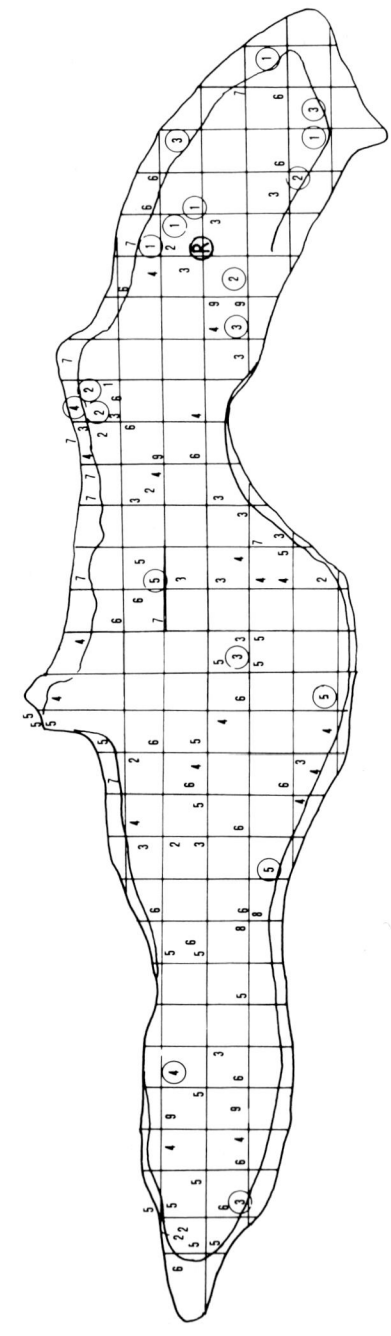

FIGURE 1. See legend on next page.

FIGURE 1. Distribution of mice trapped on Great Gull Island after release of $+/t^{w11}$ males in 1956 and 1957, at the point designated R. Each number designates a particular sample and represents a single mouse trapped and tested for $\underline{t}$-genotype. Numbers in circles represent a $+/\underline{t}$ heterozygote, uncircled numbers indicate a $+/+$ animal.

Sample No.	Date	No. mice progeny tested	No. $+/\underline{t}$ heterozygotes
1:	1959	6	5
2:	1960	12	4
3:	1961	22	5
4:	1962	20	2
5:	1966	23	3
6:	1969	22	0
7:	1970	10	0
8:	1972 (April)	2	0
9:	1972 (October)	5	0

and theoretical picture that emerges is that most populations
of mice carry t-mutations in some substantial frequency.
The question then arises whether this polymorphism for lethal
or semilethal genes confers some intrinsic adaptive advantage to mouse populations and, if so, what that advantage is.

Perhaps the most obvious possibility is some relationship
with the H-2 complex. The observation that only lethal and
semilethal t-factors have high transmission ratios and are
found in wild populations, and significantly that only these
inhibit recombination through the region of H-2, has led to
speculations (24-26) that the T/t complex and the H-2 complex may have some functional or evolutionary relationship.

Evidence for an actual physical relationship has indeed
been found. Hammerberg and Klein (7) and Hammerberg et al.
(27) have serologically examined H-2 types in t-bearing
chromosomes and found only a restricted number of H-2 types
in that sample. With a few exceptions, t-haplotypes of the
same complementation group, no matter how disparate their
geographical origin, had identical H-2 haplotypes. These
findings have been confirmed with respect to I-region genes
by McDevitt et al. (28). These findings are particularly
surprising in view of the extensive polymorphism of H-2 in
wild mouse populations (29), which makes it highly unlikely
that mice from separate populations would carry the same
H-2 region genes. Obviously, this instance of linkage disequilibrium could have either a functional or an evolutionary
interpretation. It could be assumed that particular t-haplotypes form especially advantageous gene complexes (super-
genes?) with particular H-2 haplotypes, and thus are selected
for in wild populations. Alternatively, it could be assumed
that t-haplotypes of the same complementation group must
often actually represent the same chromosome which originated
once and then was dispersed geographically by migration, with

the _t_-haplotype and the existing H-2 complex held permanently together by recombination suppression. This is perhaps the more likely explanation, since it implies that the "primordial" mutation to a _t_-factor is a very rare event; this seems indeed to be the case since none have been observed in the laboratory. The rare instances of lethal or semilethal _t_-haplotypes with "inappropriate" H-2 partners are also explainable on this hypothesis, since at present it appears that the generation of one lethal from another is not only a rare event but one that is not recombinational but mutational in nature. Thus a new lethal generated by mutation would necessarily remain associated with an H-2 type typical of the parent complementation group.

In summary, at present it is clear that lethal and semilethal _t_-factors that suppress recombination at least as far as the H-2 region and have high male transmission ratios are polymorphic in natural populations of mice. Whether they confer an adaptive advantage is not yet clear, but if they do, it is likely to be due to some as yet undefined relationship to H-2 region genes.

REFERENCES

1. Dobrovolskaia-Zavadskaia, N. (1927). Comp. Rend Soc. Biol. (Paris) 97:114.
2. Dobrovolskaia-Zavadskaia, N., and Kobozieff, N. (1932). Comp. Rend. Soc. Biol. (Paris) 110:782.
3. Chesley, P., and Dunn, L.C. (1936). Genetics 21:525.
4. Bennett, D. (1975). Cell 6:545.
5. Dooher, G.B., and Bennett, D. (1974). J. Embryol. Exp. Morph. 43:749.
6. Dooher, G.B., and Bennett, D. (1977). Biology of Reproduction 17:269.
7. Hammerberg, C., and Klein, J. (1975). Nature 258:296.

8. Dunn, L.C., and Gluecksohn-Schoenheimer, S. (1950). Proc. Natl. Acad. Sci. USA 36:233.
9. Lyon, M.F., and Mason, I. (1977). Genet. Res. 29:255.
10. Dunn, L.C., Bennett, D., and Beasley, A.B. (1962). Genetics 47:285.
11. Klein, J., and Hammerberg, C. (1977). Immunol. Rev. 33:70.
12. Schneider, H.A. (1946). Proc. Soc. Exp. Biol. Med. 63:161.
13. Dunn, L.C., and Morgan, W.C., Jr. (1952). Amer. Nat. 536:321.
14. Klein, J. (1975). "Biology of the Mouse Histocompatibility-2 Complex." Springer-Verlag, New York.
15. Anderson, P.K. (1964). Science 145:177.
16. Dunn, L.C., Beasley, A.B., and Tinker, H. (1960). J. Mammal. 41:220.
17. Allison, A.C. (1954). Brit. Med. J. 1:290.
18. Dunn, L.C., Beasley, A.B., and Tinker, H. (1958). Amer. Nat. 92:215.
19. Bruck, D. (1957). Proc. Natl. Acad. Sci. USA 43:152.
20. Dunn, L.C., and Levene, H. (1961). Evolution 15:385.
21. Lewontin, R.C., and Dunn, L.C. (1960). Genetics 45:795.
22. Anderson, P.K., Dunn, L.C., and Beasley, A.B. (1964). Amer. Nat. 98:57.
23. Bennett, D., Bruck, D., Dunn, L.C., Klyde, B., Shutsky, F., and Smith, L.J. (1967). Amer. Nat. 101:538.
24. Snell, G.D. (1968). Folia Biol. (Praha) 14:335.
25. Klein, J. (1975). Adv. Exp. Med. Biol. 64:467.
26. Artzt, K., and Bennett, D. (1975). Nature 256:545.
27. Hammerberg, C., Klein, J., Artzt, K., and Bennett, D. (1976). Transplantation 21:199.
28. McDevitt, H.O., and Levenson, J.R. (1976). J. Exp. Med. 144:834.
29. Klein, J. (1974). Ann. Rev. Genet. 8:63.

INDIVIDUAL MICE OF ONE INBRED STRAIN PRODUCE ANTI-H-2 ANTIBODIES OF DIFFERENT SPECIFICITIES[1]

Pavol Iványi

Paul de Greeve

Central Laboratory of the Netherlands Red Cross
Blood Transfusion Service, and Laboratory of
Experimental and Clinical Immunology
University of Amsterdam
Amsterdam, The Netherlands

The H-2 chart (1-6) provides information on the reaction pattern of allo-immune sera against gene products of the H-2K and H-2D regions of the major histocompatibility system (MHS) of the mouse.

The extent and arrangement of the H-2 chart is subject to continual change as new information is obtained from the study of inbred mouse strains having new H-2 haplotypes, and as new advances are made in the understanding of the genetic composition of the H-2 system. In this sense, the intrinsic supplement of the H-2 chart is the list of allo-immune sera (with the donor-recipient combinations and absorptions) used for the definition of H-2 antigens.

H-2K and H-2D antigens are operationally subdivided into private and public antigens. The public antigens are shared

---

[1] The studies were financially supported by the Foundation for Medical Research (FUNGO).

by two or more H-2 alleles. <u>Public</u> antigens illustrate the cross-reactivity pattern of the respective anti-H-2 sera. The molecular basis of the extensive cross-reactivity pattern of the MHS has not yet been sufficiently elucidated. Two possible factors that may account for cross-reactivity appear not to be mutually exclusive. In the first instance, an allo-immune serum is composed of a heterogeneous family of antibodies directed against components of a complex allo-antigen. Second, in addition to recognizing the specific target, a homogeneous antibody population can also recognize specificities that are different but closely related to those of the original immunogen. We use the term "cross-reactions" or "cross-reactivity pattern" of an anti-H-2 serum produced against one K- or D-region difference for both these possibilities.

The <u>private</u> H-2 antigens are restricted to one H-2K or H-2D allele product. They are defined by anti-donor strain antibodies produced in a recipient or F_1 hybrid and/or appropriately absorbed to reveal only anti-donor activity. Serotyping of wild mice has shown, however (7, 8), that some "monospecific" anti-private sera still exhibit cross-reactions with products of other H-2 alleles, if the panel of available targets is extended beyond those defining the present boundaries of the H-2 chart. Absorption experiments with cells from positively reacting wild mice have disclosed the cross-reactivity pattern of "monospecific" anti-private sera.

Most public antigens (coded by one K or D allele) are "parts" of one complex H-2 molecule. Therefore, it might be supposed that "privateness" is a reflection of the sum of a number of public specificities. Caution should, however, be exercised when attempts at generalization are made, and the need for this has been illustrated by the discovery of a separate locus for certain H-2D region "public" specificities (9, 10). More details and alternatives have been discussed by Snell et al. (3), Démant (5), Shreffler and David (11), and Klein (6, 12, 13). The number of thoroughly analyzed H-2 alleles is about 10. Studies on wild mice (7, 12, 14-17) have shown the polymorphism of H-2 loci to be much more extensive.

The impressive amount of H-2 serological work carried out during histocompatibility and immunogenetic studies has been inevitably and unavoidably performed by using serum pools. These serum pools were obtained by immunization of groups of recipients of one inbred strain (or F_1 hybrids) with cells of the donor strain. It is probably tacitly assumed, because all serum producers are members of one highly inbred strain, that no substantial differences in the content of individual sera might be expected. The use, as a routine procedure in H-2 laboratories, of pools of sera from several recipients is usually dictated by the need to obtain a sufficient amount of a serum. Although considerable variations must have occurred in the different laboratories in the course of the preparation of the individual serum pools, in general terms the procedures from different centers are roughly comparable. These can be characterized as being repeated intraperitoneal injections of different amounts of lymphoid cells, administered over differing time schedules, with the subsequent bleedings and pooling of positive sera to produce one batch of anti-H-2 serum. There are data about differences in the content of specificities from early and late pools during prolonged immunizations (e.g., the NIH catalogue of anti-H-2 sera), but to the best of our knowledge, there are no reports about differences in serum samples obtained from individual mice. We do not know whether this is because such differences were not found, or whether they had been noticed and then considered as trivial, or whether the possibility of divergence was not tested at all. The number of animals in one group used for the preparation of an anti-H-2 serum pool probably differs greatly from experiment to experiment, but groups of more than 20 recipients are frequently used.

Our first look at the diversity of the antibody content of individual H-2 sera (i.e., in sera from individual mice of one inbred strain) came from experiments in which these sera were tested on human cells. H-2 sera exert a strong cytotoxic effect on human cells, possibly due to cross-reactions with HLA antigens. The reaction pattern of H-2 sera on human cells depends on the mouse donor-recipient combination and the HLA phenotype of the human cells (18-21). Thus, the wide variety of HLA alleles, presented by a panel of human lymphocytes, served as a first indicator that antibody variability was present in individual H-2 sera (22).

A short summary of this work will be given here. Anti-H-2 sera were produced by immunization of B10.A(4R) ($H-2^{h4}$; kkbbbb) recipients with lymphoid cells of H-2 congenic B10.A ($H-2^a$; kkkdddd) donors. In this combination anti-D^d antibodies are primarily produced. The donor (D^d)-recipient (D^b) H-2 antigen difference, according to the H-2 chart, enables the generation of anti-H-2.4 (private) and anti-H-2.3, 13,35,36,41,42,43,44,49 (public) antibodies. Instead of one serum pool from several immunized recipients, the serum samples of each individual mouse were tested on a panel of about 100 HLA-typed human peripheral blood lymphocytes. It was found (for summary see Table 1) that individual mouse sera differed both in strength and specificity. A number of sera were negative, whilst others reacted in various titers on human lymphocytes. The reaction pattern of the positively reacting individual mouse sera was correlated with the presence of certain HLA antigens. Significant correlations

TABLE 1. Reaction Pattern of 38 Anti-D^d Sera on Lymph Node Cells of Inbred Strains

			Anti-H-2D^d (D^b ← D^d) sera													
			3	3	3	3	3	3	3	3	3	3	3	3	3	3
Cells tested			45	7	3	1	42	34	16	35	40	10	39	37	31	32
A Reaction on human cells	strength*		T	T	T	T	S	T	T	T	T	T	T	T	T	T
	major HLA-specificity		N	N	W	W	W Aw31	N	N	N	N	N	N	M	M A3	W A11
B Mice strain tested	H-2 haplotype	alleles specificity**														
B10.A (2R)	h2	K^k D^b	–	–	–	–	–	–	–	–	–	–	–	–	–	–
B10.A	a	K^k D^d 4,3,13,35,36(41,42,43,44,49)	–	±	±	±	±	++	++	+++	+++	++++	++++	++++	++++	++++
B10.AKM	m	K^k D^q 3,13, ? ? (49)	–	–	–	–	–	–	–	–	–	–	–	–	–	–
A.SW	s	K^s (42), D^s 3, 36, (42), (49)	–	–	–	–	–	–	–	–	–	–	–	–	–	++
C3H.NB	p	K^p D^p 3, 35, (41), (49)	–	–	–	–	–	–	–	–	–	–	–	–	–	–
B10.RIII	r	K^r D^r 3? (49)	–	–	–	–	–	–	–	–	–	–	–	–	–	–
C3H	k	K^k D^k 3	–	–	–	–	–	–	–	–	–	–	–	–	–	–
B10	b	K^b D^b 35,36,	–	–	–	–	–	–	±	–	–	–	–	–	–	–
HTG	g	K^d D^b d	–	–	–	–	–	–	–	–	–	–	–	–	–	–

– negative, ± very weak or inconsistent, + weak, ++ only weak reactions (+ or ++), titer 1:4–1:16, +++ strong reactions (+++ in dilution 1:64), ++++ strong reactions (++++ in dilution 1:128), +++++ strong reactions (+++++ in dilution 1:256 or higher).

* reaction strength and frequency on human cells arbitrarily classified as N = negative, W = weak, M = medium and S = strong (see ref. 22).

** H-2 specificities according to the H-2 chart; specificities reactive primarily in hemagglutination are given in parentheses (see ref. 6).

TABLE 1 continued

Cells tested				Anti-H-2D^d (D^b + D^d) sera														
				3	3	3	3x	3	3	3	3	3	3	3	3	3	3x	3	3
				38	12	47	46	30	6	48	24	36	37	45	25	18
A Reaction on human cells	strength*			T	T	T	S	T	T	S	S	T	N	S	T	S
	major HLA-specificity			N	W	M	N	N	M	S	S	N	W	M	M	S
					A11				B27	B17	B27		B27		B27	B17
															B37	
B Mice strain tested	H-2	haplotype alleles specificity**															
B10.A (2R)	h2	K^k	–	–	–	–	–	–	–	–	–	–	–	–	–
		D^b	–													
B10.A	a	K^k	–	++++	+++++	+++++	+++++	+++++	++++	++++	++++	+++++	+++++	+++++	++++	+++++
		D^d	4,3,13,35,36(41,42,43,44,49)													
B10.AKM	m	K^k	–	+	–	–	–	+++	++++	++	++	++++	++++	+++++	+++++	+++++
		D^q	3,13, ? ? (49)													
A.SW	s	K^s	(42),	+	++	++	+++++	–	++	+	++	+	+	+++	++	+++
		D^s	3, 36, (42), (49)													
C3H.NB	p	K^p	–	–	–	–	–	+	++	+	+++	++	–	–	++	++++
		D^p	3, 35, (41), (49)													
B10.RIII	r	K^r	–	–	–	–	+$\stackrel{+}{-}$	+	–	–	+	–	–	–	–
		D^r	3, 3?													
C3H	k	K^k	–	–	–	–	–	–	–	–	–	–	–	–	–
		D^k	3 (49)													
B10	b	K^b	35,36	–	–	–	–	–	–	–	+	–	–	–	+	–
		D^b	–													
HTG	g	K^d	3	–	–	–	–	–	–	–	–	–	–	–	–	–

TABLE 1 continued

A Cells tested — Anti-H-2D^d (D^b + D^d) sera

strength*	3	3	3	3	3	3	3	3	3	3x	3	3
	44	13	17	47	28	15	2	9	33	9	19	22
	T	T	T	T	S	T	T	T	T	M	T	T
Reaction on human cells	N	M	W	S	S	S	S	W	M	S	M	W
major HLA-specificity		B27	Aw31	B27	B27	B37	B27	Aw31	B27		B27	A11
			A11		B37			A11	B37			

B Mice strain tested

strain	H-2 haplotype	alleles	specificity**												
B10.A (2R)	h2	K^k D^b	−, −	−	−	−	−	−	−	−	−	−	−	−	−
B10.A	a	K^k D^d	4,3,13,35,36(41,42,43,44,49)	++++	+++++	+++++	+++++	+++	++++	+++	+++++	+++++	+++++	+++++	+++++
B10.AKM	m	K^k D^q	3,13,?,? (49)	++++	++++	++	+++++	+++	++++	−	++	++	+++++	+++++	+++++
A.SW	s	K^s D^s	3, 36, (42),(49)	++++	++++	+	++++	+	++++	++++	−	++++	+++++	++++	+
C3H.NB	p	K^p D^p	3, −, −	+	+++	+++	+	−	−	−	+	+++	−	−	−
B10.RIII	r	K^r D^r	3, 32, (49)	+++	++	++	−	+	+	+	+	++++	−	−	−
C3H	k	K^k D^k	3, −	−	−	−	−	−	−	−	−	−	−	−	−
B10	b	K^b D^b	35,36, (49), −	+	+	+++	+++++	+++++	+++	+++	++++	+++++	+++	++++	++++
HTG	g	K^d D^b	3, −			−						+++	+++	+++++	++

were found between the high cytotoxicity scoring grade (CSG) and the presence of HLA-B12, B17, B27, and B37. (All these antigens are included in the Bw4/4a supertypic specificity.) Positive correlations were also found with some HLA-A locus antigens, and this was most pronounced for A11 and Aw31. However, as in previous studies, none of the individual H-2 sera behaved as "monospecific" with regard to individual HLA antigens. It is presumed that this reaction pattern of the individual mouse sera is attributable to the presence of certain common structures shared by all these antigens; individual mice of one inbred strain might recognize, or produce, antibodies against different parts of the complex allo-antigenic difference.

During this study, it became obvious that a similar approach should also be applied to the mouse H-2 system, and preliminary results are given in this work.

Anti-H-2D^d sera from individual mice of one highly inbred strain were tested on a number of H-2 haplotypes of inbred strains which were further supplemented with a small number of wild mice. The wild mice were included in the investigation to extend the number of H-2 targets as an expedient in studying variation in the sera of individual mice. These results on mouse cells showed an analogous extent of variation in titer and specificity similar to that already observed on human cell targets.

These findings, if confirmed on further combinations, might have implications for the understanding of 1) the generation of antibody diversity during MHS allo-immunization, and 2) the target H-2 specificities that are responsible for the cytotoxic effect of H-2 sera. It has been hypothesized (23) and widely discussed that the MHS itself is involved in the generation of antibody diversity.

MATERIALS AND METHODS

Mice. The mouse strains were maintained at our Institute. The breeding nuclei were obtained from Drs. J. Colombani (Paris), J. Klein (Dallas), P. Demant (Amsterdam), and J.P. Levy (Paris), and from the Jackson Laboratory (Bar Harbor).

Serology. The cytotoxic action of H-2 sera on mouse lymph node cells (24) was tested by the microlymphocytotoxicity test (25), which was regularly performed in serial geometrical dilutions (for details, see ref. 15). Absorbed rabbit serum was used as a source of complement. Serum dilutions and cell suspensions were prepared in RPMI + 10% FCS medium. The reaction in each well was recorded as -, +, ++, +++, ++++ according to the finding of not more than 15, 30, 50, 80 and more than 80% dead cells, respectively.

Immunizations. Donor spleen + thymus + lymph node cells (in doses of one donor per ten recipients) were injected intraperitoneally into female mice 4 ($\pm$ 1) months old. In spite of possible individual variations (mostly due to differences in spleen weight) in the total number of lymphoid cells received from individual donors, no effort was made to equalize the number of cells injected each week. However, each mouse received the same aliquot from one common pool of the prepared lymphoid cell suspension. The immunizations were performed at weekly intervals with a one-week intermission after every fourth injection. Serum samples from individual mice were collected every week 3-4 days after injection. Samples obtained after the first injection were labeled A, after the second injection B, and so on. Most work was performed with serum samples of the immunization week T (i.e., after a series of 19 injections). When it was decided to terminate a given donor-recipient immunization combination, the recipients were bled 3, 7, and 9 days after the last immunization, and the three samples of sera (from one individual mouse) were pooled and designated as being the last sample of the immunization (with a letter of the last immunization week).

<u>Wild mice</u>. Wild mice nos. 1, 2, 9, and 10 were captured at different localities in or near Amsterdam, mice nos. 3-8 were captured in South Holland in one building complex.

RESULTS AND DISCUSSION

Serotyping of Inbred Strains

The anti-H-2 sera were raised by immunization of B10.A (4R) ($H-2^{h4}$; kkbbbb) recipients with B10.A ($H-2^a$; kkkdddd) donor cells (serum no. CLS-3). Serum samples of bleeds S or T from 38 individual recipients were tested for their cytotoxicity on lymph node cells of inbred mouse strains representative of the well-defined H-2 haplotypes. The reaction of all sera was negative on B10.A(2R) (kkkdd.b) cells. This shows that only anti-H-2D^d antibodies were present. The data are summarized in Table 1.

For general information, the reaction strength and major HLA antigen associations of the individual sera are given in the top part of Table 1 (for details, see ref. 22).

The following conclusions and questions emerge from the reaction pattern observed on mouse cells.

1. One serum was negative and three gave only very weak reactions with the cells of the donor strain. All other sera reacted with the donor strain cells in titers higher than 1:64 up to titers of 1:2000-4000. Eight sera, however, only reacted with the donor strain. Such sera represent monospecific anti-private (anti-H-2.4) sera. As will later

be shown, this designation is only operational because some of the "monospecific" sera cross-reacted with certain wild mice.

2. Further sera exhibited a variable extent of cross-reactivity with other H-2 haplotype products. At the right hand end of Table 1 appear those sera that cross-reacted with all H-2 haplotypes having H-2 public specificities shared with the cells of the donor strain. The only exception to the expected cross-reactions were the negative reaction of all sera with the $H-2^k$ haplotype. It is not clear why anti-D^d sera produced in D^b recipients did not react with the $H-2^k$ haplotype. Both D^d and D^k contain H-2.3. This H-2 public antigen was originally defined (by hemagglutination) with an anti-$H-2^d$ serum produced in $H-2^b$ recipients (26). H-2.3 was later described as actually being a "family" of 3-like specificities (27). The family consists of H-2.3, 13, 35, 36, 41, 42, 43, 44 specificities. Only H-2.3 is present on $H-2^k$. Demant et al. (27) noticed that the presence of $H-2^b$ in the recipient can block the formation of cytotoxic anti-H-2.3 antibodies. It was presumed that the presence of H-2.35 and H-2.36 of $H-2^b(K^b)$ are responsible for this situation. However, in our combination, only D^b (H-2.35, 36 negative) was present in the recipients. If we admit that the sera do not contain cytotoxic anti-3 antibodies, it would become difficult to explain the reactions with $H-2^r$ and K^d alleles. We therefore propose the following explanation. H-2.3 of the D^d allele is a D-end antigen with a counterpart of a similar structure at the K-end of $H-2^d$. $H-2^r$ has the $H-2D^d$-like characteristics of H-2.3. However, H-2.3 is primarily a K-end specificity on the $H-2^k$ haplotype. The formation of anti-K-end antibodies in our sera was blocked by the presence of identical K^k alleles in

the donor and recipient. In other terms, we presume that the absence of anti-H-2.3 in our sera (blocked by the presence of K^k in the recipient) is responsible for the negative reactions with $H-2^k$.

3. Between the two extremes (negative sera versus "strong" cross-reactive sera) individual sera exerted a variable reaction pattern when tested on the panel of H-2 haplotypes (q, s, p, r). Serum no. 46 only reacted strongly with D^d and $H-2^s$. Sera nos. 6, 36 and 37 only reacted strongly with D^d and $H-2^q$ (although weaker reactions also occurred with other haplotypes). Sera nos. 24 and 17 gave stronger reactions on $H-2^p$ than on $H-2^q$ and $H-2^s$ targets. Many more sera cross-reacted with $H-2^p$ when compared with $H-2^r$. No sera reacted strongly with $H-2^p$ and/or $H-2^r$, which did not have concurrent reactivity against $H-2^q$ and $H-2^s$.

There are clear-cut differences in the <u>frequency of occurrence of different cross-reacting patterns</u>. (See anti-q and anti-s versus anti-p and anti-r). Further experiments might elucidate the biological meaning of these differences.

Two interpretations emerge from these findings: 1) certain individual mice produce anti-H-2.4 (the anti-"private" antibody) with a variable cross-reactivity pattern for other specificities; and 2) individual sera represent different combinations of distinct antibody populations with variable predominance for some specificities. These predominant components (or cross-reactivity patterns) are characteristic for some individual sera, and correspond with public H-2 antigens H-2.13 (for $H-2^q$), H-2.36 (for $H-2^s$), H-2.35 (for $H-2^p$) and H-2.? (for $H-2^r$).

4. Cross-reaction of anti-D^d antibodies with K region products. a) Anti-H-2.35 and 36 antibodies can be responsible for cross-reactions of anti-D^d antibodies with $H-2K^b$

products of the $H-2^b$ haplotype. A rather low number of sera exerted a strong cross-reaction of this type (21%). In one serum (no. 9), anti-K^b activity was the strongest. There was good accord for the requirement that anti-$H-2^s$ and anti-$H-2^p$ cross-reactivity be present in sera showing anti-$H-2^b$ positivity. However, not all sera that reacted strongly with $H-2^p$ and/or $H-2^r$ reacted with $H-2^b$. Thus some as yet unknown requirement(s) must be fulfilled to achieve anti-K-end (K^b) cross-reactivity in individual anti-D-end sera.

b) An unexpected finding was the strong reactions of three sera with HTG cells ($H-2^g$; ddddd.b). These reactions point to shared specificities of D^d with K^d allele products. According to the H-2 chart, H-2.3 is the only specificity that might be responsible for these reactions. However, as mentioned under point 2, all sera were negative with $H-2^k$. Furthermore, the reactivity of at least two of the anti-K^d-positive sera was either negative or very weak with cells of other strains bearing H-2.3. Thus sera 3x-9 and 3-19 either detect unknown K^d allele products or they visualize the complex nature of H-2.3. The latter possibility seems to be more valid, because both DBA/1 ($H-2^q$) and A.SW ($H-2^s$) cells absorbed anti-HTG (K^d) activity from serum 3x-9.

Anti-H-2.3 sera have been reported as exerting very weak or inconsistent reactions on HTG cells ($H-2^g$) (27). This observation might be due to the use of serum pools in which the content of anti-$H-2^g$ (anti-K^d) might have been diluted out by the high number of negative sera.

As mentioned before for sera cross-reacting with K^b, it seems in a similar sense that some unknown requirement(s) must be fulfilled for the achievement of anti-K-end (K^d) cross-reactivity in individual anti-D-end sera. Only three out of 38 sera of individual mice from one inbred strain had this capacity.

TABLE 2. Reaction Pattern of 36 Anti-H-2D^d Sera on Lymph Node Cells of 10 Wild Mice and the Donor (B10.A) Strain

Wild mouse	3	3	3	3	3	3	3	3	3	3	3x	3	3	3	3x	3	3	3	3	3
No.	45	7	1	42	34	16	35	40	10	39	37	31	38	12	47	46	30	6	48	24
	T	T	T	S	T	T	T	T	T	T	P	T	T	T	S	T	T	T	T	S
w5	-	-	-	-	-	+	-	+	-	+	-	-	-	+		-	-	-	-	-
w6	-	-	-	-	-	-	-	-	-	+	-	-	+	++	+	+	++	++	+	-
w7	-	-	-	-	-	-	-	-	-	-	+	-	-	+	-	-	++	+	-	-
w3	-	-	-	-	-	-	-	-	-	-	-	-	-	-	-	+++	++	+	++	++
w4	-	-	-	-	-	-	-	-	-	-	+	-	-	++	-	++	+	+	+	+
w1	-	-	-	-	-	-	-	-	-	+	-	-	-	-	-	-	+++	-	-	+
w10	-	-	-	-	-	-	-	-	-	-	-	++	-	++	-	-	+	++	-	-
w8	-	-	-	-	-	-	-	+	-	-	+	+	-	+++	++++	++	+++	+++	+++	-
w2	-	-	-	-	-	-	-	-	-	-	++	++++	++	++++	++++	+	++++	++++	++++	++++
w9	-	-	-	-	-	-	-	-	-	-	+	++	-	++++	++++	+++	++++	++++	++++	++++
B10.A	-	±	±	±	++	++	+++	+++	+++	++++	++++	++++	++++	++++	++++	++++	++++	++++	++++	++++

* - negative; + weak; ++ only weak reactions (+ or ++), titer 1:4–1:16; +++ strong reaction (+++) in dilutions 1:16–1:32; ++++ strong reaction (++++) in dilution 1:64 or higher.

TABLE 2 continued

Wild mouse No.	3 36 T	3 37 S	3x 45 S	3 25 T	3 18 S	3 44 T	3 13 T	3 17 T	3 47 T	3 28 S	3 2 T	3 9 T	3 33 T	3x 9 M	3 19 T	3 22 T
w5	++		−	+	−	−	+	−	−	−	−	++	+	++		−
w6	+		+	++	+++	+	++	+	+	+	−	++	+++	+		−
w7	++		+	++	+++	+	−	++	+++	++	+++	+++	+++	+		−
w3	−		++	++	++	+	++	+	+	+	++	+	+++	−		−
w4	+		+	++	+	+	+++	+	++	+	++	++	+++	+++		−
w1		+	−	+++	−				++				+++	+++	+++++	+++
w10	++		+++	++	++	++	++++	++	++++	++	−	++	+++++	++		+
w8	++		++++	+++	+++	++	+++	+++	+++++	++	++	+++++	+++++	++++	+++++	−
w2		+++	++++	+++++	+++++				+++++				+++++	+++++	+++++	+++++
w9	++		++	++	++++	++++	+++++	+++++	+++++	+++++	+++	+++++	+++++	+++++		−
B10.A	++++	++++	++++	++++	+++++	++++	++++	++++	+++++	+++	+++	+++++	+++++	+++++	+++++	+++++

5. Although both the donor and recipient strains have the "whole" H-2.28 complex (H-2.27, 28, 29), we have tested for the presence of anti-H-2L locus antibodies (9, 10). This may occur if the donor and recipient strains differ in subtypes of the H-2.28 complex. Serotyping of BALB/c-H-2^d and BALB/cH-2^{db} cells, and absorption experiments, revealed no difference between these two cell types although it was concluded that the H-2.28 public specificity was absent from cells of the H-2^{db} mutant (28).

Serotyping of Wild Mice

These experiments were aimed at extending the available panel of H-2 haplotypes and thus to allow for a better resolution of the extent of diversity of the individual H-2 sera.

The same sera as those tested on inbred mouse strains and human cells were tested for cytotoxicity on lymph node cells of 10 wild mice. The data are summarized in Table 2 (the sera remain arranged in the same order as for Table 1). The following conclusions and questions emerged from the reaction pattern observed with wild mice cells.

1. None of the wild mice tested can be classified as H-2.4 or D^d positive. This is because none of them reacted positively with all sera containing anti-donor antibodies.

2. The <u>extent</u> of cross-reactivity of the sera on wild mice cells (strength and number of positive reactions) was similar to that found on cells from mice of an inbred strain. This is certainly true for sera grouped to the left ("narrow" sera) and for most of the sera to the right ("broad" sera) of Table 2. Sera nos. 25, 9 and 33, which reacted with all wild mice cells, were obviously noted as being the most polyvalent sera on cells of inbred strains. However, there are examples that do not exactly follow this tendency.

3. Some sera that behaved as monospecific anti-H-2.4 (sera nos. 3x-37 and 31) on inbred strains reacted with cells of certain wild mice. This shows that the monospecific nature of these sera was operational and the "hidden" polyvalency or cross-reactivity pattern only emerged when sera were tested on wild mice cells. This finding parallels previous data obtained with "monospecific" (pooled) anti-H-2-private antigen sera (7, 8).

4. The interesting finding obtained from these preliminary tests on wild mice cells was the actual extent of cross-reactivity (broadness) of some sera. While some sera were negative or only reactive with a restricted number of individuals (inbred or wild mice) a few sera reacted with a very high number of unrelated individuals with different H-2 haplotypes (sera nos. 25, 9 and 33 as extreme examples). Sera of this kind show that not only the extent of polymorphism, but also the extent of cross-reactivity among allelic products is unique for the MHS. Individuals of one inbred strain differ up to the two extreme ends of this very great individual variability; some form cytotoxic antibodies reactive only with the donor strain and possibly a few other haplotypes, whereas others produce antibodies that react with a very high proportion of the polymorphic variants present in the given species. Although illustrated only in a preliminary approach, knowledge about the extent of these variations may be important for further studies on the mechanisms involved in the regulation of this kind of individual diversity in anti-MHS antibody production.

We conclude that the observed variations among individual sera were surprising as they have not yet been reported despite the great amount of H-2 serological work. (We cannot exclude similar notions in some work that we were unable to review.) In several instances where physiological factors involved in the regulation of anti-H-2 antibody formation were studied and in studies on genetic differences in anti-H-2 antibody production among inbred strains, the data were obtained by testing serum pools. In some studies, mean values from individual sera are shown (with or without values for S.D.). Furthermore, the respective sera were mostly tested on the donor strain only. As also shown by our data, in such a situation individual variations could be considered rather as being "exceptions" and thus escaped interest for immunobiological evaluation (29-32).

Experiments are in progress to test further combinations by a similar approach. We have seen a similar kind of individual variations when B10.A(2R) recipients were immunized by B10.A donor cells. In this combination, the donor and the recipient differ only in the D region.

We should like to stress that the important point, inherent in the approach that we follow, is that a high number of individual sera should encounter a far larger variety of MHS polymorphic targets. It seems clear that the higher the number of possible targets, the greater the extent of the observed diversity. This could be put forward for consideration as a general rule when we attempt to characterize the extent of individual variations in antibody formation in the frame of MHS. The same sera which were used in this work were also tested for cytotoxicity activity on human cells which provide a much more extensive panel (variety) of target structures than the limited panel of available H-2 haplotypes. The individual mouse sera differed in strength, reaction frequency, and specificity (related to HLA antigens) (22). There was a good correlation between the broadness of individual sera on both mouse and human cells (Fig. 1). This is a further parameter that illustrates the extent of the observed differences of individual sera. While some sera were rather

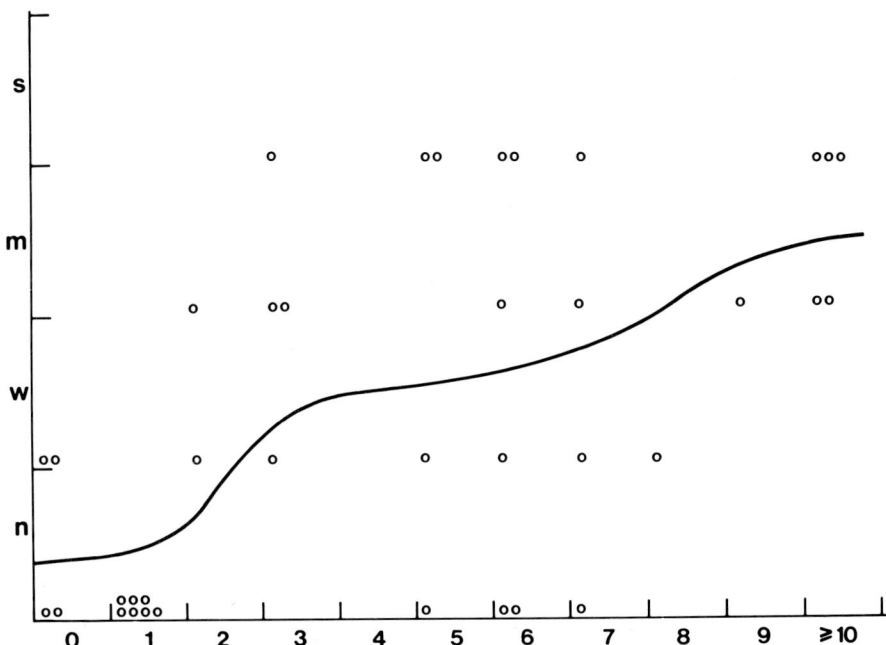

FIGURE 1. Correlation between the extent of cross-reactivity of anti-H-2D^d sera on human and mice cells. (Not corrected for differences in the number of cell types tested). x axis: number of positively reacting inbred strains of wild mice. (Only reactions stronger than + were counted.) y axis: n, negative, w, weak; m, medium; s, strong (22).

restricted when tested on the mouse panel and exerted a low reaction frequency on human cells, other sera reacted with the majority of mouse haplotype products and also with about 80% of human cells from unrelated individuals (more than 100 cells tested).

The pronounced differences in the reaction pattern of individual mouse sera are of interest for further studies of the antibody diversity that occurs during MHS allo-immunization. Probably, we are faced with an extensive, but hitherto not experienced, variability in the number of clones that became or remained specifically effective in the production of antibody to components of the H-2D region. Several studies show different aspects of individual variations in antibody response among individual mice of one inbred strain when these were immunized with heterologous antigens (bacterial, erythrocyte, hapten, etc.) (33-39). For estimates of the extent of antibody diversity in inbred mouse strains and the proportion of this potential that is being used by an individual mouse, see Kreth and Williamson (36) and Pink and Askonas (37). Some of the most restricted antisera can be predicted to be analogous to mono- or oligo-clonal antibody populations generated by techniques in vitro (40). These latter antisera are produced to unravel the complex nature of the immunological response to products of the MHS region. Experiments can now be designed to test the possible homogeneity of our restricted (with regard to target specificity) antibody populations, and to attempt to elucidate the role of biological factors (41) that contribute to the regulation in vivo of the observed antibody diversity during the course of MHS allo-antigenic immunization.

Finally, for operational and practical purposes, elucidation of the type of response occurring in an individual experimental animal might be useful before the setting up in vitro of a series of continuous lymphoid cell cultures (hybridomas) from any given MHS immunized individual.

ACKNOWLEDGMENTS

We express our gratitude to Peter van Mourik for his excellent technical assistance and to Sietse Weide and William Molenaar for their help in the preparation of the anti-H-2 sera. We are grateful to Anja Maas and Jetty Gerritsen for their secretarial assistance.

REFERENCES

1. Snell, G.D., and Stimpfling, J.H. (1966). In "Biology of the Laboratory Mouse," 2nd ed. (E.L. Green, ed.), p. 457. McGraw-Hill Book Co., New York.
2. Snell, G.D., Hoecker, G., Amos, D.B., and Stimpfling, J.H. (1964). Transplantation 2:777.
3. Snell, G.D., Cherry, M., and Demant, P. (1973). Transplant. Rev. 15:3.
4. Shreffler, D.C., and Snell, G.D. (1969). Transplantation 8:435.
5. Démant, P. (1973). Transplant. Rev. 15:162.
6. Klein, J. (1975). "Biology of the Mouse Histocompatibility-2 Complex." Springer Verlag, New York.
7. Micková, M., and Iványi, P. (1971). In "Proceedings of the Symposium on Immunogenetics of the H-2 System" (A. Lengerová and M. Vojtisková, eds.), p. 20. S. Karger AG, Basel.
8. Iványi, P., and Micková, M. (1972). Transplantation 14:802.
9. Démant, P., Snell, G.D., Hess, M., Lemonnier, F., Neauport-Sautes, C., and Kourilsky, F. (1975). J. Immunogenetics 2:263.
10. Démant, P., Neauport-Sautes, C., and Joskowitz, M. (1977). Tissue Antigens 10:252.
11. Shreffler, D.C., and David, C.S. (1975). Adv. Immunol. 20:125.
12. Klein, J. (1971). Nature 229:635.
13. Klein, J. (1973). Transplant. Proc. 5:11.
14. Iványi, P., Démant, P., Vojtisková, M., and Iványi, D. (1969). Transplant. Proc. 1:365.
15. Micková, M., and Iványi, P. (1976). Folia Biol. (Praha) 22:169.
16. Klein, J. (1970). Science 168:1362.

17. Zaleska-Rutczynska, Z., and Klein, J. (1977). J. Immunol. 119:1903.
18. Ivasková, E., Dausset, J., and Iványi, P. (1972). Folia Biol. (Praha) 18:194.
19. Ivasková, E., Iványi, P., Mickova, M., and Démant, P. (1975). Tissue Antigens 6:286.
20. Iványi, P., Pavljuková, H., and Ivasková, E. (1976). Transplantation 22:612.
21. Iványi, P. (1978). Proc. Roy. Soc., in press.
22. Iványi, P., van den Berg-Loonen, E.M., and de Greeve, P. (1978). Tissue Antigens, in press.
23. Jerne, N.K. (1971). Eur. J. Immunol. 1:1.
24. Gorer, P.A., and O'Gorman, P. (1956). Transpl. Bull. 3:142.
25. Terasaki, P.I., and McClelland, J.D. (1964). Nature 204:998.
26. Hoecker, G., Counce, S., and Smith, P. (1954). Proc. Natl. Acad. Sci. USA 40:1040.
27. Démant, P., Snell, G.D., and Cherry, M. (1971). Transplantation 11:242.
28. McKenzie, I.F.C., Morgan, G.M., Melvold, R.G., and Kohn, H.I. (1977). J. Immunogenetics 4:333.
29. Stimpfling, J.H., and Durham, T. (1972). J. Immunol. 108:947.
30. Lilly, F., Graham, H., and Coley, R. (1973). Transplant. Proc. 5:193.
31. Peters, K., Schreiber, B., Gleichmann, H., and Gleichmann, E. (1974). J. Immunogenetics 1:141.
32. Shinohara, N., Ricks, J., Hansen, T.D., and Sachs, D.H. (1977). J. Immunol. 119:1732.
33. Gershon, R.K., and Kondo, K. (1972). Immunology 23:321.
34. Silver, D.M., McKenzie, I.F.C., and Winn, H.J. (1972). J. Exp. Med. 136:1063.

35. Briles, D.E., and Krause, R.M. (1972). J. Immunol. 109: 1311.
36. Kreth, H.W., and Williamson, A.R. (1973). Eur. J. Immunol. 3:141.
37. Pink, J.R.L., and Askonas, A. (1974). Eur. J. Immunol. 4:426.
38. Eichmann, K. (1974). Eur. J. Immunol. 4:296.
39. Eichmann, K. (1975). J. Immunogenetics 2:491.
40. Lemke, H., Hammerling, G.J., Hohmann, Ch., and Rajewsky, K. (1978). Nature 271:249.
41. Pilarski, L.M., and Cunningham, A.J. (1975). Eur. J. Immunol. 5:10.
42. Iványi, P. (1970). Curr. Top. Microbiol. Immunol. 53:1.

STUDIES ON HISTOCOMPATIBILITY ANTIGENS AND
HYBRID STERILITY GENE IN WILD MICE: A SHORT SURVEY

Pavol Iványi

Central Laboratory of the Netherlands Red Cross
Blood Transfusion Service and Laboratory of
Experimental and Clinical Immunology
University of Amsterdam
Amsterdam, The Netherlands

Around 1967-1968 it became obvious that research on H-2 immunogenetics had to be extended to wild mice populations. At that time, several functional aspects of the H-2 system began to receive attention. To understand better the biological role of the H-2 system, it was desirable to have at least initial information about the degree of polymorphism of the H-2 system in wild mouse populations. Such data would also be essential for speculations about the mechanisms involved in the generation and maintenance of this kind of polymorphism and complexity. The initiation of these studies was promoted by a successful definition of MHS antigens of non-inbred populations in other species such as human, chicken, and rabbit (for review see 1, 2). In other words, if MHS serology became possible in other non-inbred populations, why should all work in mice remain restricted to inbred strains? The available inbred strains have frequently a common genetic origin. They might represent a selection of the most suitable (and interesting) genes for some specialized laboratory studies, but they are certainly not fully representative of the natural degree of variation either at the structural or the functional levels. The laboratory conditions might have enabled us to study genes which had no chance to spread (and to be found) under natural conditions.

When H-2 studies on wild mice began, there was a temptation to extend the number of thoroughly studied H-2 haplotypes (alleles at that time) because the number of "independent" inbred H-2 haplotypes was less than ten. The only information about H-2 antigens of non-inbred mice came from the work of Rubinstein and Ferrebee (3). Serotyping of 47 Swiss-Webster mice of a random-bred colony showed a high degree of polymorphism. New H-2 alleles and H-2 antigens were to be expected.

In this short survey only the summary of our work performed until 1975 will be given. All data have been published and the interested reader should consult the original articles. The extensive data of Dr. Jan Klein's group are summarized in a separate communication in this volume. The remarkable series of Dr. Klein's new $H-2^w$ haplotypes now available for laboratory work will doubtless open new horizons in H-2 immunogenetics.

An Estimate of the Degree of Heterozygosity at Histocompatibility Loci in Wild Populations of House Mouse

Wild males were mated with inbred females of the C57BL/10 strain. Skin grafts were exchanged among the progeny of one wild male. In such a situation the number of surviving grafts is proportional to the number of H loci which were in a heterozygous state in the original wild male. It was found that, while 20-30 histocompatibility loci occur in heterozygous state in F_1 hybrids of two inbred strains, only 3-7 histocompatibility loci can be supposed to occur in a heterozygous state in individual wild males. This indicates that wild mice are relatively homozygous in respect to their H loci (4-6).

This study represents a general approach for the estimation of the degree of heterozygosity at H loci in any non-inbred individual. However, two points should be considered. 1) The number of surviving grafts can be biased if there are antigens shared between the outbred individual and the inbred strain. Therefore, any one outbred individual should be tested by mating with two or more inbred strains of diverse genetic background. When a greater number of non-inbred individuals are tested with the same inbred strain, the average values thus obtained can be considered as real because it is highly improbable that different non-inbred individuals would share histocompatibility genes with the same inbred strain. 2) The number of surviving grafts is influenced by the sex of the recipients and possibly other physiological factors. Pre-immunization of the recipient with donor tissue can result in graft rejection due to weak H systems otherwise undetected. Thus, the above-mentioned estimates have only a relative value obtained in an operationally comparable methodological assay (6).

Studies on H-2 Antigens in Wild Mice

1. The H-2 phenotype of 226 wild mice was tested by the PVP hemagglutination test with anti-H-2 sera prepared by immunization of inbred strains. Mice from 16 localities were tested (13 from central Bohemia). Wild mice within any given locality (with the exception of two or three localities) did not show much similarity in their H-2 phenotypes. This contrasted sharply with the phenotypes of 100 wild mice captured at Great Gull Island (a small rocky island in the Atlantic Ocean) where only three different H-2 phenotypes were detected. In all localities the phenotypes observed

suggested the presence of new, thus far unknown, H-2 haplotypes. Some wild mice reacted with a very high number of anti-H-2 sera, others were negative for all H-2 antigenic specificities tested (4, 7, 8).

2. Sera against public H-2 specificities had a high reaction frequency on wild mice. The most frequent reactions were observed with anti-H-2 sera defining so-called long public specificities of the inbred mouse strains (anti-H-2.5, 3, 28, 11, 1). These data suggest extensive sharing of public antigens among inbred and wild mice or, in other words, they illustrate the extent of sharing of cross-reactive structures among gene products on a large sample of individuals of the given species.

Sera against private H-2 antigens also reacted positively although less frequently. Absorption experiments, however, showed that the $H-2^w$ haplotypes did not share <u>identical</u> private antigens with the inbred strains (7-9). The positive reactions of anti-H-2-inbred-private sera with wild mice were rather due to cross reactions with <u>similar</u> antigenic products of the $H-2^w$ haplotypes.

3. Three new H-2 congenic strains were prepared on the C57BL/10 background by introducing three new $H-2^w$ haplotypes by successive backcross matings (B10.W44 with $H-2^{w7p}$, B10.W67 with $H-2^{w8p}$, B10.W627 with $H-2^{w9p}$). Those were characterized for some of the inbred public specificities, Ss phenotypes, three new private specificities, and three new public specificities detected by antisera prepared against the respective $H-2^w$ haplotypes (8).

The haplotype designations must be considered as provisional because the new $H-2^{wp}$ haplotypes have not been compared with the extensive series of the $H-2^w$ haplotypes of Dr. Jan Klein. The addition of letter "p" to the haplotype symbol is proposed to designate the origin (Prague) of these new haplotypes until direct serological comparison will allow a final nomenclature.

A number of functional, structural, and genetic analogies emerge from the comparison of the H-2 and HLA system. In general, the H-2K and H-2D loci are considered as analogous with HLA-A and HLA-B loci. However, the H-2 system, after preliminary insight into the degree of its polymorphism, seems to be much more polymorphic than the HLA system. Several monospecific anti-HLA sera can be completely absorbed by cells obtained from different populations. Although several HLA antigen splits are yet to be expected, the frequencies of several HLA genes in a large area (e.g., Caucasoid populations) are estimated to be around 0.05-0.10. The H-2 gene frequency estimates are more difficult, but prediction for a large enough population sample is <0.01. To explain this difference we suggest that mutations arising in a mouse population can be maintained because of the social structure, generation time, and number of offspring in mice. Furthermore, the distribution of the mutant H-2 alleles may be affected by the alleles at the T/t region. We presume that the difference in polymorphism is not "inherent" to the MHS of the two species (e.g., different mutation rates) but the consequence of a number of circumstances "outside" the MHS of both species (8).

Identification of a Male-Hybrid-Sterility-Gene Located Between the T/t and H-2 System on Chromosome 17

When wild males were mated with inbred females, we noticed that a) some wild males mated with C57BL/10 (B10) females produced sterile sons; b) the same wild males produced fertile sons with C3H females; and c) the daughters from all matings were fertile (4, 7, 10-13).

Because both parents were fully fertile and sterile sons occurred only in certain combinations, the observed type of sterility was designated as (male)-hybrid sterility (Hst).

A major gene responsible for the observed hybrid sterility was located on chromosome 17 between the T/t and the H-2 system. This was shown by a three-point test cross (B10-TxC3H)T/+ ♀ x W ♂. The hybrid sterility gene (Hst-1) mapped 6 cM distally from dominant T.

Some wild males were heterozygous for the Hst gene. Serotyping of the two $H-2^w$ haplotypes in the male progeny from B10 ♀ x W ♂ matings showed that the recombination fraction between H-2 and Hst was 8-13%.

Fifty-three wild males were tested for the presence of Hst gene; of these, 23 males yielded only sterile sons (Hst homozygotes), 10 wild males yielded sterile and fertile sons (Hst heterozygotes), and 20 wild males sired only fertile sons. The majority of wild males with the Hst gene were captured in the Prague zoological garden, but five were captured in different localities (four in Bohemia, one in Denmark).

Wild males with the Hst gene sired sterile sons with C57BL/10, A, BALB/c, DBA/1, and AKR/J females, whereas the same wild males sired fertile males with inbred females of C3H/Di, CBA/J, P/J, I/St, and F/St strains.

The cause of the observed male-hybrid-sterility was a spermatogenesis arrest at the stage of spermatogonia or primary spermatocytes; the sterile males had small testis (<90 mg). Examination of meiotic and mitotic chromosomes of sterile hybrids did not reveal any gross chromosomal rearrangements which would point to a chromosomal type of sterility. Blood testosterone level, vesicular gland weight (an indicator of target organ sensitivity to testosterone), and gonadotropin activity were within ranges of physiological variation (13).

The sterile males were healthy individuals with signs of "hybrid vigor" for quantitative traits identical to that of fertile wild x inbred male hybrids.

The mechanism of the male-hybrid-sterility as well as its consequences for possible incipient reproductive isolation of Mus musculus subspecies is not known and its clarification will require more data.

The identification of the Hst gene clearly shows a hitherto unknown type of polymorphism on the 17th chromosome of the inbred strains (e.g., B10 versus C3H) which would have remained undetected without combining the inbred chromosome with the wild chromosome.

A nomenclature for the hybrid sterility genetic system (Hst) was proposed by Forejt and Iványi (13).

SUMMARY

A short summary of studies on histocompatibility antigens and hybrid sterility in wild mice was presented. It was found that 1) wild mice are relatively homozygous for the numerous histocompatibility loci when compared with F_1 hybrids of two inbred strains; 2) the H-2 system is highly polymorphic and serologically complex in wild mice populations; 3) three new congenic strains that contain $H-2^w$ haplotypes on the C57BL/10 background have been established and characterized; and 4) hybrid sterility gene was located on chromosome 17 between the T/t and H-2 region.

ACKNOWLEDGMENTS

The experiments described in this communication were performed at the Institute of Experimental Biology and Genetics in close cooperation with Drs. Milada Micková, Jiri

Forejt, Peter Démant, Marta Vojtisková, and Dagmar Iványi. Requests for B10.W^{wp} mice and wild mice with the Hst gene should be addressed to Drs. Milada Micková and Jiri Forejt, respectively (Institute of Molecular Genetics, Czechoslovac Academy of Sciences, Prague 4).

REFERENCES

1. Iványi, P. (1970). Current Topics Microbiol. Immunol. 53:1.
2. Götze, D. (ed.) (1977). "The Major Histocompatibility System in Man and Animals." Springer Verlag, Berlin, Heidelberg, New York.
3. Rubinstein, P., and Ferrebee, J.W. (1964). Transplantation 2:715.
4. Iványi, P., Démant, P., Vojtisková, M., and Iványi, D. (1969). Transplant. Proc. 1:365.
5. Iványi, P., and Démant, P. (1970). In "XIth European Conference on Animal Blood Groups and Biochemical Polymorphism, Warszaw 1968," p. 547. W. Junk, N.V., Publishers, The Hague.
6. Micková, M., and Iványi, P. (1972). Folia Biol. 18:350.
7. Micková, M., and Iványi, P. (1971). In "Immunogenetics of the H-2 System" (A. Lengerová and M. Vojtisková, eds.), p. 20. Karger, Basel.
8. Micková, M., and Iványi, P. (1976). Folia Biol. 22:56.
9. Iványi, P., and Micková, M. (1972). Transplantation 14:802.
10. Iványi, P., Vojtisková, M., Démant, P., and Micková, M. (1969). Folia Biol 15:401.
11. Iványi, P., and Micková, M. (1971). In "Immunogenetics of the H-2 System" (A. Lengerová and M. Vojtisková, eds.), p. 104. Karger, Basel.

12. Micková, M., and Iványi, P. (1972). In "Proceedings XIIth European Conference on Animal Blood Groups and Biochemical Polymorphism," p. 621. Akademiai Kiado, Budapest.
13. Forejt, J., and Iványi, P. (1975). Genetic. Res. 24:189.

CHARACTERIZATION OF H-2 HAPLOTYPES IN WILD MICE[1]

Jan Klein[2]
William R. Duncan
Edward K. Wakeland
Zofia Zaleska-Rutczynska[3]
Huei-Jen S. Huang
Ellen Hsu

Department of Microbiology
University of Texas Southwestern Medical School
Dallas, Texas

Department of Immunogenetics
The Max Planck Institute for Biology
Tubingen, Federal Republic of Germany

For someone who is accustomed to going almost daily to the animal room with a shopping list of mice needed for experiments, it is hard to imagine that there had been a time when no inbred strains existed. It is even more difficult to envision how the world would have looked had inbred strains never been invented. Although the resourceful

---

[1] The work was supported by grant number AI 11879 from the National Institutes of Health.

[2] Permanent address: Max Planck Institute for Biology, Tubingen, W. Germany.

[3] Permanent address: Department of Histology and Embryology, Academy of Medicine, Warsaw, Poland.

human mind would probably have found ways to eventually arrive at our present level of knowledge even without the strains, such an alternate course, no doubt, would have been much more arduous. Among the disciplines hardest hit by the lack of inbred strains would have been immunogenetics and, in particular, transplantation genetics. We would have learned about the existence of histocompatibility genes without the strains, but how we would have accumulated all the current knowledge about the individual genes, particularly the genes in the major histocompatibility complex, we cannot even guess. We are, therefore, happy to join in this tribute to the pioneers who initiated inbreeding of mouse strains. They have put into our hands a tool which for contemporary biology is as indispensable as the scintillation counter, the pH meter, or the Sephadex column. They have enabled the construction of new scientific disciplines probing into new territories of modern biology. For all this we are extremely grateful and it is our pleasure to share with them today some of the excitement that they have helped to generate.

The work that we are going to describe started with inbred-strain analysis but it soon led us to the original source of all such strains -- the wild mice. It concerns the H-2 complex, the major histocompatibility complex of the mouse (Fig. 1). When we began the work some ten years ago, Peter Gorer, Bernard Amos, Gustavo Hoecker, and George Snell had already established that among the inbred strains there were some ten alleles, or haplotypes as they are now referred to, at this complex locus (for review see ref. 1). This degree of genetic variability was surprising since only two or three alleles were known at all other loci identified at that time. Was this variability of the H-2 complex a true polymorphism or was it merely the result of chance fixation? And if the H-2 complex was more polymorphic than

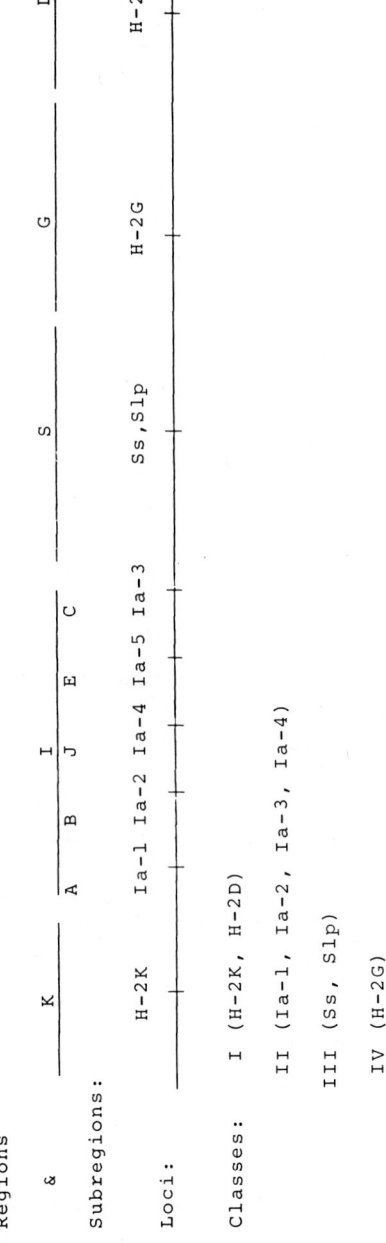

FIGURE 1. The genetic map of the H-2 complex.

other loci, what was then the meaning of the polymorphism, and how did the polymorphism relate to the function of the H-2 loci? What was the function of the H-2 complex anyway? These questions were provocative enough to stimulate our interest in the study of H-2 variability.

Wild Mice Typing with Inbred Reagents

The first step in this study was to trap wild mice in farm buildings, bring them into the laboratory, and type them with antisera defining individual antigens controlled by the H-2 complex. All such antisera available to us at that time were produced by crossimmunization of inbred strains and they defined H-2 antigens of the inbred strains. The typing of the wild mice with the antisera established two important facts, but provided only partial answers to our original questions (Fig. 2). The two facts were, first, that some antisera were broadly reactive with wild mice, while others reacted with such mice hardly at all (2), and second, that two mice from a single locality resembled each other in their H-2 antigens more closely than mice from different localities (3, 4). The broadly cross-reactive antisera were against H-2 antigens which were also shared by many inbred strains and which we therefore designated as public (2). The reactivity pattern of these antisera suggested that the wild mice carried H-2 haplotypes different from those present in the inbred strains, but we could not be completely sure about this conclusion because it is the private (the most restricted antigen) that unambiguously characterizes a haplotype. The antisera to the private H-2 antigens reacted hardly at all with the wild mice, and if they did, it often turned out that the reactivity was caused by a previously undetected public antigen. When antibodies to the new antigen were removed by absorption, the antiserum ceased to react with the particular wild mouse while it

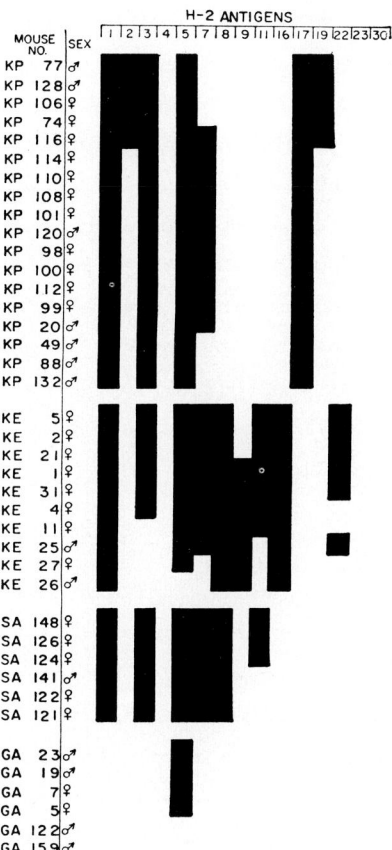

FIGURE 2. Distribution of 15 inbred-derived antigens among 40 wild mice from the area of Ann Arbor, Michigan. Solid bars indicate the presence of an antigen; blanks indicate absence of antigens. Mice from the same locality have the same capital letters. (Reproduced from ref. 3 with the permission of the American Association for the Advancement of Science.)

still reacted with the inbred strain carrying the private antigen. It was tantalizing finding: all we could get was a hint that wild mice possessed H-2 haplotypes absent in inbred strains, but we could neither prove this variability nor could we determine how many new haplotypes were present among the mice.

The finding of locality-dependent diversification of H-2 haplotypes was interesting and it fit well with observations made earlier by ecologists about the social structure of the mouse populations. Mice apparently live in small family units (demes) which are relatively isolated and migration from one unit to another is probably rare (5, 6). The similarity of H-2 haplotypes within a single locality could be, therefore, the result of the genetic isolation of the families.

Production of B10.W Congenic Lines

The ambiguity of wild-mice typing with antisera prepared against inbred H-2 haplotypes convinced us that the way of going about characterizing wild H-2 haplotypes was to produce antisera against antigens controlled by these haplotypes. This may sound like a simple thing to do, but in fact there were problems with the approach. First of all, there was the question of how to get enough material for immunization. A single wild mouse was not going to provide enough tissue for the large number of injections necessary to produce an H-2 antiserum. Then, there was the problem of how to test the antiserum when the donor of the immunizing tissue was no longer available. And finally, we feared that immunization with so genetically heterogenous a donor as the wild mouse would produce a mixture of antibodies that would take us forever to sort out.

Because of these problems we decided to use a different approach. We resolved to isolate a sample of H-2 haplotypes from wild mice and place them on the well-defined genetic background of an inbred strain. In other words, we wanted to do what George Snell did earlier with so many inbred

H-2 haplotypes -- to produce H-2 congenic lines. Because most of Snell's congenic lines were on the background of strain C57BL/1oSn, or B10 for short, we decided to use this strain for the crossing with wild mice, and to produce a series of B10.W congenic lines (where W stands for "wild"). It was then that we got the first taste of what it was like in the pioneering days of inbred and congenic strain production! For the task of producing 30 or 40 B10.W lines looks simple on paper, but in reality one has to overcome all kinds of mishaps to get the mice through the ten or 12 generations of backcrossing required for congenic line production. Now, more than ever, we can appreciate what was involved when the entire program of inbred-strain production was begun!

To produce the B10.W lines, we took advantage of the fact that in the middle of the H-2 complex is a locus coding for a serum protein, the serum serological substance or Ss (Fig. 1, ref. 7). Because most wild mice carry the Ss^h allele at this locus, while some inbred strains carry the Ss^l allele, in the B10.W-production system we would backcross repeatedly to an $Ss^l Ss^l$ strain (B10.BR) and always select for further mating animals carrying the Ss^h allele and thus presumably the wild-derived alleles at the H-2 loci on both sides of Ss (Fig. 3).

We started the project with some 80 B10.W lines; now, some ten years later, when the project is almost completed, we ended up with 35 lines. The others were all lost because of poor breeding performance.

Typing of B10.W Lines

Once the production of the B10.W lines was completed, we took the next step -- to analyze their H-2 haplotypes. To this end we tested the lines with a battery of antisera

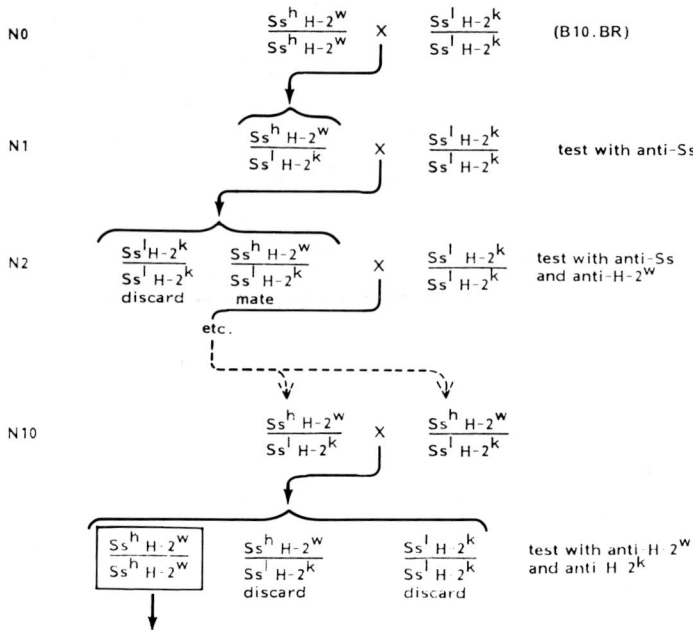

FIGURE 3. Method used for the production of B10.W lines.

defining all the known private H-2 antigens. We then immunized inbred strains [usually the (A x B10)F_1 hybrids] with B10.W tissues, produced anti-B10.W sera, and tested the antisera against all inbred strains carrying different H-2 haplotypes, and against all other B10.W lines. Finally, we did an absorption analysis with all strains and all antisera that were positive in a direct test. The analysis consisted of absorbing each positive antiserum with each positive strain and testing on all positive strains. Although we tried to avoid the production of antibodies to public H-2 antigens in the anti-B10.W sera by choosing the A strain as one of the parents of the F_1 hybrid recipient (the H-2^a haplotype of strain A codes for most of the known public antigens), we were not completely successful. The broad cross-reactivity of the anti-B10.W indicated that there were many more public antigens than were previously defined and that wild mice carried a number of public antigens absent in inbred strains. Because we were interested only in antigens that characterize a given haplotype -- that is, in private H-2 antigens -- we did not make any effort to identify and characterize the new public antigens. On the contrary, we tried to remove the anti-public antibodies from our antisera and to restrict the reactivity pattern of the antisera as much as we could.

The analysis of all the B10.W strains has not been completed yet. Thus far, however, we have been able to identify 20 new H-2 haplotypes and 25 new H-2 antigens (Table 1), all of which are absent in the inbred strains (8-13). We designated the new haplotypes $H-2^{w1}$ through $H-2^{w20}$ and the new antigens H-2.101 through H-2.125. In addition, we identified three new H-2 haplotypes in strains carrying alleles at the t locus (11) which is positioned in the same chromosome as H-2 and which affects embryonic and sperm development (14). These haplotypes we designated according to the t allele they carry, that is, $H-2^{tw1}$, $H-2^{t12}$, and $H-2^{tw5}$.

Haplotype Counting

Table 2 lists all the H-2 haplotypes -- inbred or wild -- identified thus far. The total number is 109. Some of these haplotypes, however, share alleles at both H-2K and H-2D (they differ at other loci of the H-2 complex not discussed in this communication). If we exclude these shared haplotypes, we still end up with an impressive number of 68 different haplotypes. Thus, among inbred strains and wild mice, 68 different combinations of alleles at the H-2K and H-2D loci have been already identified.

TABLE 1. H-2 Chart of B10.W and T/t Strains

Type strain	H-2 haplotype	H-2 antigen																													
		2	9	19	21	23	26	31	32	102	103	104	105	106	107	108	109	110	111	112	113	114	115	116	117	118	119	120	122	123	124	125
B10.KPA42	w1	-	-	-	21	-	-	-	-	-	-	-	-	-	-	-	-	-	-	-	-	-	-	-	-	-	-	-	-	-	-	-
B10.KPB68	w2	-	-	-	-	-	-	-	-	102	-	-	-	-	-	-	-	-	-	-	-	-	-	-	-	-	-	120	122	-	-	-
B10.SAA48	w3	-	-	-	-	-	-	-	-	-	103	-	-	-	-	-	-	-	-	-	-	-	-	-	-	-	-	-	-	1237	124	-
B10.GAA20	w4	-	-	-	-	-	-	-	-	-	-	104	-	-	-	-	-	-	-	-	-	-	-	-	-	118	119	1207	-	-	-	-
B10.KEA5	w5	-	-	-	-	-	-	31	-	-	-	-	105	-	-	-	-	-	-	-	-	-	-	-	-	-	-	?	-	-	-	125
B10.RB7	w6	-	-	-	-	-	-	-	-	-	-	-	-	-	107	-	-	-	-	-	-	-	-	-	-	-	-	-	-	-	-	-
B10.MOA1	w7	-	-	-	-	-	-	-	-	-	-	-	-	-	-	-	-	-	-	-	-	-	-	-	-	-	-	-	-	-	-	-
B10.SNA70	w8	-	-	-	-	23	-	-	-	-	-	-	-	-	-	-	-	-	-	-	-	-	-	-	-	-	-	-	-	-	-	-
B10.BUA19	w9	-	-	-	-	-	-	-	32	-	-	-	-	-	-	-	-	110	-	-	-	-	-	-	-	-	-	-	-	-	-	-
B10.KEA2	w10	-	-	-	-	-	-	-	-	-	-	-	-	-	-	-	-	-	-	-	-	-	-	-	-	-	-	-	-	-	-	-
B10.CAA2	w11	-	97	-	-	-	26	-	-	-	-	-	-	-	-	-	-	-	-	-	-	-	-	-	-	-	-	-	-	-	-	-
B10.MOL1	w12	-	-	-	-	-	-	-	-	-	-	-	-	-	-	-	-	-	-	112	-	-	-	-	-	-	-	-	-	-	124	-
B10.STA10	w13	-	-	-	-	23	-	-	-	-	-	-	-	-	-	-	-	-	-	-	113	-	-	-	-	-	-	-	-	-	-	125
B10.STC77	w14	-	-	-	-	-	-	31	-	-	-	-	-	-	-	-	-	-	-	-	-	114	-	-	-	-	-	1207	122	123	-	125
B10.STC90	w15	-	-	-	-	-	-	-	-	-	-	-	-	-	-	-	-	-	-	-	-	-	115	-	-	-	119	-	-	-	?	-
B10.BUA1	w16	-	-	-	-	-	-	-	-	-	-	-	-	-	-	-	-	-	-	-	-	-	-	116	117	-	-	-	-	-	-	-
B10.CAS2	w17	-	-	-	-	-	-	-	-	-	-	-	-	-	-	-	-	-	-	-	-	-	-	-	117	-	-	-	-	-	-	-
B10.CHR51	w18	-	-	-	-	-	-	-	-	-	-	-	-	-	-	-	-	-	111	-	-	-	-	-	-	118	119	120	-	1237	1247	-
B10.KPB128	w19	-	-	19	-	-	-	31	-	-	-	-	-	-	-	-	-	-	-	-	-	-	-	-	117	-	-	-	-	-	-	125
B10.LIB18	w20	2	-	-	-	-	-	-	-	-	-	-	-	-	-	-	-	-	-	-	-	-	-	116	-	-	-	-	-	-	-	-
T/t	t12	-	-	-	-	-	-	-	-	-	-	-	-	106	107	-	-	110	-	-	-	-	-	-	-	-	-	-	-	-	-	-
T/t	tw1	-	-	-	-	-	-	-	-	-	-	-	-	-	107	108	-	-	-	-	-	-	-	-	-	-	-	-	-	-	-	-
T/t	tw5	-	-	-	-	-	-	-	-	-	-	-	-	-	-	-	109	-	-	-	-	-	-	-	-	-	-	-	-	-	-	-

TABLE 2. Known H-2 Haplotypes and Alleles at the H-2K and H-2D Loci

Haplotype	Alleles H-2K	Alleles H-2D	Haplotype	Alleles H-2K	Alleles H-2D
a1,a2	k	d	o1,o2	d	k
an1	s	f	p	p	p
ap1,ap2,ap3,ap4,ap5	f	d	q	q	q
aq1	k	f	qp1	q	s
ar1	f	?	qq1	b	q
as1	k	s	r	r	r
at1,at2,at3,at4	s	b	s	s	s
b	b	b	sq1,sq2	s	q
ba	ba	b	t1,t2,t3,t4,t5,t6	s	d
b	bb	b	t12	t12	t12
bd	bd	b	tw1	tw1	tw1
bf	bf	b	tw5	tw5	tw5
bg1,bg2,bg3	bg	b	u	u	d
bh	bh	b	v,w1	v	q*?
bi	bi	b	w2	w2	w2
bj	bj	b	w3	w3	w3
bk	bk	b	w4	w4	w4
bp	b	?	w5	d	w5
by1,by2	q	b	w6	w6	w6
d	d	d	w7	w7	w7
da	d	da	w8	k	w8
dx	d	?	w9	w9	k
f,w11	f	f	w10	w10	w10
fa	fa	f	w12	k	w12
fb	f	fb	w13	w13	w13
g,g1,g2,g3,g4,g5,g6	d	b	w14	d	w14
h,h1,h2,h3,h4,h5,h15	k	b	w15	w15	w15
hj	k	b*	w16	w16	w16
i,i3,i5,i7,i8,i9,i18,i21	b	d	w17	w17	w17
ia1	b	da	w18	d	w18
j,ja	j	b*	w19	s	b
k	k	k	w20	w20	w20
ka	ka	k	y1,y2	q	d
m,m1	k	q	z	z	z

Some of the haplotypes in Table 2 are recombinants, that is, they possess alleles at H-2K or H-2D loci also present in other haplotypes but in different combinations. Much of the variability of the H-2 complex is thus generated by combinatorial means with alleles at the two H-2 loci existing in many different combinations. Theoretically, the number of different H-2K-H-2D combinations equals the number of alleles at the H-2K locus multiplied by the number of alleles at the H-2D locus. The alleles at the two loci are listed in Table 3: there are 37 of them at the H-2K locus and 32 at H-2D. The total number of possible H-2K-H-2D combinations generated by the known alleles is, therefore, 1,184.

As mentioned earlier, the H-2 complex contains several loci in addition to H-2K and H-2D. The total number of loci composing the complex is not yet known, but ten loci have been already identified (Fig. 1). Although most other H-2 loci are probably far less polymorphic in comparison with H-2K and H-2D, they do contain multiple alleles. And so, even if one assumes that there are just a few alleles at other H-2 loci, the number of possible H-2 haplotypes reaches almost astronomical proportions.

Determining Phenotypic Frequencies of H-2 Antigens

The development of the B10.W lines was meant to be just a springboard for our analysis of wild mouse populations. Because the private inbred H-2 antigens were so poorly represented among wild mice, we hoped to identify private antigens of wild mice by means of our B10.W lines. As Table 1 shows, we have succeeded in doing so. Once we produced antisera defining new private or semi-private H-2 antigens, we went back to wild mouse populations with this considerably enriched battery of typing reagents to determine the

TABLE 3. Known Alleles at the H-2K and H-2D Loci

H-2K	H-2D
b	b
ba	b*
bb	d
bd	da
bf	f
bg	fb
bh	k
bi	p
bj	q
bk	q*
d	r
f	s
fa	t12
ja	tw1
k	tw5
ka	w2
p	w3
q	w4
r	w5
s	w6
t12	w7
tw1	w8
tw5	w10
u	w12
v	w13
w2	w14
w3	w15
w4	w16
w6	w17
w7	w18
w9	w20
w10	z
w13	
w15	
w17	
w20	
z	

frequencies of individual H-2 antigens. This time the battery contained antisera against 36 H-2 antigens (Table 4), of which about one-half were characterized as private and the rest as semi-private (the latter being shared by a limited number of strains available to us).

So far we have typed 70 wild mice, but the work continues and data on a much larger panel should be available soon. The mice were trapped at the different localities in Texas, most of them in various farm buildings; only a few were trapped in the fields. (One should bear in mind that the wild mice from which our B10.W lines were developed were trapped in Michigan and thus represented a geographically very distant population with regard to the Texas mice.)

To calculate phenotypic frequencies of the 36 H-2 antigens defined by our battery of antisera, we also used the panel of B10.W lines and took each line as a representative of one wild mouse. The total number of wild mice typed is therefore 90 (70 trapped mice and 20 B10.W lines).

The calculations of phenotypic frequencies were carried out by dividing the number of mice typed as positive for a given antigen by the number of mice tested (90). The frequencies calculated in this manner are subject to several qualifications.

In the first place, we ignored the fact that groups of mice came from the same localities and that the presence in a given locality of several mice positive for a given antigen probably reflected close blood-relationship among such mice. The justification for ignoring the origin of the mice was our hope that in a large sample covering wide geographical area the local effects on antigen frequencies would be minimized. However, the sample described here might not have been large enough and local effects might have biased our antigen-frequency calculations.

TABLE 4. Frequencies of H-2 Antigens Among Wild Mice and B10.W Congenic Lines

H-2 antigen	No. of positive wild mice[a]	No. of positive B10.W lines[b]	Phenotypic frequency (%)[c]
2	4	1	5.5
4	2	0	2.2
9	10	1	12.2
15	0	0	-
16	2(9)	0	2.2(10.0)
17	2	0	2.2
18	0	0	-
19	1	1	2.2
20	4	0	4.4
21	1	3	4.4
23	0	1	1.1
25	7	1	8.8
26	2	1	3.3
30	3	0	3.3
31	6	3	10.0
32	0	1	1.1
33	1	0	1.1
103	10	1	12.2
105(107)	13	1	15.5
106	0	1	1.1
108	5	1	6.6
109	1	1	2.2
110	2	2	4.4
111	2	1	3.3
112	0	1	1.1
113	1	1	2.2
114	0	1	1.1
115	17	1	20.0
116	5	2	7.7
118	19	2	23.3
119	15	3	20.0
120	16	4	22.2
122	0	2	2.2
123	3	3	6.6
124	3	3	6.6
125	10	4	15.5

[a] Out of 70 mice tested.
[b] Out of 20 tested.
[c] Combined wild mice and B10.W lines (90 mice).

In the second place, we did not take into consideration the H-2 zygosity of the wild mice. A mouse typed as negative for a given antigen does not pose any problem in this respect, but a positive mouse could be either homo- or heterozygous for the antigen. We are in a process of establishing the frequency of H-2 heterozygosity among wild mice and once we have the figure, we should be able to use it as a correction factor in our frequency estimates. For the calculations in Table 4, such a factor has not yet been introduced and the data should be therefore viewed with this fact in mind. It is also important to realize that the data in Table 4 represent phenotypic and not gene frequencies.

The third qualification concerns the pooling of data on wild mice and B10.W line typing. Each B10.W line is, in fact, not equivalent to one wild mouse because it carries only one of the two H-2 haplotypes of the original wild mouse from which it was derived. Consequently, by typing wild mice we test for two haplotypes (provided that the mouse is an H-2 heterozygote), while by typing B10.W lines we test for only one haplotype. However, since in most instances the antigen frequencies calculated separately for wild mice and B10.W lines do not differ significantly, the pooling of the data is probably justified.

The last qualification concerns the specificity of the reagents in the typing battery. Each antiserum, before it was included in the battery, was extensively analyzed against panels of inbred strains and B10.W lines and absorbed so that it reacted only against a given antigen, but we could never be sure that it did not contain additional antibodies. Such antibodies might have escaped our detection when the antiserum was tested against the strain panels, but they might have emerged when the serum was tested on wild mice. In one or two instances we know that this actually did happen. For example, antiserum detecting antigen H-2.16 was monospecific in our panel tests, that is, it reacted only with strains carrying H-2.16 and no other mice. But when the antiserum was tested against our sample of 70 wild mice, it reacted with an unexpectedly high number of them. Furthermore, some mice positive with this antiserum were also positive for several other private H-2 antigens and this observation, of course, aroused our suspicion that the anti-H-2.16 might not be monospecific after all. And indeed, when the antiserum was absorbed by some of the positive wild mice, the reactivity against these mice was removed but the serum still reacted with H-2.16 positive strains. Apparently, despite all efforts to make it monospecific, the antiserum still contained at least two antibodies, one against antigen H-2.16 and another against a public antigen which is present in H-2.16-positive strains and some wild mice, but absent in all other strains in our typing panels. Once we learned about the second antibody, we made sure that for further testing an anti-H-2.16 serum was used from which this antibody was absorbed out, but we of course could not tell how many of the previously typed positive wild mice actually carried H-2.16 and how many reacted with the antiserum because of the presence of the previously unidentified antibody to the public antigen. We solved this problem by ignoring the results obtained with the antiserum before its complexity was discovered.

This example, however, raises the question of how many of our other typing antisera are complex in a similar way. To answer this question we would have to do absorption analysis with each positive mouse, but such analysis with so many reagents and so many mice to test is not humanly possible. We do, however, spot check all the antisera by absorbing them with wild mice and testing them for their monospecificity, and we check, of course, all antisera for additional antibodies when our suspicion is aroused. But despite all these precautions, one must consider the antigen-frequency data in Table 4 as possibly overestimated. The frequencies of at least some of the antigens might be actually lower than our preliminary data indicate.

Evaluation of H-2 Antigen Frequencies

Among the private H-2 antigens in Table 4 the following are detected by antisera which we have no reason to suspect as containing more than one antibody: H-2.2, 4, 15, 17, 18, 19, 20, 21, 23, 26, 30, 31, 32, 33, 106, 108, 109, 110, 111, 112, 113, and 114. The other antigens are either semi-private ones or antigens defined by antisera that might contain additional antibodies. Several points can be made about these preliminary data.

The most frequent among the "clean" private antigens is H-2.31 which was found in six wild mice and three B10.W strains (phenotypic frequency of some 10%). At the opposite end of this scale are antigens H-2.15 and H-2.18 which we have not found in any of the typed wild mice and which are also absent in all the B10.W lines. The antigens are carried by inbred strains and one might speculate that these strains were derived from wild mice unrelated to the population of wild mice that we are currently typing.

The majority of the clean private antigens occur in the Texas wild mice and in the B10.W lines with frequencies ranging from 1 to 5%. If we assume -- as our data suggest -- that aside from H-2.31, there are not very many frequently occurring H-2 antigens and that the average H-2 antigen frequency is 2%, then we may estimate that the population we are presently testing may contain some 50 antigens controlled by the H-2K locus and another 50 antigens controlled by the H-2D locus. Barring linkage disequilibrium, these numbers would then lead to the estimate of 50^2 or 2,500 haplotypes (or more precisely H-2K-H-2D allelic combinations).

It is interesting to compare these estimates with the frequencies of antigens controlled by the human homologue of the H-2 complex -- the HLA system. There are now some 15 antigens known to be controlled by the HLA-A locus and another 20 antigens controlled by the HLA-B locus (15). About half of these HLA antigens occur in frequencies similar to the phenotypic frequency of H-2.31 (i.e., 10 to 20%), while the other half of HLA antigens occurs in frequencies of 5% or less, that is, frequencies similar to those of the majority of H-2 antigens. In other words, in the HLA system there seem to be more antigens occurring in high frequencies than in the H-2 complex. This difference between the H-2 and HLA systems could mean one of two things: either some of the HLA antisera defining the more common antigens are not truly monospecific and the difference between H-2 and HLA is the consequence of differences in methodologies, or the H-2 complex is more polymorphic than the HLA system.

Some 43% of the wild mice in our sample failed to react with any of the antisera in our battery. These "blanks" indicate the presence in the wild mice of H-2 antigens against which we have not yet made antisera. Thus there are probably more antigens than we have been able to identify so far, and search for these antigens is in progress.

One can also expect that mice from other parts of the world might have still different antigens. A comparison of North American, European, and Asiatic mice might provide some interesting results in this respect.

CONCLUSION

In addition to providing, for the first time, information about H-2 antigen frequencies and about the extent of the H-2 polymorphism, our wild mice studies are proving to be useful in other areas of H-2 immunogenetics, primarily as a source of new H-2 variants. Several of our B10.W lines turned out to be H-2 recombinants, carrying haplotypes that are new combinations of H-2K or H-2D alleles present in other strains. Such an enrichment of the H-2 recombinant collection should be helpful for all kinds of H-2 mapping studies.

Other B10.W strains seem to carry alleles which are similar to but not identical with alleles carried by other strains. The difference between these alleles is very likely the consequence of recent mutations. Comparison -- particularly a biochemical one -- of such natural variants should provide much needed information as to how the variability of the H-2 loci is generated.

And so in our studies we have traveled the full circle: we began with inbred strains, extended from there to wild mice, and now with our B10.W lines we have come back to inbred strains again. And we have been able to travel the circle in such a relatively short time only because we had a solid foundation in the lines produced by the fathers of mouse inbreeding.

ACKNOWLEDGMENTS

We thank Mr. Wayne Bundy for breeding most of the strains mentioned in this communication, and Ms. Jeanne Lively for secretarial help.

REFERENCES

1. Klein, J. (1975). "Biology of the Mouse Histocompatibility-2 Complex." Springer-Verlag, New York.
2. Klein, J. (1971). Nature 229:635.
3. Klein, J. (1970). Science 168:1362.
4. Klein, J., and Bailey, D. (1971). Genetics 68:287.
5. Crowcroft, P.L. (1955). J. Mammal. 36:299.
6. Reimer, J., and Petras, M. (1967). J. Mammal. 48:88.
7. Shreffler, D. (1964). Genetics 49:973.
8. Klein, J. (1972). Transplantation 13:291.
9. Klein, J. (1975). Immunogenetics 2:297.
10. Hauptfeld, M., and Klein, J. (1976). Immunogenetics 3:603.
11. Hammerberg, C., Klein, J., Artzt, K., and Bennett, D. (1976). Transplantation 21:199.
12. Zaleska-Rutczynska, Z., and Klein, J. (1977). J. Immunol. 119:1903.
13. Klein, J., and Zaleska-Rutczynska, Z. (1977). J. Immunol. 119:1912.
14. Klein, J., and Hammerberg, C. (1977). Immunol. Rev. 33:70.
15. Snell, G., Dausset, J., and Nathenson, S.G. (1977). "Histocompatibility." Academic Press, New York.

FOUR MAJOR ENDOGENOUS RETROVIRUS CLASSES
EACH GENETICALLY TRANSMITTED IN VARIOUS SPECIES OF MUS

Robert Callahan
George J. Todaro

Laboratory of Viral Carcinogenesis
National Cancer Institute
National Institutes of Health
Bethesda, Maryland

Several properties of endogenous retroviruses (1) make them useful reagents for the study of evolutionary relationships among species. In a variety of mammalian and avian systems it has been demonstrated that endogenous viral gene sequences are transmitted through the germline and are present in the normal cellular DNA in multiple copies (2, 3). As such, they are subjected to the same evolutionary pressures as are cellular DNA sequences. Because, in some species, these viruses can be isolated and transmitted in cell culture, it is possible to readily purify their genomes as well as the proteins produced by the viral RNA. We have previously taken advantage of this property of endogenous primate type C viruses to study the evolutionary relationship between primates (4, 5); similar studies have also been performed with other endogenous mammalian type C and type B viruses (6-8). One surprising result of some of the above studies was that transmission of type C viruses does occur between the germlines of distantly related species (9, 10).

In other cases, type C viruses are found only in certain animals and in certain tissues of a species and is the result of infection.

In the latter category are the viruses isolated from one woolly monkey with a fibrosarcoma (11, 12) and from several gibbons, both normal and leukemic (13-15). DNA transcripts of the viral RNA show homology to rodent and not to primate cellular DNA; in particular, transcripts would hybridize most extensively to the cellular DNAs of various Mus species (6). To define more precisely the origin of the woolly monkey-gibbon group of infectious type C viruses, various species of Southeast Asian mice were surveyed fro their endogenous type C viruses. Since the gibbons were the primate species that showed widespread infection by this group of viruses, Southeast Asian members of the family Muridae were studied for viruses which may be present that might best account for the transmissible leukemia virus of gibbons (16).

In this report, we describe the properties of the various endogenous retroviruses of Southeast Asian species of Mus. These animals, we have found, readily release distinct groups of retroviridae. On the basis of biological and biochemical criteria, two subclasses of endogenous type C viruses could be distinguished. One subclass (C-I) is related to the woolly monkey-gibbon ape infectious primate type C viruses; the other subclass (C-II) is more closely related to laboratory murine leukemia viruses (MuLVs) that are commonly isolated from inbred strains and feral populations of M. musculus (6, 17). In addition, a novel class of endogenous retroviruses (designated M432) could be distinguished which is unrelated by immunological and molecular hybridization criteria to all other classes of retroviruses (8, 18) and type B particles have been identified in the milk of M. cervicolor mice which are morphologically and antigenically related to M. musculus mammary tumor virus (MMTV) (19).

Using viral reverse transcripts we have determined the number of endogenous retrovirus-related sequences in various members of the genus Mus. Taken together, the total complement of retroviral sequences represents a significant portion of the cellular genome, approximately 0.04%. The three classes of viral sequences we have studied (C-I, C-II, and M432) appear to be conserved to different extents, suggesting differing evolutionary pressures on each.

Phylogenetic Relationship of the Southeast Asian Species of Mus

Table 1 lists the species of Mus which have been studied as well as their origin. The mice were initially trapped by Dr. J. Marshall in different provinces of Thailand, except M. dunni from India and M. musculus molossinus from Japan (20). Recently, random-bred colonies of certain of these species have been established in laboratories in the United States.

Using M. cervicolor as the point of reference, the evolutionary relationships of the different species were determined. Radioactive nonrepetitive cellular DNA was prepared from a tissue culture cell line derived from M. cervicolor lung tissue (6). The labeled DNA was used as a probe to measure nucleic acid sequence homology between the nonrepetitive DNA of M. cervicolor and cellular DNA extracted from pooled organs of different rodents. Table 2 lists the final extent of hybridization (at a $C_o t$ of 2×10^4) and the $\Delta T_m R$ of the hybrid. The $\Delta T_m R$ is a measure of the thermal stability of the hybrids formed and is the index of the base-pair mismatch between the DNAs of the species being tested. The results reveal a broad range of homology between the M.

TABLE 1. Southeast Asian Species of Mus and Their Origin

Species	Country	Province
M. caroli	Thailand	Chantaburi[a]
M. cervicolor cervicolor	Thailand	Loei, Lampang, Chantaburi[a]
M. cervicolor popaeus	Thailand	Chantaburi, Klorat, Tak, Saraburi
M. cookii	Thailand	Tak, Loei
M. dunni	India	Unknown
M. musculus castaneus	Thailand	Unknown
M. musculus molossinus	Japan	Unknown
M. pahari	Thailand	Tak
M. shortridgei	Thailand	Tak, Lampang

[a] From Roswell Park Colony.

cervicolor cellular genome and those of the other murine species. By these criteria, M. cervicolor cervicolor, the subspecies M. cervicolor popaeus, and M. cookii are all closely related. Of the other species tested, M. caroli, M. musculus, and M. dunni are more distantly related to M. cervicolor. The two least related species are M. pahari and M. shortridgei. Labeled nonrepetitive cellular DNA from M. shortridgei, however, when hybridized to mouse, rat, and hamster cellular DNAs, showed more homology to mouse DNA than to either rat or hamster, consistent with it being a member of the genus Mus (data not shown). The evolutionary relationship of various Mus species with Mus cervicolor as the point of reference is summarized in the phylogenetic tree of the Muridae family shown in Figure 1. This is based on the final extent

TABLE 2. Nucleic Acid Homology and Thermal Stability Between Rodent Nonrepetitive Cellular DNAs

Species[a]	^{3}H-labeled nonrepetitive cellular DNA[b] M. cervicolor	
	% Hybrid	$\Delta T_m R$ (°C)[c]
Mouse		
M. cervicolor cervicolor	100	–
M. cervicolor popaeus	95	2.0
M. cookii	87	3.7
M. caroli	75	4.7
M. musculus	64	5.2
M. dunni	60	7.2
M. pahari	24	>28
M. shortridgei	11	>28
Rat		
R. norvegicus	4	ND
Chinese hamster		
Cricelulus griseus	4	ND

[a] Cellular DNA was extracted from various tissues or tissue culture cell lines of the species listed as described (33) and sonicated to yield a mean size of 6 to 8S as determined by alkaline sucrose centrifugation.

[b] The preparation of [^{3}H]thymidine-labeled nonrepetitive cellular DNA has been previously described (6, 33, 34). Hybridization reactions contained 0.7 M NaCl, 0.01 M Tris (pH 7.4), 2×10^{-3} M EDTA, 0.05% SDS, 10,000 to 20,000 cpm/ml of ^{3}H-labeled nonrepetitive cellular DNA, and 2 to 4 mg/ml cellular DNA. Hybridizations were initiated by heating the mixture to 100°C for 10 min, followed by incubation at 65°C. At a $C_o t$ value of 2×10^4, 0.025 ml or 0.05 ml aliquots were removed and frozen at -80°C until digested with single-

Legend to Table 2 continued

strand specific nuclease, S_1, as described (35, 36). The $C_o t$ values (C_o is the concentration of cellular DNA in moles of nucleotides per liter, and t is the time in seconds) were calculated as suggested by Britten and Kohne (37) and corrected to a monovalent cation concentration of 0.18 M (38).

^{c}The temperature at which 50% of the hybrids are dissociated (T_m) was 91 to 93°C for the homologous DNA:DNA hybrids. The ΔT_m is the difference in T_m between the other DNA:DNA hybrids and the T_m of the homologous hybrid. $\Delta T_m R$ is the thermal stability for all the nonrepeated DNA and is defined as the temperature at which 50% of the total labeled DNA (normalized to the homologous reaction) is in a hybrid form (39). The $T_m R$ values were derived from at least three experiments in which hybrids were melted at 3°C increments over a range of $\pm 10°C$ from the $T_m R$, using S_1 nuclease to detect hybrids.

---

of hybridization of nonrepetitive cellular DNA, since it is difficult to assign time scales to the divergence points in the absence of paleontological references.

Properties of Endogenous Retroviruses Isolated from Southeast Asian Species of Mus

Three different classes of endogenous retroviruses have been isolated from Asian species of Mus. These viruses were obtained either from tissue culture cell lines which spontaneously release virus or by induction of tissue culture cell lines with the pyrimidine analog, BUdR, followed by mixed cell cultivation with sensitive cell lines chosen for their permissiveness for the replication of retroviruses.

On the basis of biological and biochemical criteria, two subclasses of type C virus could be identified. Subclass C-I

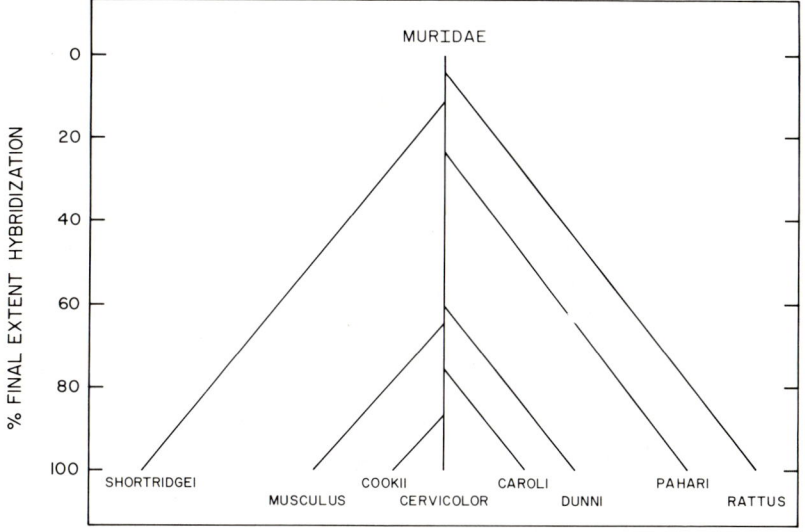

FIGURE 1. Phylogenetic tree depicting relationships among members of the family Muridae. The final extents of hybridization shown in Table 2 are depicted diagrammatically. The distance between divergent lines represents differences in nucleic acid homology which is dependent on hybridization conditions and not relative times of divergences from a common ancestor.

replicates only in nonmurine tissue culture cell lines, whereas subclass C-II can replicate in cell lines derived from different species of the genus Mus. Table 3 shows the host ranges of representatives of each subclass, isolated from M. cervicolor tissue culture cells. By immunological criteria, the viral-associated proteins of the two subclasses are readily distinguishable. The major structural protein p30 of the C-I subclass competes in a woolly monkey-gibbon ape group-specific radioimmunoassay (17), whereas the respective proteins of the C-II subclass show no competition (Fig. 2A).

TABLE 3. Host Range of Retroviruses Isolated from Mus cervicolor

Host cell line[b]	Species	Supernatant reverse transcriptase assay[a] ($[^3H]$dTMP incorporated, cpm $\times 10^{-3}$)		
		CERV C-I virus growing in SIRC cells	CERV C-II virus growing in SC-1 cells	M432 virus growing in NIH/3T3 cells
M. caroli heart	M. caroli	2.6	24.3	2.3
M. cervicolor lung	M. cervicolor cervicolor	2.0	1.8	1.2
M. dunni lung	M. dunni	2.2	458.0	1.7
NIH/3T3	M. musculus	1.2	461.0	336.3
M. shortridgei lung	M. shortridgei	1.1	11.7	1.9
NRK	Rat	3.0	1.1	1.1
SIRC	Rabbit	198.0	1.0	1.0
FEC	Cat	155.0	1.4	1.4
DBS-FRhL-1	Rhesus	86.5	2.4	2.4
A204	Human	4.0	3.6	1.9

Legend to Table 3

^aCell culture medium from the three viruses propagated in the cell lines shown was filtered and used to infect the host cell lines listed. Medium from infected cultures was assayed for viral polymerase activity at 2-week intervals after infection. The results shown are the values obtained 12 to 15 weeks after infection. Underlined numbers indicate levels of viral polymerase activity significantly above background.

^bThe derivation and source of these cell lines have been described (6, 17, 40). The M. dunni lung and M. shortridgei lung tissue culture cell lines are described elsewhere (Callahan and Todaro, in preparation). The cat cell line, FEC, and SIRC rabbit cell line (CCL 60) are from the American Type Culture Collection.

In this assay, the slope of the competition curve can be used as a measure of relatedness between the labeled test antigen and the competing protein. Thus, by this criterion, the p30 proteins of the type C-I viruses exhibit varying degrees of relatedness to the test protein derived from the woolly monkey virus, with the M. shortridgei virus being the most distantly related. Other than the various isolates of MuLV from feral and inbred strains of M. musculus, the only other members of the C-II subclass are virus isolates we have obtained from M. cervicolor and M. cookii cell cultures. The viral-associated p30 proteins of this subclass compete in a M. musculus MuLV group-specific radioimmunoassay, albeit the M. cervicolor and M. cookii viral proteins compete with a reduced slope (Fig. 2B). Although the p30 proteins of these two subclasses of viruses can be distinguished by their unique antigenic determinants, they do share other

A
B
C
% DISPLACEMENT OF 125 I
MICROGRAMS PROTEIN
WOOLLY MONKEY
CERVICOLOR C-I
CAROLI C-I
DUNNI C-I
SHORTRIDGEI C-I
MUSCULUS C-II
CERV C-II
CERVICOLOR C-II
RAT
100
80
60
40
20
0.1
1.0
10
100

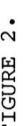

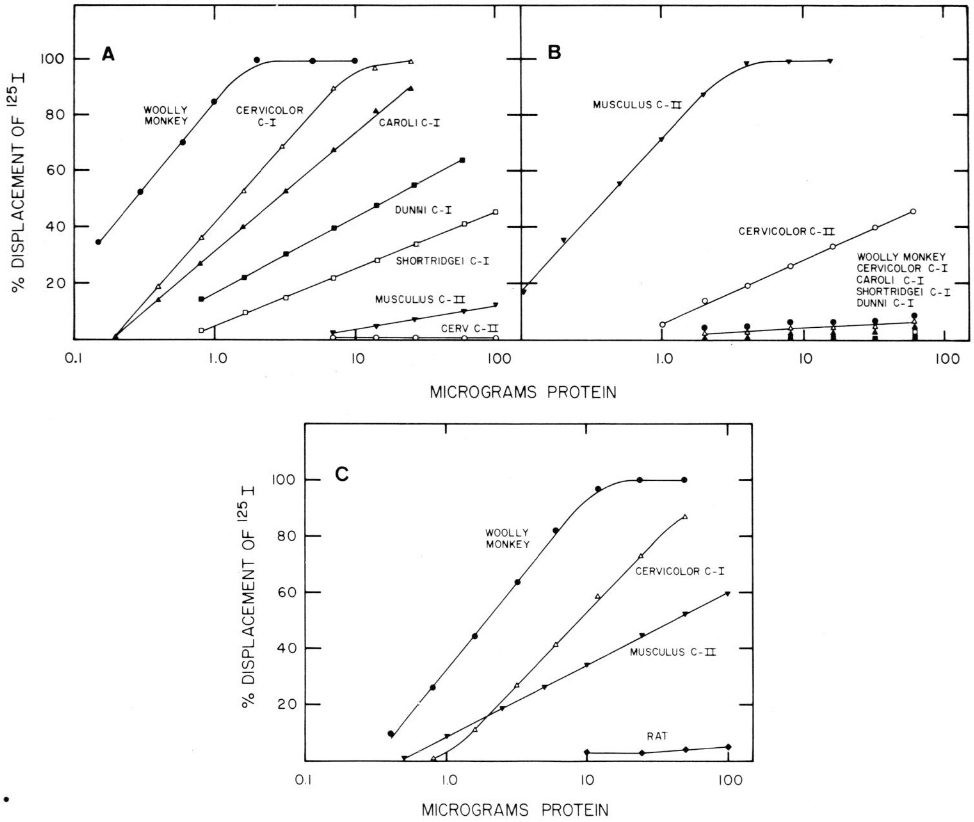

FIGURE 2.

Legend to Figure 2.

FIGURE 2. Competitive radioimmunoassays for p30 proteins. Extracts of cells producing type C virus were used as competing antigens (17). The results are plotted as percentage displacement of the labeled test antigen versus micrograms of competing protein. (A) Assay for SSAV p30 protein using antisera to p30 and ^{125}I-labeled SSAV p30. (B) Assay for murine p30 protein using antisera to M. musculus type C-II virus p30 protein (BALB/c ecotropic type C virus S2CL3) and ^{125}I-labeled S2CL3 p30 protein. (C) Interspecies assay using antisera to the p30 protein of endogenous porcine virus PK-15 and ^{125}I-labeled SSAV p30 (17, 21).

determinants in common. This can be demonstrated by using a rodent interspecies assay in which the p30 proteins of both subclasses of murine type C viruses compete with the test antigen, while the rat type C viral protein does not (Fig. 2C). A similar pattern between the two subclasses of viruses can be demonstrated by the ability of antisera against the viral-associated reverse transcriptase of the woolly monkey virus or R-MuLV to inhibit enzymatic activity of the respective subclass of viruses (6, 21).

Approximately half of the viral genome of type C viruses is believed to contain the coding sequences for the viral structural proteins (the gag proteins) and reverse transcriptase (the pol gene product) (22). Using labeled viral cDNA probes representing the entire viral genome, the relationship between the M. cervicolor C-I and C-II subclass virogenes was measured (data not shown). In agreement with the immunological data, the C-I and C-II virogenes share very little nucleic acid homology (approximately 10% under the conditions used). Moreover, the M. cervicolor C-I probe hybridizes partially with the M. cookii and M. caroli C-I viral RNAs and to a lesser extent with M. dunni and M. shortridgei as well as SSAV and GALV-1 viral RNAs. In contrast to these results, the M. cervicolor C-II probe did not hybridize appreciably with either of the viruses from M. cookii or M. musculus; this suggests that the C-II subclass of virogenes has been less evolutionarily conserved than the virogenes of the C-I subclass.

A third class of endogenous retroviruses has been isolated from two Asian rodents, M. cervicolor and M. caroli. Members of this class so far have only replicated in cell culture lines derived from M. musculus (Table 3). They share certain morphological and biochemical properties in common with M. musculus type B mammary tumor virus. Cells producing these viruses contain cytoplasmic A particles and budding forms contain doughnut-shaped nucleoids (Fig. 3). However, the mature particles lack the characteristic surface spikes of MMTV and have a centrally located nucleoid. This group of viruses contains two envelope glycoproteins (gp65, gp32), a major internal protein (p24), as well as p16 and a phosphoprotein (p12). The viral-associated reverse transcriptase, with a molecular weight of 70,000 daltons, is smaller than the M. musculus type B viral enzyme (23) although both share a preference for magnesium to fulfill their divalent cation requirement for activity (8).

Antisera raised against the M. cervicolor p24 protein show a partial cross-reactivity with the corresponding protein of the M. caroli virus (M832) (Fig. 4). In this assay, MMTV, bovine leukemia virus (BLV), Rauscher MuLV (R-MuLV), the Mason-Pfizer monkey virus (MPMV), and both M. cervicolor type C viruses all fail to compete with the test antigen. Similar results were obtained with antisera prepared against the M. cervicolor M432 viral-associated reverse transcriptase. In addition, by molecular hybridization criteria, the M. cervicolor M432 viral genes are unrelated to M. musculus MMTV,

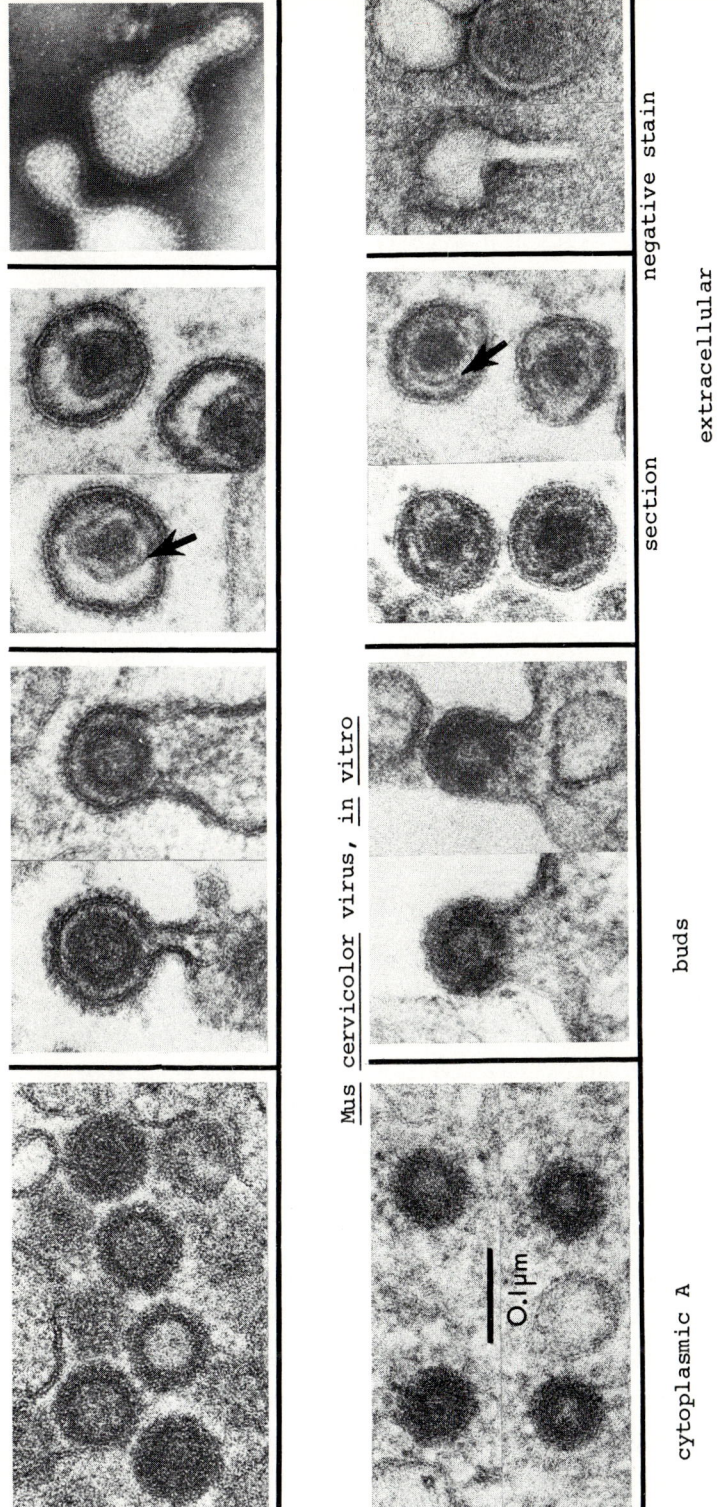

FIGURE 3. Electron micrographs of virus particles from NIH/3T3 cells infected with M. cervicolor M432 and M. musculus type B (MMTV) virus from RIII tumor cells (18). The intermediate line is indicated by the arrows. The marker bar equals 0.1 μm.

both murine type C viral subclasses, as well as MPMV and BLV (8).

Table 4 summarizes the classes of viral isolates obtained from each species. With the one exception, M. musculus, viruses of the type C subclass C-I have been isolated from every species tested. Type C subclass C-II viruses have been isolated from only M. musculus, M. cervicolor, and M. cookii. The M432-related viruses have been isolated from M. cervicolor and M. caroli, while the type B class contains viral isolates from M. musculus. A virus morphologically and antigenically related to the MTV subclass of viruses has recently been observed in the milk of M. cervicolor mice (19).

Conservation of Retroviral Genes in the Cellular Genome of Rodents

The cellular DNA of Asian species of Mus were examined for the presence of nucleic acid sequences related to the genomes of the M. cervicolor C-I, C-II, and M432 subclasses of viruses   Table 5 summarizes the final extents of hybridization and the thermal stabilities of the hybrids formed with DNA from the listed species and labeled M. cervicolor viral DNA transcripts synthesized in the endogenous polymerase reaction. As shown, viral-related nucleic acid sequences corresponding to the three subclasses could be detected in all the species of Mus as well as more distantly related rodents. All three viral DNA probes hybridized most extensively and with the highest thermal stability to cellular DNA from Mus cervicolor tissues. The extent to which the viral sequences were detected in other related Mus species depended both on the genetic distance from M. cervicolor DNA, as determined by the nonrepetitive cellular DNA hybridization studies (see Table 2) and on the viral subclass. Thus the M. cervicolor C-I and M432 subclasses of virogenes are conserved

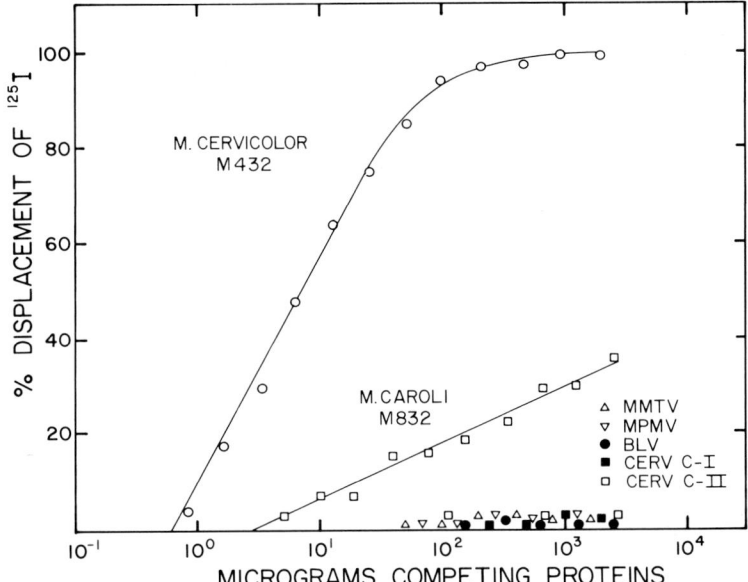

FIGURE 4. Competitive radioimmunoassay for M. cervicolor M432 p24 protein. Assay for M. cervicolor M432 p24 protein using antisera to p24 and ^{125}I-labeled p24 protein (8). Extracts from infected cells were used for competing antigen. The results are plotted as in Figure 2.

with a pattern similar to the nonrepetitive cellular DNA of this species. The fact that the final extents of hybridization and thermal stabilities of the hybrids formed decrease in parallel indicates that the partial homologies detected are not the result of the conservation of discrete, contiguous regions of the viral genome. Rather, it suggests that the differences are due to base-pair substitution throughout much of the viral genome. The virogene sequences related to the M. cervicolor C-II subclass appear much more species-specific. In species more distantly related to M. cervicolor than M. cookii, only low levels of homology can be detected. A similar result has been found using viral DNA probe from

TABLE 4. Summary of Viral Isolates from Different Mus Species

Species	Retrovirus class[a]			
	M432	Type B	Type C	
			C-I	C-II
M. caroli	+	−	+	−
M. cervicolor	+	+[b]	+	+
M. cookii	−	−	+	+
M. dunni	−	−	+	−
M. musculus	−	+	−	+
M. shortridgei	−	−	+	−

[a] The properties of the different subclasses of retrovirus are given in the text. (+) at least one viral isolate from a tissue culture cell line derived from the indicated species, (−) no viral isolates were obtained.

[b] Observed in the milk of M. cervicolor popaeus (19).

M. musculus C-II viruses (either BALB/c ecotropic or BALB/c xenotropic viruses) (6). Thus, there seems to be less rigorous evolutionary conservation of virogene sequences for this subclass of type C viruses.

Quantitation of the Endogenous Virogenes Corresponding to the Different Classes of Retroviruses

It has been shown in a number of mammalian and avian species that the genomes of endogenous retroviruses are represented in cellular DNA in multiple copies. Using labeled viral DNA, the frequency of viral-related sequences of each M. cervicolor viral subclass was measured. Both the type C-I and C-II virogenes are reiterated approximately

TABLE 5. Nucleic Acid Homology and Thermal Stability Between Retrovirus Isolated from M. cervicolor and Various Rodent DNAs

Species[b]	M. cervicolor viral ^{3}H-labeled DNA[a]								
	Type C-I			Type C-II			M432		
	% Hybrid	ΔT_m	$\Delta T_m R$	% Hybrid	ΔT_m	$\Delta T_m R$	% Hybrid	ΔT_m	$\Delta T_m R$
Mouse									
M. cervicolor cervicolor	100			100			100		
M. cervicolor popaeus	96	1.0	1.0	30			94	0.6	1.0
M. cookii	80	3.0	5.0	26			90	2.3	2.0
M. caroli	64	4.5	9.0	5			48	5.2	17
M. musculus	44	7.0	>27	11			35	9.0	>27
M. dunni	42	7.0	>27	7			31	9.0	>27
M. pahari	17	NT	NT	10			3	NT	NT
M. shortridgei	6	NT	NT	6			3	NT	NT
Rat									
R. norvegicus	7	NT	NT	4			6	NT	NT
Chinese hamster									
Cricelulus griseus	2	NT	NT	3			3	NT	NT

Legend to Table 5.

[a] The [^{3}H]thymidine-labeled DNA probes were prepared from the indicated detergent-disrupted virus in the presence of actinomycin D as described (41). The specific activity of the [^{3}H]DNA probes was 2×10^7 cpm/g. The [^{3}H]DNA probes used were approximately full-length copies of the viral genome as shown by alkaline sucrose gradient analysis. The type C-I virus was grown in the SIRC rabbit cell line. The type M432 and type C-II viruses were grown in, respectively, M. musculus NIH Swiss 3T3 or feral mouse SC-1 tissue culture cell lines (40, 42). The conditions of hybridization are described in the legend to Table 2. Percent hybridization was determined at a $C_o t$ of 2×10^4. The T_m of the homologous hybrids were 89-93°C. Values for ΔT_m and $\Delta T_m R$ were determined as described in the legend to Table 2.

[b] Cellular DNA was extracted from pooled tissue or from tissue culture cell lines of the indicated species as described (33).

NT = Not tested.

---

six times per haploid cellular genome, while the type M432 virogenes are reiterated approximately 25 times (Fig. 5). Recently, Morris et al. have estimated that there are at least 25 copies of type B virogenes (MMTV) per haploid genome in M. cervicolor (7). The summaries of our estimates of the reiteration frequency of each subclass of virogenes as well as mouse ribosomal RNA genes (24) are compared in Table 6. The approximate genetic complexity of a single viral genome is 10,000 base pairs and that of the cellular genome is 1.5×10^9 base pairs/haploid genome (25). Thus at least 0.04% of the M. cervicolor cellular genome contains the genetic information for endogenous retroviral-associated proteins.

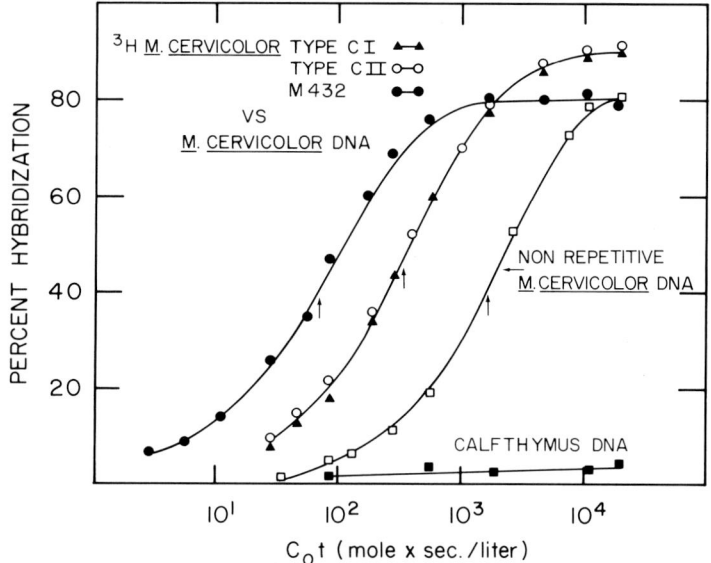

FIGURE 5. Hybridization of ^{3}H-labeled DNA probes prepared from each of the M. cervicolor retroviruses to M. cervicolor DNA. The preparation of nonrepetitive M. cervicolor cellular DNA and the ^{3}H-labeled viral DNA probes are described in Tables 2 and 4, respectively. Hybridization conditions are described in the legend to Table 2. At various time, 0.05 ml aliquots were removed from the reaction mixture and frozen at -80°C until digested with S_1 nuclease (35, 36).

This value is comparable to the genetic complexity of the structural genes coding for 28S and 18S ribosomal RNAs. It is therefore fair to conclude that the retroviral genes represent a significant portion of the cellular genome. However, these represent minimal estimates as additional subclasses of retroviruses may yet be isolated from one or more of these Mus species.

TABLE 6. Comparison of Viral Gene Frequency with Ribosomal RNA Gene Frequency

Viral genome	Copy number per haploid genome[a]	Complexity[b] (base pairs)	Percent of cellular genome[c]
C-I	6	6×10^4	4×10^{-3}
C-II	6	6×10^4	4×10^{-3}
M432	25	2.5×10^5	1.7×10^{-2}
MMTV	25	2.5×10^5	1.7×10^{-2}
Total	62	6.2×10^5	4.2×10^{-2}
Ribosomal (28S and 18S)	88	7×10^5	4.6×10^{-2}

[a] The copy number per haploid genome of the C-I, C-II, and M432 subclasses of virogenes was determined by $C_o t$ analysis (Fig. 5). The copy number for the type B class (MMTV) in M. cervicolor was reported by Morris et al. (7). The ribosomal RNA (28S and 18S) copy number as reported by Gaubatz and Cutler (24).

[b] The complexities of a single viral genome and of 28S and 18S ribosomal RNA were assumed to be, respectively, 10,000 and 8,000 base pairs (43).

[c] The complexity of the mouse cellular genome was taken as 1.5×10^9 base pairs per haploid genome (25).

CONCLUSIONS

On the basis of nucleic acid homology, it has been possible to establish the evolutionary relationships between a variety of Southeast Asian species of Mus (6, 26). We have placed the species of Mus so far examined into three groups based on the nucleic acid homology they share with the point of reference, the M. cervicolor cellular genome. On this basis, M. cervicolor cervicolor, M. cervicolor popaeus, and M. cookii are highly related, whereas M. caroli, M. dunni, and M. musculus are a somewhat more distantly related group. The M. pahari and M. shortridgei cellular genomes share the least homology with that of M. cervicolor.

The Mus cellular genome contains the genetic information for type B, at least two subclasses of type C, and a novel class (M432) of endogenous retroviral genes. Each of the four classes is present in the cellular genome in multiple copies. In M. cervicolor the total complement of endogenous retroviral sequences accounts for a minimum of 0.04% of the cellular genome. This calculation takes into account only the genetic information corresponding to the viral-associated structural proteins. Additional genetic information would be required to account for the various mechanisms the cell has evolved to regulate the expression of these virogenes. Although the absolute copy number of a given subclass of virogenes may vary with the species of Mus, it is apparent that they represent a significant portion of the cellular genome.

The evolutionary conservation of these classes of endogenous retroviral genes is not uniform. The most highly conserved is subclass C-I which has evolved at a rate comparable to the nonrepetitive cellular DNA sequences. The M432 class of virogenes, although highly conserved, seems to have evolved at a faster rate than the C-I subclass. The C-II subclass is the least conserved. By nucleic hybridization there is little homology between the M. cervicolor type C-II virogenes within members of this species and even less in more distantly related species. Similarly, the various viral isolates of type C-II from M. musculus also share varying degrees of nucleic acid homology between their genomes (27, 28). The relationship between members of this subclass is more easily demonstrated by the shared unique antigenic determinants of the p30 and reverse transcriptase proteins.

The existence of two subclasses of endogenous type C virogenes in the murine cellular genome is not unique to the genus <u>Mus</u>. By the criteria of nucleic acid hybridization, the rat cellular genome contains two subclasses of unrelated endogenous type C virogenes (29, 30). The domestic cat and the Old World monkey genomes (5, 31) also have been shown to harbor multiple copies of two or more distantly related retroviral genomes. The extensive genetic and immunologic data available in the mouse (<u>Mus musculus</u>) and its close relatives (20, 32) make it especially attractive for studies on the evolution and function of these viral gene sequences.

REFERENCES

1. Dalton, A.J., Melnick, J.L., Bauer, H., Beaudreau, G., Bentvelzen, P., Bolognesi, D., Gallo, R., Graffi, A., Haguenau, F., Heston, W., Huebner, R., Rowe, W., and Todaro, G.J. (1974). Intervirology 4:201.
2. Todaro, G.J., Benveniste, R.E., Callahan, R., Lieber, M.M., and Sherr, C.J. (1974). Cold Spring Harbor Symp. Quant. Biol. 39:1159.
3. Aaronson, S.A., and Stephenson, J.R. (1976). Biochim. Biophys. Acta 458:323.
4. Benveniste, R.E., and Todaro, G.J. (1976). Nature 261:101.
5. Todaro, G.J., Benveniste, R.E., Sherwin, S.A., and Sherr, C.J. Cell, in press.
6. Benveniste, R.E., Callahan, R., Sherr, C.J., Chapman, V., and Todaro, G.J. (1977). J. Virol. 21:849.

7. Morris, V.L., Medeiros, E., Ringold, G.M., Bishop, J.M., and Varmus, H.E. (1977). J. Mol. Biol. 114:73.
8. Callahan, R., Sherr, C.J., and Todaro, G.J. (1977). Virology 80:401.
9. Benveniste, R.E., and Todaro, G.J. (1974). Nature 252: 456.
10. Benveniste, R.E., Sherr, C.J., and Todaro, G.J. (1975). Science 190:886.
11. Theilen, G.H., Gould, D., Fowler, M., and Dungworth, D.L. (1971). J. Natl. Cancer Inst. 47:881.
12. Wolfe, L.G., Deinhardt, F., Theilen, G.H., Rabin, H., Kawakami, T., and Bustad, L.K. (1971). J. Natl. Cancer Inst. 47:1115.
13. Kawakami, T., Huff, S.D., Buckley, P.M., Dungworth, D.L., Snyder, S.P., and Gilden, R.V. (1972). Nature New Biol. 235:170.
14. Kawakami, T.G., and Buckley, P.M. (1974). Transplant. Proc. 6:193.
15. Lieber, M.M., Benveniste, R.E., Sherr, C.J., and Todaro, G.J. (1975). Virology 66:616.
16. Charman, H.P., Kim, N., White, M., Marquardt, H., Gilden, R.V., and Kawakami, T. (1975). J. Natl. Cancer Inst. 55:1419.
17. Lieber, M.M., Sherr, C.J., Todaro, G.J., Benveniste, R.E., Callahan, R., and Coon, H.G. (1975). Proc. Natl. Acad. Sci. USA 72:2315.
18. Callahan, R., Benveniste, R.E., Sherr, C.J., Schidlovsky, G., and Todaro, G.J. (1976). Proc. Natl. Acad. Sci. USA 73:3579.
19. Schlom, J., Hand, P., Teramoto, Y.A., Callahan, R., and Todaro, G. (1978). J. Natl. Cancer Inst., in press.
20. Marshall, J.T. (1977). Bull. Amer. Mus. Nat. Hist. 158:173.

21. Sherr, C.J., Fedele, L.A., Benveniste, R.E., and Todaro, G.J. (1975). J. Virol. 15:1440.
22. Baltimore, D. (1974). Cold Spring Harbor Symp. Quant. Biol. 39:1187.
23. Dion, A., Vaidya, A., Font, G.S., and Moore, D.H. (1974). J. Virol. 14:40.
24. Gaubatz, J., and Cutler, R.G. (1975). Biochemistry 14:760.
25. Laird, C. (1971). Chromosoma 32:378.
26. Rice, N.R., and Strauss, N.A. (1973). Proc. Natl. Acad. Sci. USA 70:3546.
27. Callahan, R., Benveniste, R.E., Lieber, M.M., and Todaro, G.J. (1974). J. Virol. 14:1394.
28. Callahan, R., Lieber, M.M., and Todaro, G.J. (1975). J. Virol. 15:1378.
29. Scolnick, E.M., Goldberg, R.J., and Williams, D. (1976). J. Virol. 18:559.
30. Scolnick, E.M., Williams, D., Maryak, J., Vass, W., Goldberg, R.J., and Parks, W.P. (1976). J. Virol. 20:570.
31. Benveniste, R.E., and Todaro, G.J. Primate and Human Cancer Meeting, San Antonio, Texas, in press.
32. Chapman, V.M., and Shows, T.B. (1976). Nature 259:665.
33. Benveniste, R.E., Heinemann, R., Wilson, G.L., Callahan, R., and Todaro, G.J. (1974). J. Virol. 14:56.
34. Benveniste, R.E., and Todaro, G.J. (1974). Proc. Natl. Acad. Sci. USA 71:4513.
35. Ando, T. (1966). Biochim. Biophys. Acta 114:158.
36. Benveniste, R.E., and Scolnick, E.M. (1973). Virology 51:370.
37. Britten, R.J., and Kohne, D.E. (1968). Science 161:529.
38. Britten, R.J., and Smith, J. (1970). Carnegie Inst. Washington Yearbook 68:378.

39. Kohne, D.E. (1970). Q. Rev. Biophys. 3:327.
40. Jainchill, J.L., Aaronson, S.A., and Todaro, G.J. (1969). J. Virol. 4:549-553.
41. Benveniste, R.E., and Todaro, G.J. (1977). Proc. Natl. Acad. Sci. USA 74:4557.
42. Hartley, J.W., and Rowe, W.P. (1975). Virology 65:128.
43. Birnstiel, M., Chipehase, M., and Speirs, J. (1971). Progr. Nucleic Acid Res. Mol. Biol. 11:351.

Index

A

Andervont, Howard B., 99, 103, 110
  chemical carcinogenesis, 105
  mammary tumor virus, 100
Altrock, Bruce W., 321
Antigen
  FMR, 394
  G_{IX}, 394
  gross cell surface, 394
  Thy-1, 394
  TL, 394, 399
  tumor specific, 403

B

Bagg, Halsey, 37, 46, 186
Bailey, Donald W., 197, 481
Bennett, Dorothea, 615
Bittner, John J., 38, 110, 322
Boggs, Sallie S., 463, 465
Bonhomme, Francois, 569
Bussey Institute, 7, 35, 55, 179

C

Callahan, Robert, 343, 689
Cancer Research Institute, 23
Cardiff, Robert D., 321
Castle, William E., 6, 35, 46, 180
  family tree, 8
Chapman, Verne M., 555
Cherry, Marianna, 481
Chromosome banding, 591
  C-banding, 599
  G-banding, 592
  Q-banding, 592
  wild mouse, 596
Chused, Thomas M., 297
Cloudman, Arthur M., 38
Colcher, D., 343
Cold Spring Harbor, 35, 46, 185

D

de Greeve, Paul, 633
Deringer Margaret K., 99, 112
Dickie, Margaret M., 175
Drohan, W., 343
Duncan, William R., 667
Dunn, L. C., 37

E

Enzyme
  β-Glucuronidase, 559
    hidden variant, 574
    induction gene
      Gur, 264
      Tfm, 265
    NADH diaphorase, 536
    processing gene
      architectural, *Eg*, 260
      modification, *Map-1*, *Map-2*, 259
      organelle, function
        bg, 261
        le, 261
    turnover, *Ce*, 259
    regulatory gene
      Bgs, 263
      Lv, 262
    strain distribution, 238
    structural gene, 236
      Gus^a, Gus^b, Gus^h, 258
    temporal gene
      Gut, 268
      Int, 271
      Rec, 271
      Tem, 271
    thermostability, 577

F

Fekete, Elizabeth, 38
Flaherty, Lorraine, 409
Fluorescence activated cell sorter, 302
Furth, Jacob, 69

G

Gautsch, J., 343
Gene
  Ah, 267, 287
  Akv-1, 291
  Fgv-1, 291
  C, mutation, 245
  deletion, 244
  Got-1, 243
  Got-2, 243
  Gpi-1, 248
  Gv-1, 398
  Gv-2, 398
  Hbb   Hst, 662
  leukemia, 287
  linkage map, 238
  Lps, 457
  null allele, Qa locus, 420
  pseudoallele-*t*, 617
  recombination suppression, T/t complex, 618
  tandem duplication, 419
    Amy-1, Amy-2, 242
    Car-1, Car-2, 242
    Es-1,2,5,6,7,9,11, 242
    Hbb, Hby, 245
  Tla, 399
Graff, Ralph J. 371
Green, Earl L., 167
Green, Margaret C., 157
Gross, Ludwik, 82

H

Hand, P., 343
Hartley, Janet W., 289
Held, Joe R., 29
Heston, Walter E., 109, 172, 323
Histocompatibility antigen, 409
  congenic strain
    listing, 124
    mating system, 122
  H-2
    alleles at *H-2K, H-2D*, 679
    genetic map, 669
  H-2 complex
    private specificity, 634
    public specificity, 634
    serologic reactivity, 642
  immunogenicity, 380
  linkage map, 377
  listing, 375
  wild mouse, H-2 complex, 670

Hoffman, Harold A., 217
Howard, D., 343
Hsu, Ellen, 667
Huang, Huei-Jen S., 667
Hutton, John J., 235

I

Inbred mouse
  AKR, origin, 69
  RF, 71
Inbred strain
  genetic quality control, 217
  nomenclature, 213, 445
    congenic, 148
  subline differences, 129, 441, 490
    A, 210, 443, 489
    AKR, 467
    BALB/c, 210, 415, 483
    CBA, 211, 471, 475, 489
    CE, 490
    contamination, 230
    C3H, 211, 443, 457, 461, 463, 465, 489
    C57, 443, 486
    C57BL, 212, 481, 483
    DBA, 212, 489
    MA, 490
    origin of, 197
    SJL, 490
Iványi, Pavol, 633, 657

J

Jackson Laboratory, 39, 57, 71, 109, 159
Japanese waltzing mouse, 36

K

Keeler, Clyde, 4, 179, 497
Klein, Jan, 667
Krause, Richard M., 25
Kufe, D., 343

L

Lathrop, Abbie E. C., 10, 37, 497
Lerner, R., 343
Leukemia, carcinogen-induced, 80
Liebelt, Annabel G., 475
Lilly, Frank, 281
Little, Clarence, 33, 46, 70, 109, 171, 181
Loeb, Leo, 13, 497
Lung tumor, 25, 110
Lynch, Clara, 25, 74, 172
Lyon, Mary F., 445

INDEX

M

McDowell, E. Carlton, 35, 186
Mammary tumor, 114
  clonal origin, 333
  endogenous virus, 323
    immune response, 327
  exogenous virus, 323
    immune response, 327
  foster-nursing, 64, 105
  genetics, 13, 325
  hormone effect
    estrogen, 327
    glucocorticoid, 328
    insulin, 328
    prolactin, 327
  hyperplastic nodule, 323
  pregnancy effect, 327
  strain difference, 345
Mammary tumor virus, 321
  endogenous virus, 354
  exogenous virus, unique DNA sequence, 358
  gp52 antigen
    group specificity, 346
    peptide map, 351
    type specificity, 346
  GR mouse, 358
  immune response, 326
  p10 antigen, peptide map, 351
  p28 antigen
    group specificity, 348
    peptide map, 351
    type specificity, 350
  rat, 360
  type B, 364
Manhattan Project, 113
Marshall, Joe T., 500, 511
Mendel, Gregor, 34
Miller, Dorothy A., 591
Miller, Orlando J., 591
Moisset, Beatriz, 485
Morgan, Thomas H., 47
Morse, Herbert C. III, 3, 297, 441
Mosier, Donald E., 471
Mouse gene
  furless, 174
  linkage map, 161
  linkage testing stock, 159
  opossum, 175
  pale ear, 175
  shambling, 175
  short ear, 158

Mühlbock, Otto, 324
Murine leukemia virus
  amphotropic virus, 293, 297
  ecotropic virus, 297
    integrated genome, 291
    reintegration, 291
    strain difference, 290
    structural gene, 283
  genome, 298
  gp70 antigen $G_{(AKSL2)}$, 396
    $G_{(ERLD)}$, 396
    G_{IX}, 396
    $G_{(RADA1)}$, 396
    polymorphism, 299
    X.1, 286
    XenCSA, 303
      strain distribution, 307
  immune response
    cell-mediated, 286
    humoral, 286
  MCF virus, 294, 299
  regulation of expression, *Fv-1* locus, 284
  relationship to primate virus, 690
  reverse transcriptase, 86
  structure, 297
    C-I subclass, 690
    C-II subclass, 690
  xenotropic virus, 284, 297, 442
    Fv-1 effect, 285, 313
    induction loci, 293
    structural gene, 315
Murphy, James, 52, 76
Murray, Joseph M., 37
Murray, William S., 38
Mus caroli, 500, 692
Mus cervicolor, 500, 692
Mus cookii, 692
Mus dunni, 692
Mus leggada, 499
Mus m. castaneus, 692
Mus m. domesticus, 545
Mus m. molossinus, 498, 547, 692
Mus m. musculus, 583
Mus pahari, 692
Mus poschiavinus, 499
Mus shortridgei, 692

N

Nauts, Helen C., 23
Nucleolus organizer, 592
  silver stain, 603

O

Old, Lloyd J., 391

P

Paigen, Kenneth, 255
Population genetics
  allelic variation, 542
    electromorph, 571
    mutation–equilibrium theory, 571
    neutral theory, 571
  genic heterozygosity, 570
Potter, Michael, 497

Q

Qa antigen, 409
  Qa-1, 412
  Qa-2, 414
  Qa-3, 414

R

Radiation
  atomic bomb, 73
  chromosome break, 607
  leukemia in man, 72
  leukemia induction, 286
  leukemogenesis, 80
Recombinant inbred strain, 423
  gene mapping, 432
  linkage detection, 426
  Mup, 266
  strain distribution pattern, 424
  XenCSA, 309
Robertsonian fusion, 499, 548
Robertsonian translocation, 596
  Q-banding, 592
Rockefeller University, 25
Roderick, Thomas H., 485
Rosenstreich, David L., 457
Rowe, Wallace P., 289
Russell, Elizabeth S., 33, 172

S

Sage, Richard D., 519
Sawin, Paul B., 170
Schidlovsky, G., 343
Schlom, J., 343
Selander, Robert K., 569
Shimkin, Michael B., 103, 111
Slye, Maude, 35, 46
Snell, George D., 40, 119, 172, 446, 481
Somatic cell genetics
  hamster–mouse hybrid, 607
  mouse–human hybrid, 605

Stockert, Elisabeth, 391
Strong, Leonell C., 38, 45, 168, 186
  C3H mouse, early distribution, 60
  enzyme differences in strains, 255
  origin of strain C3H, CBA, 53
Swiss mouse, 26

T

Tagliabue, Aldo, 461
Taylor, Benjamin A., 423
Teramoto, Y., 343
Thymus
  hormone, 83
  leukemia, 79
Tla, 409
Todaro, George J., 343, 689
T/t complex, 615
  H-2 antigen, wild mouse, 675
  t complementation group, 617
Tyzzer, E. E., 35, 46

W

Wakeland, Edward K., 667
Wild mouse
  Africa, 499
  Asia, 499
  C banding, 504
  contribution to inbreds, 497
  Europe, taxonomy, 512
  genetic polymorphism, 556
  geographic distribution, 522
  geographic origin, 506
  H-2 antigen, serotyping, 648
  H-2 complex
    frequency of alleles, 684
    polymorphism, 660
  H-2 congenic, 672
  immunoglobulin, 502
  karyotype, 506
  mammary tumor virus, 360
  MUP polymorphism, 504
  Mus booduga fulvidentris, 596
  Mus caroli, 360, 511, 548, 596
  Mus cervicolor, 360, 511, 596, 691
    Q-banding, 593
  Mus cervicolor popaeus, type B virus, 364
  Mus cookii, 596
  Mus dunni, 596
  Mus hispanicus, 521
  Mus m. brevirostris, 520, 582
  Mus m. castaneus, 499, 511, 557, 581
  Mus m. domesticus, 513, 582
  Mus m. homourus, 548

INDEX

Mus m. molossinus, 360, 557, 581, 596
   waltzing mouse, 502
Mus m. musculus, 499, 512, 582, 596
Mus m. Peru Atteck, 360
Mus m. spretus, 520
Mus poschiavinus, 548, 596
phylogenetic tree, 695
Spanish house mouse, 519
   genetic relationship, 543

Switzerland, 499
T/t complex, 621
Woolley, George, 40
Wright, Sewall, 35, 111

Z

Zaleska-Rutczynska, Zofia, 667
Zatz, Marion M., 467

This book is due on the last date st
below. Fines will be charged o
due books.